Interaction between Nervous Tissue Cells

Interaction between Nervous Tissue Cells

Edited by Emma Hixon

www.statesacademicpress.com

States Academic Press,
109 South 5th Street,
Brooklyn, NY 11249, USA

Visit us on the World Wide Web at:
www.statesacademicpress.com

ISBN: 978-1-63989-779-7

Cataloging-in-Publication Data

Interaction between nervous tissue cells / edited by Emma Hixon.
   p. cm.
Includes bibliographical references and index.
ISBN 978-1-63989-779-7
1. Neurons. 2. Nerve tissue. 3. Stem cells. I. Hixon, Emma.
QP363 .I58 2023
591.188--dc23

Table of Contents

Preface

The nervous system is responsible for the control and regulation of body activities and functions. It has been classified into two different parts including the central nervous system (CNS), which consists of spinal cord and the brain, and the peripheral nervous system (PNS), which comprises the branching peripheral nerves. Cells of the nervous system are responsible for transmitting electrical impulses around the human body. Neurons and glial cells are the major types of cells found in the nervous tissues. Neurons interact and compute information provided by the nervous system. Glial cells are the supporting cells of the nervous tissue and are smaller in size as compared to neurons. This book aims to understand the interactions between nervous tissue cells. It strives to provide a fair idea about this topic to help develop a better understanding of the latest advances in the study of nervous tissue cells. The extensive content of this book provides the readers with a thorough understanding of the subject.

This book is a comprehensive compilation of works of different researchers from varied parts of the world. It includes valuable experiences of the researchers with the sole objective of providing the readers (learners) with a proper knowledge of the concerned field. This book will be beneficial in evoking inspiration and enhancing the knowledge of the interested readers.

In the end, I would like to extend my heartiest thanks to the authors who worked with great determination on their chapters. I also appreciate the publisher's support in the course of the book. I would also like to deeply acknowledge my family who stood by me as a source of inspiration during the project.

Editor

Astroglial Modulation of Hydromineral Balance and Cerebral Edema

Yu-Feng Wang¹* and Vladimir Parpura²*

¹Department of Physiology, School of Basic Medical Sciences, Harbin Medical University, Harbin, China, ²Department of Neurobiology, The University of Alabama at Birmingham, Birmingham, AL, United States

*Correspondence:
Yu-Feng Wang
yufengwang@ems.hrbmu.edu.cn
Vladimir Parpura
vlad@uab.edu

Maintenance of hydromineral balance (HB) is an essential condition for life activity at cellular, tissue, organ and system levels. This activity has been considered as a function of the osmotic regulatory system that focuses on hypothalamic vasopressin (VP) neurons, which can reflexively release VP into the brain and blood to meet the demand of HB. Recently, astrocytes have emerged as an essential component of the osmotic regulatory system in addition to functioning as a regulator of the HB at cellular and tissue levels. Astrocytes express all the components of osmoreceptors, including aquaporins, molecules of the extracellular matrix, integrins and transient receptor potential channels, with an operational dynamic range allowing them to detect and respond to osmotic changes, perhaps more efficiently than neurons. The resultant responses, i.e., astroglial morphological and functional plasticity in the supraoptic and paraventricular nuclei, can be conveyed, physically and chemically, to adjacent VP neurons, thereby influencing HB at the system level. In addition, astrocytes, particularly those in the circumventricular organs, are involved not only in VP-mediated osmotic regulation, but also in regulation of other osmolality-modulating hormones, including natriuretic peptides and angiotensin. Thus, astrocytes play a role in local/brain and systemic HB. The adaptive astrocytic reactions to osmotic challenges are associated with signaling events related to the expression of glial fibrillary acidic protein and aquaporin 4 to promote cell survival and repair. However, prolonged osmotic stress can initiate inflammatory and apoptotic signaling processes, leading to glial dysfunction and a variety of brain diseases. Among many diseases of brain injury and hydromineral disorders, cytotoxic and osmotic cerebral edemas are the most common pathological manifestation. Hyponatremia is the most common cause of osmotic cerebral edema. Overly fast

Abbreviations: Ang II, angiotensin II; ANP, atrial natriuretic peptide; AQP 4, aquaporin 4; BBB, blood-brain barrier; cAMP, cyclic adenosine 3′,5′-monophosphate; CVOs, circumventricular organs; ECM, extracellular matrix; ER, endoplasmic reticulum; ERK 1/2, extracellular signal-regulated kinase 1/2; FAK, focal adhesion kinase; GABA, γ-aminobutyric acid; GFAP, glial fibrillary acidic protein; HB, hydromineral balance; I/R, ischemia/reperfusion; MAPKs, mitogen-activated protein kinases; MNCs, magnocellular neuroendocrine cells; NCX, sodium-calcium exchanger; NKCC1, Na⁺, K⁺, 2 Cl⁻ and water cotransporter 1; NO, nitric oxide; NTS, nucleus of the tractus solitarii; OT, oxytocin; OVLT, organum vasculosum of the lamina terminalis; pERK 1/2, phosphorylated ERK 1/2; PKA, protein kinase A; PTN, protein tyrosine nitration; PVN, paraventricular nucleus; RAAS, renin-androgen-aldosterone system; RVD, regulatory volume decrease; RVI, regulatory volume increase; SFO, subfornical organ; SON, supraoptic nucleus; TRP channels, transient receptor potential channels; TRPC, TRP canonical channel; TRPV, TRP vallinoid channel; VP, vasopressin; VRAC, volume-regulated anion channel.

correction of hyponatremia could lead to central pontine myelinolysis. Ischemic stroke exemplifies cytotoxic cerebral edema. In this review, we summarize and analyze the osmosensory functions of astrocytes and their implications in cerebral edema.

Keywords: astrocytes, cerebral edema, osmosensation, osmotransduction, vasopressin

INTRODUCTION

Homeostasis of the internal environment is the prerequisite for normal activity of an organism and heavily depends on the hydromineral balance (HB) of the extracellular fluid. This balance is based on equivalent amounts of water drinking and salt intake vs. their excretion, and is commonly measured by the volume and osmolality of the extracellular fluid (Muhsin and Mount, 2016). Many factors have been implicated in regulation of HB, such as thirst, along with water- and salt-regulating hormones. Among hormones, vasopressin (VP, also called as antidiuretic hormone, ADH) released by hypothalamic neuroendocrine cells has been considered the most sensitive and powerful factor regulating HB (Kinsman et al., 2017). In response to increased osmotic pressure or reduced blood volume, VP release into the brain and blood increases significantly to surge reabsorption of water, thereby maintaining relative fidelity of the osmolality and volume of the brain and blood (Brown, 2016). However, maladapted response of this VP secretion can cause cerebral edema and threaten the life of patients, e.g., due to high intracranial pressure and the resultant brain herniation in hyponatremia (Wang et al., 2011) or ischemic stroke (Jia et al., 2016). Thus, understanding the mechanisms underlying neurohumoral regulation of VP secretion is critically important.

In neurohumoral regulation of VP neuronal activity, one of the most sensitive and important factors is astrocytic plasticity in the osmotic regulatory system (Wang and Zhu, 2014; Jiao et al., 2017). Astrocytes adjacent to neurons in the osmosensory system are sensitive to VP and can also influence VP secretion through their morphological and functional plasticity and produce other hormones that also regulate osmolality, while directly regulating local HB; of note, a variety of astrocytes in other brain areas are sensitive to VP (Simard and Nedergaard, 2004), a subject that is beyond the scope of the present review, however. Nonetheless, under pathological conditions, malfunctions of osmosensory mechanisms in astrocytes worsen hydromineral disorders by affecting the activity of VP neurons and other osmotic regulatory factors; these events lead to brain edema, neural degeneration and irreversible brain damages. In this review, we sum up our current understanding of astrocyte-associated osmosensation, osmotransduction and the resultant changes in the activity of the osmotic regulatory system, along with their implication in diseases involving disorders of HB.

OSMOTIC BALANCE AND ITS NEUROHUMORAL REGULATION

Homeostasis of the internal environment is under constant challenge of the external environment. Factors that can markedly change extracellular and intracellular fluid volume include ingestion and elimination of water and salts, hydromineral distribution across capillaries (e.g., within the blood vs. brain parenchyma/other organs) and, at the cellular level, across the plasma membrane. In response to these challenges, the endocrine system and the autonomic nervous system respond promptly to adaptively modulate the HB by regulating water and salts intake and elimination, as well as their distribution among different compartments of the body.

General Regulation of HB

Hydromineral regulation requires activation of the osmosensory system in conjunction with volemic regulation and is largely determined by a variety of neurohumoral factors. An increase in plasma osmolality draws water from cells and interstitium into the blood, causing dehydration, which activates specific brain osmoreceptors to stimulate drinking and release of VP from VP neurons. VP reduces water loss via increasing water reabsorption in the kidneys. In contrast, hypoosmotic challenge causes opposite reaction by VP neurons and the hydromineral regulation. Hypovolemia, a state of decreased blood volume, stimulates vascular volume sensors that signal brain centers to initiate drinking and VP release; it also stimulates baro/volume receptors in the kidneys to activate the renin-angiotensin (Ang)-aldosterone system (RAAS), which initiates drinking and VP release while increasing Na^+ reabsorption. By contrast, blood volume expansion or hypervolemia can lead to significant increase in plasma concentrations of atrial natriuretic peptide (ANP), oxytocin (OT), prolactin and corticosterone, to eliminate Na^+ and its bound water while suppressing hyperosmotic stimulation of VP secretion. In addition, osmotic challenge can lead to the production of gaseous neurotransmitters, which can affect neuronal secretion; e.g., nitric oxide (NO) inhibits (Stern and Ludwig, 2001), while carbon monoxide stimulates (Reis et al., 2012) VP secretion. The aforementioned factors and events, and likely many more, all work in concert so that the HB is maintained.

To achieve electroneutrality and volume stability of a cell, the sum of osmotically active particles, i.e., osmolytes, in the intracellular space must be equal to that in the extracellular space. As the major extracellular osmolyte, Na^+ in optimal extracellular concentration becomes the reference point for osmoreceptors to control thirst and VP secretion, RAAS activity, levels of natriuretic factors as well as the cell volume. The control over the cell volume is achieved by a chloride shift and by modulating the activity of Na^+/K^+-ATPase (Kurbel, 2008). Movement of water across the cell membranes occurs when an osmotic pressure gradient forms between the intracellular compartment and interstitial fluid. Large amounts of Na^+, Cl^- and HCO_3^- ions are present in the extracellular space whereas K^+, Mg^{2+} and PO_4^{3-} ions are the major ions in the intracellular compartment. In most cells at rest, the plasma membrane exhibits relatively high permeability for K^+ and water, but not for Na^+ and Cl^-. Consequently, the distribution of fluid across the plasma membrane is determined mainly by the osmotic effect of Na^+ and Cl^-. Whenever osmotic gradients form, following disturbance of the electrochemical balance across the membranes, water will diffuse to the side of higher solute concentration, thereby maintaining intracellular fluid isotonic. The transport of ions and water across cell membranes has been reviewed recently elsewhere (Jia et al., 2016) and, hence, is not further discussed here.

Along with the effect on water balance, VP is also a critical modulator of the arterial diameter, which along with water reabsorption/distribution, determines the effective arterial volume. For instance, in the rat hypothalamic supraoptic nucleus (SON), hyperosmotic stimulation mainly led to vasodilatation followed by vasoconstriction (Du et al., 2015). The initial vasodilatation could be the result of both glial retraction and NO action at the blood vessels, while the later vasoconstriction phase results from VP-mediated activation of V1a VP receptor. By changing the arterial contractility and the blood pressure, VP can maintain irrigation pressure to provide sufficient nutrients to the brain and other important organs (**Figure 1**).

VP Neuronal Activity and its Neural Regulation

VP is an essential regulator of the HB. In the hypothalamus, there are several neuroendocrine nuclei that contain VP neurons. These cells are primarily located in the SON and the paraventricular nucleus (PVN); they are mainly magnocellular neuroendocrine cells (MNCs) releasing VP into the blood and brain. In addition, there are also parvocellular VP neurons in the parvocellular section of the PVN that release VP to regulate the activity of the autonomic nervous system and a variety of brain activities (Hou et al., 2016). In patients with central diabetes insipidus, the lack of VP production causes thirst (polydipsia) and excessive excretion of urine (polyuria) because of a dysfunction in the thirst mechanism, development of hyponatremia and the ensuing complications (Sailer et al., 2017).

The effects of osmotic challenges on VP neuronal activity and VP secretion depend on both external factors and the intrinsic features of these neurons (**Figure 1**). Classically, neural regulation is considered the major factor modulating VP neuronal activity during osmotic challenges (**Figure 1A**). Brain regions around the circumventricular organs (CVOs) and the nucleus of the tractus solitarii (NTS) are the major source of neural inputs and regulation of VP neuronal activity (Wang et al., 2011; McKinley et al., 2015). The CVOs are highly vascularized brain areas lacking a normal/tight blood-brain barrier (BBB). These organs/structures include (but are not limited to) the subfornical organ (SFO), the organum vasculosum of the lamina terminalis (OVLT) and the area postrema. They are the sites allowing communication between the blood plasma and the interstitium of the brain parenchyma, as their fenestrated capillaries are freely permeable to ions and water. Neural inputs

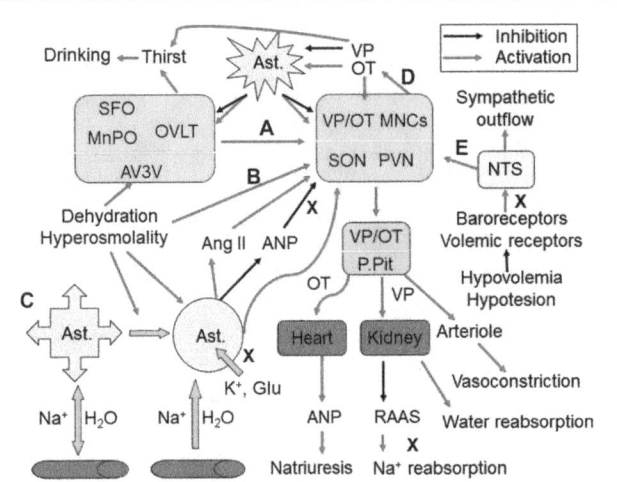

FIGURE 1 | Extrinsic modulation of hydromineral balance (HB). The sketch shows that hyperosmotic stress modulates HB by the following approaches. **(A)** Activating the subfornical organ (SFO), organum vasculosum of the lamina terminalis (OVLT) and the medial preoptic nucleus (MnPO) in the anteroventral-third ventricle (AV3V) area, which in turn activates magnocellular cells. i.e., vasopressin (VP) and oxytocin (OT) neurons in the supraoptic (SON) and paraventricular (PVN) nuclei. **(B)** Changing activity of VP neurons after the osmolality is increased in brain parenchyma. **(C)** Causing retraction of astrocyte (Ast.) processes from the areas surrounding the osmosensory neurons, which increases the excitability of VP neurons by removing the physical barriers between adjacent neurons and by increasing extracellular levels of K^+, glutamate (Glu) and angiotensin II (Ang II), and by reducing extracellular levels of atrial natriuretic peptide (ANP) or brain natriuretic peptide. **(D)** Resultant increases in intrahypothalamic release of OT can activate VP neurons through eliciting the retraction of astrocytic processes, while VP inhibits this retraction by increasing aquaporin expression. **(E)** In addition, hypovolemia or hypotension can increase VP release from the posterior pituitary (P.Pit) by reducing the baroreflex through the nucleus of the tractus solitarii (NTS) and by decreasing activity of the renin–angiotensin–aldosterone system (RAAS), while increasing the sympathetic output. As a result, serum osmotic pressure gets reduced and blood volume/pressure rises through increasing water reabsorption and natriuresis. See text for details.

from the NTS can relay hypovolemic/hypotensive information to the CVOs while directly modulating VP neuronal activity to correct fluid deficits (Miyata, 2015). However, CVOs and the NTS do not regulate the HB directly; rather, they convey osmotic and volemic messages to VP neurons. Among the CVOs, the SFO can sense and integrate hydromineral signals in responses to osmotic, volumetric and cardiovascular challenges, and integrate information from circulating signals of the metabolic status (Hindmarch and Ferguson, 2016). Moreover, VP neurons can also be regulated by other brain regions involved in the integration of circulatory and fluid information (McKinley and McAllen, 2007). Activation of neurons in these brain areas, which include the lateral parabrachial nucleus, the midbrain raphe nuclei, the medial preoptic nucleus and the septum, increases VP neuronal activity in the SON and PVN, and VP secretion that modulates kidney functions and arteriole contractility, along with activation of the anterior cingulate cortex and insula to cause thirst and resultant water drinking (Macchione et al., 2015). The efficiency of the above neuronal inputs on VP neuronal activity depends on astrocytic plasticity and the neurochemical environment surrounding VP neurons (Wang and Zhu, 2014).

Astrocytic Regulation of HB

In the extrinsic modulation of VP neuronal activity, astrocytic plasticity plays an important role in local and systemic HB (Wang et al., 2011). Astrocytes are the sensors of neuronal activity while detecting humoral information from the blood at the CVOs (Farmer et al., 2016). The responsive changes of astrocytes, i.e., their plasticity, in turn modulate VP neuronal activity, VP secretion and thirst.

Similar to the modulatory role of astrocytes in neuronal activity (Simard and Nedergaard, 2004), astrocytic morphological plasticity is also essential for the integrity of the neurovascular unit. Of note, the neurovascular unit is a functional unit comprised of capillary endothelial cells, along with surrounding neurons, and non-neuronal cells such as pericytes and astrocytes. Malfunction of astrocytes can change the structural integrity of this unit and subsequently disturb the HB across the BBB and the transport of metabolites, ions, varieties of organic molecules and water between the brain and the blood (Wang and Parpura, 2016).

Astrocytic Morphological Plasticity and VP Neuronal Activity

In osmosensation, astrocytes exhibit remarkable morphological plasticity (**Figure 1C**). It was observed that after exposing cell culture to a hypoosmotic solution, astrocytes swell within 30 s and then undergo a regulatory volume decrease (RVD; Eriksson et al., 1992) following the opening of anion channels and release of organic osmolytes. Conversely, when exposed to a hyperosmotic solution, astrocytes shrink first and then exhibit a regulatory volume increase (RVI; Eriksson et al., 1992). This dual morphological plasticity, occurring during acute hypo- or hyper-osmolality, provides astrocytes with unique capacity to keep the extracellular volume stable while differentially regulating extracellular osmolality. Regulatory volume changes could occur in a microdomain specific manner or as a result of intracellular volume redistribution; that is, an expansion of astrocytic processes with a reduced size of astrocytic somata and vice versa (Choe et al., 2016).

Different from the RVI and RVD mentioned above, when exposed to a gradual osmolality decrease, some cells use the mechanism of isovolumetric regulation to adapt to the hypoosmotic challenge without dramatically changing their volume. The apparent isovolumetry is due to the coordinated extrusion of osmolytes, whereby following an early efflux of taurine (2-aminoethanesulfonate$^-$) and Cl$^-$, a delayed K$^+$ efflux occurs, as shown in cerebellar astrocytes (Ordaz et al., 2004). These diverse volemic regulatory machineries endow astrocytes with the ability to directly sense and regulate HB in local brain parenchyma.

An example of astrocytic involvement in osmotic regulation is the osmotic response of the SON during osmotic challenges. Plastic change in the morphology of the SON and PVN is a responsive feature to hydromineral disturbance, which is particularly dramatic in the SON during dehydration. As previously reviewed (Hatton, 2002), during dehydration or salt (hypertonic) water drinking, astrocytes show retraction of their processes along with hypertrophic changes in VP neurons. This retraction of astrocytic processes from the surroundings of VP neurons increases the interaction between neurons through interneuronal/dendritic gap junctions and preponderance of synaptic connections, thereby increasing the excitability of VP neurons and promoting synchronous release of VP. To the contrary, hypoosmotic challenges can inhibit VP neuronal activity by evoking the expansion of astrocytic processes, wedging them in between neuronal dendrites and lessening dendritic gap junctional connectivity. These changes reduce direct interactions between adjacent VP neurons, while increasing astrocytic uptake of extracellular K$^+$ and glutamate; opposite effects are seen during hyperosmotic stress. Correspondingly, VP neurons become more excitable during hyperosmotic stress and inhibited during acute hypoosmotic challenges.

It is worth to notice that during chronic hyperosmotic stimulation, VP release remains at high levels. This is likely due to increased VP synthesis as well as increased release probability of VP packaged in secretory vesicles, along with increased OT-mediated excitation (**Figure 1D**). It has been reported that during chronic hypernatremia, VP and OT mRNA levels increase twofold along with hypertrophy of VP neurons and increased release of VP (Glasgow et al., 2000). Moreover, this condition also causes an activity-dependent depolarizing drift in the chloride reversal potential and abolition of inhibitory pathways, which is prevented by blocking central OT receptors (Kim et al., 2011), suggesting that hyperosmotic stress increased OT receptor signaling that is responsible for the reversal of GABAergic inhibition. Consistently, OT can excite VP neurons (Wang and Hatton, 2006) while exerting autoregulatory effect on OT neuronal activity (Wang et al., 2006; Wang and Hatton, 2007a,b). Thus, the maintenance of high level of VP release during chronic hyperosmotic stimulation is at least partially supported by an increase of OT release. Noteworthy is that the excitatory effect of OT on VP neurons takes minutes to occur, and thus is highly unlikely a direct effect on OT receptors on VP neurons, but most likely an indirect effect due to a retraction of astrocyte processes in the SON caused by OT (Wang and Hatton, 2009; Wang et al., 2017).

Astrocytes of the SON have the ability to change the osmosensory threshold within the local neural circuitry during chronic hypoosmotic challenges, as previously reviewed elsewhere (Wang et al., 2011). That is, the sensitivity of osmosensors to hypoosmotic challenge gradually wears off during continuous stimulation; correspondingly, the initial inhibitory reaction vanishes during prolonged presence of hypoosmotic environment. This notion is partly supported by the recovery of firing rate of VP neurons after initial hypoosmotic inhibition, which is based on dual astrocytic morphological plasticity, i.e., extension and subsequent retraction of astroglial processes during the initial decrease in firing rate of VP neurons followed by its recovery, respectively (Wang et al., 2013a,b).

Taken together, astrocytic morphological plasticity can modulate the activity of osmosensory neurons (OS neurons), which in turn changes astrocytic plasticity, thereby forming a functional heterocellular network in central osmosensation (**Figure 1**).

Astrocytic Functional Plasticity and VP Neuronal Activity

Accumulating evidence reveals that astrocytes are not just a passive buffer of extracellular environment, but also an active modulator of neuronal activity (**Figure 1C**). Astrocytic modulation of VP neuronal activity is closely related to its ability to take-up neurotransmitters from the extracellular space (Wang et al., 2013a,b) as well as to produce, release and take-up gliotransmitters to modulate interneuronal communication (Parpura et al., 2004; Montana et al., 2006; Ni et al., 2007). As the source of glutamate in the SON (Ponzio et al., 2006), astrocytes can excite OS neurons, including VP neurons during hyperosmotic stimulation. Moreover, the presumed reduction in uptake of β-alanine by astrocytes in the SON can inhibit GABA transporters and subsequently increase extracellular GABA levels (Park et al., 2006; Wang et al., 2013a). Therefore, astrocytes can actively alter the neurochemical environment and modulate the activity of OS neurons. Of note, cultured astrocytes from different brain regions can release various other gliotransmitters using multiple underlying mechanisms and in response not only to hypoosmotic challenge but also to an array of neurotransmitters and modulators (Malarkey and Parpura, 2009). Whether a palette of gliotransmitters, released by astrocytes elsewhere, could be released by astrocytes in the SON in times of osmotic challenge and whether that would lead to modulation of synaptic transmission in the SON remain to be determined.

ASTROCYTIC INTERACTION WITH HORMONES THAT REGULATE OSMOLALITY

While astrocytes can directly regulate thirst and maintain the integrity of neurovascular units, they also directly interact with hormones that regulate osmolality, including VP, Ang II, aldosterone, natriuretic peptide, OT and hormones in the hypothalamic-pituitary-adrenal axis. Thus, astrocytes can modulate the effect of hormones on osmotic regulation of HB (**Figure 1C**).

Astrocytes and VP

It has been reported that hyperosmotic stimulation increases VP release in the brain (Ludwig et al., 2005). VP in the brain can evoke thirst and drinking (Abrão Saad et al., 2004), while increased levels of VP in the blood increase water reabsorption in the kidney by changing water channel activity (Tamma et al., 2017). In addition, VP can antagonize diuresis and natriuretic role of ANP (Lipari et al., 2007), thereby limiting ANP's natriuretic effect. As a result, VP is both an antagonizing agent of hypovolemia or hyperosmolality and a facilitator of hyponatremia; excess release of VP may result in water retention and pathophysiological hyponatremia.

VP is a major facilitator of cell swelling in the brain, which involves several local pathways. VP can increase intracellular osmolality by uptake of Na^+, K^+, $2 Cl^-$ or by lack of extrusion of Na^+ via activating Na^+, K^+, $2 Cl^-$ and water cotransporter 1 (NKCC1) or via inhibiting the activity of Na^+/K^+-ATPase, respectively (Hertz et al., 2017). In addition, VP is also a major activator of astrocytic water channel protein, aquaporin 4 (AQP4) by activating V1a type of VP receptor (Niermann et al., 2001). Inhibition of VP1a receptor leads to decrease in AQP-4 expression and prevents brain edema after port-traumatic injury (Marmarou et al., 2014). Thereby, VP could facilitate cell swelling when intracellular osmotic pressure is higher than that of the extracellular fluid.

Astrocytes and RAAS

Along with water reabsorption/excretion mediated by VP secretion, the sodium-retaining function of the RAAS makes a critical contribution to extracellular Na^+ levels and osmolality. In the central nervous system, the RAAS acts mainly through the sensory CVOs, in particular the area postrema, to activate brain neural pathways that elevate blood pressure, release VP and aldosterone, increase renal sympathetic nerve activity, and increase the ingestion of water and Na^+ to restore Na^+ loss to the environment (Geerling and Loewy, 2008). In these processes, Ang II binds to the brain Ang type 1 receptor to stimulate thirst, Na^+ appetite and secretion of VP and OT (Felgendreger et al., 2013; Almeida-Pereira et al., 2016). The prolonged Ang type 1 receptor blockade caused rebound increase in levels of Ang II and VP secretion to compensate for hypovolemia in association with reduced ANP plasma concentrations (Araujo et al., 2013). These findings are in agreement with the classical function of the RAAS in Na^+ balance.

Astrocytes are a major source of the Ang II precursor protein angiotensinogen and Ang II in the brain (Hermann et al., 1988); expression of receptors for Ang was also identified in the rat and monkey astrocytes (Garrido-Gil et al., 2017). Thus, astrocytes could exert self-modulation during osmotic challenges. This notion is supported by the observation that following 7 and 14 days of 2% NaCl (N.B., isotonic solution contains 0.9% NaCl) in drinking water, a significant increase in Ang II precursor, preproangiotensinogen mRNA was detected in astrocytes in regions of the anterior hypothalamus, including the PVN, the medial preoptic area and medial preoptic nucleus, while a decrease was observed in astrocytes in the SON (Ryan and Gundlach, 1997). These results, consistent with a recent report (Dominguez-Meijide et al., 2017), indicate that astrocytes could increase levels of extracellular Ang II to regulate thirst, while Ang II facilitation of VP release is reduced during prolonged hyperosmotic stimulation.

Astrocytes and Natriuretic System

In contrast with the sodium-retaining functions of the RAAS, natriuretic peptides, such as ANP, brain natriuretic peptide, and OT, belong to the "natriuretic system." OT neurons in the hypothalamus can be activated simultaneously with VP neurons during dehydration or hypertonic stimulation. However, OT exerts a natriuretic function, which plays a synergistic role in maintaining HB with VP-increased water reabsorption following hyperosmotic stimulation. Similar to the modulation of VP neurons in hydromineral disturbance, astrocytic plasticity also strongly influences OT secretion from the SON and PVN (Yuan et al., 2010). In addition, during blood volume expansion, OT is secreted from the posterior pituitary into circulation to activate atrial OT receptors in the heart and promote ANP release. ANP diminishes VP-induced water reabsorption while exerting natriuretic and diuretic actions in the kidney (Gutkowska et al., 2014; Theilig and Wu, 2015), thereby reducing blood volume.

In the brain, ANP-mRNA is present in the hypothalamic suprachiasmatic nucleus, and ANP appears in both the SON and suprachiasmatic nucleus (Lipari et al., 2007). At the cellular level, ANP is present in and exocytotically released from membrane-bound vesicles as shown in cultured rat cortical astrocytes (Krzan et al., 2003; Chatterjee and Sikdar, 2013). Centrally released ANP can inhibit osmotically evoked VP and OT release through presynaptic inhibition of glutamate release from the OVLT. It has been reported that ANP application to rat hypothalamic explants did not affect depolarizing responses of VP neurons to local hypertonicity. However, ANP reversibly abolished the synaptic excitation of MNCs after hypertonic or electrical stimulation of the OVLT (Richard and Bourque, 1996). The inhibition of VP release following activation of ANP neurons can lead to diuresis, and decreased adrenocorticotropin release and blood pressure (Gutkowska et al., 1997). In addition, central ANP release is also increased by glucocorticoids (Lauand et al., 2007) and estradiol (Vilhena-Franco et al., 2011) in response to osmotic/volemic stimulation. Thus, astrocytic ANP can contribute to reducing hydromineral overload under various physiological states.

Other Effects

Hormonal cross talks and receptor-receptor interactions are common phenomena in brain control of HB. For instance, estradiol acts mainly on the VP neurons in response to water deprivation, potentiating VP neuronal activation and VP secretion without altering VP mRNA expression (Vilhena-Franco et al., 2016). Bradykinin, thyrotropin-releasing hormone, neurotensin and opioids (Irazusta et al., 2001), as well as secretin (Bai et al., 2017) and prolactin (Seale et al., 2012), are also implicated in neurohumoral regulation of HB and their expressions are likely regulated by astrocytes. Lastly, astrocytes can participate in the synthesis of many hormones that, under different osmotic conditions, regulate osmolality by differential expression of peptidases involved in the maturation and degradation of peptide (pro)hormones and neuropeptides. For example, in water-loaded rats, prolyl endopeptidase was decreased in the brain cortex (Irazusta et al., 2001). Thus, astrocytes can participate in HB by catabolizing humoral factors as well. These lines of evidence indicate that astrocytes could contribute to the HB through multiple hormones.

OSMOSENSATION BY ASTROCYTES

Homeostasis of the internal environment largely depends on neurohumoral regulation of the central osmosensory system.

As reviewed recently (Jiao et al., 2017), the key feature of osmosensation is the activation of mechanoreceptors, particularly vallinoid and canonical types of transient receptor potential channels (TRPV and TRPC, respectively), both of which are highly permeable to Ca^{2+}. Indeed, the activation of these TRP channels increases cytosolic Ca^{2+} levels in osmosensory cells, including VP neurons, and triggers a series of secondary reactions. For example, hypotonic stimulus induces intracellular Ca^{2+} elevations through TRP channels, which then trigger AQP1 translocation and activation (Conner et al., 2012). The activation of TRP channels relies on changes in cell volume, membrane stretch and cytoskeletal reorganization as well as the hydration status of the extracellular matrix (ECM) and activity of integrins in a spatiotemporal dependent manner. In this process, astrocytic plasticity plays a key role since acute hyperosmotic stimulus-induced Fos expression in neurons depends on activation of astrocytes in the SON of rats (Yuan et al., 2010). Moreover, disabling astrocytic plasticity also blocked the rebound excitation of VP neurons in response to hypoosmotic challenges (Wang et al., 2013a,b). Thus, astrocytes could be a prominent component of the osmosensory system and could be more sensitive to the osmotic challenges than neurons are.

Characteristics of Osmosensation

Osmosensation is not a simple result of activation or inactivation of the stretch-activated channels (Cheng et al., 2017). To fully understand the participation of astrocytes in osmosensation, it is necessary to analyze the characteristics of osmosensation by the central osmosensory system.

Osmosensation Is Time-Dependent Process

In an organism, osmosensation is a dynamic process in temporal domain, as outlined below using two examples. (1) Chronic hyperosmotic stress weakens $GABA_A$ receptor-mediated synaptic inhibition of VP and OT secretion in rat hypothalamic MNCs (Kim et al., 2011). That is, hyperosmotic stress caused a profound depolarizing shift in the reversal potential of GABAergic response (E_{GABA}) in MNCs. This E_{GABA} shift was associated with increased expression of NKCC1 in MNCs and was blocked by the NKCC inhibitor bumetanide. (2) In response to hyposmolality challenge, VP neurons in the SON show a transient inhibition of firing activity followed by a rebound excitation (Wang et al., 2013a,b), which can partially account for the rebound increase in VP release after initial inhibition in response to hyponatremic challenges (Yagil and Sladek, 1990). It is likely that the initial osmosensory response involves Ca^{2+}-activated K^+ channels and, the BK and SK types of K^+-selective ion channels (Ohbuchi et al., 2010), leading to elimination of action potentials and to hyperpolarization of membrane potentials, while the latter phase could be a result of RVD-related activation of the volume-regulated anion channel (VRAC) and Cl^- efflux (Muraki et al., 2007; Hübel and Ullah, 2016). Taken together, these findings indicate the presence of a diversity of mechanisms underlying temporal dynamics in osmosensation.

Osmosensation Is Related to the Spatial Location of Osmosensory Cells

Mechanosensitive ion channels could function differently at different loci. For instance, excitation of the OVLT neurons elicited by hyperosmotic stress was not affected by the deletion of TRPV 4 but was abolished in cells lacking TRPV 1 (Ciura et al., 2011). Thus, TRPV1 likely plays a dominant role in OVLT sensation of hyperosmotic stress; however, hyperosmolality-sensitive TRPV4 (Liedtke and Friedman, 2003) could carry out the same function at other loci. Moreover, same TRPV channels at different loci could have different sensitivity to osmotic challenges. For example, cell and cell nuclear sizes in the SON were increased or decreased with hyperosmolality or hypoosmolality, respectively. However, both conditions did not affect those sizes in the medial habenular nucleus *in vivo* (Zhang et al., 2001), although this brain area also expresses TRPV1 and TRPV4 (Ishikura et al., 2015). Furthermore, hyperosmotic glucose or urea solutions activate VP neurons but not CVO neurons (Ho et al., 2007). These findings indicate existence of spatial diversity in osmosensation.

Roles of Various Ion Channels in Osmosensation Might be Interchangeable

In classical view, sensation of osmotic pressure is the matter of activation or inhibition of stretch-activated cation channels, (Prager-Khoutorsky and Bourque, 2015). Besides the TRP channels discussed above, other mechanical ion channels have been implicated in osmosensation, such as TMEM63 proteins found in *Arabidopsis* (Zhao et al., 2016), acid-sensing ion channel 3 in the intervertebral disc (Uchiyama et al., 2007), mammalian TRP ankyrin-1 and TRP melastatin-8 channels (Soya et al., 2014). It is possible that once one ion channel is dysfunctional (Taylor et al., 2008; Ciura et al., 2011), other ion channels with similar responsive features could easily compensate this deficit and thus maintain the osmosensory ability.

Osmoreceptors in Astrocytes

As stated above, the traditional theory of neuron dominance in osmosensation could not entirely explain findings presented above and, thus, the osmosensory ability of astrocytes needs to be considered in the operation of the osmosensory system.

AQP4 and Water Transport

Astrocytic volume change and subsequent morphological plasticity are essential for astrocytic regulation of local HB and VP neuronal activity (Theodosis et al., 2008; Wang and Zhu, 2014), the latter of which heavily relies on AQP4 expression and activity in astrocytes (Wang and Parpura, 2016).

In the rodent brain, AQP4 is the predominant form of specific water channel protein expressed in astrocytes; it is essential for quick water transport between intracellular and extracellular compartments when osmotic gradients are present (Xu et al., 2017). As previously reported, volemic change is essential for osmosensation as it changes the interaction between cytoskeletal elements and osmosensitive TRP channels (Prager-Khoutorsky and Bourque, 2015). Moreover, the expanded astrocyte processes promote releasing

gliotransmitters and neuromodulators, metabolizing neural active substance, and transporting content of the interstitial fluid to the blood vessels (Parpura and Verkhratsky, 2012; Zorec et al., 2012) and thus exert the effect of osmosensation while maintaining the HB in the brain. The osmolality-associated AQP4 permeability determines astrocytic morphological plasticity, which can change the activity of their adjacent OS neurons independent of the activity of TRP channels, thereby making individual TRP channel type replaceable in osmosensation.

The participation of AQP4 in osmotic regulation is a dynamic process. It has been observed that the expression of AQP4 increased during chronic hyperosmotic stress (Yang et al., 2013). By contrast, hypoosmotic stimulation increases water influx through AQP4 due to increased osmotic gradient across cell membranes. This latter effect is not due to increase in AQP4 membrane installation but in its permeability to water (Potokar et al., 2013). It is possible, however, that there is a transient increase in AQP4 expression, or a reduction in its catabolism, at the early stage of hypoosmotic stimulation because levels of AQP4 are found to rapidly increase or decrease; these changes are accompanied by expansion and retraction of astrocyte processes in rat SON, respectively (Wang and Hatton, 2009).

Ion Channel Activity

Neurons in the CVOs and the MNCs in the SON and PVN, all important for thirst and urination, are also the target of astrocytic plasticity, which is related to alteration of astrocytic ion channel activity. In addition to the extensively identified astrocyte involvement in osmotic regulation of VP neuronal activity in the SON, astrocytes are also Na^+ sensors. Namely, Na_x, an atypical Na^+ channel and Na^+-level sensor, is found located in perineuronal lamellate processes extending from ependymal cells and astrocytes in the CVOs; astrocytes isolated from the SFO were sensitive to an increase in the extracellular Na^+ level (Noda, 2007).

Another important osmosensory machinery in astrocytes is the expression of TRP channels. Astrocytes are known to express many types of osmosensitive TRP channels. For example, TRPV1 has been identified at thick cellular processes of astrocytes in the CVOs, and intravenous administration of a TRPV1 agonist resiniferatoxin induced Fos expression in astrocytes in the OVLT, SFO and area postrema (Mannari et al., 2013). The resultant Ca^{2+} influx following activation of TRP channels is not only associated with astrocytic volemic change but also related to neuromodulation through releasing glutamate, and this transmitter can modulate neuronal activity at extrasynaptic sites (Parpura et al., 2017). Thus, Na_x channels and TRPV1 channels in astrocytes can sense Na^+ levels in bodily fluids (Noda, 2007), particularly in the blood, as well as in the interstitium of the osmotic regulatory system, and in turn convert hydromineral message to neuronal activity, thereby maintaining the HB. The different types of TRP channel expression in astrocytes and its association with different types of OS neurons endow some of the spatial features of the osmosensory system.

Osmolyte Transport and its Coupling to AQP4

Transporting osmolytes is essential for HB between intra- and extracellular spaces and involves the redistribution of osmolytes through ion transporters and channels.

The function of AQP4 is coordinated with a variety of closely associated osmolyte transporters. As reviewed recently (Wang and Parpura, 2016), AQP4 is colocalized and/or assembled with the TRPV4, Kir4.1/Kir5.1, connexin 43, glutamate transporter-1, metabotropic glutamate receptor 5, and Na^+/K^+-ATPase, mainly at astrocytic endfeet. Along with changes in AQP4 activity, there are also alterations in non-selective cation currents, K^+ and Na^+ transmembrane activity, and glutamate uptake and metabolism. In addition, many other molecules are also involved in this process, including: (1) channels, such as Ca^{2+} release-activated Ca^{2+} channel, voltage gated K^+ channels, sulfonylurea 1/TRP melastatin-4 ion channel; (2) carriers including Na^+, Cl^- cotransporter, NKCC, Na^+/H^+ exchangers, Cl^-/HCO_3^- exchanger, acid-sensing ion channel 1a, Na^+-glucose cotransporter, and glutamate transporters; (3) receptor-coupled channels such as ionotropic AMPA and NMDA types of glutamate receptors, and $GABA_A$ receptor; and (4) Na^+/K^+-ATPase or sodium pump, as reviewed recently (Jia et al., 2016). Lastly, epithelial Na^+ channel (Miller and Loewy, 2013) and VRAC (Hussy et al., 2000) are also involved in astrocyte regulation of HB.

In astrocytes, there is NKCC-mediated $Na^+/K^+/Cl^-$ inward flux evoked by elevated extracellular K^+. The increase in clearance of extracellular K^+ during neuronal excitation is initially performed by Na^+/K^+-ATPase-mediated K^+ uptake into astrocytes; at K^+ concentrations above ~10 mM this process is aided by uptake of Na^+, K^+ and 2 Cl^- via the cotransporter NKCC1. This activity allows subsequent K^+ release via the inward rectifying K^+ channel Kir4.1, perhaps after syncytial buffering (Walz, 1992), i.e., trans-astrocytic connexin-mediated K^+ transfer. Since the Na^+/K^+-ATPase exchanges 3 Na^+ with 2 K^+, it creates extracellular hypertonicity and cell shrinkage; however, hypertonicity stimulates NKCC1 and can also cause RVI to minimize ionic disequilibrium due to asymmetric Na^+/K^+-ATPase fluxes (Hertz et al., 2013).

Astrocytic ECM

The ECM represents an essential component of osmosensation (Jiao et al., 2017). In response to hydromineral disbalance, the ECM can quickly detect changes in extracellular cation levels and then transmit this message to other components of the osmosensors, including integrins, cell adhesion molecules and cytoskeletal elements. Integrins can transduce hydromineral signals from extracellular to intracellular compartments by activation of TRP channels via integrin-linked kinase and actin, as previously discussed (Jiao et al., 2017). In the SON and PVN, neural cell adhesion molecule, which gets decorated with various amounts of sialic acid, can significantly influence cell adhesion and thus allows astrocyte processes to expand or retract (Theodosis et al., 2008).

The ECM-integrin bonding is modulated by both cell volume and hydration state of the ECM in osmotic stress. Under different osmotic conditions, signaling process modulated by integrin-ECM contacts is either activated or inhibited. As observed in the SON, hyperosmotic stress caused by increasing NaCl concentration can decrease polysialic acid-neural cell adhesion molecule complex content (Schmelzer and Miller, 2002) and increases extracellular volume (Perkins et al., 2017). The changes in the astrocytic ECM (Pierre et al., 2001) could further activate astrocytic TRP channels through integrins (Song and Dityatev, 2018). In contrast, their decrease promotes reactive gliosis or expansion of astrocytic processes that reduces neuronal excitability (Theodosis et al., 2008), at least transiently (Jiao et al., 2017). These features endow astrocytes with the ability to detect osmotic changes in the internal environment. It is likely that changes in extracellular osmolality activate TRP channels, which increases intracellular Ca^{2+}, changes glial fibrillary acidic protein (GFAP) polymerization state and AQP4 distribution at the plasma membrane, leading to astrocytic morphological and functional plasticity.

GFAP Plasticity

The typical morphological feature of astrocytes during hydromineral disturbance is the plastic changes in the expression of GFAP, which determines not only the morphology, but also the function of astrocytes (Wang and Parpura, 2016). As the retraction and expansion of astrocytic processes take place, depolymerization/disassembling and polymerization/assembly of GFAP also occur, correspondingly; these events translocate GFAP-associated functional molecules, such as glutamine synthetase, AQP4 and actin, to (GFAP assembly) and from (GFAP disassembly) astrocyte processes surrounding adjacent neurons, thereby changing the uptake rate of extracellular neurochemical, their transport, conversion and/or release. As a result, neuronal activity is also changed as discussed below.

Hypotonic stimulation of the SON caused transient inhibition of VP neuronal activity (Wang et al., 2013a,b) and VP release (Yagil and Sladek, 1990), which were followed by recovery of the firing activity of VP neurons and rebound of VP secretion. These phenomena were accompanied with an initial increase in GFAP expression followed by a decrease in its expression, and were blocked by gliotoxin, L-aminoadipic acid. Importantly, GFAP increase was accompanied with simultaneous increase in the expression of glutamine synthetase and redistribution of this enzyme toward peripheral processes. This subcellular relocalization of glutamine synthetase to astrocytic processes (vs. soma) reflects upon an increased astrocytic ability to cause a shift in excitation-inhibition by decreasing extracellular concentrations of glutamate (as per increases conversion of glutamate to glutamine in astrocytes) and resulting in suppression of VP neuronal activity. Conversely a subsequent decrease in GFAP was associated with a decrease in membrane installation of GFAP-associated proteins and their functions (Wang et al., 2013a,b), including that of glutamine synthetase, resulting in an increase of extracellular levels of glutamate. Through this GFAP plasticity, astrocytes can adaptively exert their influence on the activity of VP and other OS neurons via differently localizing their functional proteins and changing neurochemical environment.

MECHANISMS UNDERLYING OSMOTRANSDUCTION IN ASTROCYTES

Activation of osmosensory cells is only the beginning of the osmosensory reaction to hydromineral disturbance. Osmosensation in the osmosensory system initiates osmotransduction that is responsible for osmotic adaptation process. The osmotransduction involves a complex set of signaling processes that are responsible for astrocytic plasticity under osmotic stress (**Figures 2A–C**). In this process, changes in GFAP and AQP4 expression are critical events, linked to diverse signaling cascades.

Transduction of Hyperosmotic Signals

Hyperosmotic stress can activate a variety of signaling molecules (**Figure 2B**). We provide a contemporary analysis of the spatiotemporal distribution of signaling molecules and their interactions; this analysis can aid better understanding of signaling cascades responsible for the transduction of different osmotic information in astrocytes. Several lines of evidence reveal that cellular effects of osmotic stress are commonly associated with intracellular Ca^{2+} mobilization and activation of mitogen-activated protein kinases (MAPKs) including c-Jun NH2-terminal kinase, p38 MAPK and extracellular signal-regulated kinase 1/2 (ERK 1/2; **Figures 2A–C**).

ECM-Integrin and Kinase Pathways

When osmotic stress acts on cell surface glycocalyx, integrin-mediated cell-matrix adhesion forms, cytoskeleton reorganizes, and focal adhesion kinase (FAK)-ERK1/2-c-Jun signaling axis gets activated, leading to the upregulation of interstitial collagenase, i.e., matrix metalloproteinase 13 (MMP-13) expression in rats (Horiguchi et al., 2011).

Hyperosmotic activation of the above pathway has dual effect on cellular events. The protective effect of integrins is clearly associated with phosphorylation/activation of FAK (Lunn and Rozengurt, 2004). In astrocytes, it has been reported that ECM binding to integrins can activate FAK, which is colocalized with actin stress fibers at sites of focal adhesion complexes (Rutka et al., 1999). An important downstream signal of FAK is phosphorylated ERK 1/2 (pERK 1/2) that is both essential for the activation of TRP channels during hyperosmotic stimulation (Prager-Khoutorsky and Bourque, 2015) and for GFAP polymerization. As reported, integrin-mediated ECM adhesions and cytoskeleton organization activate the FAK-ERK signaling axis, leading to upregulation of MMP expression and increased cell motility (Shi et al., 2011), a condition favoring astrocyte process retraction during hyperosmotic stimulation (Theodosis et al., 2008). Consistently, FAK signaling can promote organizing filamentous actin into parallel bundles (Rutka et al., 1999), which can interact with GFAP through calponin 3 (Plantier et al., 1999) and fascin (Mondal et al., 2010). Since an increase in this molecular association occurs during astrocyte retraction along with GFAP reduction (Wang and Hatton, 2009), the activation of FAK signaling could account for the astrocyte retraction following hyperosmotic stimulation, at least at the initial stage.

It is worth noting that increase in pERK 1/2 is essential for GFAP polymerization in the SON; however, pERK 1/2 is likely expressed in an undetectable level in astrocytes in the SON during hyperosmotic stimulation (Dine et al., 2014). Thus, the promoting effect of pERK 1/2 on GFAP polymerization could

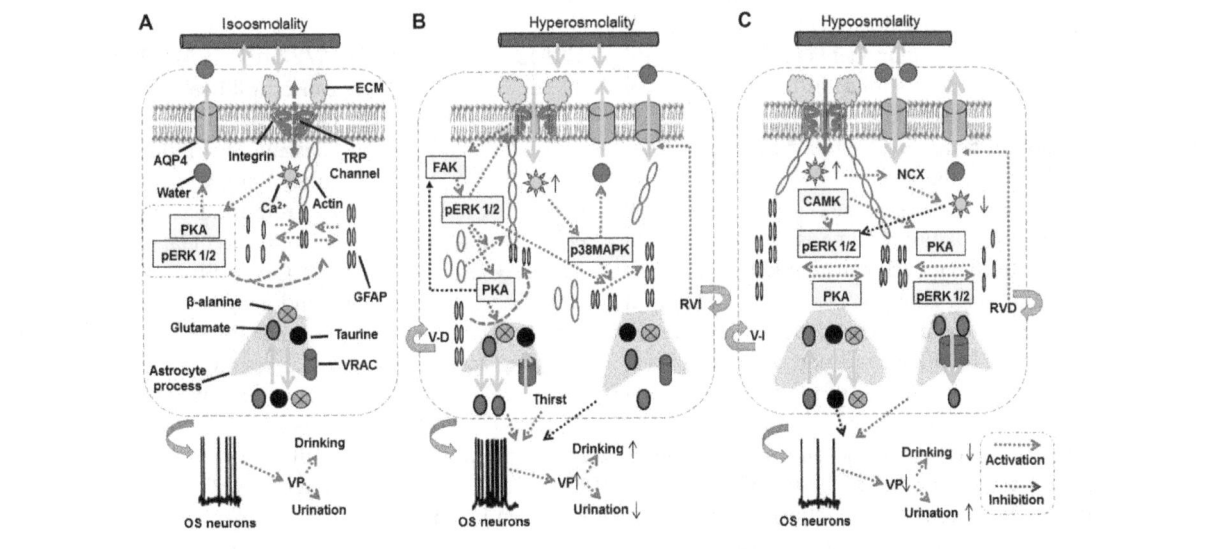

FIGURE 2 | The proposed mechanisms underlying astrocytic sensation and transduction of osmotic signals. **(A–C)** The signaling processes for isosmotic **(A)**, hyperosmotic **(B)** and hypoosmotic **(C)** conditions, respectively. The abbreviations are: AQP4, aquaporin4; CAMK, calmodulin kinase; ECM, extracellular matrix; FAK, focal adhesion kinase; GFAP, glial fibrillary acidic protein; MAPKs, mitogen-activated protein kinases; NCX, sodium-calcium exchanger; OS neuron, osmosensory neurons; pERK 1/2, phosphorylated extracellular signal-regulated kinase 1/2; PKA, protein kinase A; RVD, regulatory volume decrease; RVI, regulatory volume increase; TRP, transient receptor potential; V-D, volume decrease; V-I, volume increase; VRAC, volume-regulated anion channel. See text for details.

be minimal. Alternatively, activation of FAK-ERK 1/2 signaling could be restrained to the somata of astrocytes since pERK 1/2 promotion of GFAP polymerization has strong feature of microdomain-specificity (Wang et al., 2017). Lastly, pERK 1/2 signaling could be an upstream signal of protein kinase A (PKA) that can inhibit FAK activation (Padmanabhan et al., 1999) while increasing AQP4 permeability (Song and Gunnarson, 2012) and GFAP depolymerization in the SON (Wang et al., 2017).

It has been reported that cyclic adenosine 3′,5′-monophosphate (cAMP)-associated signaling is a component of the hyperosmotic signals. Salt-loaded rats had elevated levels of cAMP in the SON within 2 days of drinking 2% NaCl saline. Similarly, increasing medium osmolality from 290 to 325 mOsm/kg increased cAMP levels in the SON in rat explant culture (Carter and Murphy, 1989). Furthermore, increased cAMP promotes bovine VP gene expression (Pardy et al., 1992) through cAMP-PKA-CREB signaling (McCabe and Burrell, 2001). Albeit not directly demonstrated, one might expect that increased levels of cAMP would lead to increased expression-polymerization/assembly of GFAP in the SON, as it has been demonstrated in cultured astrocytes from mouse visual cortex (Gottipati et al., 2012). Taken together, it is very likely that a microdomain specific expression of pERK 1/2 and PKA is responsible for the GFAP-associated astrocytic plasticity during osmotic stress.

Cytosolic Ca^{2+} Levels and Associated Signaling Molecules

A direct effect of activating ECM-integrin signaling is the opening of TRPV channels, which causes increases in intracellular Ca^{2+} level. In turn, Ca^{2+} increase can activate MAPKs through Rho-type small G-protein (Kino et al., 2010), and osmotic stress further promotes couplings of activated Rho-type small G-proteins with c-Jun NH2-terminal kinase-interacting protein 4 to cascade components of the p38 MAPK signaling pathway. This signaling stimulates the expression of nuclear factor of activated T-cells 5, a transcription factor which regulates the expression of genes involved in the osmotic stress (Lee et al., 2011). Moreover, the activation of p38 MAPK also increases AQP4 expression (Salman et al., 2017), particularly following hyperosmotic stimulation (Arima et al., 2003), which facilitates water efflux from intracellular compartment to the extracellular space. Ca^{2+} signaling is also associated with GFAP plasticity (Chatterjee and Sikdar, 2013), which could also been mediated by p38 MAPK in addition to pERK 1/2 (Li et al., 2017). Indeed, in a mouse model of the middle cerebral artery occlusion, p38 MAPK activation was observed in the glial scar area, while in culture hypoxia and scratch injury-induced astrogliosis was attenuated by both p38 inhibition and knockout of p38 MAPK (Roy Choudhury et al., 2014); p38MAPK activation was associated with increased GFAP expression. Since increase in GFAP expression along with its associated astrocytic processes expansion does not occur at the initial stage of hyperosmotic stimulation, the GFAP increase could be a secondary reaction occurring before the RVI. Lastly, Ca^{2+} signaling can contribute to the RVI by antagonizing

a PKA-mediated increase in AQP4 permeability (Song and Gunnarson, 2012), which is supposed to further reduce water loss in the presence of higher extracellular osmolality. However, detailed regulation of GFAP reduction and astrocyte retraction following hyperosmotic stimulation remains to be explored.

Pro-inflammatory and Apoptotic Pathway

Chronic and severe hyperosmotic challenges can cause maladaptation of astrocytes and consequently result in detrimental responses of the osmosensory system. Uncontrolled hyperosmotic stress-elicited TRPV1 channel activation increases release of cytokines interleukin-6 and interleukin-8 which could result from transactivation of epidermal growth factor receptor, MAPK, and nuclear factor kappa-light-chain-enhancer of activated B cells activation (Cavet et al., 2011). Moreover, hyperosmotic saline also acts via MMP9 that binds to the low-density lipoprotein receptor-related protein-1, triggers the phosphorylation of ERK 1/2, and induces down-regulation of the perineurial barrier-forming tight junction protein claudin-1 (Hackel et al., 2012). The activation of MAPKs can also evoke cell apoptosis. For example, hyperosmotic stress-induced apoptosis is mediated by p38 MAPK, c-Jun NH2-terminal kinase and activating AP-1 transcription complex, and manifests with acceleration of cytochrome c release and caspase-3 activation (Ben Messaoud et al., 2015). Thus, hyperosmotic stress could cause systemic inflammation and apoptosis if dehydration or hypertonic stress is not corrected promptly.

Hypoosmotic Signals

Under hypoosmotic challenges, cell swelling and the ensuing RVD could involve a series of cellular and molecular events, which lead to adaptive changes in astrocytic morphology (**Figure 2C**).

Cytosolic Ca^{2+}, pERK1/2 and GFAP Plasticity

Hypoosmolality initially causes mild cell swelling that activates Ca^{2+}-permeable cation channels such as TRPV 4 (Benfenati et al., 2007) and leads to transient Ca^{2+} influx into the cells (Sato et al., 2013). The Ca^{2+} influx along with increased membrane stretch triggers Ca^{2+} release from intracellular stores (Borgdorff et al., 2000). This elevation in intracellular Ca^{2+} is essential for an increase in formation of GFAP filaments (González et al., 2007; Salazar et al., 2008) and a hypoosmotic volume increase (Ebner et al., 2006). Hypoosmolality-elicited Ca^{2+}/calmodulin signaling can increase the expression of pERK1/2 that is likely responsible for Ca^{2+}-associated cell swelling because chelation of extracellular Ca^{2+} also abolished pERK 1/2 increases under hypoosmotic conditions, while blocking the phosphorylation of ERK 1/2 significantly reduced cell swelling and the ensuing RVD, as shown in trout hepatocytes (Ebner et al., 2006). Importantly, prolonged hypoosmotic challenge *in vitro* was found to evoke only transient intracellular Ca^{2+} increase that was followed by a long decrease in Ca^{2+} levels (Sánchez-Olea et al., 1993), likely due to the activation of plasmalemmal Na$^+$-Ca^{2+} exchangers as shown in mouse odontoblasts (Sato et al., 2013). In the SON, the pERK1/2-associated GFAP polymerization and the elongation of astrocytic processes (Wang et al., 2017) appears as a short-lasting (minutes) event, which could be reversed by some subsequent event that causes GFAP depolymerization, such as PKA activation (Wang et al., 2017). The causal association between GFAP depolymerization and hypoosmolality-evoked RVD in chronic hypoosmotic challenges remains to be elucidated.

Cellular Signaling and RVD

Following Ca^{2+} influx and mobilization of intracellular Ca^{2+} store, high cytosolic Ca^{2+} levels lead to further cell swelling and RVD (Mola et al., 2015). In this process, calmodulin kinase plays a key role since intracellular application of monoclonal anti-calmodulin antibody blocked hypoosmotic activation of VRAC opening in rat cerebral astrocytes (Olson et al., 2004). Moreover, p38 MAPK is involved in the occurrence of RVD in hypoosmotic environment (Ebner et al., 2006). Whether these signals are directly linked to RVD remains to be explored.

Interestingly, PKA activation is associated with increased activation of VRAC and efflux of taurine and Cl^- from astrocytes with redistribution of actin network, i.e., its disruption at the somata, while concentrated at foci corresponding to the tips of the cell projections retracted by swelling (Moran et al., 2001). Moreover, actin cytoskeleton and microtubule are important for astrocyte swelling and they may form a barrier for the efflux of taurine and Cl^- since the presence of dense actin network can impede the flow of ions and water (Lange, 2000; Platonova et al., 2015).

Effects of Hypoosmotic Signals on Neuronal Activity

Accompanying the above signaling events and the resultant morphological changes, astrocytes could change neuronal activity through releasing gliotransmitters. The initial hypoosmotic inhibition of VP neuronal activity is associated with the increased availability of extracellular β-alanine, presumably through the reduction in its uptake (Johnston and Stephanson, 1976); the increased level of this non-essential amino acid then inhibits plasmalemmal GABA transporters and subsequently increases extracellular GABA levels (Park et al., 2006). The rebound excitation of VP neurons during hypoosmolality-evoked RVD is associated with the failure of GABAergic inhibition (Wang et al., 2013a) and coordinated D-serine signaling between astrocytes and MNCs in the SON to increase both NMDA receptor activation and VP neuronal activity (Wang et al., 2013b). In addition, the release of ATP from astrocytes during cell swelling could also contribute to the rebound excitation of neurons that are initially inhibited by hypoosmolality (Förster and Reiser, 2016) by activating ATP-sensitive K^+ channel (Thomzig et al., 2003). Lastly, hypoosmotic challenge-evoked taurine release through VRAC is also involved in the transmission of osmotic information. In the SON, taurine is released from astrocytes in an osmolality-dependent manner (Choe et al., 2012) and acts on glycine receptors to inhibit VP neuronal activity (Hussy et al., 2000). However, this effect is reversible during prolonged hypoosmotic challenges (Song and Hatton, 2003). The summary of potential mechanisms underlying astrocyte sensation and transduction of osmotic signals are outlined in **Figure 2**.

ASTROCYTES AND INTRACELLULAR CEREBRAL EDEMA

Cerebral edema is a common outcome when excess fluid is accumulated in the intracellular or extracellular space of the brain due to brain trauma or hydromineral disorder. According to the extent of the BBB disruption, cerebral edema can be divided into two types: (1) Cerebral edema with BBB disruption, which includes both vasogenic and interstitial edemas, featuring extracellular retention of water and Na^+; and (2) Cerebral edema without the BBB disruption, which includes cytotoxic and osmotic edemas, both of which feature intracellular swelling. Astrocytic plasticity is involved in brain edemas. Pathogenesis of vasogenic and interstitial edemas was recently reviewed (Jia et al., 2016; Wang and Parpura, 2016), so was the involvement of maladaptive ion transport in the occurrence of cerebral edema (Stokum et al., 2016); ionic edema is usually associated with cytotoxic edema, where water and ions exit capillaries into the brain interstitium. Thus, in this review we further explore the mechanisms underlying astrocyte-associated cytotoxic edema and osmotic edema.

Hyponatremia and Cerebral Edema

Hyponatremia is the most common clinical electrolyte disorder. When Na^+ plasma concentration falls more than several mmol/L below normal level, hyponatremia occurs (Kleindienst et al., 2016), and is associated with decompensated heart failure, hepatic cirrhosis with ascites formation, renal failure, inappropriate intravenous transfusion, and cancer-associated syndrome of inappropriate antidiuretic hormone secretion. Hyponatremia due to loss of NaCl occurs as a consequence to diarrhea and vomiting, overuse of diuretics, certain types of Na^+-wasting kidney diseases and Addison's disease with reduced aldosterone (Muñoz et al., 2010; Raimann et al., 2017). Hyponatremia creates an abnormal osmotic pressure gradient across the BBB and plasma membrane, which drives movement of water into the brain parenchyma and leads to osmotic cerebral edema (Walcott et al., 2012; Voets and Maas, 2018). The subsequent water intoxication leads to astrocytic swelling and RVD, which further disturb the activity of osmotic regulating system.

Hyponatremia-evoked cerebral edema likely delivers signals through the following process. As reported, excessive Ca^{2+} influx through TRPV4 predisposes Müller glial cells of the retina to activation of Ca^{2+}-dependent proapoptotic signaling pathways (Ryskamp et al., 2011). In glial cells of nematodes, it was also identified that osmotic stress induces massive protein aggregation coupled with unfolded protein response and endoplasmic reticulum (ER) stress (Gankam-Kengne et al., 2017); adequate protein folding is a tightly regulated process that requires proper intracellular ionic strength and it is necessary for normal cell function. Prolonged hyponatremia along with the activation of proapoptotic pathways could similarly cause damages to mammalian astrocytes. During RVD, the reversal of glutamate transporters and anion channel opening could increase glutamate release (Milanese et al., 2009). A sustained low extracellular sodium ion concentration decreased glutamate uptake by astrocytes and elevated extracellular glutamate concentration (Fujisawa et al., 2016). The release of glutamate from astrocytes can activate extrasynaptic NMDA receptors (Araque et al., 1998) to increase spontaneous synaptic transmission which can lead to an increase in production of NO (Kenny et al., 2018). Prolonged accumulation of NO could cause protein tyrosine nitration (PTN), damage the oxidative phosphorylation and trigger apoptosis in astrocytes (Raju et al., 2015). As a result, astrocyte uptake mechanism, e.g., for glutamate, can be disrupted, leading to the accumulation of the extracellular glutamate, and consequently increased VP neuronal activity and VP secretion as previously reviewed (Wang et al., 2011; Jia et al., 2016). However, since PTN is only possible with NO levels way in excess of what can be expected from either neuronal or epithelial NOS under strong hyperosmotic stimulus, the PTN-associated apoptosis in astrocytes is most likely a result of activation of inducible NOS as discussed in the pathogenesis of cardiac injury (Wang et al., 2018).

It is important to note that MNCs in the SON remain functional during dehydration and ischemic stroke because of down-regulation of NMDA glutamate receptor (Currás-Collazo and Dao, 1999), and neuroprotection by glia and vascular/perivascular cells (Currás-Collazo et al., 2002).

Central Pontine Myelinolysis

Central pontine myelinolysis is an osmotic demyelization syndrome often occurring during overly rapid correction of chronic hyponatremia with hypertonic saline infusion or using VP receptor antagonist. Astrocytes play an essential role in its development. As reported in rat models, massive astrocyte death occurred after rapid correction of hyponatremia. Astrocyte death caused a disruption of the astrocyte-oligodendrocyte network, rapidly upregulated inflammatory cytokines genes, and increased serum S100β, which predicted clinical manifestations and outcome of osmotic demyelination (Gankam Kengne et al., 2011). Treatment of hyponatremia with non-peptide VP receptor antagonists (vaptans) or hypertonic saline resulted in a higher mortality rate, when compared to treatment with urea, because of osmotic breakdown of the BBB, microglia activation, astrocyte demise and demyelination (Gankam Kengne et al., 2015). Moreover, astrocytes in central pontine myelinolysis were small, and exhibited fewer and shorter processes than perilesional astrocytes (Popescu et al., 2013) in addition to the loss of AQP4 in astrocytes within demyelinating lesions (Takagi et al., 2014). The damage of astrocytes is associated with imbalance between protein synthesis and degradation which can induce ER stress and cell death. As reported, rapid correction of chronic hyponatremia induces severe alterations in proteostasis, i.e., the biogenesis, folding, trafficking and degradation of proteins present within and outside the cell. The alterations are characterized by diffuse protein aggregation, ubiquitination and ER stress accompanied by increased autophagic activity and apoptosis in astrocytes within regions previously shown to be demyelinated in later stages of this syndrome (Gankam-Kengne et al., 2017). These results indicate that osmotic demyelination during severe osmotic stress might be a consequence of the

failure of proteostasis. Furthermore, these results support a model for the pathophysiology of osmotic brain injury in which rapid correction of hyponatremia triggers apoptosis in astrocytes followed by a loss of trophic communication between astrocytes and oligodendrocytes, secondary inflammation, microglial activation and finally demyelination.

Ischemic Cerebral Edema

The cytotoxic cerebral edema results from disruption in cellular metabolism and cellular retention of sodium and water due to poisoning or hypoxia. Hypoxia-associated cytotoxic cerebral edema is often observed in ischemic cerebral stroke (Jia et al., 2016), traumatic brain injury (Burda et al., 2016), cerebral palsy (Thomas et al., 2004) and cardiac arrest (Hirko et al., 2008). Astrocytes are critical contributors in this process, as typically observed in ischemic stroke (Wang and Parpura, 2016).

In the ischemic cerebral edema, a disruption of ionic and neurotransmitter homeostasis plays a pivotal role, particularly the accumulation of K^+ and glutamate in the extracellular space. The activation of astrocytic NKCC1 by highly increased extracellular K^+ concentrations and hypoxic inhibition of Na^+/K^+-ATPase form conditions promoting influx of $Na^+/K^+/Cl^-$, while inhibiting efflux of K^+ and Cl^- through K^+ and Cl^- co-transporter (Wilson and Mongin, 2018). During reperfusion following initial ischemia in the stroke, with the activation of Na^+/K^+-ATPase, more Na^+ release than K^+ uptake occurs in a 3:2 ratio, which drives NKCC1 to move more ions into the cell, leading to the development of cytotoxic cell swelling (Wang and Parpura, 2016). All these pathological processes are associated with malfunctioned uptake machinery for K^+ and glutamate (Steiner et al., 2012; Yan et al., 2016) and fluid volume transfer through AQP4 (Anderova et al., 2014) by astrocytes during ischemic stroke. Additionally, there is an increased level of extracellular VP during ischemic stroke (Jia et al., 2016). This notion is supported by a clinical observation that continuous intravenous infusion with conivaptan, a non-peptide antagonist of V1a and V2 VP receptors, for 48 h after experimental stroke reduces brain edema and BBB disruption (Zeynalov et al., 2015). Thus, in this process, damaged astrocytes and malfunctioned VP neurons could form a vicious cycle to worsen ischemic brain damages.

The disruption of ionic/neurotransmitter homeostasis is derived from reduced delivery of nutrients to the brain, in particular glucose and oxygen. In this process, disorders of many interacting molecular pathways in astrocytes, particularly TRPV2 activation-associated Ca^{2+} overload, are implicated. As reported in rat cortical astrocytes, oxygen-glucose deprivation followed by reoxygenation treatment enhanced the expression of TRPV2 and increased intracellular Ca^{2+} level (Zhang et al., 2017), which could disrupt normal functions involving mitochondria, ER, and nucleus, leading to cell damage.

Under physiological conditions, mitochondrial activity depends on increases in cytosolic Ca^{2+}, from the extracellular space or from the ER, which is taken up by mitochondria through specialized contact sites between the ER and mitochondria known as mitochondrial-associated ER membranes. The coordination of these Ca^{2+} pools is required to synchronize

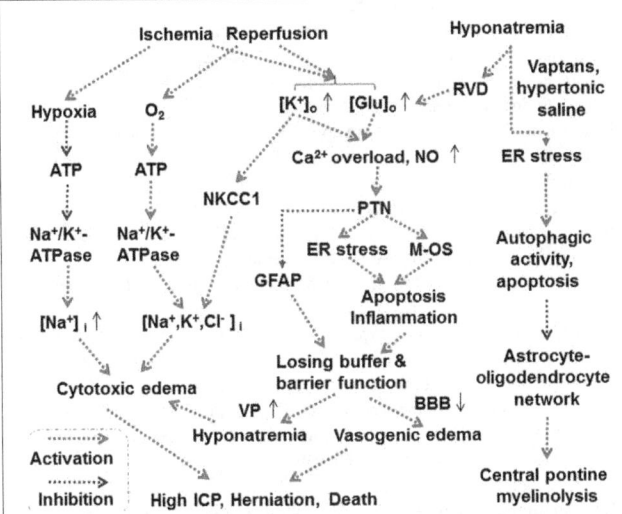

FIGURE 3 | Scheme depicting proposed astrocytes-associated mechanisms underlying cerebral edema. Abbreviations: BBB, blood-brain barrier; ER, endoplasmic reticulum; Glu, glutamate; ICP, intracranial pressure; M-OS, mitochondrial oxidative stress; NKCC1, Na+, K+, 2 Cl− and water cotransporter 1; NO, nitric oxide; PTN, protein tyrosine nitration. See text for details. Other annotations refer to **Figures 1, 2**.

post-traumatic edema (Shi et al., 2012; Wu et al., 2014) while the others just the opposite Hirt et al. (2017). It is possible that AQP4 is differentially involved in different types of cerebral water balance. By slowing the rate of water entry into brain, AQP4-null mice could be protected from cytotoxic brain edema induced by water intoxication, brain ischemia, or meningitis. By contrast, by reducing the rate of water outflow from brain parenchyma, AQP4 deletion could aggravate vasogenic brain edema caused by tumor, cortical freeze, intraparenchymal fluid infusion, or brain abscess (Papadopoulos and Verkman, 2007). However, the detailed mechanisms underlying this controversy remain to be explored. **Figure 3** summarizes the mechanisms underlying astrocytes-associated cerebral edema.

CONCLUSION

Malfunctions of astrocytes have been observed in many diseases in addition to those that are directly associated with hydromineral disorders, such as epilepsy (Losi et al., 2016), neuroinflammatory demyelinating disease (Verkman, 2012), Canavan disease (Clarner et al., 2014), Huntington's disease (Hsiao et al., 2014), amyotrophic lateral sclerosis (Ramírez-Jarquín et al., 2017), and post-traumatic syringomyelia (Najafi et al., 2016). Understanding of the involvement of astrocytes in neurohumoral modulation of HB and astrocytic mechanisms underlying cerebral edema will provide important references for clarification of all the pathogenesis of these diseases. Consequently, specific therapeutic targets might be identified with the perspectives of prevention of disease development and improvement of their prognosis.

AUTHOR CONTRIBUTIONS

Y-FW wrote the first draft. Y-FW and VP designed this topic and edited the manuscript.

mitochondrial respiration rates and ATP synthesis to meet physiological demands (Schäfer et al., 2014). However, serious hypoxia and glucose deprivation during ischemic stroke could cause strong mobilization of intracellular Ca2+ stores, ER stress, mitochondrial Ca2+ overload and oxidative stress in astrocytes (Hori et al., 2002; Ouyang et al., 2011), leading to inflammatory reaction and apoptosis. In this process, abnormal RVD following initial astrocytic swelling could create a hyperosmotic environment (Wang and Parpura, 2016) to cause inflammation (Cavet et al., 2011) and astrocytic apoptosis (Ben Messaoud et al., 2015), leading to irreversible brain damage.

It is worth to notice that a knockout of AQP4 yielded contradictory reports, some claiming reduction in

REFERENCES

Abrão Saad, W., Antonio De Arruda Camargo, L., Sérgio Cerri, P., Simões, S., Abrão Saad, G., Garcia, G., et al. (2004). Influence of arginine vasopressin receptors and angiotensin receptor subtypes on the water intake and arterial blood pressure induced by vasopressin injected into the lateral septal area of the rat. *Auton. Neurosci.* 111, 66–70. doi: 10.1016/j.autneu.2003.08.013

Almeida-Pereira, G., Coletti, R., Mecawi, A. S., Reis, L. C., Elias, L. L., and Antunes-Rodrigues, J. (2016). Estradiol and angiotensin II crosstalk in hydromineral balance: role of the ERK1/2 and JNK signaling pathways. *Neuroscience* 322, 525–538. doi: 10.1016/j.neuroscience.2016.02.067

Anderova, M., Benesova, J., Mikesova, M., Dzamba, D., Honsa, P., Kriska, J., et al. (2014). Altered astrocytic swelling in the cortex of α-syntrophin-negative GFAP/EGFP mice. *PLoS One* 9:e113444. doi: 10.1371/journal.pone.0113444

Araque, A., Sanzgiri, R. P., Parpura, V., and Haydon, P. G. (1998). Calcium elevation in astrocytes causes an NMDA receptor-dependent increase in the frequency of miniature synaptic currents in cultured hippocampal neurons. *J. Neurosci.* 18, 6822–6829. doi: 10.1523/JNEUROSCI.18-17-06822.1998

Araujo, I. G., Elias, L. L., Antunes-Rodrigues, J., Reis, L. C., and Mecawi, A. S.

(2013). Effects of acute and subchronic AT1 receptor blockade on cardiovascular, hydromineral and neuroendocrine responses in female rats. *Physiol. Behav.* 122, 104–112. doi: 10.1016/j.physbeh.2013.08.018

Arima, H., Yamamoto, N., Sobue, K., Umenishi, F., Tada, T., Katsuya, H., et al. (2003). Hyperosmolar mannitol simulates expression of aquaporins 4 and 9 through a p38 mitogen-activated protein kinase-dependent pathway in rat astrocytes. *J. Biol. Chem.* 278, 44525–44534. doi: 10.1074/jbc.M304368200

Bai, J. J., Tan, C. D., and Chow, B. K. C. (2017). Secretin, at the hub of water-salt homeostasis. *Am. J. Physiol. Renal Physiol.* 312, F852–F860. doi: 10.1152/ajprenal.00191.2015

Ben Messaoud, N., Katzarova, I., and López, J. M. (2015). Basic properties of the p38 signaling pathway in response to hyperosmotic shock. *PLoS One* 10:e0135249. doi: 10.1371/journal.pone.0135249

Benfenati, V., Amiry-Moghaddam, M., Caprini, M., Mylonakou, M. N., Rapisarda, C., Ottersen, O. P., et al. (2007). Expression and functional characterization of transient receptor potential vanilloid-related channel 4 (TRPV4) in rat cortical astrocytes. *Neuroscience* 148, 876–892. doi: 10.1016/j.neuroscience.2007.06.039

Borgdorff, A. J., Somjen, G. G., and Wadman, W. J. (2000). Two mechanisms that raise free intracellular calcium in rat hippocampal neurons during hypoosmotic and low NaCl treatment. *J. Neurophysiol.* 83, 81–89. doi: 10.1152/jn.2000.83.1.81

Brown, C. H. (2016). Magnocellular neurons and posterior pituitary function. *Compr. Physiol.* 6, 1701–1741. doi: 10.1002/cphy.c150053

Burda, J. E., Bernstein, A. M., and Sofroniew, M. V. (2016). Astrocyte roles in traumatic brain injury. *Exp. Neurol.* 275, 305–315. doi: 10.1016/j.expneurol.2015.03.020

Carter, D. A., and Murphy, D. (1989). Cyclic nucleotide dynamics in the rat hypothalamus during osmotic stimulation: *in vivo* and *in vitro* studies. *Brain Res.* 487, 350–356. doi: 10.1016/0006-8993(89)90839-1

Cavet, M. E., Harrington, K. L., Vollmer, T. R., Ward, K. W., and Zhang, J. Z. (2011). Anti-inflammatory and anti-oxidative effects of the green tea polyphenol epigallocatechin gallate in human corneal epithelial cells. *Mol. Vis.* 17, 533–542.

Chatterjee, S., and Sikdar, S. K. (2013). Corticosterone treatment results in enhanced release of peptidergic vesicles in astrocytes via cytoskeletal rearrangements. *Glia* 61, 2050–2062. doi: 10.1002/glia.22576

Cheng, B., Lin, M., Huang, G., Li, Y., Ji, B., Genin, G. M., et al. (2017). Cellular mechanosensing of the biophysical microenvironment: a review of mathematical models of biophysical regulation of cell responses. *Phys. Life Rev.* 22–23, 88–119. doi: 10.1016/j.plrev.2017.06.016

Choe, K. Y., Olson, J. E., and Bourque, C. W. (2012). Taurine release by astrocytes modulates osmosensitive glycine receptor tone and excitability in the adult supraoptic nucleus. *J. Neurosci.* 32, 12518–12527. doi: 10.1523/JNEUROSCI.1380-12.2012

Choe, K. Y., Prager-Khoutorsky, M., Farmer, W. T., Murai, K. K., and Bourque, C. W. (2016). Effects of salt loading on the morphology of astrocytes in the ventral glia limitans of the rat supraoptic nucleus. *J. Neuroendocrinol.* 28:4. doi: 10.1111/jne.12370

Ciura, S., Liedtke, W., and Bourque, C. W. (2011). Hypertonicity sensing in organum vasculosum lamina terminalis neurons: a mechanical process involving TRPV1 but not TRPV4. *J. Neurosci.* 31, 14669–14676. doi: 10.1523/JNEUROSCI.1420-11.2011

Clarner, T., Wieczorek, N., Krauspe, B., Jansen, K., Beyer, C., and Kipp, M. (2014). Astroglial redistribution of aquaporin 4 during spongy degeneration in a Canavan disease mouse model. *J. Mol. Neurosci.* 53, 22–30. doi: 10.1007/s12031-013-0184-4

Conner, M. T., Conner, A. C., Bland, C. E., Taylor, L. H., Brown, J. E., Parri, H. R., et al. (2012). Rapid aquaporin translocation regulates cellular water flow: mechanism of hypotonicity-induced subcellular localization of aquaporin 1 water channel. *J. Biol. Chem.* 287, 11516–11525. doi: 10.1074/jbc.M111.329219

Currás-Collazo, M. C., and Dao, J. (1999). Osmotic activation of the hypothalamo-neurohypophysial system reversibly downregulates the NMDA receptor subunit, NR2B, in the supraoptic nucleus of the hypothalamus. *Mol. Brain Res.* 70, 187–196. doi: 10.1016/s0169-328x(99)00129-1

Currás-Collazo, M. C., Patel, U. B., and Hussein, M. O. (2002). Reduced susceptibility of magnocellular neuroendocrine nuclei of the rat hypothalamus to transient focal ischemia produced by middle cerebral artery occlusion. *Exp. Neurol.* 178, 268–279. doi: 10.1006/exnr.2002.8032

Dine, J., Ducourneau, V. R., Fenelon, V. S., Fossat, P., Amadio, A., Eder, M., et al. (2014). Extracellular signal-regulated kinase phosphorylation in forebrain neurones contributes to osmoregulatory mechanisms. *J. Physiol.* 592, 1637–1654. doi: 10.1113/jphysiol.2013.261008

Dominguez-Meijide, A., Rodriguez-Perez, A. I., Diaz-Ruiz, C., Guerra, M. J., and Labandeira-Garcia, J. L. (2017). Dopamine modulates astroglial and microglial activity via glial renin-angiotensin system in cultures. *Brain Behav. Immun.* 62, 277–290. doi: 10.1016/j.bbi.2017.02.013

Du, W., Stern, J. E., and Filosa, J. A. (2015). Neuronal-derived nitric oxide and somatodendritically released vasopressin regulate neurovascular coupling in the rat hypothalamic supraoptic nucleus. *J. Neurosci.* 35, 5330–5341. doi: 10.1523/JNEUROSCI.3674-14.2015

Ebner, H. L., Fiechtner, B., Pelster, B., and Krumschnabel, G. (2006). Extracellular signal regulated MAP-kinase signalling in osmotically stressed trout hepatocytes. *Biochim. Biophys. Acta* 1760, 941–950. doi: 10.1016/j.bbagen.2006.03.017

Eriksson, P. S., Nilsson, M., Wågberg, M., Rönnbäck, L., and Hansson, E. (1992). Volume regulation of single astroglial cells in primary culture. *Neurosci. Lett.* 143, 195–199. doi: 10.1016/0304-3940(92)90264-8

Farmer, W. T., Abrahamsson, T., Chierzi, S., Lui, C., Zaelzer, C., Jones, E. V., et al. (2016). Neurons diversify astrocytes in the adult brain through sonic hedgehog signaling. *Science* 351, 849–854. doi: 10.1126/science.aab3103

Felgendreger, L. A., Fluharty, S. J., Yee, D. K., and Flanagan-Cato, L. M. (2013). Endogenous angiotensin II-induced p44/42 mitogen-activated protein kinase activation mediates sodium appetite but not thirst or neurohypophysial secretion in male rats. *J. Neuroendocrinol.* 25, 97–106. doi: 10.1111/j.1365-2826.2012.02376.x

Förster, D., and Reiser, G. (2016). Nucleotides protect rat brain astrocytes against hydrogen peroxide toxicity and induce antioxidant defense via P2Y receptors. *Neurochem. Int.* 94, 57–66. doi: 10.1016/j.neuint.2016.02.006

Fujisawa, H., Sugimura, Y., Takagi, H., Mizoguchi, H., Takeuchi, H., Izumida, H., et al. (2016). Chronic hyponatremia causes neurologic and psychologic impairments. *J. Am. Soc. Nephrol.* 27, 766–780. doi: 10.1681/ASN.2014121196

Gankam-Kengne, F., Couturier, B. S., Soupart, A., Brion, J. P., and Decaux, G. (2017). Osmotic stress-induced defective glial proteostasis contributes to brain demyelination after hyponatremia treatment. *J. Am. Soc. Nephrol.* 28, 1802–1813. doi: 10.1681/ASN.2016050509

Gankam Kengne, F., Couturier, B. S., Soupart, A., and Decaux, G. (2015). Urea minimizes brain complications following rapid correction of chronic hyponatremia compared with vasopressin antagonist or hypertonic saline. *Kidney Int.* 87, 323–331. doi: 10.1038/ki.2014.273

Gankam Kengne, F., Nicaise, C., Soupart, A., Boom, A., Schiettecatte, J., Pochet, R., et al. (2011). Astrocytes are an early target in osmotic demyelination syndrome. *J. Am. Soc. Nephrol.* 22, 1834–1845. doi: 10.1681/ASN.2010111127

Garrido-Gil, P., Rodriguez-Perez, A. I., Fernandez-Rodriguez, P., Lanciego, J. L., and Labandeira-Garcia, J. L. (2017). Expression of angiotensinogen and receptors for angiotensin and prorenin in the rat and monkey striatal neurons and glial cells. *Brain Struct. Funct.* 222, 2559–2571. doi: 10.1007/s00429-016-1357-z

Geerling, J. C., and Loewy, A. D. (2008). Central regulation of sodium appetite. *Exp. Physiol.* 93, 177–209. doi: 10.1113/expphysiol.2007.039891

Glasgow, E., Murase, T., Zhang, B., Verbalis, J. G., and Gainer, H. (2000). Gene expression in the rat supraoptic nucleus induced by chronic hyperosmolality versus hyposmolality. *Am. J. Physiol. Regul. Integr. Comp. Physiol.* 279, R1239–1250. doi: 10.1152/ajpregu.2000.279.4.r1239

González, A., Pariente, J. A., and Salido, G. M. (2007). Ethanol stimulates ROS generation by mitochondria through Ca^{2+} mobilization and increases GFAP content in rat hippocampal astrocytes. *Brain Res.* 1178, 28–37. doi: 10.1016/j.brainres.2007.08.040

Gottipati, M. K., Kalinina, I., Bekyarova, E., Haddon, R. C., and Parpura, V. (2012). Chemically functionalized water-soluble single-walled carbon nanotubes modulate morpho-functional characteristics of astrocytes. *Nano Lett.* 12, 4742–4747. doi: 10.1021/nl302178s

Gutkowska, J., Jankowski, M., and Antunes-Rodrigues, J. (2014). The role of oxytocin in cardiovascular regulation. *Braz. J. Med. Biol. Res.* 47, 206–214. doi: 10.1590/1414-431x20133309

Gutkowska, J., Jankowski, M., Lambert, C., Mukaddam-Daher, S., Zingg, H. H., and Mccann, S. M. (1997). Oxytocin releases atrial natriuretic peptide by combining with oxytocin receptors in the heart. *Proc. Natl. Acad. Sci. U S A* 94, 11704–11709. doi: 10.1073/pnas.94.21.11704

Hackel, D., Krug, S. M., Sauer, R. S., Mousa, S. A., Böcker, A., Pflücke, D., et al. (2012). Transient opening of the perineurial barrier for analgesic drug delivery. *Proc. Natl. Acad. Sci. U S A* 109, E2018–E2027. doi: 10.1073/pnas.1120800109

Hatton, G. I. (2002). Glial-neuronal interactions in the mammalian brain. *Adv. Physiol. Educ.* 26, 225–237. doi: 10.1152/advan.00038.2002

Hermann, K., Phillips, M. I., Hilgenfeldt, U., and Raizada, M. K. (1988). Biosynthesis of angiotensinogen and angiotensins by brain cells in primary culture. *J. Neurochem.* 51, 398–405. doi: 10.1111/j.1471-4159.1988.tb01052.x

Hertz, L., Chen, Y., and Song, D. (2017). Astrocyte cultures mimicking brain astrocytes in gene expression, signaling, metabolism and K^+ uptake and showing astrocytic gene expression overlooked by immunohistochemistry and in situ hybridization. *Neurochem. Res.* 42, 254–271. doi: 10.1007/s11064-016-1828-x

Hertz, L., Xu, J., Song, D., Yan, E., Gu, L., and Peng, L. (2013). Astrocytic and neuronal accumulation of elevated extracellular K^+ with a 2/3 K^+/Na^+ flux ratio-consequences for energy metabolism, osmolarity and higher brain function. *Front. Comput. Neurosci.* 7:114. doi: 10.3389/fncom.2013. 00114

Hindmarch, C. C., and Ferguson, A. V. (2016). Physiological roles for the subfornical organ: a dynamic transcriptome shaped by autonomic state. *J. Physiol.* 594, 1581–1589. doi: 10.1113/JP270726

Hirko, A. C., Dallasen, R., Jomura, S., and Xu, Y. (2008). Modulation of inflammatory responses after global ischemia by transplanted umbilical cord matrix stem cells. *Stem Cells* 26, 2893–2901. doi: 10.1634/stemcells. 2008-0075

Hirt, L., Fukuda, A. M., Ambadipudi, K., Rashid, F., Binder, D., Verkman, A., et al. (2017). Improved long-term outcome after transient cerebral ischemia in aquaporin-4 knockout mice. *J. Cereb. Blood Flow Metab.* 37, 277–290. doi: 10.1177/0271678x15623290

Ho, J. M., Zierath, D. K., Savos, A. V., Femiano, D. J., Bassett, J. E., McKinley, M. J., et al. (2007). Differential effects of intravenous hyperosmotic solutes on drinking latency and c-Fos expression in the circumventricular organs and hypothalamus of the rat. *Am. J. Physiol. Regul. Integr. Comp. Physiol.* 292, R1690–R1698. doi: 10.1152/ajpregu.00547.2006

Hori, O., Ichinoda, F., Tamatani, T., Yamaguchi, A., Sato, N., Ozawa, K., et al. (2002). Transmission of cell stress from endoplasmic reticulum to mitochondria: enhanced expression of Lon protease. *J. Cell Biol.* 157, 1151–1160. doi: 10.1083/jcb.200108103

Horiguchi, K., Fujiwara, K., Ilmiawati, C., Kikuchi, M., Tsukada, T., Kouki, T., et al. (2011). Caveolin 3-mediated integrin β1 signaling is required for the proliferation of folliculostellate cells in rat anterior pituitary gland under the influence of extracellular matrix. *J. Endocrinol.* 210, 29–36. doi: 10.1530/joe-11-0103

Hou, D., Jin, F., Li, J., Lian, J., Liu, M., Liu, X.-N., et al. (2016). Model roles of the hypothalamo-neurohypophysial system in neuroscience study. *Biochem. Pharmacol.* 5:211. doi: 10.4172/2167-0501.1000211

Hsiao, H.-Y., Chiu, F.-L., Chen, C.-M., Wu, Y.-R., Chen, H.-M., Chen, Y.-C., et al. (2014). Inhibition of soluble tumor necrosis factor is therapeutic in Huntington's disease. *Hum. Mol. Genet.* 23, 4328–4344. doi: 10.1093/hmg/ddu151

Hübel, N., and Ullah, G. (2016). Anions govern cell volume: a case study of relative astrocytic and neuronal swelling in spreading depolarization. *PLoS One* 11:e0147060. doi: 10.1371/journal.pone.0147060

Hussy, N., Deleuze, C., Desarménien, M. G., and Moos, F. C. (2000). Osmotic regulation of neuronal activity: a new role for taurine and glial cells in a hypothalamic neuroendocrine structure. *Prog. Neurobiol.* 62, 113–134. doi: 10.1016/s0301-0082(99)00071-4

Irazusta, J., Silveira, P. F., Gil, J., Varona, A., and Casis, L. (2001). Effects of hydrosaline treatments on prolyl endopeptidase activity in rat tissues. *Regul. Pept.* 101, 141–147. doi: 10.1016/s0167-0115(01)00277-4

Ishikura, T., Suzuki, H., Shoguchi, K., Koreeda, Y., Aritomi, T., Matsuura, T., et al. (2015). Possible involvement of TRPV1 and TRPV4 in nociceptive stimulation-induced nocifensive behavior and neuroendocrine response in mice. *Brain Res. Bull.* 118, 7–16. doi: 10.1016/j.brainresbull.2015.08.004

Jia, S. W., Liu, X. Y., Wang, S. C., and Wang, Y. F. (2016). Vasopressin hypersecretion-associated brain edema formation in ischemic stroke: underlying mechanisms. *J. Stroke Cerebrovasc. Dis.* 25, 1289–1300. doi: 10.1016/j.jstrokecerebrovasdis.2016.02.002

Jiao, R., Cui, D., Wang, S. C., Li, D., and Wang, Y. F. (2017). Interactions of the mechanosensitive channels with extracellular matrix, integrins, and cytoskeletal network in osmosensation. *Front. Mol. Neurosci.* 10:96. doi: 10.3389/fnmol.2017.00096

Johnston, G. A., and Stephanson, A. L. (1976). Inhibitors of the glial uptake of β-alanine in rat brain slices. *Brain Res.* 102, 374–378. doi: 10.1016/0006-8993(76)90895-7

Kenny, A., Plank, M. J., and David, T. (2018). The role of astrocytic calcium and TRPV4 channels in neurovascular coupling. *J. Comput. Neurosci.* 44, 97–114. doi: 10.1007/s10827-017-0671-7

Kim, J. S., Kim, W. B., Kim, Y. B., Lee, Y., Kim, Y. S., Shen, F. Y., et al. (2011). Chronic hyperosmotic stress converts GABAergic inhibition into excitation in vasopressin and oxytocin neurons in the rat. *J. Neurosci.* 31, 13312–13322. doi: 10.1523/JNEUROSCI.1440-11.2011

Kino, T., Segars, J. H., and Chrousos, G. P. (2010). The guanine nucleotide exchange factor brx: a link between osmotic stress, inflammation and organ physiology and pathophysiology. *Expert Rev. Endocrinol. Metab.* 5, 603–614. doi: 10.1586/eem.10.3

Kinsman, B. J., Nation, H. N., and Stocker, S. D. (2017). Hypothalamic signaling in body fluid homeostasis and hypertension. *Curr. Hypertens. Rep.* 19:50. doi: 10.1007/s11906-017-0749-7

Kleindienst, A., Hannon, M. J., Buchfelder, M., and Verbalis, J. G. (2016). Hyponatremia in neurotrauma: the role of vasopressin. *J. Neurotrauma* 33, 615–624. doi: 10.1089/neu.2015.3981

Krzan, M., Stenovec, M., Kreft, M., Pangrsic, T., Grilc, S., Haydon, P. G., et al. (2003). Calcium-dependent exocytosis of atrial natriuretic peptide from astrocytes. *J. Neurosci.* 23, 1580–1583. doi: 10.1523/JNEUROSCI.23-05-01580.2003

Kurbel, S. (2008). Are extracellular osmolality and sodium concentration determined by Donnan effects of intracellular protein charges and of pumped sodium? *J. Theor. Biol.* 252, 769–772. doi: 10.1016/j.jtbi.2008.02.022

Lange, K. (2000). Regulation of cell volume via microvillar ion channels. *J. Cell. Physiol.* 185, 21–35. doi: 10.1002/1097-4652(200010)185:1<21::aid-jcp2>3.3. co;2-4

Lauand, F., Ruginsk, S. G., Rodrigues, H. L., Reis, W. L., de Castro, M., Elias, L. L., et al. (2007). Glucocorticoid modulation of atrial natriuretic peptide, oxytocin, vasopressin and Fos expression in response to osmotic, angiotensinergic and cholinergic stimulation. *Neuroscience* 147, 247–257. doi: 10.1016/j.neuroscience.2007.04.021

Lee, S. D., Choi, S. Y., and Kwon, H. M. (2011). Distinct cellular pathways for resistance to urea stress and hypertonic stress. *Am. J. Physiol. Cell Physiol.* 300, C692–C696. doi: 10.1152/ajpcell.00150.2010

Li, D., Liu, N., Zhao, H. H., Zhang, X., Kawano, H., Liu, L., et al. (2017). Interactions between Sirt1 and MAPKs regulate astrocyte activation induced by brain injury *in vitro* and *in vivo*. *J. Neuroinflammation* 14:67. doi: 10.1186/s12974-017-0841-6

Liedtke, W., and Friedman, J. M. (2003). Abnormal osmotic regulation in *trpv4$^{-/-}$* mice. *Proc. Natl. Acad. Sci. U S A* 100, 13698–13703. doi: 10.1073/pnas.1735416100

Lipari, E. F., Lipari, A., Dieli, F., and Valentino, B. (2007). ANP presence in the hypothalamic suprachiasmatic nucleus of developing rat. *Ital. J. Anat. Embryol.* 112, 19–25.

Losi, G., Marcon, I., Mariotti, L., Sessolo, M., Chiavegato, A., and Carmignoto, G. (2016). A brain slice experimental model to study the generation and the propagation of focally-induced epileptiform activity. *J. Neurosci. Methods* 260, 125–131. doi: 10.1016/j.jneumeth.2015.04.001

Ludwig, M., Bull, P. M., Tobin, V. A., Sabatier, N., Landgraf, R., Dayanithi, G., et al. (2005). Regulation of activity-dependent dendritic vasopressin release from rat supraoptic neurones. *J. Physiol.* 564, 515–522. doi: 10.1113/jphysiol. 2005.083931

Lunn, J. A., and Rozengurt, E. (2004). Hyperosmotic stress induces rapid focal adhesion kinase phosphorylation at tyrosines 397 and 577. Role of Src family kinases and Rho family GTPases. *J. Biol. Chem.* 279, 45266–45278. doi: 10.1074/jbc.M314132200

Macchione, A. F., Beas, C., Dadam, F. M., Caeiro, X. E., Godino, A., Ponce, L. F., et al. (2015). Early free access to hypertonic NaCl solution induces a long-term effect on drinking, brain cell activity and gene expression of adult rat offspring. *Neuroscience* 298, 120–136. doi: 10.1016/j.neuroscience. 2015.04.004

Malarkey, E. B., and Parpura, V. (2009). "Mechanisms of transmitter release from astrocytes," in *Astrocytes in (Patho)Physiology of the Nervous System*, eds V. Parpura and P. G. Haydon (New York, NY: Springer), 301–350.

Mannari, T., Morita, S., Furube, E., Tominaga, M., and Miyata, S. (2013). Astrocytic TRPV1 ion channels detect blood-borne signals in the sensory circumventricular organs of adult mouse brains. *Glia* 61, 957–971. doi: 10.1002/glia.22488

Marmarou, C. R., Liang, X., Abidi, N. H., Parveen, S., Taya, K., Henderson, S. C., et al. (2014). Selective vasopressin-1a receptor antagonist prevents brain edema, reduces astrocytic cell swelling and GFAP, V1aR and AQP4 expression after focal traumatic brain injury. *Brain Res.* 1581, 89–102. doi: 10.1016/j.brainres. 2014.06.005

McCabe, J. T., and Burrell, A. S. (2001). Alterations of AP-1 and CREB protein DNA binding in rat supraoptic and paraventricular nuclei by acute and repeated hyperosmotic stress. *Brain Res. Bull.* 55, 347–358. doi: 10.1016/s0361-9230(01)00520-2

McKinley, M. J., and McAllen, R. M. (2007). Neuroendocrine self-control: dendritic release of vasopressin. *Endocrinology* 148, 477–478. doi: 10.1210/en.2006-1531

McKinley, M. J., Yao, S. T., Uschakov, A., McAllen, R. M., Rundgren, M., and Martelli, D. (2015). The median preoptic nucleus: front and centre for the regulation of body fluid, sodium, temperature, sleep and cardiovascular homeostasis. *Acta Physiol.* 214, 8–32. doi: 10.1111/apha.12487

Milanese, M., Bonifacino, T., Zappettini, S., Usai, C., Tacchetti, C., Nobile, M., et al. (2009). Glutamate release from astrocytic gliosomes under physiological and pathological conditions. *Int. Rev. Neurobiol.* 85, 295–318. doi: 10.1016/s0074-7742(09)85021-6

Miller, R. L., and Loewy, A. D. (2013). ENaC γ-expressing astrocytes in the circumventricular organs, white matter, and ventral medullary surface: sites for Na$^+$ regulation by glial cells. *J. Chem. Neuroanat.* 53, 72–80. doi: 10.1016/j.jchemneu.2013.10.002

Miyata, S. (2015). New aspects in fenestrated capillary and tissue dynamics in the sensory circumventricular organs of adult brains. *Front. Neurosci.* 9:390. doi: 10.3389/fnins.2015.00390

Mola, M. G., Sparaneo, A., Gargano, C. D., Spray, D. C., Svelto, M., Frigeri, A., et al. (2015). The speed of swelling kinetics modulates cell volume regulation and calcium signaling in astrocytes: a different point of view on the role of aquaporins. *Glia* 64, 139–154. doi: 10.1002/glia.22921

Mondal, S., Dirks, P., and Rutka, J. T. (2010). Immunolocalization of fascin, an actin-bundling protein and glial fibrillary acidic protein in human astrocytoma cells. *Brain Pathol.* 20, 190–199. doi: 10.1111/j.1750-3639.2008.00261.x

Montana, V., Malarkey, E. B., Verderio, C., Matteoli, M., and Parpura, V. (2006). Vesicular transmitter release from astrocytes. *Glia* 54, 700–715. doi: 10.1002/glia.20367

Moran, J., Morales-Mulia, M., and Pasantes-Morales, H. (2001). Reduction of phospholemman expression decreases osmosensitive taurine efflux in astrocytes. *Biochim. Biophys. Acta* 1538, 313–320. doi: 10.1016/s0167-4889(01)00082-9

Muhsin, S. A., and Mount, D. B. (2016). Diagnosis and treatment of hypernatremia. *Best Pract. Res. Clin. Endocrinol. Metab.* 30, 189–203. doi: 10.1016/j.beem.2016.02.014

Muñoz, A., Riber, C., Trigo, P., and Castejón, F. (2010). Muscle damage, hydration, electrolyte balance and vasopressin concentrations in successful and exhausted endurance horses. *Pol. J. Vet. Sci.* 13, 373–379. doi: 10.1111/j.2042-3306.2010.00211.x

Muraki, K., Shigekawa, M., and Imaizumi, Y. (2007). "Chapter 28-A new insight into the function of TRPV2 in circulatory organs," in *TRP Ion Channel Function in Sensory Transduction and Cellular Signaling Cascades*, eds W. B. Liedtke and S. Heller (Boca Raton, FL: CRC Press/Taylor & Francis).

Najafi, E., Stoodley, M. A., Bilston, L. E., and Hemley, S. J. (2016). Inwardly rectifying potassium channel 4.1 expression in post-traumatic syringomyelia. *Neuroscience* 317, 23–35. doi: 10.1016/j.neuroscience.2016.01.001

Ni, Y., Malarkey, E. B., and Parpura, V. (2007). Vesicular release of glutamate mediates bidirectional signaling between astrocytes and neurons. *J. Neurochem.* 103, 1273–1284. doi: 10.1111/j.1471-4159.2007.04864.x

Niermann, H., Amiry-Moghaddam, M., Holthoff, K., Witte, O. W., and Ottersen, O. P. (2001). A novel role of vasopressin in the brain: modulation of activity-dependent water flux in the neocortex. *J. Neurosci.* 21, 3045–3051. doi: 10.1523/JNEUROSCI.21-09-03045.2001

Noda, M. (2007). Hydromineral neuroendocrinology: mechanism of sensing sodium levels in the mammalian brain. *Exp. Physiol.* 92, 513–522. doi: 10.1113/expphysiol.2006.035659

Ohbuchi, T., Yokoyama, T., Saito, T., Suzuki, H., Fujihara, H., Katoh, A., et al. (2010). Modulators of BK and SK channels alter electrical activity *in vitro* in single vasopressin neurons isolated from the rat supraoptic nucleus. *Neurosci. Lett.* 484, 26–29. doi: 10.1016/j.neulet.2010.08.010

Olson, J. E., Li, G. Z., Wang, L., and Lu, L. (2004). Volume-regulated anion conductance in cultured rat cerebral astrocytes requires calmodulin activity. *Glia* 46, 391–401. doi: 10.1002/glia.20014

Ordaz, B., Tuz, K., Ochoa, L. D., Lezama, R., Peña-Segura, C., and Franco, R. (2004). Osmolytes and mechanisms involved in regulatory volume decrease

under conditions of sudden or gradual osmolarity decrease. *Neurochem. Res.* 29, 65–72. doi: 10.1023/b:nere.0000010434.06311.18

Ouyang, Y. B., Xu, L. J., Emery, J. F., Lee, A. S., and Giffard, R. G. (2011). Overexpressing GRP78 influences Ca^{2+} handling and function of mitochondria in astrocytes after ischemia-like stress. *Mitochondrion* 11, 279–286. doi: 10.1016/j.mito.2010.10.007

Padmanabhan, J., Clayton, D., and Shelanski, M. L. (1999). Dibutyryl cyclic AMP-induced process formation in astrocytes is associated with a decrease in tyrosine phosphorylation of focal adhesion kinase and paxillin. *J. Neurobiol.* 39, 407–422. doi: 10.1002/(sici)1097-4695(19990605)39:3<407::aid-neu7>3.0.co;2-s

Papadopoulos, M. C., and Verkman, A. S. (2007). Aquaporin-4 and brain edema. *Pediatr. Nephrol.* 22, 778–784. doi: 10.1007/s00467-006-0411-0

Pardy, K., Adan, R. A., Carter, D. A., Seah, V., Burbach, J. P., and Murphy, D. (1992). The identification of a cis-acting element involved in cyclic 3′,5′-adenosine monophosphate regulation of bovine vasopressin gene expression. *J. Biol. Chem.* 267, 21746–21752.

Park, J. B., Skalska, S., and Stern, J. E. (2006). Characterization of a novel tonic γ-aminobutyric acidA receptor-mediated inhibition in magnocellular neurosecretory neurons and its modulation by glia. *Endocrinology* 147, 3746–3760. doi: 10.1210/en.2006-0218

Parpura, V., and Verkhratsky, A. (2012). Astrocytes revisited: concise historic outlook on glutamate homeostasis and signaling. *Croat. Med. J.* 53, 518–528. doi: 10.3325/cmj.2012.53.518

Parpura, V., Fisher, E. S., Lechleiter, J. D., Schousboe, A., Waagepetersen, H. S., Brunet, S., et al. (2017). Glutamate and ATP at the interface between signaling and metabolism in astroglia: examples from pathology. *Neurochem. Res.* 42, 19–34. doi: 10.1007/s11064-016-1848-6

Parpura, V., Scemes, E., and Spray, D. C. (2004). Mechanisms of glutamate release from astrocytes: gap junction "hemichannels", purinergic receptors and exocytotic release. *Neurochem. Int.* 45, 259–264. doi: 10.1016/s0197-0186(03)00290-0

Perkins, K. L., Arranz, A. M., Yamaguchi, Y., and Hrabetova, S. (2017). Brain extracellular space, hyaluronan, and the prevention of epileptic seizures. *Rev. Neurosci.* 28, 869–892. doi: 10.1515/revneuro-2017-0017

Pierre, K., Bonhomme, R., Dupouy, B., Poulain, D. A., and Theodosis, D. T. (2001). The polysialylated neural cell adhesion molecule reaches cell surfaces of hypothalamic neurons and astrocytes via the constitutive pathway. *Neuroscience* 103, 133–142. doi: 10.1016/s0306-4522(00)00536-4

Plantier, M., Fattoum, A., Menn, B., Ben-Ari, Y., Der Terrossian, E., and Represa, A. (1999). Acidic calponin immunoreactivity in postnatal rat brain and cultures: subcellular localization in growth cones, under the plasma membrane and along actin and glial filaments. *Eur. J. Neurosci.* 11, 2801–2812. doi: 10.1046/j.1460-9568.1999.00702.x

Platonova, A., Ponomarchuk, O., Boudreault, F., Kapilevich, L. V., Maksimov, G. V., Grygorczyk, R., et al. (2015). Role of cytoskeleton network in anisosmotic volume changes of intact and permeabilized A549 cells. *Biochim. Biophys. Acta* 1848, 2337–2343. doi: 10.1016/j.bbamem.2015.07.005

Ponzio, T. A., Ni, Y., Montana, V., Parpura, V., and Hatton, G. I. (2006). Vesicular glutamate transporter expression in supraoptic neurones suggests a glutamatergic phenotype. *J. Neuroendocrinol.* 18, 253–265. doi: 10.1111/j.1365-2826.2006.01410.x

Popescu, B. F., Bunyan, R. F., Guo, Y., Parisi, J. E., Lennon, V. A., and Lucchinetti, C. F. (2013). Evidence of aquaporin involvement in human central pontine myelinolysis. *Acta Neuropathol. Commun.* 1:40. doi: 10.1186/2051-5960-1-40

Potokar, M., Stenovec, M., Jorgacevski, J., Holen, T., Kreft, M., Ottersen, O. P., et al. (2013). Regulation of AQP4 surface expression via vesicle mobility in astrocytes. *Glia* 61, 917–928. doi: 10.1002/glia.22485

Prager-Khoutorsky, M., and Bourque, C. W. (2015). Mechanical basis of osmosensory transduction in magnocellular neurosecretory neurones of the rat supraoptic nucleus. *J. Neuroendocrinol.* 27, 507–515. doi: 10.1111/jne.12270

Raimann, J. G., Tzamaloukas, A. H., Levin, N. W., and Ing, T. S. (2017). Osmotic pressure in clinical medicine with an emphasis on dialysis. *Semin. Dial.* 30, 69–79. doi: 10.1111/sdi.12537

Raju, K., Doulias, P. T., Evans, P., Krizman, E. N., Jackson, J. G., Horyn, O., et al. (2015). Regulation of brain glutamate metabolism by nitric oxide and S-nitrosylation. *Sci. Signal.* 8:ra68. doi: 10.1126/scisignal.aaa4312

Ramírez-Jarquín, U. N., Rojas, F., van Zundert, B., and Tapia, R. (2017). Chronic infusion of SOD1G93A astrocyte-secreted factors induces spinal motoneuron degeneration and neuromuscular dysfunction in healthy rats. *J. Cell. Physiol.* 232, 2610–2615. doi: 10.1002/jcp.25827

Reis, W. L., Biancardi, V. C., Son, S., Antunes-Rodrigues, J., and Stern, J. E. (2012). Enhanced expression of heme oxygenase-1 and carbon monoxide excitatory effects in oxytocin and vasopressin neurones during water deprivation. *J. Neuroendocrinol.* 24, 653–663. doi: 10.1111/j.1365-2826.2011.02249.x

Richard, D., and Bourque, C. W. (1996). Atrial natriuretic peptide modulates synaptic transmission from osmoreceptor afferents to the supraoptic nucleus. *J. Neurosci.* 16, 7526–7532. doi: 10.1523/JNEUROSCI.16-23-07526.1996

Roy Choudhury, G., Ryou, M. G., Poteet, E., Wen, Y., He, R., Sun, F., et al. (2014). Involvement of p38 MAPK in reactive astrogliosis induced by ischemic stroke. *Brain Res.* 1551, 45–58. doi: 10.1016/j.brainres.2014.01.013

Rutka, J. T., Muller, M., Hubbard, S. L., Forsdike, J., Dirks, P. B., Jung, S., et al. (1999). Astrocytoma adhesion to extracellular matrix: functional significance of integrin and focal adhesion kinase expression. *J. Neuropathol. Exp. Neurol.* 58, 198–209. doi: 10.1097/00005072-199902000-00009

Ryan, M. C., and Gundlach, A. L. (1997). Differential regulation of angiotensinogen and natriuretic peptide mRNAs in rat brain by osmotic stimulation: focus on anterior hypothalamus and supraoptic nucleus. *Peptides* 18, 1365–1375. doi: 10.1016/s0196-9781(97)00192-7

Ryskamp, D. A., Witkovsky, P., Barabas, P., Huang, W., Koehler, C., Akimov, N. P., et al. (2011). The polymodal ion channel transient receptor potential vanilloid 4 modulates calcium flux, spiking rate, and apoptosis of mouse retinal ganglion cells. *J. Neurosci.* 31, 7089–7101. doi: 10.1523/JNEUROSCI.0359-11.2011

Sailer, C., Winzeler, B., and Christ-Crain, M. (2017). Primary polydipsia in the medical and psychiatric patient: characteristics, complications and therapy. *Swiss Med. Wkly.* 147:w14514. doi: 10.4414/smw.2017.14514

Salazar, M., Pariente, J. A., Salido, G. M., and Gonzalez, A. (2008). Ebselen increases cytosolic free Ca^{2+} concentration, stimulates glutamate release and increases GFAP content in rat hippocampal astrocytes. *Toxicology* 244, 280–291. doi: 10.1016/j.tox.2007.12.002

Salman, M. M., Sheilabi, M. A., Bhattacharyya, D., Kitchen, P., Conner, A. C., Bill, R. M., et al. (2017). Transcriptome analysis suggests a role for the differential expression of cerebral aquaporins and the MAPK signalling pathway in human temporal lobe epilepsy. *Eur. J. Neurosci.* 46, 2121–2132. doi: 10.1111/ejn.13652

Sánchez-Olea, R., Pasantes-Morales, H., and Schousboe, A. (1993). Neurons respond to hyposmotic conditions by an increase in intracellular free calcium. *Neurochem. Res.* 18, 147–152. doi: 10.1007/bf01474677

Sato, M., Sobhan, U., Tsumura, M., Kuroda, H., Soya, M., Masamura, A., et al. (2013). Hypotonic-induced stretching of plasma membrane activates transient receptor potential vanilloid channels and sodium-calcium exchangers in mouse odontoblasts. *J. Endod.* 39, 779–787. doi: 10.1016/j.joen.2013.01.012

Schäfer, M. K., Pfeiffer, A., Jaeckel, M., Pouya, A., Dolga, A. M., and Methner, A. (2014). Regulators of mitochondrial Ca^{2+} homeostasis in cerebral ischemia. *Cell Tissue Res.* 357, 395–405. doi: 10.1007/s00441-014-1807-y

Schmelzer, A. E., and Miller, W. M. (2002). Effects of osmoprotectant compounds on NCAM polysialylation under hyperosmotic stress and elevated pCO_2. *Biotechnol. Bioeng.* 77, 359–368. doi: 10.1002/bit.10175

Seale, A. P., Watanabe, S., and Grau, E. G. (2012). Osmoreception: perspectives on signal transduction and environmental modulation. *Gen Comp Endocrinol* 176, 354–360. doi: 10.1016/j.ygcen.2011.10.005

Shi, W. Z., Qi, L. L., Fang, S. H., Lu, Y. B., Zhang, W. P., and Wei, E. Q. (2012). Aggravated chronic brain injury after focal cerebral ischemia in aquaporin-4-deficient mice. *Neurosci. Lett.* 520, 121–125. doi: 10.1016/j.neulet.2012.05.052

Shi, Z. D., Wang, H., and Tarbell, J. M. (2011). Heparan sulfate proteoglycans mediate interstitial flow mechanotransduction regulating MMP-13 expression and cell motility via FAK-ERK in 3D collagen. *PLoS One* 6:e15956. doi: 10.1371/journal.pone.0015956

Simard, M., and Nedergaard, M. (2004). The neurobiology of glia in the context of water and ion homeostasis. *Neuroscience* 129, 877–896. doi: 10.1016/j.neuroscience.2004.09.053

Song, I., and Dityatev, A. (2018). Crosstalk between glia, extracellular matrix and neurons. *Brain Res. Bull.* 136, 101–108. doi: 10.1016/j.brainresbull.2017.03.003

Song, Y., and Gunnarson, E. (2012). Potassium dependent regulation of astrocyte water permeability is mediated by cAMP signaling. *PLoS One* 7:e34936. doi: 10.1371/journal.pone.0034936

Song, Z., and Hatton, G. I. (2003). Taurine and the control of basal hormone release from rat neurohypophysis. *Exp. Neurol.* 183, 330–337. doi: 10.1016/s0014-4886(03)00105-5

Soya, M., Sato, M., Sobhan, U., Tsumura, M., Ichinohe, T., Tazaki, M., et al. (2014). Plasma membrane stretch activates transient receptor potential vanilloid and ankyrin channels in Merkel cells from hamster buccal mucosa. *Cell Calcium* 55, 208–218. doi: 10.1016/j.ceca.2014.02.015

Steiner, E., Enzmann, G. U., Lin, S., Ghavampour, S., Hannocks, M. J., Zuber, B., et al. (2012). Loss of astrocyte polarization upon transient focal brain ischemia as a possible mechanism to counteract early edema formation. *Glia* 60, 1646–1659. doi: 10.1002/glia.22383

Stern, J. E., and Ludwig, M. (2001). NO inhibits supraoptic oxytocin and vasopressin neurons via activation of GABAergic synaptic inputs. *Am. J. Physiol. Regul. Integr. Comp. Physiol.* 280, R1815–R1822. doi: 10.1152/ajpregu.2001.280.6.r1815

Stokum, J. A., Gerzanich, V., and Simard, J. M. (2016). Molecular pathophysiology of cerebral edema. *J. Cereb. Blood Flow Metab.* 36, 513–538. doi: 10.1177/0271678X15617172

Takagi, H., Sugimura, Y., Suzuki, H., Iwama, S., Izumida, H., Fujisawa, H., et al. (2014). Minocycline prevents osmotic demyelination associated with aquaresis. *Kidney Int.* 86, 954–964. doi: 10.1038/ki.2014.119

Tamma, G., Di Mise, A., Ranieri, M., Geller, A., Tamma, R., Zallone, A., et al. (2017). The V2 receptor antagonist tolvaptan raises cytosolic calcium and prevents AQP2 trafficking and function: an *in vitro* and *in vivo* assessment. *J. Cell. Mol. Med.* 21, 1767–1780. doi: 10.1111/jcmm.13098

Taylor, A. C., McCarthy, J. J., and Stocker, S. D. (2008). Mice lacking the transient receptor vanilloid potential 1 channel display normal thirst responses and central Fos activation to hypernatremia. *Am. J. Physiol. Regul. Integr. Comp. Physiol.* 294, R1285–R1293. doi: 10.1152/ajpregu.00003.2008

Theilig, F., and Wu, Q. (2015). ANP-induced signaling cascade and its implications in renal pathophysiology. *Am. J. Physiol. Renal Physiol.* 308, F1047–F1055. doi: 10.1152/ajprenal.00164.2014

Theodosis, D. T., Poulain, D. A., and Oliet, S. H. (2008). Activity-dependent structural and functional plasticity of astrocyte-neuron interactions. *Physiol. Rev.* 88, 983–1008. doi: 10.1152/physrev.00036.2007

Thomas, R., Salter, M. G., Wilke, S., Husen, A., Allcock, N., Nivison, M., et al. (2004). Acute ischemic injury of astrocytes is mediated by Na-K-Cl cotransport and not Ca^{2+} influx at a key point in white matter development. *J. Neuropathol. Exp. Neurol.* 63, 856–871. doi: 10.1093/jnen/63.8.856

Thomzig, A., Prüss, H., and Veh, R. W. (2003). The Kir6.1-protein, a pore-forming subunit of ATP-sensitive potassium channels, is prominently expressed by giant cholinergic interneurons in the striatum of the rat brain. *Brain Res.* 986, 132–138. doi: 10.1016/s0006-8993(03)03222-0

Uchiyama, Y., Cheng, C. C., Danielson, K. G., Mochida, J., Albert, T. J., Shapiro, I. M., et al. (2007). Expression of acid-sensing ion channel 3 (ASIC3) in nucleus pulposus cells of the intervertebral disc is regulated by p75NTR and ERK signaling. *J. Bone Miner. Res.* 22, 1996–2006. doi: 10.1359/jbmr.070805

Verkman, A. S. (2012). Aquaporins in clinical medicine. *Annu. Rev. Med.* 63, 303–316. doi: 10.1146/annurev-med-043010-193843

Vilhena-Franco, T., Mecawi, A. S., Elias, L. L., and Antunes-Rodrigues, J. (2011). Oestradiol potentiates hormone secretion and neuronal activation in response to hypertonic extracellular volume expansion in ovariectomised rats. *J. Neuroendocrinol.* 23, 481–489. doi: 10.1111/j.1365-2826.2011.02133.x

Vilhena-Franco, T., Mecawi, A. S., Elias, L. L., and Antunes-Rodrigues, J. (2016). Oestradiol effects on neuroendocrine responses induced by water deprivation in rats. *J. Endocrinol.* 231, 167–180. doi: 10.1530/joe-16-0311

Voets, P. J. G. M., and Maas, R. P. P. W. M. (2018). Extracellular volume depletion and resultant hypotonic hyponatremia: a novel translational approach. *Math. Biosci.* 295, 62–66. doi: 10.1016/j.mbs.2017.11.005

Walcott, B. P., Kahle, K. T., and Simard, J. M. (2012). Novel treatment targets for cerebral edema. *Neurotherapeutics* 9, 65–72. doi: 10.1007/s13311-011-0087-4

Walz, W. (1992). Role of Na/K/Cl cotransport in astrocytes. *Can. J. Physiol. Pharmacol.* 70, S260–S262. doi: 10.1139/y92-270

Wang, Y. F., and Hatton, G. I. (2006). Mechanisms underlying oxytocin-induced excitation of supraoptic neurons: prostaglandin mediation of actin polymerization. *J. Neurophysiol.* 95, 3933–3947. doi: 10.1152/jn.01267.2005

Wang, Y. F., and Hatton, G. I. (2007a). Dominant role of βγ subunits of G-proteins in oxytocin-evoked burst firing. *J. Neurosci.* 27, 1902–1912. doi: 10.1523/JNEUROSCI.5346-06.2007

Wang, Y. F., and Hatton, G. I. (2007b). Interaction of extracellular signal-regulated protein kinase 1/2 with actin cytoskeleton in supraoptic oxytocin neurons and astrocytes: role in burst firing. *J. Neurosci.* 27, 13822–13834. doi: 10.1523/JNEUROSCI.4119-07.2007

Wang, Y. F., and Hatton, G. I. (2009). Astrocytic plasticity and patterned oxytocin neuronal activity: dynamic interactions. *J. Neurosci.* 29, 1743–1754. doi: 10.1523/JNEUROSCI.4669-08.2009

Wang, Y.-F., Liu, L.-X., and Yang, H.-P. (2011). Neurophysiological involvement in hypervolemic hyponatremia-evoked by hypersecretion of vasopressin. *Transl. Biomed.* 2:3. doi: 10.3823/425

Wang, S. C., Meng, D., Yang, H.-P., Wang, X., Shuwei, J., Wang, P., et al. (2018). Pathological basis of cardiac arrhythmias: vicious cycle of immune-metabolic dysregulation. *Cardiovasc. Disord. Med.* 3, 1–7. doi: 10.15761/cdm.1000158

Wang, Y. F., and Parpura, V. (2016). Central role of maladapted astrocytic plasticity in ischemic brain edema formation. *Front. Cell. Neurosci.* 10:129. doi: 10.3389/fncel.2016.00129

Wang, Y. F., Ponzio, T. A., and Hatton, G. I. (2006). Autofeedback effects of progressively rising oxytocin concentrations on supraoptic oxytocin neuronal activity in slices from lactating rats. *Am. J. Physiol. Regul. Integr. Comp. Physiol.* 290, R1191–R1198. doi: 10.1152/ajpregu.00725.2005

Wang, P., Qin, D., and Wang, Y. F. (2017). Oxytocin rapidly changes astrocytic GFAP plasticity by differentially modulating the expressions of pERK 1/2 and protein kinase A. *Front. Mol. Neurosci.* 10:262. doi: 10.3389/fnmol.2017.00262

Wang, Y. F., Sun, M. Y., Hou, Q., and Hamilton, K. A. (2013a). GABAergic inhibition through synergistic astrocytic neuronal interaction transiently decreases vasopressin neuronal activity during hypoosmotic challenge. *Eur. J. Neurosci.* 37, 1260–1269. doi: 10.1111/ejn.12137

Wang, Y. F., Sun, M. Y., Hou, Q., and Parpura, V. (2013b). Hyposmolality differentially and spatiotemporally modulates levels of glutamine synthetase and serine racemase in rat supraoptic nucleus. *Glia* 61, 529–538. doi: 10.1002/glia.22453

Wang, Y. F., and Zhu, H. (2014). Mechanisms underlying astrocyte regulation of hypothalamic neuroendocrine neuron activity. *Sheng Li Ke Xue Jin Zhan* 45, 177–184.

Wilson, C. S., and Mongin, A. A. (2018). The signaling role for chloride in the bidirectional communication between neurons and astrocytes. *Neurosci. Lett.* doi: 10.1016/j.neulet.2018.01.012 [Epub ahead of print].

Wu, Q., Zhang, Y. J., Gao, J. Y., Li, X. M., Kong, H., Zhang, Y. P., et al. (2014). Aquaporin-4 mitigates retrograde degeneration of rubrospinal neurons

by facilitating edema clearance and glial scar formation after spinal cord injury in mice. *Mol. Neurobiol.* 49, 1327–1337. doi: 10.1007/s12035-013-8607-3

Xu, M., Xiao, M., Li, S., and Yang, B. (2017). Aquaporins in nervous system. *Adv. Exp. Med. Biol.* 969, 81–103. doi: 10.1007/978-94-024-1057-0_5

Yagil, C., and Sladek, C. D. (1990). Osmotic regulation of vasopressin and oxytocin release is rate sensitive in hypothalamoneurohypophysial explants. *Am. J. Physiol.* 258, R492–R500. doi: 10.1152/ajpregu.1990.258.2.r492

Yan, W., Zhao, X., Chen, H., Zhong, D., Jin, J., Qin, Q., et al. (2016). β-Dystroglycan cleavage by matrix metalloproteinase-2/-9 disturbs aquaporin-4 polarization and influences brain edema in acute cerebral ischemia. *Neuroscience* 326, 141–157. doi: 10.1016/j.neuroscience.2016.03.055

Yang, M., Gao, F., Liu, H., Yu, W. H., Zhuo, F., Qiu, G. P., et al. (2013). Hyperosmotic induction of aquaporin expression in rat astrocytes through a different MAPK pathway. *J. Cell. Biochem.* 114, 111–119. doi: 10.1002/jcb.24308

Yuan, H., Gao, B., Duan, L., Jiang, S., Cao, R., Xiong, Y. F., et al. (2010). Acute hyperosmotic stimulus-induced Fos expression in neurons depends on activation of astrocytes in the supraoptic nucleus of rats. *J. Neurosci. Res.* 88, 1364–1373. doi: 10.1002/jnr.22297

Zeynalov, E., Jones, S. M., Seo, J. W., Snell, L. D., and Elliott, J. P. (2015). Arginine-vasopressin receptor blocker conivaptan reduces brain edema and blood-brain barrier disruption after experimental stroke in mice. *PLoS One* 10:e0136121. doi: 10.1371/journal.pone.0136121

Zhang, B., Glasgow, E., Murase, T., Verbalis, J. G., and Gainer, H. (2001). Chronic hypoosmolality induces a selective decrease in magnocellular neurone soma and nuclear size in the rat hypothalamic supraoptic nucleus. *J. Neuroendocrinol.* 13, 29–36. doi: 10.1111/j.1365-2826.2001.00593.x

Zhang, Y., Hong, G., Lee, K. S., Hammock, B. D., Gebremedhin, D., Harder, D. R., et al. (2017). Inhibition of soluble epoxide hydrolase augments astrocyte release of vascular endothelial growth factor and neuronal recovery after oxygen-glucose deprivation. *J. Neurochem.* 140, 814–825. doi: 10.1111/jnc.13933

Zhao, X., Yan, X., Liu, Y., Zhang, P., and Ni, X. (2016). Co-expression of mouse TMEM63A, TMEM63B and TMEM63C confers hyperosmolarity activated ion currents in HEK293 cells. *Cell Biochem. Funct.* 34, 238–241. doi: 10.1002/cbf.3185

Zorec, R., Araque, A., Carmignoto, G., Haydon, P. G., Verkhratsky, A., and Parpura, V. (2012). Astroglial excitability and gliotransmission: an appraisal of Ca^{2+} as a signalling route. *ASN Neuro* 4:e00080. doi: 10.1042/AN20110061

2

Caffeine and Modafinil Ameliorate the Neuroinflammation and Anxious Behavior in Rats During Sleep Deprivation by Inhibiting the Microglia Activation

Meetu Wadhwa, Garima Chauhan, Koustav Roy, Surajit Sahu[†], Satyanarayan Deep[†], Vishal Jain, Krishna Kishore, Koushik Ray, Lalan Thakur and Usha Panjwani[*]

Defence Institute of Physiology & Allied Sciences (DIPAS), Defence Research and Development Organization (DRDO), New Delhi, India

*Correspondence:
Usha Panjwani
neurophysiolab.dipas@gmail.com

†Present address:
Surajit Sahu,
INSERM U901 INMED, Marseille,
France
Satyanarayan Deep,
Department of Neurology, University
of New Mexico, Albuquerque, NM,
United States

Background: Sleep deprivation (SD) plagues modern society due to the professional demands. It prevails in patients with mood and neuroinflammatory disorders. Although growing evidence suggests the improvement in the cognitive performance by psychostimulants during sleep-deprived conditions, the impending involved mechanism is rarely studied. Thus, we hypothesized that mood and inflammatory changes might be due to the glial cells activation induced modulation of the inflammatory cytokines during SD, which could be improved by administering psychostimulants. The present study evaluated the role of caffeine/modafinil on SD-induced behavioral and inflammatory consequences.

Methods: Adult male Sprague-Dawley rats were sleep deprived for 48 h using automated SD apparatus. Caffeine (60 mg/kg/day) or modafinil (100 mg/kg/day) were administered orally to rats once every day during SD. Rats were subjected to anxious and depressive behavioral evaluation after SD. Subsequently, blood and brain were collected for biochemical, immunohistochemical and molecular studies.

Results: Sleep deprived rats presented an increased number of entries and time spent in closed arms in elevated plus maze test and decreased total distance traveled in the open field (OF) test. Caffeine/modafinil treatment significantly improved these anxious consequences. However, we did not observe substantial changes in immobility and anhedonia in sleep-deprived rats. Caffeine/modafinil significantly down-regulated the pro- and up-regulated the anti-inflammatory cytokine mRNA and protein expression in the hippocampus during SD. Similar outcomes were observed in blood plasma cytokine levels. Caffeine/modafinil treatment significantly decreased the microglial immunoreactivity in DG, CA1 and CA3 regions of the hippocampus during SD, however, no significant increase in immunoreactivity of astrocytes was observed. Sholl analysis

---

Abbreviations: DAB, Diaminobenzidine; ELISA, Enzyme-linked immunosorbent assay; EPM, Elevated plus maze; FST, Forced swim test; GFAP, Glial fibrillary acidic protein; Iba-1, Ionized calcium-binding adapter molecule I; OF, Open field; PBS, Phosphate buffered saline; PFA, Paraformaldehyde; PBST, Phosphate buffered saline containing tween-20/tritonX-100; SD, Sleep deprivation.

signified the improvement in the morphological alterations of astrocytes and microglia after caffeine/modafinil administration during SD. Stereological analysis demonstrated a significant improvement in the number of ionized calcium binding adapter molecule I (Iba-1) positive cells (different states) in different regions of the hippocampus after caffeine or modafinil treatment during SD without showing any significant change in total microglial cell number. Eventually, the correlation analysis displayed a positive relationship between anxiety, pro-inflammatory cytokines and activated microglial cell count during SD.

Conclusion: The present study suggests the role of caffeine or modafinil in the amelioration of SD-induced inflammatory response and anxious behavior in rats.

Highlights

- SD induced mood alterations in rats.
- Glial cells activated in association with the changes in the inflammatory cytokines.
- Caffeine or modafinil improved the mood and restored inflammatory changes during SD.
- SD-induced anxious behavior correlated with the inflammatory consequences.

Keywords: sleep deprivation, mood changes, microglia, cytokines, neuroinflammation, caffeine, modafinil

INTRODUCTION

Insufficient sleep is one of the most common and significant health problem worldwide associated with the immune system modulation (Dworak et al., 2011) and mood decline (Babson et al., 2010; Alkadhi et al., 2013). Documented evidence support the high prevalence of anxiety and depression with an associated link to inflammation in several pathological conditions like rheumatoid arthritis, kidney disease and bowel disease conditions (Kang et al., 2011). Cytokines play a crucial role in inflammation, neurobehavioral and emotional deficits. During the inflammatory challenge, microglial cells get activated and affect the release of cytokines (pro-inflammatory cytokines increase and anti-inflammatory cytokines decrease), often coinciding with behavioral manifestations (Kang et al., 2014; Wohleb et al., 2014b).

There are growing lines of evidence showing bi-directional communication between the sleep and immune system. Sleep influences the immune system and vice versa (Zielinski and Krueger, 2011). The pro-inflammatory cytokines such as interleukin-1β (IL-1β), TNF-α, IL-6 are found to be increased upon sleep deprivation (SD) in humans as well as experimental animals. Glial cells comprise the innate immune cells of the brain. Once activated, these cells imbalance the cytokine levels leading to behavioral abnormalities. However, their role under sleep-deprived conditions is remained unclear (Wisor et al., 2011; Alkadhi et al., 2013). It had reported that insufficient sleep decreases the mental performance and increases the risk of immune dysfunctions (Carey et al., 2011). Pro-inflammatory cytokines are associated with SD and mood disorders, however, the underlying mechanism is poorly understood (Rönnbäck and Hansson, 2004; Abelaira et al., 2013; Hong et al., 2016).

Caffeine and modafinil are widely consumed psychoactive drugs in the world showing beneficial effects on cognitive performance. Caffeine acts as a non-selective adenosine antagonist showing stimulant activity and prevents the deterioration of the cognitive performance (Daly, 2007; Nehlig, 2010; Sanday et al., 2013; Cappelletti et al., 2015). Concurrently, modafinil acts as a cognitive enhancer after directly binding to dopamine transporter and elevates the level of serotonin (Minzenberg and Carter, 2008; Rasetti et al., 2010). Caffeine or modafinil are thought to improve the mood (Cunha and Agostinho, 2010; Boele et al., 2013). Available literature suggests the dose-dependent effect of caffeine on mood showing an anxiolytic effect at low or moderate doses and anxiogenic effects at higher doses (Rusconi et al., 2014; Yamada et al., 2014).

Human and animal studies have shown improved memory performance after SD upon taking caffeine and modafinil (McGaugh et al., 2009; Cunha and Agostinho, 2010). Caffeine/modafinil administration prevents the neuroinflammation mediating memory disturbance in animal models of Alzheimer's, Parkinson's, stress, diabetes, convulsions, or alcohol-induced amnesia (Brothers et al., 2010; Raineri et al., 2012; Gyoneva et al., 2014a). However, neither the detailed mechanisms underlying neuroinflammation mediated emotional regulation during SD nor the effectiveness of the psychostimulants agents has been established yet. Therefore, we selected the two well-known psychostimulants viz., caffeine and modafinil; to assess the mood changes, astrocytes and microglial cells investigation along with the inflammatory cytokine levels during SD. Additionally, the predicted mechanism was investigated by the correlation analysis between the behavioral and inflammatory test parameters.

MATERIALS AND METHODS

Animals

Adult male Sprague-Dawley rats of 6–8 weeks old and approximately 220 ± 10 g body weight were used for the present study. Rats were housed in the clean cages made up of plexiglass material in the animal house at a temperature of 25 ± 2°C and humidity of 55 ± 2% RH with 12 h light and dark cycles with food and water *ad libitum*. All the experimental protocols were approved by the Institutional Animal Ethics Committee (IAEC, IAEC/DIPAS/2015-19) of Government of India, in accordance with the Committee for the Purpose of Control and Supervision of Experiments on Animals (CPCSEA) guidelines. Animal handling was done regularly to make them habituate to the experimenter. Experiments were conducted during the light period of the day. All efforts were done to minimize the number of rats used and to avoid unnecessary pain to the animal.

Chemicals and Reagents

Analytical grade chemicals were procured from Sigma Chemicals (Sigma-Aldrich, St. Louis, MO, USA) unless otherwise mentioned. The Enzyme-Linked Immunosorbent Assay (ELISA) kits were purchased from BD Biosciences Laboratory Ltd. (USA) and R and D Systems, Minneapolis, MN, USA. Antibodies (primary and secondary) were procured from Sigma-Aldrich, St. Louis, MO, USA, Abcam, Cambridge, MA, USA and Millipore, CA, USA.

Experimental Design

Initially, behavioral screening of rats was done, in which the body weight, food intake, aggressiveness and stereotype behavior

were evaluated. This was done to ensure that rats were not suffering from any impairment, after that the animals were divided randomly into different groups: cage control with vehicle treatment (CC+Veh); cage control with caffeine treatment (CC+Caf); cage control with modafinil treatment (CC+Mod); sleep deprived for 48 h with vehicle treatment (SD+Veh); sleep deprived for 48 h and caffeine treatment (SD+Caf); sleep deprived for 48 h and modafinil treatment (SD+Mod). The rats underwent the vehicle and drugs treatment during the control and sleep deprived conditions for 48 h. Each group had five rats, and the behavioral study took place between 8:00 AM and 11:00 AM. We used a different set of rats in each behavioral paradigm. Animals were euthanized immediately after the behavioral test during the light period of the day and evaluated for the biochemical and immunohistochemical analyses. The schematic experimental design of the present study is shown in **Figure 1**.

Sleep Deprivation Procedure

Briefly, the male Sprague-Dawley rats were sleep deprived for 48 h in the automated SD apparatus according to the well-established SD protocol of our lab (Wadhwa et al., 2015; Chauhan et al., 2016). The exposure paradigm made the rats awake on providing the SD stimulus such as sound, light, and vibration. There was a proper provision of ventilation, food and water during the SD exposure. The control rats were kept under controlled conditions (temperature, humidity and light) in the animal house facility.

Drug Administration

Caffeine (Sigma-Aldrich, St. Louis, MO, USA, 60 mg/kg/day, dissolved in physiological saline, administered orally) and

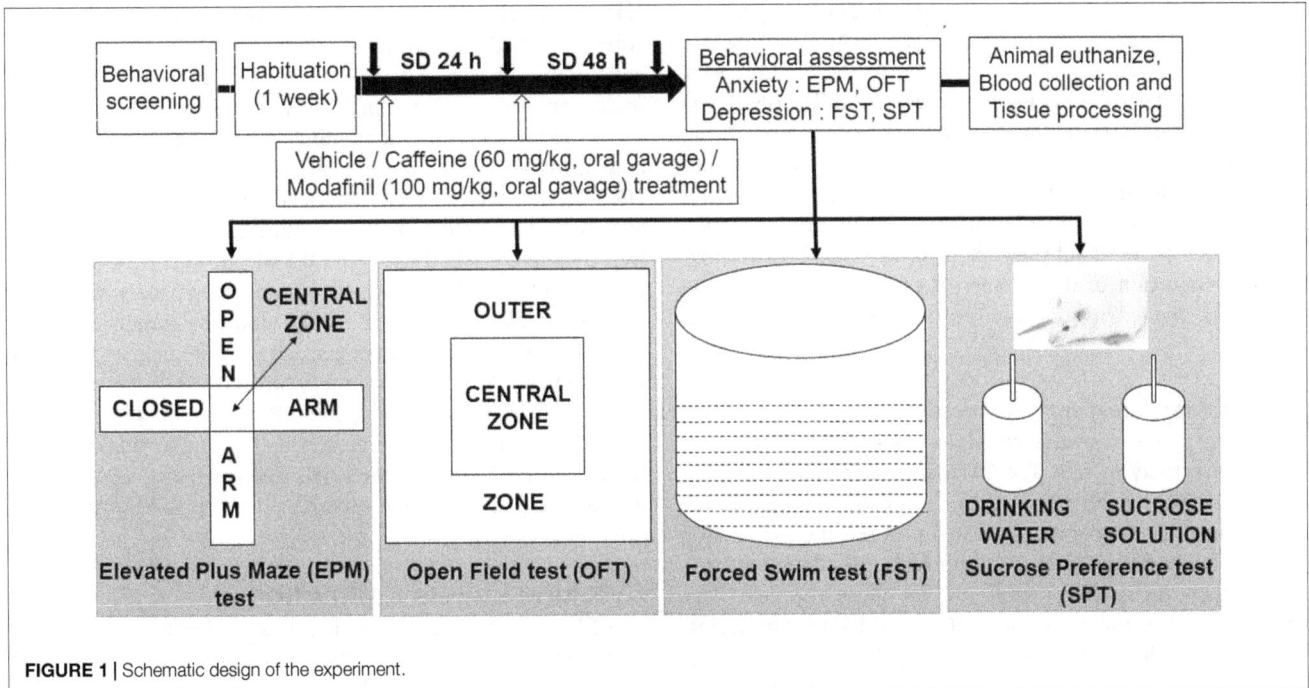

FIGURE 1 | Schematic design of the experiment.

modafinil (Modalert, Sun Pharma, India; 100 mg/kg/day, suspended in physiological saline, treated orally) dose were based upon our previous study (Sahu et al., 2013; Wadhwa et al., 2015), however, the tested doses of caffeine were higher and did not mimic habitual caffeine consumption. Caffeine or modafinil was given to the rats as an oral gavage, once a day in the morning time, during 48 h of SD.

Body Weight and Food Intake

To evaluate the physiological consequences, we monitored the body weight and food intake of both control and experimental rats regularly in the morning (1 week before the initiation of the experiments). The food was maintained at a constant amount (150 g) per animal, and every morning (8:00–9:00 AM), the remaining food was measured. At the same time, the body weight of each animal was noted down.

Behavioral Testing

We utilized a battery of behavioral tests that measure anxiety and depressive-like behavior including the open field (OF, locomotor activity, exploratory behavior), elevated plus maze (EPM, locomotor activity, exploratory behavior, anxiety), forced swim test (FST, behavioral "despair") and sucrose preference test (SPT, anhedonia). Each behavioral analysis was carried out with 15 rats in each experimental group.

Elevated Plus Maze Apparatus and Test Procedure

The EPM test is used to assess the level of anxiety in rodents. The EPM test apparatus consisted of plus shape design with four arms (two open and two closed, perpendicular to each other) having an open roof. The apparatus was about 40–70 cm elevated from the floor. Briefly, the rat was placed at the junction of the open and closed arms, facing the open arm opposite to the experimenter for 5 min. At the end of the test, rats were removed from the apparatus and placed back to their home cage. The test apparatus was properly cleaned with alcohol and dried with cotton before testing another rat. An overhead camera in association with ANY-maze software (Stoelting Co, Wood Dale, IL, USA) was properly arranged for the tracking and automatically recording the number of entries and time spent by the rat in the open and closed arms. The anxious behavior was evaluated by calculating the proportion of the time spent (time spent into the open arms divided by the total time spent in the open/closed arms) and the proportion of the number of entries (entries into the open arms divided by the total entries into the open/closed arms).

Open field apparatus and test procedure

OF test is a well-known method to assess the spontaneous locomotor activity in rats. The OF maze was divided into two zones: central and peripheral zone, using the square drawn on the maze. The apparatus consisted of a rectangular area of 81 × 81 cm surrounded by a 28 cm high wall. The field was lit with white light (23W) fixed 100 cm above the field. The rat was placed in the center of the OF, and its activity during the subsequent 5 min was recorded using ANY-Maze tracking software (Stoelting Co, Wood Dale, IL, USA). The test apparatus was properly cleaned with alcohol and dried with cotton before testing another rat to exclude any cues and smell.

Forced swim apparatus and test procedure

The FST was used to assess the depression in rodents. It is based on the assumption that an animal will try to escape, if the rat fails, the animal eventually stops trying and gives up. The FST apparatus is a vertical plexiglas cylinder (40 cm high; 20 cm in diameter) filled with 30 cm deep water (24–30°C). Briefly, the rat was placed in a cylindrical container of water from which it cannot escape, for 5 min. The rat was properly dried after removal from the water with a clean towel. The water was replaced regularly with fresh water to avoid the accumulation of the urine and fecal material. ANY-maze software (Stoelting Co, Wood Dale, IL, USA) was used to determine the test parameters.

Sucrose preference test procedure

The anhedonia, indicator of depression, means the lack of interest in rewarding stimuli. In this task, we assessed the animal interest in seeking out a sweet, rewarding drink in plain drinking water. This test was carried out in the animal's home cage. Briefly, rats were initially habituated to the presence of two bottles; one containing 2% sucrose solution and another drinking water for 2 days in their home cage. During this phase, rats had the free access to both bottles. The intake of normal drinking water and sucrose solution was measured daily, and the positions of bottles were regularly interchanged to reduce biases. On the completion of 48 h SD, the rats were presented the same two bottles (one containing water and another containing sucrose solution) and measured the intake of water and sucrose solution. Sucrose preference index was calculated as a ratio of the volume of sucrose intake over the total volume of fluid intake.

Blood Collection and Tissue Processing

After the scheduled period of SD exposure and the behavioral assessment, the blood was collected from left ventricle under anesthesia (ketamine 80 mg/kg-xylazine 20 mg/kg) in the vacutainer tube containing the sodium heparin as an anticoagulant. The blood was centrifuged at 3500 rpm at 4°C for 10–15 min, and plasma was separated. The rats were euthanized, and the brain was extracted out immediately. The hippocampi were isolated, washed with cold 0.1 M phosphate buffer saline (PBS) solution. The hippocampus was snap frozen in liquid nitrogen and then stored at −80°C until the time of analysis. Later, the samples were homogenized with the help of Polytron homogenizer (Remi Pvt. Limited) with 1× PBS and protease inhibitor cocktail containing inhibitors with broad specificity for serine, cysteine, acid proteases and aminopeptidases. After homogenization, the solution was centrifuged at 10,000 rpm for 10 min at 4°C, and the supernatant was isolated out for the cytokines assay.

Cytokine Levels Estimation

The ELISA is a specific and highly sensitive method for the quantification of cytokines. Plasma samples and hippocampal supernatant (100 μl, 1:50 dilution in assay buffer) were assayed for ILs (IL-1β, IL-6, IL-4, IL-10) and TNF-α using commercial ELISA kits. The assays were performed as per the manufacturer's protocols.

Evaluation of RNA Expression Levels of Secretory Cytokines in Hippocampus by Real-Time PCR (RT-PCR)

Isolation of the total RNA from hippocampal tissue was done using TRIZOL reagent (Sigma-Aldrich, St. Louis, MO, USA) according to the previously described protocol (Rio et al., 2010). RNA level was quantified using a Nanodrop (Thermo Fisher Scientific, Waltham, MA, USA) by measuring absorbance at 260 and 280 nm. The purity of RNA was checked by denaturing agarose gel electrophoresis and ethidium bromide staining. RNA was reverse-transcribed to cDNA using an RT2 first strand cDNA Synthesis Kit (QIAGEN Sciences, Germantown, MD, USA), according to the manufacturer's instruction. Relative quantitative analysis of the gene expression of interleukins and TNF for each group was done by employing RT^2 Profiler inflammatory cytokines and receptor array (QIAGEN Sciences, Germantown, MD, USA) using RT^2 SYBR® Green qPCR master mix (QIAGEN Sciences, Germantown, MD, USA). The analysis was performed by the comparative $2^{-\Delta\Delta CT}$ method as previously described. The gene expression analysis was done using software available online at www.sabiosciences.com, after normalization of each gene (Ct) to the housekeeping genes.

Immunohistochemistry

Transcardial perfusion and fixation were performed using 4% paraformaldehyde (PFA) in 0.1 M PBS (pH = 7.4). Brains were cryosectioned after processing with graded sucrose solution (10%, 20% and 30%) respectively dissolved in PBS (0.1 M, pH 7.4) using cryostat (Leica, Germany). Coronal sections of 30 μm thickness were taken in tissue culture plate and stored at 4°C in sodium azide solution to prevent fungal growth. The sections were processed for immunoreactivity of glial fibrillary acidic protein (GFAP) and ionized calcium binding adapter molecule I (Iba-1) proteins. Briefly, the sections were washed in PBS containing 0.1% Tween-20 or Triton X-100 (PBST) twice for 5 min each, subsequently; the antigen retrieval was done by incubating the sections with sodium citrate buffer for 10–15 min in boiling water bath. Sections were incubated with blocking buffer (5% goat serum for GFAP; 3% bovine serum albumin (BSA) for Iba-1) diluted in PBS for 2 h at room temperature, followed by washing with PBST. Prior to primary antibody labeling of an Iba-1 protein, there was an additional step of permeabilization in which the sections were treated with 0.25% Triton X-100 for 20–30 min followed by PBST washing thrice for 5 min each. The sections were then probed with rabbit anti-GFAP antibody and goat anti-Iba-1 antibody prepared in blocking solution for 40 h at 4°C. Sections were subsequently incubated with biotinylated goat anti-rabbit and rabbit anti-goat antibody for 2 h at room temperature, followed by three washings in PBST (5 min each). Finally, the sections were developed with diaminobenzidine (DAB) tetrahydrochloride solution.

Imaging and Analysis

The immune-stained sections were observed under Olympus (Melville, NY, USA) BX51TF microscope and images were taken from the DG, CA1 and CA3 regions of the dorsal hippocampus of the brain. We performed sholl analysis for the morphological evaluation of astrocytes and microglial cells. The cell quantification was performed using stereo investigator program. Similarly, immunoreactivity of astrocytes and microglial cells was quantified with the help of ImageJ software.

Statistical Analysis

All the data are expressed as Means ± SEM. Physiological, behavioral, biochemical, immunohistochemical and molecular data were analyzed by Two-way ANOVA followed by Tukey *post hoc* test with multiple comparisons. Pearson's correlation test was applied for correlation analysis. Data presented as mean percentage of control value used for graphical representation has been mentioned with graphs. All statistical analysis was done using GraphPad Prism 7.03 Software. The significance level of $p < 0.05$ was considered to be statistically significant.

RESULTS

Caffeine or Modafinil Treatment Improved the Physiological Consequences during SD

To assess the physiological changes in rats during SD, their body weight and food intake were recorded. We did not notice changes in body weight gain in the caffeine or modafinil treated control groups compared to vehicle-treated control rats, but a significant decrease in body weight was observed in vehicle-treated sleep-deprived rats as compared to vehicle treated control group rats. Administration of caffeine or modafinil to SD exposed rats significantly improved the body weight as compared to vehicle-treated SD group ($F_{(dFn, dFd); (2,114)} = 76.28$; $p < 0.0001$; **Supplementary Figure S1A**). However, changes in food intake among the different groups was non-significant ($F_{(2,114)} = 0.8429$; $p = 0.4331$; **Supplementary Figure S1B**).

Caffeine/Modafinil Administration Produced Anxiolytic Effect during SD

Sleep deprived rats showed anxious behavior while caffeine or modafinil treatment to SD exposed rats improved the anxious behavior of sleep-deprived rats as shown in the track plot of rats during EPM and OF test (**Figures 2A,D**). The proportion of the number of entries and the proportion of the time spent in the open arms were significantly reduced in sleep-deprived rats treated with vehicle compared to vehicle treated control rats Caffeine or modafinil treatment during SD significantly improved the proportion of the number of entries ($F_{(2,114)} = 17.4$; $p < 0.0001$) and the proportion of the time spent in the open arms ($F_{(2,114)} = 25.19$; $p < 0.0001$) compared to SD exposed rats (**Figures 2B,C**).

Similarly, in the OF test, the total distance traveled in the OF was significantly reduced in sleep-deprived rats. However, caffeine or modafinil administration significantly

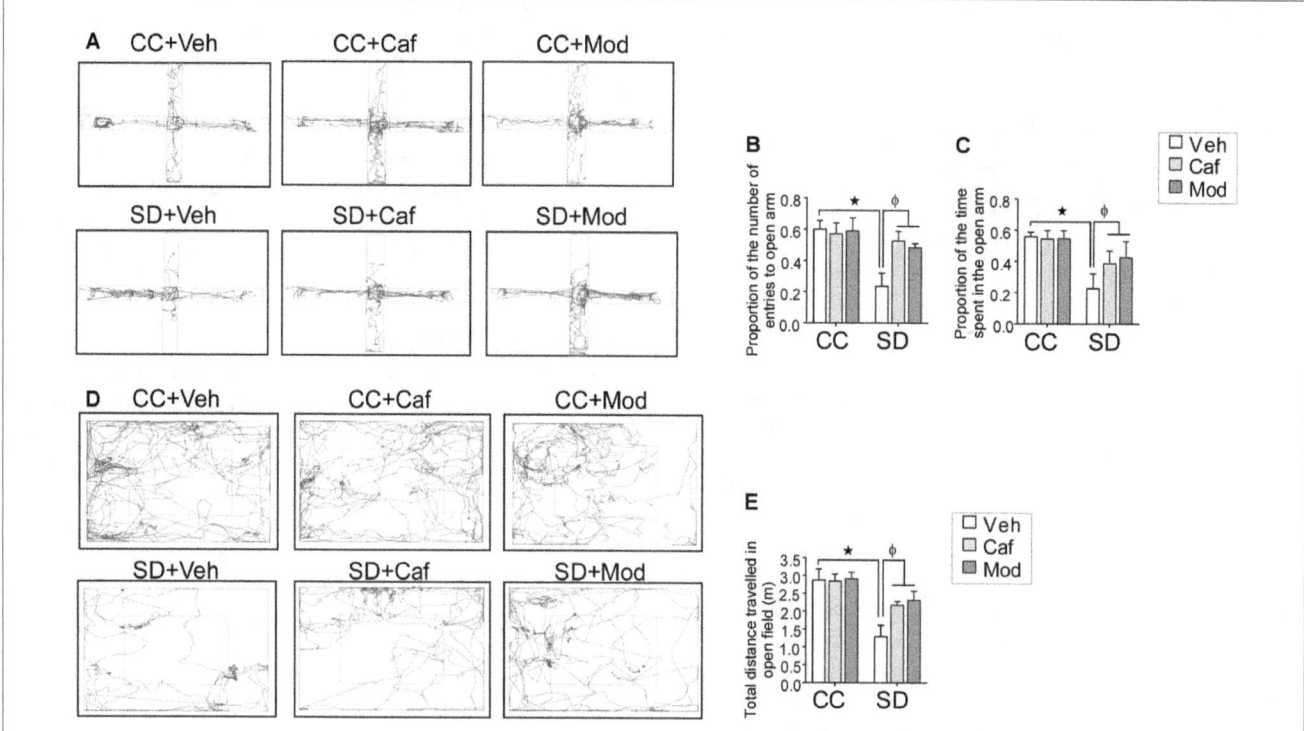

FIGURE 2 | Assessment of the anxious behavior following caffeine/modafinil treatment during sleep deprivation (SD). **(A)** Track plot of the rats during elevated plus maze (EPM) test and the study parameters: **(B)** proportion of the number of entries in the open arms; **(C)** proportion of the time spent in the open arms. **(D)** Track plot of the rats during open field (OF) test; **(E)** total distance traveled in the OF. *$p < 0.05$ when compared to control treated with vehicle; $^\phi p < 0.05$ when compared to sleep deprived treated with vehicle. Two way ANOVA followed by Tukey *post hoc* test with multiple comparison was used.

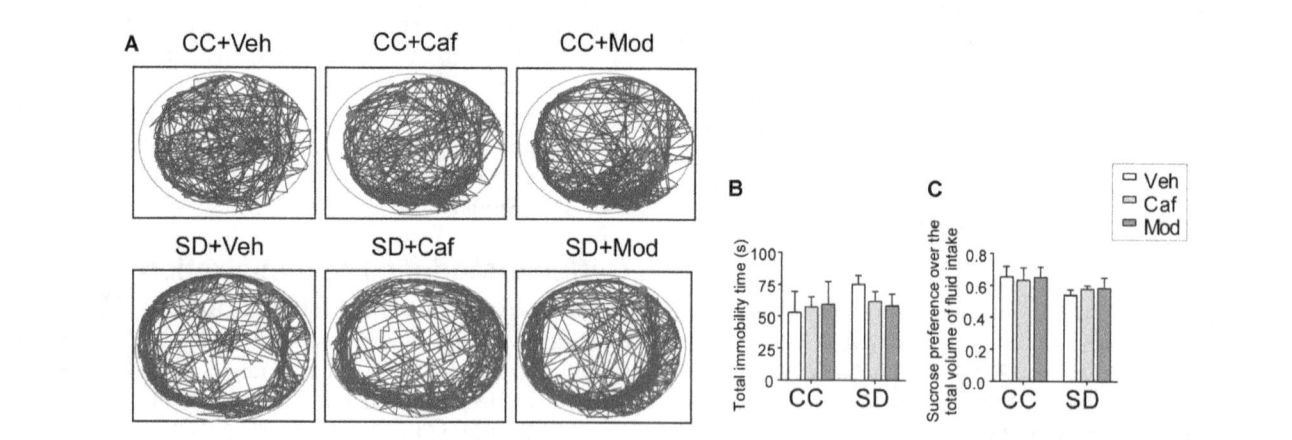

FIGURE 3 | Assessment of the depressive behavior following caffeine/modafinil treatment during SD. **(A)** Track plot of the rats during forced swim test (FST) test; **(B)** total immobility time; **(C)** sucrose preference over the total volume of fluid intake. Two way ANOVA followed by Tukey *post hoc* with multiple comparison test was applied for statistical comparison between groups. Blue dot represented the starting point and the red dot represented the end point of the test.

improved/increased the total distance traveled in the OF in SD exposed animals compared to sleep-deprived rats ($F_{(2,114)}$ = 14.91; p = 0.0001; **Figure 2E**). However, change in vehicle or caffeine, or modafinil treated control rats in EPM and OF test was comparable.

Caffeine or Modafinil Treatment Recovered the Depressive Behavior during SD

A non-significant increase in the immobility time was observed in SD exposed animals compared to control animals. Data showed that caffeine/modafinil treatment during SD

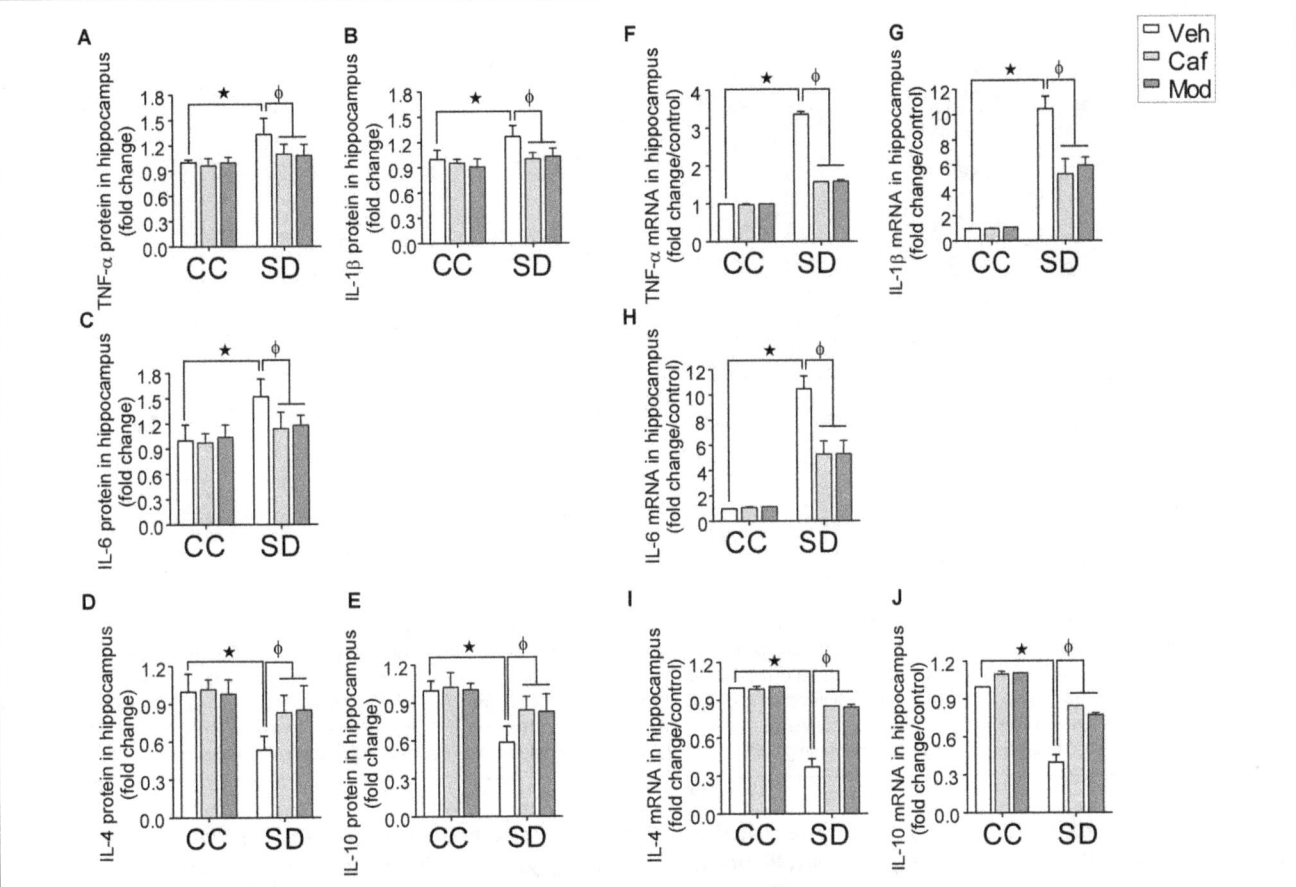

FIGURE 4 | Fold changes in the inflammatory cytokines in hippocampus during caffeine/modafinil administration following SD. The concentration of cytokine levels were measured in picograms per milliliter and expressed as the fold changes in TNF-α **(A,F)**; interleukin-1β (IL-1β) **(B,G)**; IL-6 **(C,H**; pro-inflammatory cytokines), IL-4 **(D,I)**; IL-10 **(E,J**; anti-inflammatory cytokines) in the hippocampus. *$p < 0.05$ when compared to control treated with vehicle; $^\phi p < 0.05$ when compared to sleep deprived treated with vehicle. Two way ANOVA followed by Tukey *post hoc* test with multiple comparison were used for the statistical evaluation.

non-significantly improved the immobility time in SD exposed rats ($F_{(2,54)} = 1.274$; $p = 0.2881$; **Figures 3A,B**).

Similar to FST, sucrose preference index was non-significantly reduced in rats subjected to SD compared to control, while caffeine/modafinil administration following SD exposure non-significantly improved the sucrose solution intake compared to the sleep-deprived group ($F_{(2,54)} = 0.6609$; $p = 0.5205$; **Figure 3C**). Also, we did not find a significant difference in the vehicle or caffeine or modafinil-treated control rats.

Caffeine/Modafinil Administration Maintained the Cytokines Profiling during SD

A significant fold increase in the pro-inflammatory cytokines (TNF-α, IL-1β and IL-6) and decrease in the anti-inflammatory cytokines (IL-4 and IL-10) in the hippocampus of rats subjected to SD as compared to control was observed. However, caffeine or modafinil administered during SD significantly decreased the pro-inflammatory: TNF-α ($F_{(2,84)} = 10.17$; $p = 0.0001$;

Figure 4A); IL-1β ($F_{(2,84)} = 3.693$; $p = 0.0290$; **Figure 4B**); IL-6 ($F_{(2,84)} = 4.168$; p = $P = 0.0188$; **Figure 4C**) and increased the anti-inflammatory cytokines: IL-4 ($F_{(2,84)} = 22.09$; $p < 0.0001$; **Figure 4D**); IL-10 ($F_{(2,84)} = 5.933$; $p = 0.0039$; **Figure 4E**) in the hippocampus as compared to vehicle-treated sleep deprived rats.

Subsequently, in plasma, we found a significant fold increase in the pro-inflammatory cytokines and decrease in the anti-inflammatory cytokines insleep deprived rats as compared to control, that were restored by caffeine or modafinil treatment during SD. The respective figures were: TNF-α ($F_{(2,84)} = 20.32$; $p < 0.0001$; **Supplementary Figure S2A**); IL-1β ($F_{(2,84)} = 27.67$; $p < 0.0001$; **Supplementary Figure S2B**); IL-6 ($F_{(2,84)} = 15.39$; $p < 0.0001$; **Supplementary Figure S2C**); IL-4 ($F_{(2,84)} = 18.52$; $p < 0.0001$; **Supplementary Figure S2D**); IL-10 ($F_{(2,84)} = 15.2$; $p < 0.0001$; **Supplementary Figure S2E**).

Real time PCR study showed that caffeine or modafinil administration during SD significantly down-regulated the mRNA expression of TNF-α ($F_{(2,12)} = 2007$; $p < 0.0001$; **Figure 4F**); IL-1β ($F_{(2,12)} = 27.41$, $p < 0.0001$; **Figure 4G**); IL-6

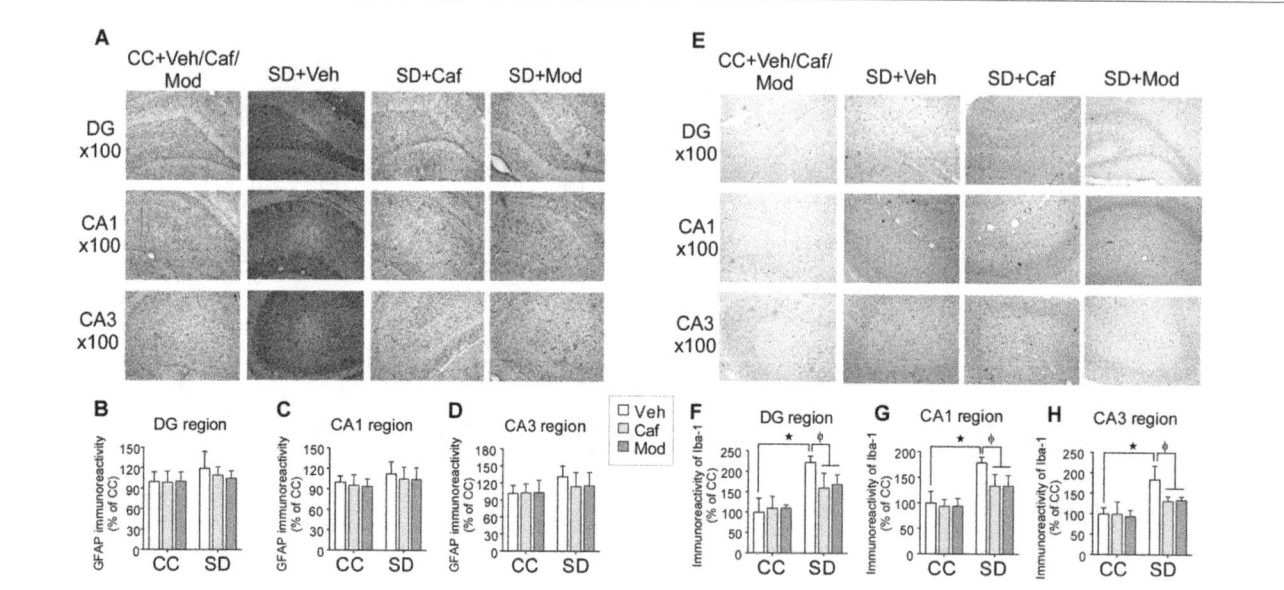

FIGURE 5 | Caffeine or modafinil treatment inhibited the glial cell immunoreactivity in rat hippocampus following SD. **(A)** Representable image of astrocytes expression in DG, CA1 and CA3 regions of the hippocampus. Glial fibrillary acidic protein (GFAP) immunoreactivity quantification in **(B)** DG region; **(C)** CA1 region; **(D)** CA3 region of the hippocampus. **(E)** Representable image of microglial cells expression in DG, CA1 and CA3 regions of the hippocampus. Ionized calcium binding adapter molecule I (Iba-1) cell immunoreactivity quantification in **(F)** DG region; **(G)** CA1 region; **(H)** CA3 region of the hippocampus. $*p < 0.05$ when compared to control treated with vehicle; $^{\phi}p < 0.05$ when compared to sleep deprived treated with vehicle. Two way ANOVA followed by Tukey *post hoc* multiple comparison test was applied for statistical comparison between groups and for the graphical representation, values expressed mean percentage of Control ± SEM.

$(F_{(2,12)} = 28.36;\ p < 0.0001;$ **Figure 4H**), and up-regulated the mRNA expression of IL-4 $(F_{(2,12)} = 153.9; p < 0.0001;$ **Figure 4I**), IL-10 $(F_{(2,12)} = 76.56; p < 0.0001;$ **Figure 4J**) during SD.

Caffeine or Modafinil Treatment Down-regulated the Astrocyte and Microglial Cells Immunoreactivity Following SD

Astrocytes and microglial cells activation were evaluated by studying the immunohistochemical changes in the expression of GFAP and Iba-1 protein in different regions of the dorsal hippocampus. Changes in the expression of astrocytes among different groups in DG, CA1 and CA3 region of the dorsal hippocampus was shown in **Figure 5A**. We found no significant increase in the relative mean pixel intensity of GFAP positive cells (astrocytes immunoreactivity) in DG, CA1 and CA3 region of the hippocampus, belonged to SD exposed rats compared to control rats. Caffeine or modafinil treatment non-significantly decrease the astrocytes immunoreactivity in DG $(F_{(2,84)} = 0.6826;$ $p = 0.5081;$ **Figure 5B**); CA1 $(F_{(2,84)} = 0.2293;\ p = 0.7956;$ **Figure 5C**), and CA3 $(F_{(2,84)} = 0.1089; p = 0.8970;$ **Figure 5D**) region of dorsal hippocampus compared to sleep deprived vehicle-treated rats.

Similarly, the vehicle and drugs treated rats showed changes in the expression of microglial cell in different hippocampal regions as shown in **Figure 5E**. We found a significant increase in the relative mean pixel intensity of microglial cell in SD exposed rats compared to control rats. Caffeine or modafinil treatment during SD significantly decreased the immunoreactivity of

microglial cell in DG $(F_{(2,84)} = 44.32;\ p < 0.0001;$ **Figure 5F**); CA1 $(F_{(2,84)} = 30.26;\ p < 0.0001;$ **Figure 5G**), and CA3 $(F_{(2,84)} = 25.97;\ p < 0.0001;$ **Figure 5H**) region of the hippocampus in comparison with vehicle-treated SD rats. Furthermore, no significant change in the immunoreactivity of GFAP and Iba-1 positive cells was observed following treatment of caffeine or modafinil to control rats compared to vehicle-treated control rats.

Caffeine or Modafinil Administration Efficiently Improved the Astrocyte and Microglial Cells Morphology Following SD

Representable intersectional and segmented images of the resting, intermediate and activated stage of astrocyte and the microglial cell was shown in **Figures 6A,B**. Astrocyte and microglial cells morphology were investigated by the following parameters such as soma density, soma area, sum inters, mean inters, ramification index and glial cell size/length among different groups in DG, CA1 and CA3 region of the dorsal hippocampus.

Administration of caffeine/modafinil significantly improved the SD-induced changes in the morphology of astrocytes in DG, CA1 and CA3 region of the hippocampus (**Figures 6C–H; Table 1**). Subsequently, microglial cell showed significantly increase in soma density, soma area and a decrease in sum inters, mean inters, ramification index, microglia length following SD while after caffeine or modafinil treatment during SD, the above consequences were significantly improved (**Figures 6I–N; Table 1**). Statistically non-significant morphological changes

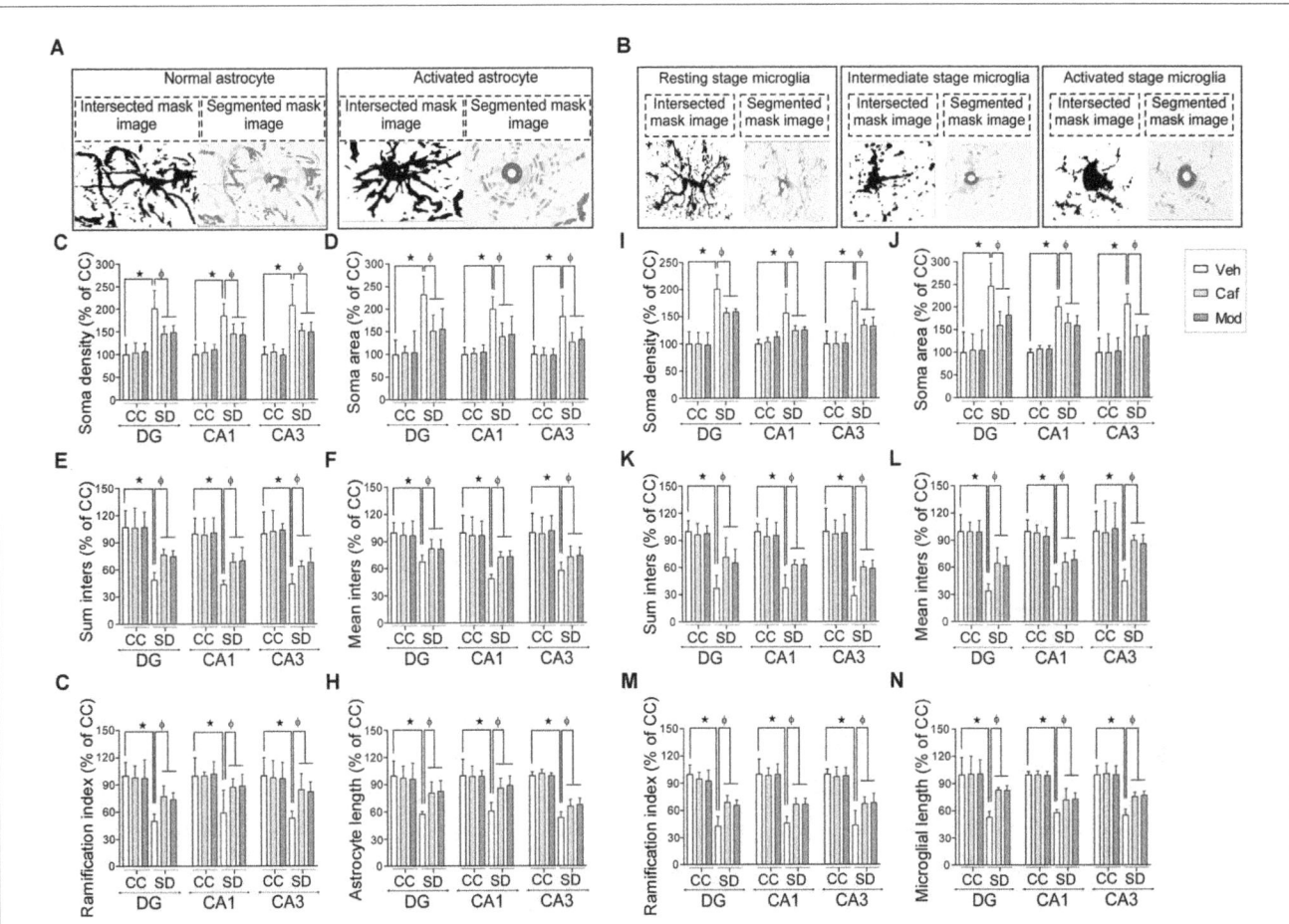

FIGURE 6 | Caffeine or modafinil administration altered the morphology of astrocytes and microglia towards the resting state in rat hippocampus during SD. Representable image (intersection and segmented mask) of **(A)** normal and activated astrocyte; **(B)** resting, intermediate and activated stage microglial cell. Changes in the **(C)** soma density; **(D)** soma area; **(E)** sum inters; **(F)** mean inters; **(G)** ramification index; **(H)** astrocyte length of astrocytes in DG, CA1 and CA3 regions of the hippocampus. Changes in the **(I)** soma density; **(J)** soma area; **(K)** sum inters; **(L)** mean inters; **(M)** ramification index; **(N)** microglial length of microglia cells. *$p < 0.05$ when compared to control treated with vehicle; $^{\phi}p < 0.05$ when compared to sleep deprived treated with vehicle. Two way ANOVA followed by Tukey *post hoc* test with multiple comparison was applied for statistical comparison between groups and for the graphical representation, values expressed mean percentage of Control ± SEM.

were observed in the vehicle and drugs treated control rats hippocampus.

Caffeine or Modafinil Treatment Maintained the Microglial Cells Numbers during SD

We found a significant decrease in resting stage microglial cell in DG and CA3 region of the hippocampus, while the activated microglial cell numbers were significantly increased in DG, CA1 and CA3 region of the hippocampus. No significant change was observed in the intermediate state microglia cell count during SD. Caffeine or modafinil treatment during SD significantly increased the resting stage microglial cell and decreased activated microglial cell count, given during SD in DG (resting ($F_{(2,84)}$ = 22.65; $p < 0.0001$; Boia et al., 2017); intermediate ($F_{(2,84)}$ = 0.1197; $p = 0.8874$); activated ($F_{(2,84)}$ = 33.82; $p < 0.0001$; **Figure 7A**), CA1 (resting ($F_{(2,84)}$ = 4.741; $p = 0.0112$); intermediate ($F_{(2,84)}$ = 0.03004;

$p = 0.9704$); activated ($F_{(2,84)}$ = 44.26; $p < 0.0001$; **Figure 7B**), and CA3 (resting ($F_{(2,84)}$ = 11.15; $p < 0.0001$); intermediate ($F_{(2,84)}$ = 0.09492; $p = 0.9095$); activated ($F_{(2,30)}$ = 3.636; $p < 0.0001$; **Figure 7C**).

Although, trivial increase in total microglial cell count in SD-vehicle group was observed, caffeine or modafinil treatment also showed in consequential improvement in DG ($F_{(2,84)}$ = 0.1095; $p = 0.8964$; **Figure 7D**); CA1 ($F_{(2,84)}$ = 0.5845; $p = 0.5597$; **Figure 7E**), and CA3 ($F_{(2,84)}$ = 0.8691; $p = 0.4231$; **Figure 7F**) region of the hippocampus.

Changes in Mood Was Correlated to Microglial Activation Induced Up-regulated Level of Pro-inflammatory Cytokines during SD

The interaction between the anxiety parameters, pro-inflammatory cytokines and activated microglia cell

TABLE 1 | Astrocyte and microglia morphology during sleep deprivation (SD) and caffeine/modafinil administration.

Parameter	$F_{(dFn,dFd)}$; $p < 0.0001$		
	DG region	CA1 region	CA3 region
Astrocyte morphology			
Soma density	$F_{(2,84)} = 30.53^*$	$F_{(2,84)} = 22.83^*$	$F_{(2,84)} = 24.37^*$
Soma area	$F_{(2,84)} = 18.27^*$	$F_{(2,84)} = 24.6^*$	$F_{(2,84)} = 19.5^*$
Sum inters	$F_{(2,84)} = 14.53^*$	$F_{(2,84)} = 13.03^*$	$F_{(2,84)} = 10.18^*$
Mean inters	$F_{(2,84)} = 20.42^*$	$F_{(2,84)} = 13.4^*$	$F_{(2,84)} = 12.76^*$
Ramification index	$F_{(2,84)} = 14.24^*$	$F_{(2,84)} = 11.37^*$	$F_{(2,84)} = 16.81^*$
Astrocyte length	$F_{(2,84)} = 20.14^*$	$F_{(2,84)} = 28.86^*$	$F_{(2,84)} = 23.1^*$
Microglia morphology			
Soma density	$F_{(2,84)} = 9.216^*$	$F_{(2,84)} = 54.27^*$	$F_{(2,84)} = 25.22^*$
Soma area	$F_{(2,84)} = 18.24^*$	$F_{(2,84)} = 47.28^*$	$F_{(2,84)} = 27.67^*$
Sum inters	$F_{(2,84)} = 17.21^*$	$F_{(2,84)} = 30.28^*$	$F_{(2,84)} = 30.85^*$
Mean inters	$F_{(2,84)} = 21.58^*$	$F_{(2,84)} = 45.18^*$	$F_{(2,84)} = 18.76^*$
Ramification index	$F_{(2,84)} = 62.23^*$	$F_{(2,84)} = 12.55^*$	$F_{(2,84)} = 27.44^*$
Microglial length	$F_{(2,84)} = 19.53^*$	$F_{(2,84)} = 32.63^*$	$F_{(2,84)} = 34.45^*$

Two way ANOVA followed by Tukey post hoc test with multiple comparison was applied for statistical comparison. $^*p < 0.05$ is significantly different compared to control.

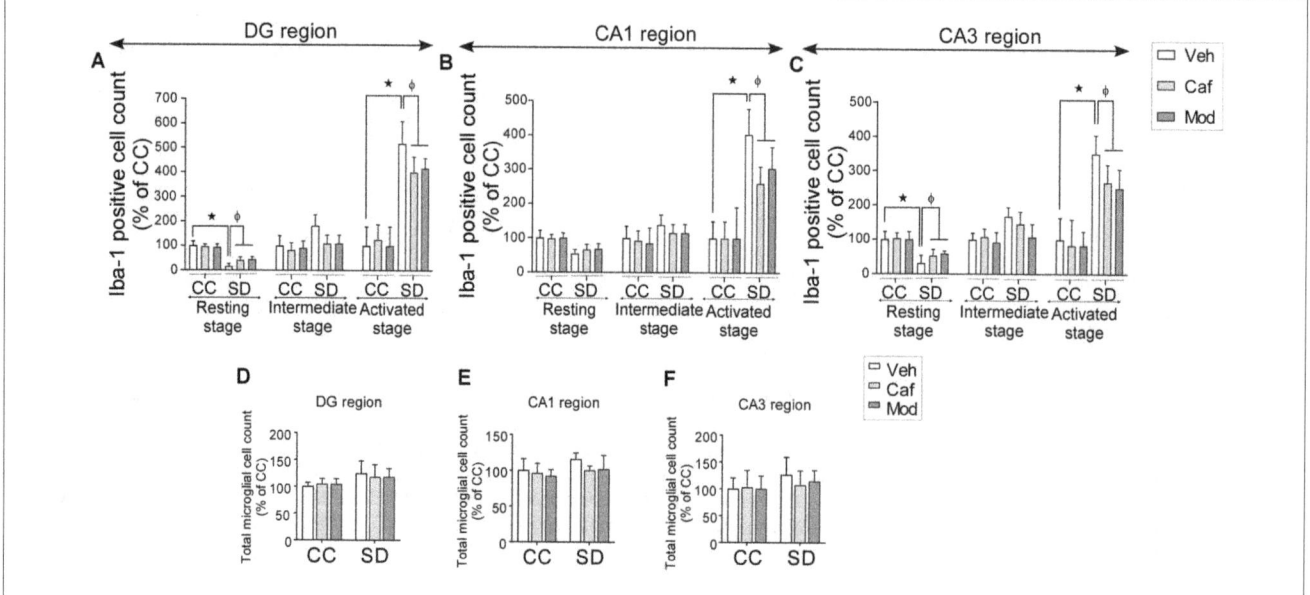

FIGURE 7 | Caffeine or modafinil improved the microglial cell numbers following SD. Changes in the microglial cell count at different stages (resting, intermediate and activated) in **(A)** DG region; **(B)** CA1 region; **(C)** CA3 region of the hippocampus. Total microglial cell count in **(D)** DG region; **(E)** CA1 region; **(F)** CA3 region of the hippocampus. $^*p < 0.05$ when compared to control treated with vehicle; $^\phi p < 0.05$ when compared to sleep deprived treated with vehicle. Two way ANOVA followed by Tukey post hoc test with multiple comparison was applied for statistical comparison between groups and for the graphical representation, values expressed mean percentage of Control ± SEM.

count was evaluated by the correlation analysis to validate the findings.

There was a significant correlation between the proportion of the number of entries in the open arms and the number of activated microglial cell count ($r^2 = 0.5634$; $p < 0.0001$; **Figure 8A**); proportion of the time spent in the open arms and the number of activated microglial cell count ($r^2 = 0.5958$; $p < 0.0001$; **Figure 8B**), and the total distance traveled in the OF and the number of activated microglial cell count ($r^2 = 0.3979$; $p < 0.0001$; **Figure 8C**).

A significant correlation was observed between the proportion of the number of entries in the open arms and

TNF-α level in hippocampus ($r^2 = 0.331$; $p < 0.0001$; **Figure 8D**); proportion of the time spent in the open arms and TNF-α level in hippocampus ($r^2 = 0.4472$; $p < 0.0001$; **Figure 8E**), and the total distance traveled in the OF and TNF-α level in hippocampus ($r^2 = 0.3734$; $p < 0.0001$; **Figure 8F**). Similarly, correlation analysis showed significant correlation between the proportion of the number of entries in the open arms and IL-1β level in hippocampus ($r^2 = 0.2514$; $p < 0.0001$; **Figure 8G**); proportion of the time spent in the open arms and IL-1β level in hippocampus ($r^2 = 0.2457$; $p < 0.0001$; **Figure 8H**) and the total distance traveled in the OF and IL-1β level in hippocampus ($r^2 = 0.26$; $p < 0.0001$; **Figure 8I**).

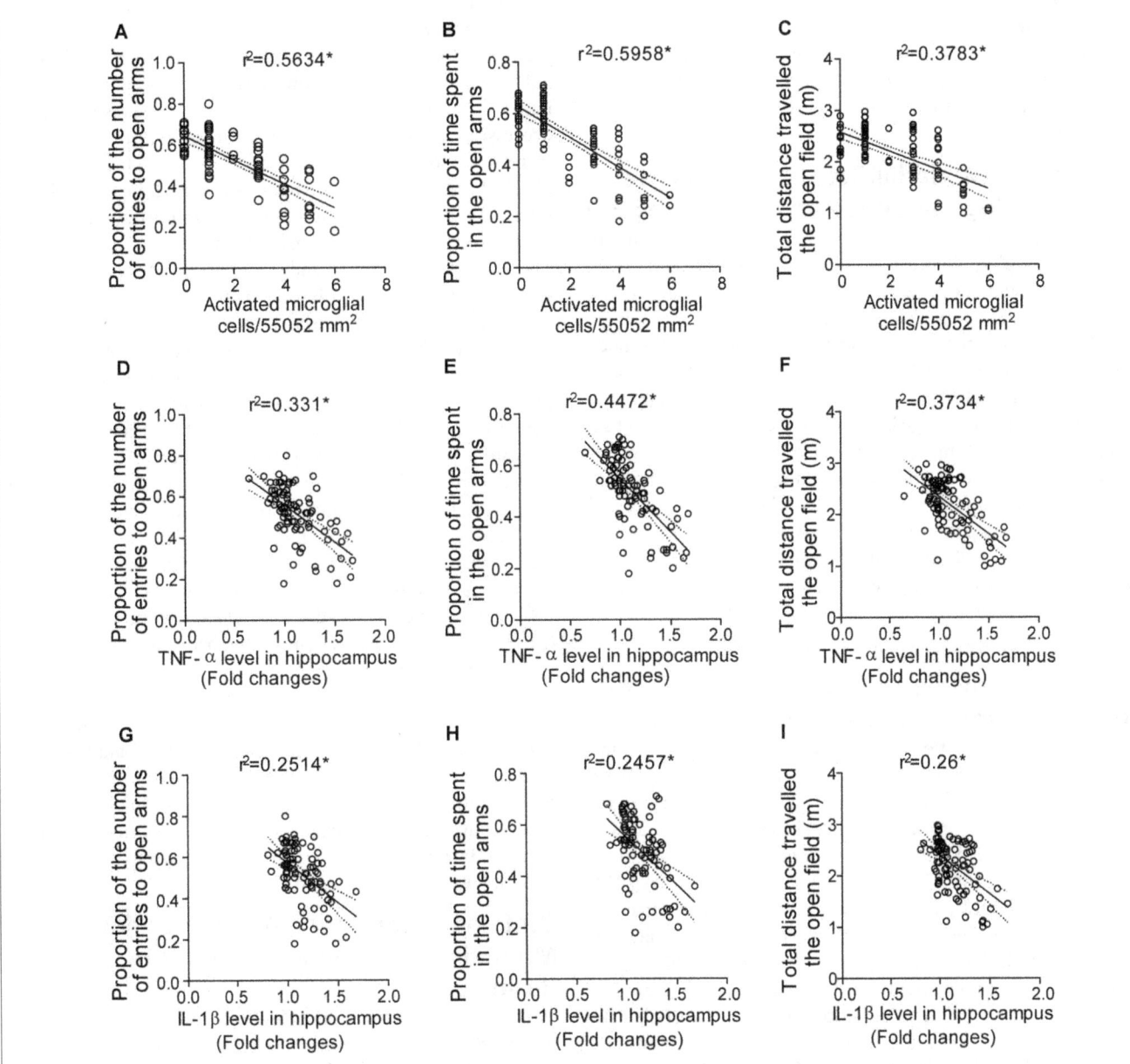

FIGURE 8 | Correlation analysis predicting the interaction between the anxious behavior, pro-inflammatory cytokines and activated microglia cell during SD. Interaction between the anxious behavior and activated microglial cell as shown by correlation between **(A)** proportion of the number of entries in the open arms and activated microglial cell count in the hippocampus; **(B)** proportion of the time spent in the open arms and activated microglial cell count in the hippocampus; **(C)** total distance traveled in the OF and activated microglial cell count in the hippocampus. Finally, the interaction between the anxious behavior and pro-inflammatory cytokine levels as shown by correlation between **(D)** proportion of the number of entries in the open arms and TNF-α level in hippocampus; **(E)** proportion of the time spent in the open arms and TNF-α level in hippocampus; **(F)** total distance traveled in the OF and TNF-α level in hippocampus; **(G)** proportion of the number of entries in the open arms and IL-1β level in hippocampus; **(H)** proportion of the time spent in the open arms and IL-1β level in hippocampus; **(I)** total distance traveled in the OF and IL-1β level in hippocampus. Pearson's test was applied for correlation analysis. $p < 0.05$ was considered to be statistically significant.

DISCUSSION

SD affected depressive/anxiety-like behavior and neuroinflammatory reactivity. Sleep deprived rats were more anxious in the EPM and OF test. However, caffeine or modafinil treatment increased OF activity, and thus reversed the effects of SD. Moreover, caffeine or modafinil decreased the time in closed but increased the time in open arms. Finally, pro-inflammatory cytokine levels were increased following SD and caffeine/modafinil decreased the pro-inflammatory and increased anti-inflammatory cytokine levels. As anticipated, the hippocampus of SD rats had higher glial immunoreactivity and altered morphology, and this effect was more pronounced in the microglial cells number, while psychostimulant drugs improved the above consequences. However, the tested doses of caffeine were higher and did not mimic habitual caffeine consumption. Together, our data highlight the influence of SD on neuroinflammatory responsiveness, and the importance of considering these factors in animal tests of depression/anxiety behaviors.

Caffeine or Modafinil Treatment Improved the Physiological Consequences during SD

Normal sleep is necessary for health and sleep disruption influences the physiological function (Kumar and Kalonia, 2007). Food intake and mood are related to each other, depending on the stress, food intake may decrease or increase (Singh, 2014). There is a controversial relationship between the body weight, food intake and sleep. There are reports on body weight reduction during SD (Mavanji et al., 2013), weight gain attenuation with no change in food intake during SD (Barf et al., 2012), at the same time increase in food intake with decreases in body weight (Koban and Stewart, 2006). Animal studies had shown that increased cytokine levels reduce food intake and body weight during neuroinflammatory conditions (Park et al., 2011a; Zombeck et al., 2013). In contrast to previous studies, we observed a decrease in body weight of sleep-deprived rats, which was improved after caffeine/modafinil treatment.

Caffeine or Modafinil Administration Produced Anxiolytic Effect and Recovered from the Depressive Behavior during SD

Under stressful conditions, the anxiety index has been found to be increased (Pechlivanova et al., 2012). Additionally, there are various reports of anxiogenic behavior in rodents as shown by EPM (decrease in time spent in open arms) and OF (impaired locomotor activity) test with body weight loss during SD (Silva et al., 2004; Garcia et al., 2011; Alkadhi et al., 2013; Matzner et al., 2013). Caffeine has a dose-dependent effect on anxious behavior. The previous findings revealed the behavioral observations that low (10 mg/kg) or moderate (20 mg/kg) dose of caffeine administration had shown reduction in anxiety behavior revealed by increased locomotor activity (Poletaeva and Oleinik, 1975; Antoniou et al., 1998), while increased dose of caffeine showed anxiogenic effect (Kayir and Uzbay, 2006) as predicted by the decrease or increase time and entry in open arms in rodents. Caffeine had shown to decrease the anxiety level during bright light stress conditions (Hughes et al., 2014) and chronic unpredictable stress conditions in rodents (Kaster et al., 2015). Like rodent studies, caffeine administration showed mood improvement in humans also (Smith et al., 2006). There are also reports available showing no effects of caffeine on anxiety (Khor et al., 2013). Similar to caffeine, modafinil showed anxiolytic effect shown by increased locomotor activity in humans (van Vliet et al., 2006) as well as rodents (Siwak et al., 2000; Quisenberry et al., 2013). This data is in consistent with the previous reports showing anxiogenic behavior during SD and caffeine or modafinil treatment improved the anxiety state during SD in rats, although the exact mechanism is still unclear.

There is a close association of the anxiety and depression with the disturbance in the normal sleep-wake cycle (Grønli et al., 2004; Jakubcakova et al., 2012; Kostyalik et al., 2014), immune system activation, assessed by decreased preference for a sweetness increased immobility time (Park et al., 2011b; Jangra et al., 2014). Imbalance in the cytokines level upon immune system activation has been found to be responsible for depression, predicted by a decrease in sucrose preference and increase in immobility (Ballok and Sakic, 2008; Braun et al., 2012). Inflammatory stimulation such as lipopolysaccharide model (Sayd et al., 2014) and Poly I:C administration (Missault et al., 2014), increased the pro-inflammatory cytokine (IL-1β, TNF-α and IL-6) levels and induced depressive symptoms. Furthermore, sickness behavior had also been found to be associated with the increased levels of the pro-inflammatory cytokine (Bluthé et al., 1995; Konsman et al., 2002; Dantzer, 2004; Vichaya et al., 2016, 2017). The incidence of anhedonia (decreased intake of sweet solution) had been reported in humans after SD (Petrovsky et al., 2014). Previous studies indicate a positive effect of caffeine on depressive symptoms shown by reduced immobility in rats (Vieira et al., 2008; Rusconi et al., 2014). Caffeine administration had been reported to alleviate the depressive behavior and memory dysfunction during chronic stress in a study by Kaster group (Kaster et al., 2015). Furthermore, inability in the reversal of the mood deficits in helpless mice during caffeine intake was reported (Machado et al., 2017). Human studies had shown improvement after modafinil treatment given during depression (Price and Taylor, 2005; Frye et al., 2007). Similar to humans, modafinil showed improvement in body weight along with mobility during stress in rats (Regenthal et al., 2009). In the paradigm used for our study, SD rats showed an increase in immobility and decrease the preference for sucrose solution, but the changes observed were not significant. Our data revealed non-significant decrease in immobility and increased sucrose preference after caffeine/modafinil treatment during SD.

Caffeine or Modafinil Administration Maintained the Inflammatory Profile during SD

The increase in the cytokine levels with decreased food and water intake had been reported in neuroinflammatory models of rodents (Zombeck et al., 2013). The anxiogenic and depressive effect of cytokines shown by a preference for closed arms than open arms reduced immobility and decreased sucrose preference validated by knock-out and cytokines administration studies (Pan et al., 2013). One rodent study on maternal obesity reported an elevated level of pro-inflammatory cytokines along with mood disorder (Kang et al., 2014); another neuroinflammatory study on lupus-prone mice described anxious behavior along with an increased pro-inflammatory cytokine level (Ballok and Sakic, 2008). The previous report suggested the beneficial effects of caffeine on MDMA induced a behavioral and neuroinflammatory response (Ruiz-Medina et al., 2013). Caffeine administration showed the ability to control the behavioral alterations in neuroinflammatory disease models such as Parkinson's (Chen et al., 2001; Joghataie et al., 2004) and Alzheimer's (Arendash et al., 2006; Dall'Igna et al., 2007). Caffeine treatment also restored the memory performance and glial cells reactivity in a rodent model of diabetes (Duarte et al., 2012) and Machado-Joseph disease (Gonçalves et al., 2013). Our data showed the increased level of pro-inflammatory and decreased levels of anti-inflammatory cytokine in hippocampus and plasma during SD. Caffeine or modafinil treatment improved the cytokine levels in the hippocampus and plasma during SD.

Astrocytes, immunocompetent cells of the brain, becomes activated and secretes several neurotoxic substances along with an enhanced GFAP protein expression (astrogliosis protein marker). This enhanced GFAP expression relates to the astrocytes activation severity. Astrocytes activation assessed by GFAP expression was increased along with increased cytokine levels during inflammatory stimulation by LPS administration (Brahmachari et al., 2006; Park et al., 2011a; Norden et al., 2016). Microglial activation marker (Iba-1) showed increased expression associated with anxiety in maternal obesity model (Kang et al., 2014), pollutants exposure (Bolton et al., 2012) in rodents. The association between the mood alterations and glial cells reactivity under the perspective of the purinergic neuromodulation had been recently studied (Rial et al., 2016). Caffeine decreases the glial cells activation along with the reducing production of pro-inflammatory cytokines due to the localization of the adenosine receptors on microglia (Sonsalla et al., 2012). In rodents, caffeine treatment has been reported to decrease the astrocytes activity (Ardais et al., 2014) and microglial activation during rodent models of neuroinflammation such as maneb and paraquat (Yadav et al., 2012) and MDMA (Ruiz-Medina et al., 2013). Our results also highlight the influence of SD on the immunoreactivity of astrocytes and microglia (increased) and caffeine or modafinil treatment down-regulated the immunoreactivity of astrocytes and microglia during SD.

We also observed the morphological changes in the astrocytes and microglia in the hippocampus of sleep-deprived rats and caffeine or modafinil administration during SD improved these changes during SD. Results of this study get supported from the previous studies of social defeat, showing elevated cytokine levels in the brain associated with anxiety (Wohleb et al., 2014a,b). Along with morphological changes, we observed increased cell numbers of activated microglial cells and decreased cell count of resting stage microglial cell during SD, while caffeine or modafinil treatment following SD improved the resting and activated microglial cells count. Previous findings dictated the modulatory role of adenosine A2A receptor system in neuroprotection by controlling the microglia inducing neuroinflammation (Dai et al., 2010; Rebola et al., 2011; Gyoneva et al., 2014b). In a previous study on stroke model, increased activated microglial cells, morphological alteration of microglia along with anxiety and depression-like behavior was reported (Nemeth et al., 2014). A recent study enlightened an interplay between the microglial cells alterations and anxiety disorders, regulated by adenosine A2A receptor (Caetano et al., 2017). Stressful conditions trigger the increased release of adenosine and ATP, in which the adenosine modulation system in association with glial cells afford maximum neuroprotection (Cunha, 2016). Caffeine attenuated the activated microglial cell count in the hippocampus on LPS stimulation (Brothers et al., 2010) and high

cholesterol diet-induced model of neuroinflammation (Chen et al., 2008). Caffeine administration prevented the microglia activation induced neuroinflammation in the transient retinal ischemic model in rodents (Boia et al., 2017). Similarly, modafinil treatment prevented the glial cells activation shown by increased resting and decrease activated stage cell with improvement in immunoreactivity in a neuroinflammatory model in rodents (Raineri et al., 2012). It was observed that microglial cells activation induced the increased level of pro-inflammatory cytokines, which further influenced the normal mood of rats towards anxious state during SD. The use of only male Sprague Dawley rats is a limitation of the present study as recent survey suggested the relation of sleep patterns with caffeine is different in males and females (Frozi et al., 2018).

CONCLUSION

The present study demonstrated that inhibition of microglial cells by caffeine or modafinil treatment modulated the cytokine levels with increased the anti-inflammatory cytokines and ameliorates the anxious behavior) during SD. Our data suggested that caffeine/modafinil are the effective therapeutic agents against SD-induced neuroinflammation and anxiety behavior.

AUTHOR CONTRIBUTIONS

UP and MW designed the study and wrote the manuscript. MW performed the experiments and analyzed the data. KRay, LT, GC, KRoy, SS, SD and VJ helped in the manuscript writing. GC, KRoy and VJ helped in the immunohistochemistry of glial proteins and cytokines level measurement. SS and SD helped in the behavioral and RT-PCR experiments. UP, KRay, KK and LT contributed in procurement and facilitated the instruments and other facilities. All authors read and approved the final manuscript.

ACKNOWLEDGMENTS

We are thankful to Director, DIPAS, for supporting and facilitating the research work. Dr. Dipti Prasad and Dr. Ekta Kohli, Neurobiology division are acknowledged for extending the behavioral facility for the study. We are thankful to Dr. Zahid Mohammad Ashraf, Genomics Division, for extending us to use RT-PCR instrument facility.

SUPPLEMENTARY MATERIAL

FIGURE S1 | Caffeine or modafinil treatment improved the sleep deprivation (SD)-induced physiological changes in rats. Changes in **(A)** body weight; **(B)** food intake. $*p < 0.05$ when compared to control treated with vehicle; $^{\phi}p < 0.05$, when compared to sleep deprived treated with vehicle. Two way ANOVA followed by Tukey post hoc test with multiple comparison was applied for statistical comparison between groups and for the graphical representation, values expressed mean percentage of Control ± SEM.

FIGURE S2 | Fold changes in the inflammatory cytokines in plasma during caffeine/modafinil administration following SD. Fold changes in **(A)** TNF-α; **(B)** IL-1β; **(C)** IL-6 (pro-inflammatory cytokines); **(D)** IL-4; **(E)** IL-10 (anti-inflammatory cytokines) in plasma. *$p < 0.05$ when compared to control treated with vehicle; $^{\phi}p < 0.05$ when compared to sleep deprived treated with vehicle. Two way

ANOVA followed by Tukey *post hoc* test with multiple comparison was applied for statistical analysis.

REFERENCES

Abelaira, H. M., Reus, G. Z., and Quevedo, J. (2013). Animal models as tools to study the pathophysiology of depression. *Rev. Bras. Psiquiatr.* 35, S112–S120. doi: 10.1590/1516-4446-2013-1098

Alkadhi, K., Zagaar, M., Alhaider, I., Salim, S., and Aleisa, A. (2013). Neurobiological consequences of sleep deprivation. *Curr. Neuropharmacol.* 11, 231–249. doi: 10.2174/1570159x11311030001

Antoniou, K., Kafetzopoulos, E., Papadopoulou-Daifoti, Z., Hyphantis, T., and Marselos, M. (1998). D-amphetamine, cocaine and caffeine: a comparative study of acute effects on locomotor activity and behavioural patterns in rats. *Neurosci. Biobehav. Rev.* 23, 189–196. doi: 10.1016/s0149-7634(98)00020-7

Ardais, A. P., Borges, M. F., Rocha, A. S., Sallaberry, C., Cunha, R. A., and Porciúncula, L. O. (2014). Caffeine triggers behavioral and neurochemical alterations in adolescent rats. *Neuroscience* 270, 27–39. doi: 10.1016/j.neuroscience.2014.04.003

Arendash, G. W., Schleif, W., Rezai-Zadeh, K., Jackson, E. K., Zacharia, L. C., Cracchiolo, J. R., et al. (2006). Caffeine protects Alzheimer's mice against cognitive impairment and reduces brain β-amyloid production. *Neuroscience* 142, 941–952. doi: 10.1016/j.neuroscience.2006.07.021

Babson, K. A., Trainor, C. D., Feldner, M. T., and Blumenthal, H. (2010). A test of the effects of acute sleep deprivation on general and specific self-reported anxiety and depressive symptoms: an experimental extension. *J. Behav. Ther. Exp. Psychiatry* 41, 297–303. doi: 10.1016/j.jbtep.2010.02.008

Ballok, D. A., and Sakic, B. (2008). Purine receptor antagonist modulates serology and affective behaviors in lupus-prone mice: evidence of autoimmune-induced pain? *Brain Behav. Immun.* 22, 1208–1216. doi: 10.1016/j.bbi.2008.06.002

Barf, R. P., Van Dijk, G., Scheurink, A. J., Hoffmann, K., Novati, A., Hulshof, H. J., et al. (2012). Metabolic consequences of chronic sleep restriction in rats: changes in body weight regulation and energy expenditure. *Physiol. Behav.* 107, 322–328. doi: 10.1016/j.physbeh.2012.09.005

Bluthé, R. M., Beaudu, C., Kelley, K. W., and Dantzer, R. (1995). Differential effects of IL-1ra on sickness behavior and weight loss induced by IL-1 in rats. *Brain Res.* 677, 171–176. doi: 10.1016/0006-8993(95)00194-u

Boele, F. W., Douw, L., De Groot, M., Van Thuijl, H. F., Cleijne, W., Heimans, J. J., et al. (2013). The effect of modafinil on fatigue, cognitive functioning and mood in primary brain tumor patients: a multicenter randomized controlled trial. *Neuro Oncol.* 15, 1420–1428. doi: 10.1093/neuonc/not102

Boia, R., Elvas, F., Madeira, M. H., Aires, I. D., Rodrigues-Neves, A. C., Tralhão, P., et al. (2017). Treatment with A$_{2A}$ receptor antagonist KW6002 and caffeine intake regulate microglia reactivity and protect retina against transient ischemic damage. *Cell Death Dis.* 8:e3065. doi: 10.1038/cddis.2017.451

Bolton, J. L., Smith, S. H., Huff, N. C., Gilmour, M. I., Foster, W. M., Auten, R. L., et al. (2012). Prenatal air pollution exposure induces neuroinflammation and predisposes offspring to weight gain in adulthood in a sex-specific manner. *FASEB J.* 26, 4743–4754. doi: 10.1096/fj.12-210989

Brahmachari, S., Fung, Y. K., and Pahan, K. (2006). Induction of glial fibrillary acidic protein expression in astrocytes by nitric oxide. *J. Neurosci.* 26, 4930–4939. doi: 10.1523/jneurosci.5480-05.2006

Braun, T. P., Grossberg, A. J., Veleva-Rotse, B. O., Maxson, J. E., Szumowski, M., Barnes, A. P., et al. (2012). Expression of myeloid differentiation factor 88 in neurons is not requisite for the induction of sickness behavior by interleukin-1β. *J. Neuroinflammation* 9:229. doi: 10.1186/1742-2094-9-229

Brothers, H. M., Marchalant, Y., and Wenk, G. L. (2010). Caffeine attenuates lipopolysaccharide-induced neuroinflammation. *Neurosci. Lett.* 480, 97–100. doi: 10.1016/j.neulet.2010.06.013

Caetano, L., Pinheiro, H., Patrício, P., Mateus-Pinheiro, A., Alves, N. D., Coimbra, B., et al. (2017). Adenosine A$_{2A}$ receptor regulation of microglia morphological remodeling-gender bias in physiology and in a model of chronic anxiety. *Mol. Psychiatry* 22, 1035–1043. doi: 10.1038/mp.2016.173

Cappelletti, S., Piacentino, D., Sani, G., and Aromatario, M. (2015). Caffeine: cognitive and physical performance enhancer or psychoactive drug? *Curr. Neuropharmacol.* 13, 71–88. doi: 10.2174/1570159x13666141210215655

Carey, M. G., Al-Zaiti, S. S., Dean, G. E., Sessanna, L., and Finnell, D. S. (2011). Sleep problems, depression, substance use, social bonding, and quality of life in professional firefighters. *J. Occup. Environ. Med.* 53, 928–933. doi: 10.1097/jom.0b013e318225898f

Chauhan, G., Ray, K., Sahu, S., Roy, K., Jain, V., Wadhwa, M., et al. (2016). Adenosine A1 receptor antagonist mitigates deleterious effects of sleep deprivation on adult neurogenesis and spatial reference memory in rats. *Neuroscience* 337, 107–116. doi: 10.1016/j.neuroscience.2016.09.007

Chen, X., Gawryluk, J. W., Wagener, J. F., Ghribi, O., and Geiger, J. D. (2008). Caffeine blocks disruption of blood brain barrier in a rabbit model of Alzheimer's disease. *J. Neuroinflammation* 5:12. doi: 10.1186/1742-2094-5-12

Chen, J. F., Xu, K., Petzer, J. P., Staal, R., Xu, Y. H., Beilstein, M., et al. (2001). Neuroprotection by caffeine and A$_{2A}$ adenosine receptor inactivation in a model of Parkinson's disease. *J. Neurosci.* 21:RC143.

Cunha, R. A. (2016). How does adenosine control neuronal dysfunction and neurodegeneration? *J. Neurochem.* 139, 1019–1055. doi: 10.1111/jnc.13724

Cunha, R. A., and Agostinho, P. M. (2010). Chronic caffeine consumption prevents memory disturbance in different animal models of memory decline. *J. Alzheimers Dis.* 20, S95–S116. doi: 10.3233/jad-2010-1408

Dai, S. S., Zhou, Y. G., Li, W., An, J. H., Li, P., Yang, N., et al. (2010). Local glutamate level dictates adenosine A$_{2A}$ receptor regulation of neuroinflammation and traumatic brain injury. *J. Neurosci.* 30, 5802–5810. doi: 10.1523/JNEUROSCI.0268-10.2010

Dall'Igna, O. P., Fett, P., Gomes, M. W., Souza, D. O., Cunha, R. A., and Lara, D. R. (2007). Caffeine and adenosine A$_{2A}$ receptor antagonists prevent β-amyloid (25–35)-induced cognitive deficits in mice. *Exp. Neurol.* 203, 241–245. doi: 10.1016/j.expneurol.2006.08.008

Daly, J. W. (2007). Caffeine analogs: biomedical impact. *Cell Mol. Life Sci.* 64, 2153–2169. doi: 10.1007/s00018-007-7051-9

Dantzer, R. (2004). Cytokine-induced sickness behaviour: a neuroimmune response to activation of innate immunity. *Eur. J. Pharmacol.* 500, 399–411. doi: 10.1016/j.ejphar.2004.07.040

Duarte, J. M., Agostinho, P. M., Carvalho, R. A., and Cunha, R. A. (2012). Caffeine consumption prevents diabetes-induced memory impairment and synaptotoxicity in the hippocampus of NONcZNO10/LTJ mice. *PLoS One* 7:e21899. doi: 10.1371/journal.pone.0021899

Dworak, M., Kim, T., Mccarley, R. W., and Basheer, R. (2011). Sleep, brain energy levels and food intake: relationship between hypothalamic ATP concentrations, food intake and body weight during sleep-wake and sleep deprivation in rats. *Somnologie* 15, 111–117. doi: 10.1007/s11818-011-0524-y

Frozi, J., De Carvalho, H. W., Ottoni, G. L., Cunha, R. A., and Lara, D. R. (2018). Distinct sensitivity to caffeine-induced insomnia related to age. *J. Psychopharmacol.* 32, 89–95. doi: 10.1177/0269881117722997

Frye, M. A., Grunze, H., Suppes, T., McElroy, S. L., Keck, P. E. Jr., Walden, J., et al. (2007). A placebo-controlled evaluation of adjunctive modafinil in the treatment of bipolar depression. *Am. J. Psychiatry* 164, 1242–1249. doi: 10.1176/appi.ajp.2007.06060981

Garcia, A. M., Cardenas, F. P., and Morato, S. (2011). The effects of pentylenetetrazol, chlordiazepoxide and caffeine in rats tested in the elevated plus-maze depend on the experimental illumination. *Behav. Brain Res.* 217, 171–177. doi: 10.1016/j.bbr.2010.09.032

Gonçalves, N., Simões, A. T., Cunha, R. A., and De Almeida, L. P. (2013). Caffeine and adenosine A$_{2A}$ receptor inactivation decrease striatal neuropathology in a lentiviral-based model of Machado-Joseph disease. *Ann. Neurol.* 73, 655–666. doi: 10.1002/ana.23866

Grønli, J., Murison, R., Bjorvatn, B., Sørensen, E., Portas, C. M., and Ursin, R. (2004). Chronic mild stress affects sucrose intake and sleep in rats. *Behav. Brain Res.* 150, 139–147. doi: 10.1016/s0166-4328(03)00252-3

Gyoneva, S., Davalos, D., Biswas, D., Swanger, S. A., Garnier-Amblard, E., Loth, F., et al. (2014a). Systemic inflammation regulates microglial responses to tissue damage *in vivo*. *Glia* 62, 1345–1360. doi: 10.1002/glia.22686

Gyoneva, S., Shapiro, L., Lazo, C., Garnier-Amblard, E., Smith, Y., Miller, G. W., et al. (2014b). Adenosine A$_{2A}$ receptor antagonism reverses inflammation-induced impairment of microglial process extension in a model of Parkinson's disease. *Neurobiol. Dis.* 67, 191–202. doi: 10.1016/j.nbd.2014.03.004

Hong, H., Kim, B. S., and Im, H. I. (2016). Pathophysiological role of neuroinflammation in neurodegenerative diseases and psychiatric disorders. *Int. Neurourol. J.* 20, S2–S7. doi: 10.5213/inj.1632604.302

Hughes, R. N., Hancock, N. J., Henwood, G. A., and Rapley, S. A. (2014). Evidence for anxiolytic effects of acute caffeine on anxiety-related behavior in male and female rats tested with and without bright light. *Behav. Brain Res.* 271, 7–15. doi: 10.1016/j.bbr.2014.05.038

Jakubcakova, V., Flachskamm, C., Landgraf, R., and Kimura, M. (2012). Sleep phenotyping in a mouse model of extreme trait anxiety. *PLoS One* 7:e40625. doi: 10.1371/journal.pone.0040625

Jangra, A., Lukhi, M. M., Sulakhiya, K., Baruah, C. C., and Lahkar, M. (2014). Protective effect of mangiferin against lipopolysaccharide-induced depressive and anxiety-like behaviour in mice. *Eur. J. Pharmacol.* 740, 337–345. doi: 10.1016/j.ejphar.2014.07.031

Joghataie, M. T., Roghani, M., Negahdar, F., and Hashemi, L. (2004). Protective effect of caffeine against neurodegeneration in a model of Parkinson's disease in rat: behavioral and histochemical evidence. *Parkinsonism Relat. Disord.* 10, 465–468. doi: 10.1016/j.parkreldis.2004.06.004

Kang, A., Hao, H., Zheng, X., Liang, Y., Xie, Y., Xie, T., et al. (2011). Peripheral anti-inflammatory effects explain the ginsenosides paradox between poor brain distribution and anti-depression efficacy. *J. Neuroinflammation* 8:100. doi: 10.1186/1742-2094-8-100

Kang, S. S., Kurti, A., Fair, D. A., and Fryer, J. D. (2014). Dietary intervention rescues maternal obesity induced behavior deficits and neuroinflammation in offspring. *J. Neuroinflammation* 11:156. doi: 10.1186/s12974-014-0156-9

Kaster, M. P., Machado, N. J., Silva, H. B., Nunes, A., Ardais, A. P., Santana, M., et al. (2015). Caffeine acts through neuronal adenosine A_{2A} receptors to prevent mood and memory dysfunction triggered by chronic stress. *Proc. Natl. Acad. Sci. U S A* 112, 7833–7838. doi: 10.1073/pnas.1423088112

Kayir, H., and Uzbay, I. T. (2006). Nicotine antagonizes caffeine- but not pentylenetetrazole-induced anxiogenic effect in mice. *Psychopharmacology* 184, 464–469. doi: 10.1007/s00213-005-0036-1

Khor, Y. M., Soga, T., and Parhar, I. S. (2013). Caffeine neuroprotects against dexamethasone-induced anxiety-like behaviour in the Zebrafish (Danio rerio). *Gen. Comp. Endocrinol.* 181, 310–315. doi: 10.1016/j.ygcen.2012.09.021

Koban, M., and Stewart, C. V. (2006). Effects of age on recovery of body weight following REM sleep deprivation of rats. *Physiol. Behav.* 87, 1–6. doi: 10.1016/j.physbeh.2005.09.006

Konsman, J. P., Parnet, P., and Dantzer, R. (2002). Cytokine-induced sickness behaviour: mechanisms and implications. *Trends Neurosci.* 25, 154–159. doi: 10.1016/s0166-2236(00)02088-9

Kostyalik, D., Vas, S., Kátai, Z., Kitka, T., Gyertyán, I., Bagdy, G., et al. (2014). Chronic escitalopram treatment attenuated the accelerated rapid eye movement sleep transitions after selective rapid eye movement sleep deprivation: a model-based analysis using Markov chains. *BMC Neurosci.* 15:120. doi: 10.1186/s12868-014-0120-8

Kumar, A., and Kalonia, H. (2007). Protective effect of *Withania somnifera* Dunal on the behavioral and biochemical alterations in sleep-disturbed mice (Grid over water suspended method). *Indian J. Exp. Biol.* 45, 524–528.

Machado, N. J., Simões, A. P., Silva, H. B., Ardais, A. P., Kaster, M. P., Garção, P., et al. (2017). Caffeine reverts memory but not mood impairment in a depression-prone mouse strain with up-regulated adenosine A_{2A} receptor in hippocampal glutamate synapses. *Mol. Neurobiol.* 54, 1552–1563. doi: 10.1007/s12035-016-9774-9

Matzner, P., Hazut, O., Naim, R., Shaashua, L., Sorski, L., Levi, B., et al. (2013). Resilience of the immune system in healthy young students to 30-hour sleep deprivation with psychological stress. *Neuroimmunomodulation* 20, 194–204. doi: 10.1159/000348698

Mavanji, V., Teske, J. A., Billington, C. J., and Kotz, C. M. (2013). Partial sleep deprivation by environmental noise increases food intake and body weight in obesity-resistant rats. *Obesity* 21, 1396–1405. doi: 10.1002/oby.20182

McGaugh, J., Mancino, M. J., Feldman, Z., Chopra, M. P., Gentry, W. B., Cargile, C., et al. (2009). Open-label pilot study of modafinil for methamphetamine dependence. *J. Clin. Psychopharmacol.* 29, 488–491. doi: 10.1097/JCP.0b013e3181b591e0

Minzenberg, M. J., and Carter, C. S. (2008). Modafinil: a review of neurochemical actions and effects on cognition. *Neuropsychopharmacology* 33, 1477–1502. doi: 10.1038/sj.npp.1301534

Missault, S., Van den Eynde, K., Vanden Berghe, W., Fransen, E., Weeren, A., Timmermans, J. P., et al. (2014). The risk for behavioural deficits is determined by the maternal immune response to prenatal immune challenge in a neurodevelopmental model. *Brain Behav. Immun.* 42, 138–146. doi: 10.1016/j.bbi.2014.06.013

Nehlig, A. (2010). Is caffeine a cognitive enhancer? *J. Alzheimers Dis.* 20, S85–S94. doi: 10.3233/JAD-2010-091315

Nemeth, C. L., Reddy, R., Bekhbat, M., Bailey, J., and Neigh, G. N. (2014). Microglial activation occurs in the absence of anxiety-like behavior following microembolic stroke in female, but not male, rats. *J. Neuroinflammation* 11:174. doi: 10.1186/s12974-014-0174-7

Norden, D. M., Trojanowski, P. J., Villanueva, E., Navarro, E., and Godbout, J. P. (2016). Sequential activation of microglia and astrocyte cytokine expression precedes increased Iba-1 or GFAP immunoreactivity following systemic immune challenge. *Glia* 64, 300–316. doi: 10.1002/glia.22930

Pan, W., Wu, X., He, Y., Hsuchou, H., Huang, E. Y., Mishra, P. K., et al. (2013). Brain interleukin-15 in neuroinflammation and behavior. *Neurosci. Biobehav. Rev.* 37, 184–192. doi: 10.1016/j.neubiorev.2012.11.009

Park, S. E., Dantzer, R., Kelley, K. W., and McCusker, R. H. (2011a). Central administration of insulin-like growth factor-I decreases depressive-like behavior and brain cytokine expression in mice. *J. Neuroinflammation* 8:12. doi: 10.1186/1742-2094-8-12

Park, S. E., Lawson, M., Dantzer, R., Kelley, K. W., and McCusker, R. H. (2011b). Insulin-like growth factor-I peptides act centrally to decrease depression-like behavior of mice treated intraperitoneally with lipopolysaccharide. *J. Neuroinflammation* 8:179. doi: 10.1186/1742-2094-8-179

Pechlivanova, D. M., Tchekalarova, J. D., Alova, L. H., Petkov, V. V., Nikolov, R. P., and Yakimova, K. S. (2012). Effect of long-term caffeine administration on depressive-like behavior in rats exposed to chronic unpredictable stress. *Behav. Pharmacol.* 23, 339–347. doi: 10.1097/FBP.0b013e3283564dd9

Petrovsky, N., Ettinger, U., Hill, A., Frenzel, L., Meyhöfer, I., Wagner, M., et al. (2014). Sleep deprivation disrupts prepulse inhibition and induces psychosis-like symptoms in healthy humans. *J. Neurosci.* 34, 9134–9140. doi: 10.1523/JNEUROSCI.0904-14.2014

Poletaeva, I. I., and Oleinik, V. M. (1975). The effect of phenamine and caffeine on the ability to extrapolate in rats. *Zh. Vyssh. Nerv. Deiat. Im. I P Pavlova* 25, 529–534.

Price, C. S., and Taylor, F. B. (2005). A retrospective chart review of the effects of modafinil on depression as monotherapy and as adjunctive therapy. *Depress. Anxiety* 21, 149–153. doi: 10.1002/da.20075

Quisenberry, A. J., Prisinzano, T. E., and Baker, L. E. (2013). Modafinil alone and in combination with low dose amphetamine does not establish conditioned place preference in male Sprague-Dawley rats. *Exp. Clin. Psychopharmacol.* 21, 252–258. doi: 10.1037/a0031832

Raineri, M., Gonzalez, B., Goitia, B., Garcia-Rill, E., Krasnova, I. N., Cadet, J. L., et al. (2012). Modafinil abrogates methamphetamine-induced neuroinflammation and apoptotic effects in the mouse striatum. *PLoS One* 7:e46599. doi: 10.1371/journal.pone.0046599

Rasetti, R., Mattay, V. S., Stankevich, B., Skjei, K., Blasi, G., Sambataro, F., et al. (2010). Modulatory effects of modafinil on neural circuits regulating emotion and cognition. *Neuropsychopharmacology* 35, 2101–2109. doi: 10.1038/npp.2010.83

Rebola, N., Simões, A. P., Canas, P. M., Tomé, A. R., Andrade, G. M., Barry, C. E., et al. (2011). Adenosine A_{2A} receptors control neuroinflammation and consequent hippocampal neuronal dysfunction. *J. Neurochem.* 117, 100–111. doi: 10.1111/j.1471-4159.2011.07178.x

Regenthal, R., Koch, H., Köhler, C., Preiss, R., and Krügel, U. (2009). Depression-like deficits in rats improved by subchronic modafinil. *Psychopharmacology* 204, 627–639. doi: 10.1007/s00213-009-1493-8

Rial, D., Lemos, C., Pinheiro, H., Duarte, J. M., Gonçalves, F. Q., Real, J. I., et al. (2016). Depression as a glial-based synaptic dysfunction. *Front. Cell. Neurosci.* 9:521. doi: 10.3389/fncel.2015.00521

Rio, D. C., Ares, M. Jr., Hannon, G. J., and Nilsen, T. W. (2010). Purification of RNA using TRIzol (TRI reagent). *Cold Spring Harb. Protoc.* 2010:pdb.prot5439. doi: 10.1101/pdb.prot5439

Rönnbäck, L., and Hansson, E. (2004). On the potential role of glutamate transport in mental fatigue. *J. Neuroinflammation* 1:22. doi: 10.1186/1742-2094-1-22

Ruiz-Medina, J., Pinto-Xavier, A., Rodríguez-Arias, M., Miñarro, J., and Valverde, O. (2013). Influence of chronic caffeine on MDMA-induced behavioral and neuroinflammatory response in mice. *Psychopharmacology* 226, 433–444. doi: 10.1007/s00213-012-2918-3

Rusconi, A. C., Valeriani, G., Carluccio, G. M., Majorana, M., Carlone, C., Raimondo, P., et al. (2014). [Coffee consumption in depressive disorders: it's not one size fits all]. *Riv. Psichiatr.* 49, 164–171. doi: 10.1708/1600.17452

Sahu, S., Kauser, H., Ray, K., Kishore, K., Kumar, S., and Panjwani, U. (2013). Caffeine and modafinil promote adult neuronal cell proliferation during 48 h of total sleep deprivation in rat dentate gyrus. *Exp. Neurol.* 248, 470–481. doi: 10.1016/j.expneurol.2013.07.021

Sanday, L., Zanin, K. A., Patti, C. L., Fernandes-Santos, L., Oliveira, L. C., Longo, B. M., et al. (2013). Role of state-dependent learning in the cognitive effects of caffeine in mice. *Int. J. Neuropsychopharmacol.* 16, 1547–1557. doi: 10.1017/S1461145712001551

Sayd, A., Antón, M., Alén, F., Caso, J. R., Pavón, J., Leza, J. C., et al. (2014). Systemic administration of oleoylethanolamide protects from neuroinflammation and anhedonia induced by LPS in rats. *Int. J. Neuropsychopharmacol.* 18:pyu111. doi: 10.1093/ijnp/pyu111

Silva, R. H., Kameda, S. R., Carvalho, R. C., Takatsu-Coleman, A. L., Niigaki, S. T., Abilio, V. C., et al. (2004). Anxiogenic effect of sleep deprivation in the elevated plus-maze test in mice. *Psychopharmacology* 176, 115–122. doi: 10.1007/s00213-004-1873-z

Singh, M. (2014). Mood, food, and obesity. *Front. Psychol.* 5:925. doi: 10.3389/fpsyg.2014.00925

Siwak, C. T., Gruet, P., Woehrlé, F., Schneider, M., Muggenburg, B. A., Murphey, H. L., et al. (2000). Behavioral activating effects of adrafinil in aged canines. *Pharmacol. Biochem. Behav.* 66, 293–300. doi: 10.1016/s0091-3057(00)00188-x

Smith, A. P., Christopher, G., and Sutherland, D. (2006). Effects of caffeine in overnight-withdrawn consumers and non-consumers. *Nutr. Neurosci.* 9, 63–71. doi: 10.1080/10284150600582927

Sonsalla, P. K., Wong, L. Y., Harris, S. L., Richardson, J. R., Khobahy, I., Li, W., et al. (2012). Delayed caffeine treatment prevents nigral dopamine neuron loss in a progressive rat model of Parkinson's disease. *Exp. Neurol.* 234, 482–487. doi: 10.1016/j.expneurol.2012.01.022

van Vliet, S. A., Jongsma, M. J., Vanwersch, R. A., Olivier, B., and Philippens, I. H. (2006). Behavioral effects of modafinil in marmoset monkeys. *Psychopharmacology* 185, 433–440. doi: 10.1007/s00213-006-0340-4

Vichaya, E. G., Molkentine, J. M., Vermeer, D. W., Walker, A. K., Feng, R., Holder, G., et al. (2016). Sickness behavior induced by cisplatin chemotherapy

and radiotherapy in a murine head and neck cancer model is associated with altered mitochondrial gene expression. *Behav. Brain Res.* 297, 241–250. doi: 10.1016/j.bbr.2015.10.024

Vichaya, E. G., Vermeer, D. W., Christian, D. L., Molkentine, J. M., Mason, K. A., Lee, J. H., et al. (2017). Neuroimmune mechanisms of behavioral alterations in a syngeneic murine model of human papilloma virus-related head and neck cancer. *Psychoneuroendocrinology* 79, 59–66. doi: 10.1016/j.psyneuen.2017.02.006

Vieira, C., De Lima, T. C., Carobrez Ade, P., and Lino-de-Oliveira, C. (2008). Frequency of climbing behavior as a predictor of altered motor activity in rat forced swimming test. *Neurosci. Lett.* 445, 170–173. doi: 10.1016/j.neulet.2008.09.001

Wadhwa, M., Sahu, S., Kumari, P., Kauser, H., Ray, K., and Panjwani, U. (2015). Caffeine and modafinil given during 48 h sleep deprivation modulate object recognition memory and synaptic proteins in the hippocampus of the rat. *Behav. Brain Res.* 294, 95–101. doi: 10.1016/j.bbr.2015.08.002

Wisor, J. P., Schmidt, M. A., and Clegern, W. C. (2011). Evidence for neuroinflammatory and microglial changes in the cerebral response to sleep loss. *Sleep* 34, 261–272. doi: 10.1093/sleep/34.3.261

Wohleb, E. S., McKim, D. B., Shea, D. T., Powell, N. D., Tarr, A. J., Sheridan, J. F., et al. (2014a). Re-establishment of anxiety in stress-sensitized mice is caused by monocyte trafficking from the spleen to the brain. *Biol. Psychiatry* 75, 970–981. doi: 10.1016/j.biopsych.2013.11.029

Wohleb, E. S., Patterson, J. M., Sharma, V., Quan, N., Godbout, J. P., and Sheridan, J. F. (2014b). Knockdown of interleukin-1 receptor type-1 on endothelial cells attenuated stress-induced neuroinflammation and prevented anxiety-like behavior. *J. Neurosci.* 34, 2583–2591. doi: 10.1523/JNEUROSCI.3723-13.2014

Yadav, S., Gupta, S. P., Srivastava, G., Srivastava, P. K., and Singh, M. P. (2012). Role of secondary mediators in caffeine-mediated neuroprotection in maneb- and paraquat-induced Parkinson's disease phenotype in the mouse. *Neurochem. Res.* 37, 875–884. doi: 10.1007/s11064-011-0682-0

Yamada, K., Kobayashi, M., and Kanda, T. (2014). Involvement of adenosine A_{2A} receptors in depression and anxiety. *Int. Rev. Neurobiol.* 119, 373–393. doi: 10.1016/B978-0-12-801022-8.00015-5

Zielinski, M. R., and Krueger, J. M. (2011). Sleep and innate immunity. *Front. Biosci.* 3, 632–642. doi: 10.2741/s176

Zombeck, J. A., Fey, E. G., Lyng, G. D., and Sonis, S. T. (2013). A clinically translatable mouse model for chemotherapy-related fatigue. *Comp. Med.* 63, 491–497.

3

Brain-Resident Microglia and Blood-Borne Macrophages Orchestrate Central Nervous System Inflammation in Neurodegenerative Disorders and Brain Cancer

Lisa Sevenich*

Georg-Speyer-Haus, Institute for Tumor Biology and Experimental Therapy, Frankfurt am Main, Germany

*Correspondence:
Lisa Sevenich
sevenich@gsh.uni-frankfurt.de

Inflammation is a hallmark of different central nervous system (CNS) pathologies. It has been linked to neurodegenerative disorders as well as primary and metastatic brain tumors. Microglia, the brain-resident immune cells, are emerging as a central player in regulating key pathways in CNS inflammation. Recent insights into neuroinflammation indicate that blood-borne immune cells represent an additional critical cellular component in mediating CNS inflammation. The lack of experimental systems that allow for discrimination between brain-resident and recruited myeloid cells has previously halted functional analysis of microglia and their blood-borne counterparts in brain malignancies. However, recent conceptual and technological advances, such as the generation of lineage tracing models and the identification of cell type-specific markers provide unprecedented opportunities to study the cellular functions of microglia and macrophages by functional interference. The use of different "omic" strategies as well as imaging techniques has significantly increased our knowledge of disease-associated gene signatures and effector functions under pathological conditions. In this review, recent developments in evaluating functions of brain-resident and recruited myeloid cells in neurodegenerative disorders and brain cancers will be discussed and unique or shared cellular traits of microglia and macrophages in different CNS disorders will be highlighted. Insight from these studies will shape our understanding of disease- and cell-type-specific effector functions of microglia or macrophages and will open new avenues for therapeutic intervention that target aberrant functions of myeloid cells in CNS pathologies.

Keywords: neuroinflammation, tissue-resident macrophages, microglia, neurodegeneration, cancer

INTRODUCTION

The brain has long been regarded as an immunologically privileged site in which the presence of the blood–brain barrier (BBB) restricts the entry of blood-borne immune and inflammatory cells to the central nervous system (CNS) [for review, see Ref. (1)]. Consequently, key functions in tissue homeostasis and immune defense were attributed to brain-resident cell types, such as microglia or astrocytes (2, 3). Microglia are regarded as the innate immune cell of the CNS. As part of their routine surveillance, microglia continuously monitor their surrounding with motile protrusions to sense and resolve any disturbance (4). Along with their well-established role as immediate responders to injury and infection (5, 6), there has been an increasing appreciation of the importance of microglia for normal CNS development and function, including developmentally regulated neuronal apoptosis, neurogenesis, myelogenesis, and synaptic pruning (7–9). Given their central role in CNS inflammation, it is not surprising that dysregulation of microglial activation and microglia-induced inflammation is observed in virtually all brain malignancies, including neurodegenerative disorders as well as primary and metastatic brain cancers. Blood-borne immune and inflammatory cells have recently emerged as an important component of the disease-associated microenvironment in the brain and are regarded as critical mediators of progression in neurodegenerative disease and brain cancers. However, the lack of experimental systems that distinguish between recruited and brain-resident myeloid cells has previously halted analysis of cell-type-specific functions in CNS inflammation. The development of new methodologies provides unprecedented opportunities for comprehensive in-depth analyses of the immune landscape of the CNS under steady-state and pathological conditions. Single-cell RNAseq or mass cytometry (CyTOF) allow for an unbiased view on the immune milieu of the brain parenchyma and adjacent boundaries. In addition to the well-characterized macrophage populations of non-parenchymal areas of the brain (10), it is increasingly recognized that various immune cell populations including a large diversity of lymphoid and myeloid subpopulations are present in particular in the meninges and the choroid plexus (11–14). Analysis of parenchymal myeloid cells also revealed high cellular heterogeneity. The existence of distinct myeloid cell phenotypes may reflect functional diversity, different ontological origins, or various cell differentiation states already at steady state (11). The question how environmental cues in different brain malignancies sculpt transcriptional profiles and epigenetic states of microglia and recruited myeloid cell populations during disease progression has recently gained attention. A growing number of studies seek to unravel the heterogeneity of the disease-associated immune landscape to functionally link different cell states to disease progression. Detailed knowledge of the impact of individual cell populations or activation states across different CNS malignancies is critical for the development of improved therapeutic strategies to target dysfunctional cells without affecting essential physiological or beneficial functions. The aim of this review is to discuss recent insights into the cellular and molecular identity of the heterogeneous population of cerebral myeloid cells in different CNS disorders to highlight common and unique features of the distinct subpopulations in the respective CNS pathologies.

ONTOLOGICAL ORIGIN OF MYELOID CELLS IN THE CNS IN HEALTH AND DISEASE

Microglia, the brain-resident macrophages, represent the largest population of myeloid cells in the CNS and are localized in the brain parenchyma. The term microglia was first coined by Pio

del Rio-Hortega to describe the non-neural, non-astrocytic "third element of the nervous system" that is distinct from neuroectodermal oligodendroglia and oligodendrocytes. Del Rio-Hortega's findings indicated a mesodermal origin of microglia [for historical review, see Ref. (15)]. However, there was a long-lasting debate on the ontological origin of microglia. An alternative hypothesis proposed that microglia originate from neuro-ectodermal-derived glioblasts (16). This theory was seemingly supported by the findings that donor bone marrow cells failed to contribute to the adult microglia population in either newborn (17) or adult rodents (18). Hickey and Kimura demonstrated that in bone marrow chimera only perivascular microglia derived from the bone marrow (19). The authors used the term perivascular microglia for the cell population that to date is referred to as perivascular macrophages that are located in the Robin-Virchow space. Further evidence that resident microglia are not replaced by cells from the bone marrow was provided by Lassmann et al. (20). The definitive proof for a mesodermal origin of microglia was achieved through a genetic study that showed that mice lacking the crucial transcription factor for myeloid cells, PU.1, are devoid of microglia (21, 22).

Even after the myeloid origin of microglia was proven, debate about the nature of microglia progenitors remained. Controversy was mainly caused by the fact that there are two major sites of hematopoiesis during embryogenesis: the yolk sac and the fetal liver. As depicted in **Figure 1**, primitive hematopoiesis in mice is initiated in the yolk sac at around E7.0, which leads predominantly to the generation of macrophages and erythrocytes (23). Yolk sac-derived primitive macrophages enter the embryo proper after the circulatory system has been established (from E8.5 to E10) (24) and populate various organs that contain tissue-resident macrophage populations, including the brain. Population of the fetal brain by primitive macrophages takes place before the onset of monocyte production by the fetal liver and before the establishment of the BBB. A second wave of "definitive" hematopoiesis is initiated by hematopoietic progenitors that are generated in the yolk sac and the AGM (aorta, gonads, and mesonephros) region of the embryo proper and that migrate into the fetal liver around E10.5. After E11.5, the fetal liver serves as the major hematopoietic organ and generates all hematopoietic linages including monocytes (25). In contrast to primitive hematopoiesis, definitive hematopoiesis depends on the transcription factor Myb (26). Around birth, hematopoiesis starts to be restricted to the bone marrow (27). It further remained elusive if under physiological conditions, monocytes contribute to the establishment of the post-natal and adult microglia population. Fate mapping studies using Runx1MerCreMer lineage tracing model, in which exclusively yolk sac-derived progenitors and their progeny are fluorescently labeled following a tamoxifen pulse at E7.25, have now established that microglia are derived from yolk-sac progenitors that generate a long-living population with self-renewal capacity (28). It was further demonstrated that microglia develop from erythro-myeloid progenitors (EMP) in a stepwise PU.1 and IRF8-dependent manner (29, 30) (**Figure 1**). The development of microglia and primitive yolk sac macrophages is completely dependent on colony-stimulating factor 1 receptor (Csf1r) signaling (28). Microglia are absent in

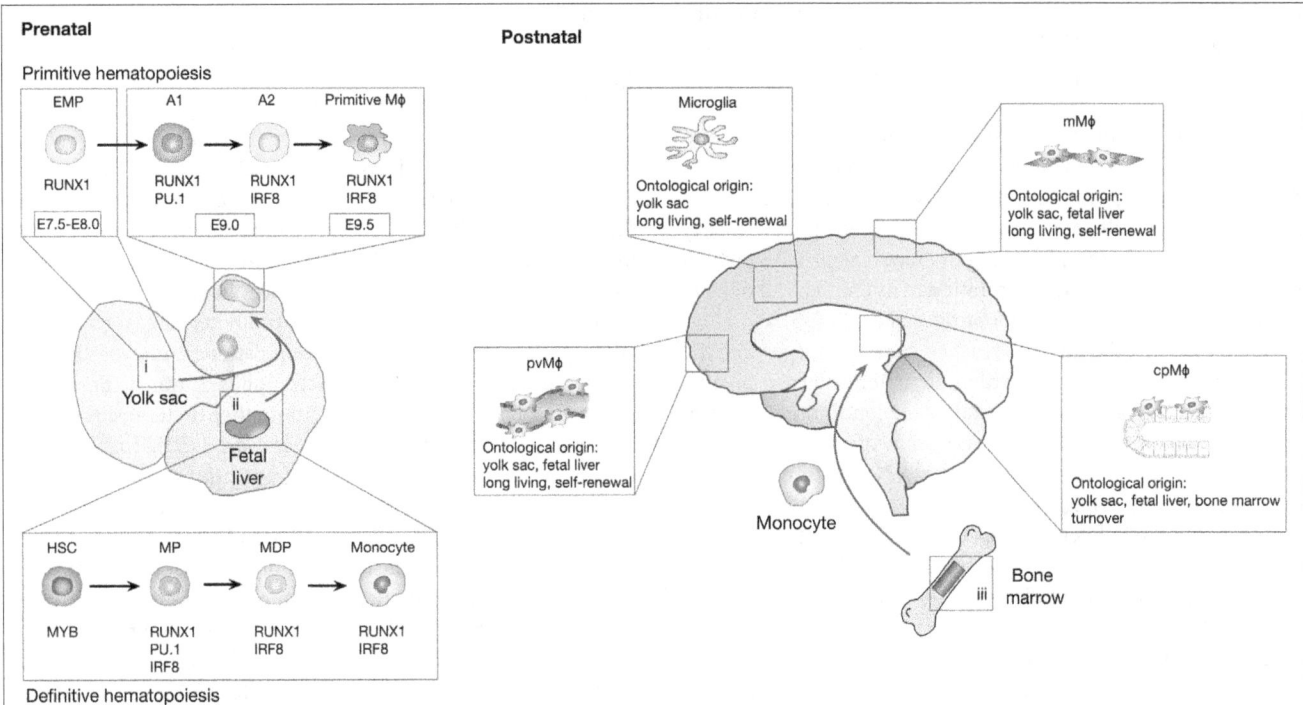

FIGURE 1 | Ontological origin of macrophage subpopulations in the central nervous system (CNS). A first wave of myeloid cell development takes place in the yolk sac (i) between E7.0 and E8.0 in a process known as primitive hematopoiesis that leads to the generation of erythro-myeloid progenitor (EMP) cells. EMP cells give rise to A1 (cKit⁺Cx3cr1⁻) cells followed by A2 (Cx3cr1⁺) cells that differentiate into microglia, perivascular macrophages (pvMφ), meningeal macrophages (mMφ), and choroid plexus macrophages (cpMφ). Microglia originate exclusively from yolk sac-derived progenitors, while non-parenchymal CNS macrophages are replenished with fetal liver-derived progenitor cells (ii) as part of definitive hematopoiesis. Perinatally, hematopoiesis starts to be restricted to the bone marrow (iii). Among the CNS macrophages, cpMφ are the only population with substantial constitution from bone marrow progenitors. Microglia, pvMφ, and mMφ are considered to be long-living cells that regenerate through self-renewal.

Csf1r knock-out mice, while mice lacking functional Csf1 did not show the same severe phenotype (31, 32). This observation was later explained by the existence of a second ligand for Csf1r, namely IL34 (33) that is highly expressed in the brain (34). Microglia represent the only tissue-resident macrophages that are exclusively derived from yolk sac-derived progenitors. By contrast, tissue-resident macrophages in other organs such as Kupffer cells in the liver, alveolar macrophages in the lung, or Langerhans cells in the skin comprise mixed populations and are repopulated by cells originating from the fetal liver during definitive hematopoiesis (27, 35). In light of recent experimental insight, it became apparent that previous findings that indicated a contribution of blood-borne monocytes to the adult microglia pool were confounded by experimental caveats that conditioned the brain for engraftment of peripheral myeloid cells, such as irradiation or parabiosis bias (36). Mildner et al. demonstrated that the use of head shields during myoablative irradiation prior to bone marrow transplantation prevented the recruitment of bone marrow-derived cells into the brain (37). These findings were further supported by studies using parabiosis in mice without the need for irradiation (38). Although chimerism in the periphery reached about 50%, there was no evidence for recruitment of peripheral monocytes to the brain. Moreover, even in the context of inflammation, when monocytes contribute to the inflammatory milieu, blood-borne cells did not

integrate into the long-term resident microglia pool (39). The microglia compartment seemed to recover from an internal pool instead. These findings are in line with previous observations demonstrating that peripheral macrophages do not transform and replace microglial cells in EAE models (20). In contrast to these findings, it was shown that under experimental conditions in which the microglial niche is completely vacant in response to microglia depletion strategies, bone marrow-derived cells enter the brain and differentiate into microglia (40, 41). Bruttger et al. recently took advantage of a Cx3cr1CreER-based system (Cx3cr1-iDTR mice) (42) that allows for conditional depletion of microglia without the necessity of generating bone marrow chimera (43). The authors demonstrated that the repopulating microglia arose exclusively from an internal CNS-resident pool. A contribution of bone marrow-derived cells was only observed in mice that were irradiated and additionally received a bone marrow transfer. Moreover, it was demonstrated that microglia self-renewal is dependent on IL1 signaling, while reconstitution from bone marrow precursor is IL1 independent. However, until recently the actual turnover rate of microglia in the brain remained elusive. Employing a multicolor fate-mapping model, the microfetti mouse [a microglia-restricted modification of the confetti mouse (44)], Tay et al. recently analyzed the rate of self-renewal of microglia in steady state, after induced CNS pathology and during the subsequent recovery phase. This study revealed heterogeneous rates of microglia replenishment in different brain regions (45). Following CNS damage, the authors found a shift from a pattern of random self-renewal within the microglial network toward a rapid expansion of selected microglia clones. This finding provides important insight into the question if microglia are recruited from adjacent regions to sites of CNS damage, or if clonal expansion results in microglial accumulation. Results obtained in the Microfetti mouse clearly favor the latter hypothesis. During the recovery phase in which microgliosis is resolved, the restoration of microglial cell density occurred through egress and apoptotic cell death (45).

Taken together, the field has reached consensus regarding the origin of microglia and the contribution of bone marrow precursors to the microglia pool under steady-state conditions. However, the debate on the functional contribution of yolk sac-derived microglia and blood-borne monocytes in CNS inflammation and their functional interplay is still in its infancy. As discussed in more detail in the following paragraphs, there is evidence that in response to inflammatory conditions associated with, e.g., irradiation, neurodegenerative disorders, or CNS cancer, the recruitment of monocytes or other bone marrow-derived progenitors can supplement the microglial pool. However, it remains unclear if the recruited cells persist and become an integral part of the microglial population, or if those cells represent a transient population that vanishes once the inflammatory stimulus is resolved. Another question that still needs to be addressed in more detail is, if yolk sac-derived microglia and bone marrow-derived macrophages (BMDM) exert redundant or cell type-specific functions in CNS pathologies and if the ontological origin determines responses against therapeutic intervention.

SHAPING OF CELLULAR IDENTITY BY THE TISSUE ENVIRONMENT

To understand the imprinting of disease-associated states on microglia and monocyte-derived macrophage identity in more detail, it is important to first consider the effects of specialized tissue environments on tissue-resident macrophages. It is increasingly recognized that in addition to the ontological origin, environmental factors play a critical role in defining functionality of tissue-resident macrophages and determine the fate and persistence of cells in tissues. Consistent with their diverse locations and functions, tissue-resident macrophages in different organs display distinct gene expression profiles (46, 47). Several studies have already dissected the genetic and epigenetic imprinting of specific tissue-resident macrophages and identified a range of transcription factors that are essential for cell type restricted gene expression profiles, e.g., SpiC for red pulp macrophages (48, 49) and GATA6 for peritoneal macrophages (46, 50). Two recent studies undertook the effort to systematically characterize the genetic and epigenetic imprinting of tissue-resident macrophages in specific organ environments. Both studies used RNA sequencing in combination with chromatin immune-precipitation (Chip)-Seq (51) and assay for transposase-accessible chromatin (ATAC)-Seq (52) to identify enhancer regions that are coupled to gene expression and accessible chromatin (53, 54). The studies by Lavin and Gosselin indicate that tissue-resident

macrophages share epigenetic structures and gene expression with other myeloid cell populations. Similarities within the lineage are largely determined by collaborating transcription factors (CTFs) such as PU.1 and lineage-determining transcription factors, including interferon regulatory factor family members and CCAAT/Enhancer-Binding-Protein (Cebp)-a (55–57). However, each tissue additionally has its unique gene expression profile that is controlled by changes in enhancer landscapes in response to environment-specific signals (**Figure 2**). Interestingly, both studies describe pronounced differences in enhancer landscapes among macrophage subtypes, while promoters are largely shared across different macrophage subpopulations and even between macrophages, monocytes, and neutrophils. It was demonstrated that microglia are most distinct from other tissue-resident macrophages in terms of their genetic landscape (53). This comparison also revealed that macrophage populations that are exposed to similar environmental cues converged to similar expression patterns. For example, Kupffer cells and splenic macrophages were shown to share a cluster of highly expressed genes that are enriched for gene ontology (GO) annotations, such as heme and porphyrin metabolism, indicating their role in erythrocyte turnover (48, 49). Similarly, small and large intestinal macrophages were shown to express genes enriched for GO annotations that reflect exposure to microbiota, such as response to bacteria and antigen processing. A more detailed comparison between microglia and peritoneal macrophages identified tissue-specific signals that determine the epigenetic and genetic imprinting of microglia and peritoneal macrophages. The genetic landscape of microglia is known to be strongly driven by the presence of TGFβ and IL34 (58, 59), while retinoic acid is a well-characterized environmental factor that dictates genetic imprinting of peritoneal macrophages and is essential for their development and function via GATA6 activation (60). The extent of tissue-specific cues on enhancer landscapes was further proven by transplantation experiments in which peritoneal macrophages were transferred to the lungs. Interestingly, the transferred tissue-resident macrophages lost most of their original tissue marks and acquired a tissue program based on their new host tissue (54).

In summary, identification of enhancer landscapes that are imprinted by specific tissue environments together with the notion that environmental cues can override ontological imprinting ultimately leads to the question, how blood-borne monocytes and macrophages are affected by the host tissue upon recruitment to sites of injury, inflammation, neurodegeneration, and neoplastic transformation and also, to which extent, the disease status dominates the imprinting of resident and recruited cell populations. The next paragraph will discuss recent findings from the field of neurodegenerative disorders, with a focus on Alzheimer's disease (AD), and brain cancers that provide critical insight into the heterogeneity of disease-associated myeloid cells.

MOLECULAR IDENTITIES OF MICROGLIA AND MACROPHAGES IN BRAIN MALIGNANCY

The local tissue environment has been shown to sculpt macrophage transcriptional profiles and epigenetic states under

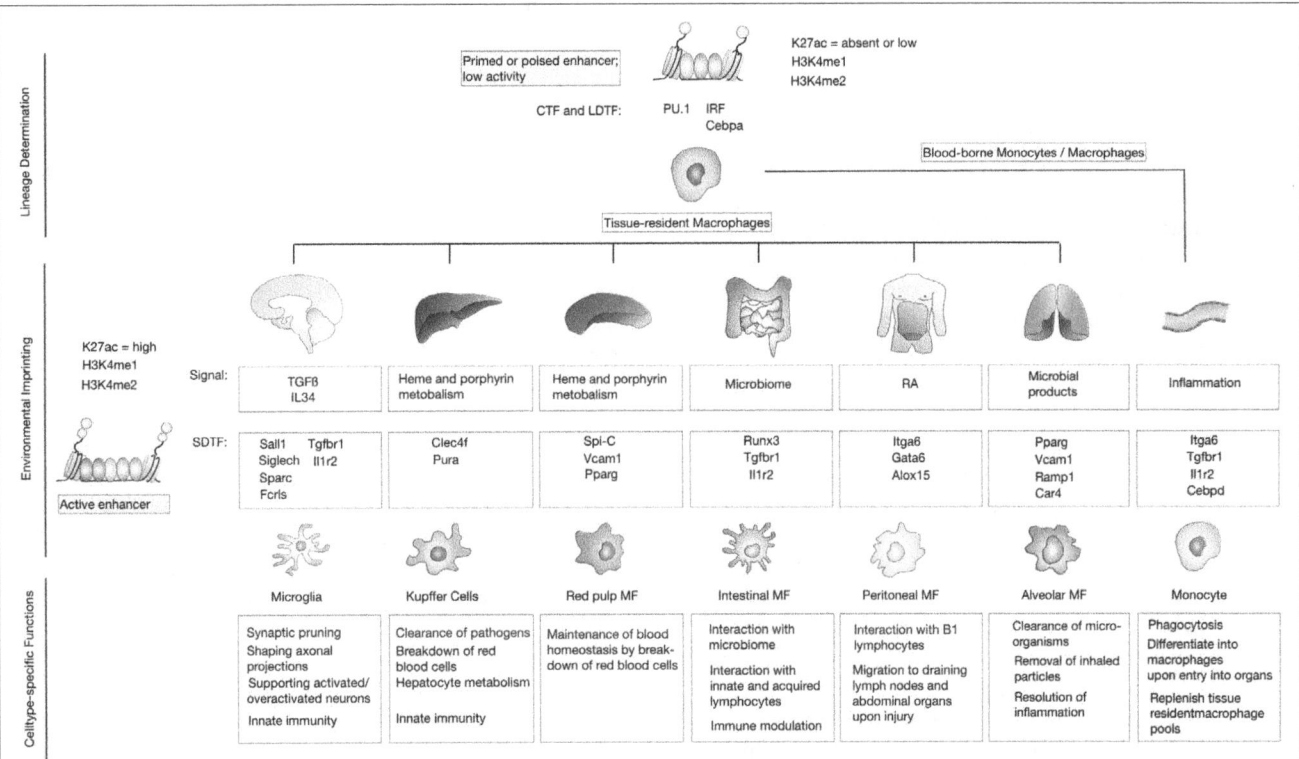

FIGURE 2 | Environmental imprinting of tissue-resident macrophages. Differentiation of precursor cells into specific lineages is determined by binding of lineage-determining transcription factors (LDTFs) and collaborating transcription factors (CTFs) to closely spaced recognition patterns on the DNA. Binding of LDTFs and CTFs selects enhancers as primed or poised. Primed enhancers are marked with characteristic histone modifications such as histone lysine 4 monomethylation (H3K4me1) or dimethylation (H3K4me2). Poised enhancers are defined by the presence of histone H3 lysine 27 trimethylation (H3K27me3). Primed or enhancers show low activity due to the lack of enhancer RNA production or the presence of H3K27me3, that has to be removed to induce an active enhancer state (upper panel). Tissue-resident macrophage populations are exposed to unique environmental cues that lead to genetic and epigenetic imprinting based on signal-dependent transcription factors (SDTF) that bind and activate primed or poised enhancers. Active enhancers are marked by H3K4m1 or H3K4me2 and histone H3 lysine 27 acetylation (H3K27ac) (middle panel). Environmental imprinting induces cell-type-specific functions of different tissue-resident macrophage populations (lower panel).

steady state conditions. However, it remained unclear whether an inflammatory tissue environment may affect differently macrophage populations of distinct ontogenies. To answer this question, it is first essential to determine the extent of peripheral recruitment of myeloid cells to the CNS under distinct pathological conditions. A critical contribution of recruited myeloid cell populations has been proposed for a long time (61). However, given the number of experimental caveats including the requirement of radiation for bone marrow transplantation (36, 37) as well as the route for tumor cell implantation, which often relies on intracranial injection, it remained unclear if the observed infiltration is due to experimental manipulation or represents an integral part of disease progression. Another important aspect that has to be taken into account is species differences that impact the extent of infiltration of cells from the periphery. It was shown that rat models for ischemia or brain cancer show lower rates of infiltration of blood-borne cells than observed in mouse models (62, 63). There is also evidence that recruitment of blood-borne inflammatory cells in human brain cancers is less pronounced than in mouse models (64). New approaches that employ lineage tracing models or single-cell RNAseq provide an unbiased view

on the extent of infiltration from the periphery and the disease-associated imprinting on different myeloid subpopulations. Moreover, a growing number of studies addresses these questions on patient-derived samples thereby excluding the possible effects of experimental artifacts and provide important insight into the clinical relevance of experimental data (65–68).

Neurodegenerative Disorders

Neurodegenerative disorders share common features including neuronal loss that ultimately leads to cognitive decline and motor dysfunction, which is associated with the establishment of an inflammatory environment. The immune milieu is comprised predominately of brain-resident microglia, which is supplemented by infiltrating immune cells. While AD, Parkinson's disease (PD), Amyotrophic Lateral Sclerosis (ALS), and multiple sclerosis (MS) are considered as neurodegenerative diseases, it is important to appreciate differences in the extent of peripheral involvement, a critical parameter to define neuroinflammation (69). ALS and MS are autoimmune inflammatory disorders of the CNS that lead to irreversible axonal damage and progressive neurological disability (70, 71). In case of ALS and MS, immune cell infiltration is causative. For example, in MS, acute demyelinating white matter lesions show myelin breakdown accompanied by infiltration of innate immune cells (i.e., monocytes) and adaptive immune cells (T- and B-lymphocytes) (72, 73). By contrast, symptoms of AD and PD should rather be regarded as innate immune reactions (74). Infiltration of lymphocytes only occurs at late disease stages when integrity of the BBB is lost. The critical contribution of CNS inflammation to disease progression in neurodegenerative disorders has been appreciated since the first description of the pathological parameters in particular due to the manifestation of microgliosis (15, 75). However, to date, it still remains controversial if CNS inflammation, in particular under conditions that trigger pronounced recruitment of myeloid cells from the periphery, is associated rather with disease amelioration or acceleration.

Alzheimer's disease represents the most common form of dementia and AD pathology is characterized by extracellular deposition of amyloid-β peptides that leads to β-amyloid plaques, formation of neurofibrillary tangles composed of hyperphosphorylated tau protein, neuroinflammation, and neuronal loss (76). An accumulating number of studies seek to evaluate the functional role of brain-resident microglia and to dissect the contribution of recruited myeloid and lymphoid cells. While some studies propose a protective role of microglia in AD (77–79), other reports show that under disease conditions, microglia acquire pro-inflammatory properties that have been associated with disease acceleration (80–83). These conflicting data can at least in part be attributed to a high phenotypic and functional heterogeneity of the disease-associated myeloid cell population. For example, Mildner et al. reported distinct and non-redundant roles of microglia and other brain-associated myeloid cells in AD mouse models. This study revealed a dominant role of CCR2-expressing myeloid cells in β-amyloid clearance (84). Previous studies that aimed at dissecting the functional contribution of myeloid subpopulations in disease progression relied on analyses of bulk cell populations categorized by a limited number of markers and single time points for analyses often at end stage disease. Importantly, analysis of bulk populations limits the capacity in resolving the heterogeneity and complexity of the immune milieu within the CNS (85–87). To date, single-cell transcriptomics allows for an unbiased characterization of immune cell types and states, thus systematically resolving the complex heterogeneity of the disease-associated immune landscape in comparison to normal tissue or in response to therapy (30, 88–91). Two recent studies used single-cell RNA seq analysis to characterize the immune landscape at different stages of disease progression in AD using the 5XFAD model (92) or the CKp25 model (93). Using the CKp25 AD-like mouse model, Mathys et al. identified early- and late-response states that differ significantly from homeostatic microglia. The early-response microglia show enrichment of cell cycle genes and genes involved in DNA replication and repair indicating that microglia expansion occurs at early disease stages. At late stages of disease progression, immune-related pathways were dominant and an enrichment of interferon-related response genes was detected. Interestingly, the late stage clusters comprised heterogeneous populations. Based on gene expression signatures of

different clusters identified at late stages, the authors conclude that those subpopulations could reflect exposure to type 1 or type 2 interferon, respectively (93). Using the 5XFAD model, Keren-Shaul et al. also identified two distinct microglia states [cluster II and III, referred to as disease-associated microglia (DAM)] in AD that were absent in normal brain. Compared to normal microglia, DAMs showed reduced expression of microglia core genes including the purinergic receptors P2ry12, P2ry13, Cx3cr1, and Tmem119 (58, 94). Concomitantly, DAMs showed enrichment of genes that are known as common risk factors for AD, including Apoe, Ctsd, Lpl, Tyrobp and Trem2. Gene set enrichment analysis indicated induction of lysosomal/phagocytic pathways, endocytosis and regulation of immune response. Interestingly, temporal resolution of the DAM phenotype manifestation indicated a two-step process. The first step appears to be accompanied by suppression of key regulators of microglial phenotype and function, such as Cx3cr1. The second stage was shown to be dependent on Trem2 and Tyrobp/Dap12. Analysis of the spatial localization of DAMs revealed close association to amyloid plaques. Given the enrichment of phagocytic and lipid metabolism pathways, the authors propose that DAM are involved in plaque clearance. The presence of DAM-like cells in AD patients has been demonstrated by histology as well as transcriptomic analysis (66, 92).

In addition to the investigation on DAMs in AD, Keren-Shaul et al. interrogated if DAMs are also present in other neurodegenerative pathologies, including a mouse model for ALS and aging. Interestingly, distinct myeloid subpopulations that showed similarities to DAMs in AD were observed also in response to aging and in ALS (92). Given the finding that progressive neurodegeneration leads to the induction of similar gene signatures in DAMs it is very interesting to compare the results by Keren-Shaul and Mathys. Both studies describe the occurrence of two distinct microglia subpopulations during disease progression that are distinct from the microglia state in the healthy brain. However, early and late disease-stage associated populations are not completely unrelated. Earlier stages rather represent a transient intermediate activation state as part of the reprogramming of homeostatic microglia in response to neurodegeneration. Interestingly, late-response microglia express increased levels of many genes that were also observed to be upregulated in DAM, suggesting a substantial similarity between the expression profiles of DAM and late-response microglia (92, 93). This observation is consistent with the idea that the DAM program may be a primed set of genes that is expressed in response to varied conditions of altered homeostasis. This is further supported by gene expression similarities between AD, ALS and aging. However, the identified populations also show important differences. For instance, Mathys et al. observed that many antiviral and interferon response genes were significantly upregulated in late-response microglia but not in DAM. Moreover, significant differences in the expression of both stage 1 and stage 2 DAM enriched genes were observed in late-response microglia. This was less pronounced in early-response microglia. Differences in gene signatures can be in part be explained by model-specific characteristics. However, this observation might also indicate that the distinct microglia states represent intermediate stages on a continuum of microglia reprogramming that ultimately converts protective/beneficial functions into neurotoxic functions (92, 93). A comparison of the observed signatures and phenotypes suggests that the early- and late-stage microglia represent the most naïve and most advanced population, respectively, while the DAM stage 1 and stage 2 might occur along the transition from early to late stage microglia. Mrdjen et al. took a similar approach as recently employed by Korin et al. for steady state conditions (11) to investigate the immune landscape in CNS inflammation using mass cytometry (95). The combination of CyTOF together with lineage tracing allowed the authors to identify different subsets of myeloid cells and the phenotypic changes in CNS immune cells during aging, AD and MS with definitive proof of the ontological origin. Microglia phenotypes observed by Mrdjen in an EAE model reflected an inflammatory phenotype that showed similarities to the phenotypes observed in aging and AD mouse models. This indicates a potential universal disease-associated microglial signature as recently proposed (96).

Although the studies by Keren-Shaul and Mathys both provide important insight into the molecular basis of DAMs or early- and late stage microglia, the beneficial or detrimental role of the respective subpopulations remains to be studied in more detail. Keren-Shaul et al. propose a protective function of DAMs due to their contribution in plaque clearance (92). Tyrobp and Trem2 are known to form a signaling complex associated with

phagocytosis (97). Trem2 expression is known to be critical for clearance of neuronal debris and loss of function of Trem2 or Tyrbp (Dap12) are associated with dementias characteristic of neocortical degeneration observed in AD. In line with this interpretation, it has previously been reported that Trem2 is critical for microglia clustering and expansion around amyloid plaques and that the Trem2-mediated early microglial response limits diffusion and toxicity of amyloid plaques (79, 82). By contrast, Jay et al. demonstrated that Trem2 deficiency resulted in reduced infiltration of inflammatory myeloid cells and thereby ameliorated AD pathology at early stages (77) and exacerbated it at later stage (98). Hence, one possible explanation might be that DAM function has a transient beneficial impact during the initial phase of AD onset while later stages might be associated with rather detrimental effects. In light of recent insight on microglial stages in AD, Hansen et al. proposed a dichotomous role of microglia, with the detrimental microglia population occurring later in disease course at the time when synapse loss is observed and symptoms manifest (99). As depicted in **Figure 3**, microglia in steady state are protective and AD is prevented by constant scavenging of aβ peptides. Once the equilibrium is lost and Aβ levels accumulate, microglia phagocytose and clear Aβ aggregates. These protective activities involve activation of microglia to a DAM state in a Trem2-dependent manner. Genetic susceptibility or aging can lead to impaired microglia function. Accumulation of toxic amyloid plaques leads to tau pathology in stressed or damaged

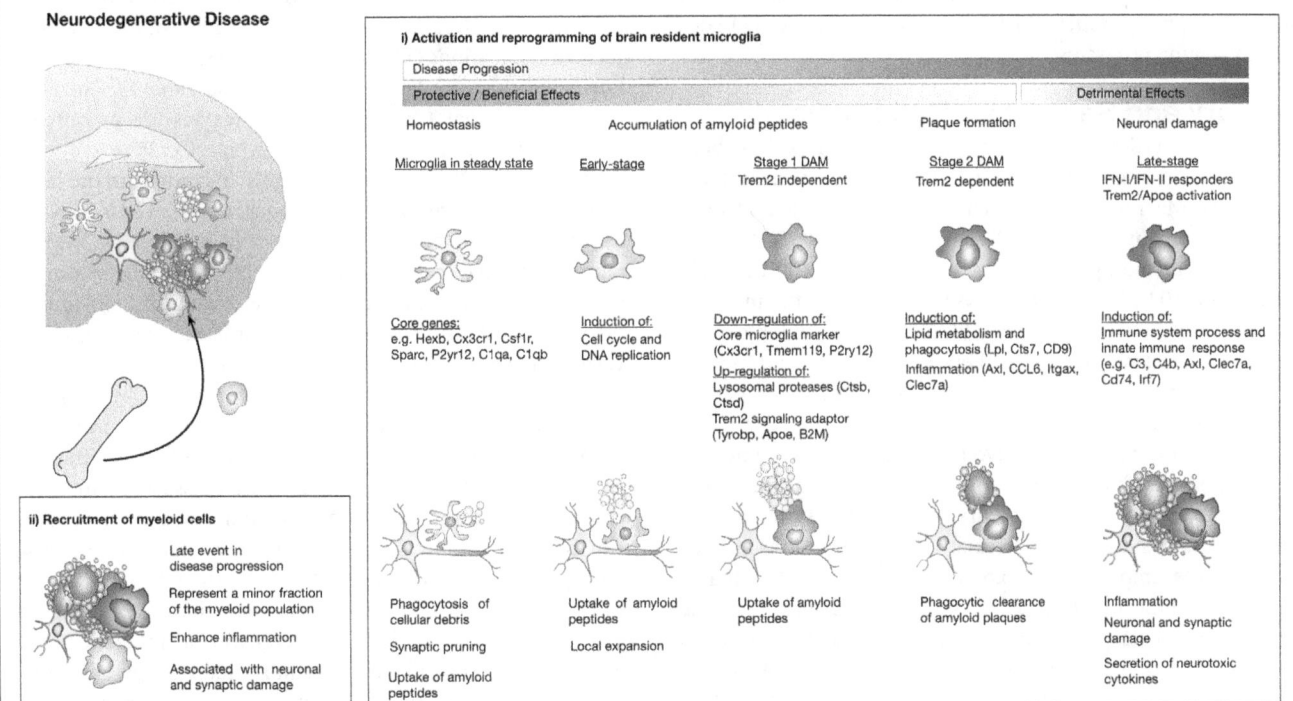

FIGURE 3 | Microglia and macrophage activation states and effector functions at different stages of disease progression in neurodegenerative disease. (i) Microglia exert protective functions including phagocytosis of cellular debris, uptake of Aβ peptides, and clearance of amyloid plaques at early disease stages. Disease progression leads to changes in microglia functions that limit their ability to confine disease manifestation or even induces inflammatory activation states that cause neuronal and synaptic damage. Induction of Trem2/Apoe signaling was shown to mediate conversion of protective microglia into tissue damaging ones. Recruitment of macrophages from the periphery appears to occur at late disease stages and contributes to disease acceleration due to enhanced inflammation.

neurons, which induces an unconstructive and inflammatory state in microglia that causes deleterious neuronal damage (99) (**Figure 3**). This hypothesis is further supported by a recent study using the APP-PS1 model for AD, the ALS model SOD1G93A and an EAE model to investigate the role of the Trem2–Apoe complex in microglial dysfunction in neurodegeneration (96). In contrast to previous studies, Krasemann et al. proposed a neurodegenerative role of microglia in which the Trem2–Apoe pathway regulates the phenotypic switch from neuroprotective to neurodegenerative microglia. A negative feedback loop of Apoe to TGFβ suppressed homeostatic microglia concomitant with an induction of Bhlhe40 response genes including Clec7a, Lgals3, Gpnmb, Itgax, Spp1, Cl2, and Fabp5. Those recent data are in line with previous findings on the presence of functionally distinct or opposing microglia/monocytes populations in EAE models (83, 100). It was demonstrated that monocyte-derived macrophages initiate demyelination, while microglia rather clear debris. Gene expression analysis confirmed that macrophages are highly phagocytic and inflammatory, while microglia showed globally suppressed metabolism at early disease stages (83). Gao et al. proposed a dichotomy of the function for Tnfr2 in myeloid cells. It was shown that microglia-derived Tnfr2 signaling is associated with protective effects, while monocytic/macrophagic Tnfr2 stimulated immune activation and EAE initiation (100). Moreover, as discussed in the following paragraph on brain cancers, reprogramming of pro-inflammatory macrophages with anti-tumor functions into immune-suppressive, tumor promoting macrophages has been described for a comprehensive number of cancer entities which is often regarded as de-regulated wound healing programs.

Primary and Secondary Brain Cancer

Massive infiltration of macrophages into tumors has been reported for a large proportion of primary tumor entities and metastasis. Macrophages represent the most abundant stromal cell type in many cancers and comprise up to 30–50% of the tumor mass including CNS cancers such as GBM (101) and brain metastases (102). Functional analyses indicate that the presence of tumor-associated macrophages (TAMs) fosters tumor growth, regulates metastasis and affects therapeutic response (103–105). Accumulation of TAMs is often associated with poor patient prognosis (106, 107).

To date only few studies have looked at the origin and fate of macrophages during cancer progression at the primary site and in metastasis. One of the first reports that systematically dissected the cellular and molecular origin of tumor-associated macrophages employed parabiosis experiments in the PyMT mouse model for breast cancer (108). The authors demonstrated that tumor development triggers a unique innate immune response that is characterized by the differentiation of inflammatory monocytes into tumor-associated macrophages (TAMs). Terminal differentiation of monocytes into TAMs occurred in a notch-dependent manner *via* the recombination signal binding protein for immunoglobulin regulator (Rbpj) (109). Monocyte-derived TAMs showed pronounced transcriptional differences compared to resident mammary-tissue-macrophages (MTM). TAM expansion during breast cancer progression led to a loss of

MTMs (108). By contrast, it was recently demonstrated in PDAC models for pancreatic cancer that tissue-resident macrophages persist and undergo significant expansion. TAMs in PDAC tissues adopted a transcriptional program that is associated with cell proliferation (110). This effect was shown to be enhanced by tissue-resident macrophages derived from the yolk sac or fetal liver, but not by HSC-derived monocytes/macrophages. Consequently, macrophages of different ontological origins had different impact on tumor progression in the PDAC model. Loss of monocyte-derived macrophages only showed marginal effects on tumor progression, while depletion of tissue-resident macrophages significantly reduced tumor progression (110). The ontological origin of TAMs in primary brain cancer was investigated in several recent studies using different glioblastoma (GBM) mouse models (101, 111–113). The study by Muller et al. used bone marrow transplantation strategies with head protected irradiation (HPI) in the GL261 model in direct comparison to total body irradiation (TBI) (113). Avoiding previously reported irradiation bias in the head region, the authors demonstrate that recruited macrophages contribute only at later stages to the tumor mass and constitute around 25% of the myeloid population. Interestingly, tumor progression in TBI-treated mice was accelerated compared to the HPI cohort, suggesting that recruited macrophages contribute in promoting tumor growth, yet their infiltration might be predominately caused by impact of IR (113). In order to fully circumvent the necessity of IR, two recent studies employed a genetically engineered model for proneural GBM and the GL261 model in combination with different lineage tracing models to discriminate ontologically distinct subpopulations (111, 112). Using the $Cx3cr1^{GFP/wt}$:$Ccr2^{RFP/wt}$ double knock-in model (114, 115) (see also **Table 1**), Chen et al. demonstrated that $Cx3cr1^{lo}Ccr2^{hi}$ monocytes are recruited to GBM, where they differentiate into $Cx3cr1^{hi}Ccr2^{lo}$ macrophages and $Cx3cr1^{hi}Ccr2^{neg}$ microglia-like cells. In contrast to the results by Muller et al., recruitment of bone marrow-derived monocytes/macrophages was reported to occur at early stages of GBM initiation. Recruited macrophages were predominantly localized to perivascular areas, while microglia were found in peri-tumoral regions. Quantification of the extent of infiltration suggested that recruited macrophages constitute up to 85% of the TAM population, with the remaining 15% being represented by microglia. A possible explanation for the discrepancy of the observed influx in both studies might be differences in tissue harvest strategies. Chen et al. focused their analysis on macro-dissected tumor areas while Muller et al. processed the entire tumor-bearing hemisphere with considerable involvement of adjacent non-tumor-bearing brain parenchyma. Moreover, it is important to note that the study by Muller et al. provided evidence that GBM-associated microglia upregulate CD45 and represent an inherent part of the CD45^{hi} population in the tumor context. Upregulation of CD45 expression on microglia was previously demonstrated in response to different inflammatory stimuli (116) and underlines the fact that CD45 levels as proposed by Ford et al. (117) can only be used to discriminate macrophages and microglia under steady-state conditions while it is not suitable under inflammatory conditions. It was also demonstrated that activated microglia including GBM-associated microglia downregulate Cx3cr1 (112).

TABLE 1 | Lineage tracing models and marker to distinguish microglia and monocyte-derived macrophages in the brain.

Approach	Cell type specificity	Principle	Advantages	Limitations	Reference
(a) Transplantation models					
BMT; TBI	BMDM	HSC source of blood monocytes is replaced with modified/labeled HSCs	High chimerism No time consuming crossing into genetic disease models	Variability in myeloablation and reconstitution Artificial engraftment of BM cells in the CNS	(113)
BMT; HPI	BMDM	HSC source of blood monocytes is replaced with modified/labeled HSCs	High chimerism No time consuming crossing into genetic disease models	Variability in myelo-ablation and reconstitution	(37, 113)
BMT; Busulfan	BMDM	HSC source of blood monocytes is replaced with modified/labeled HSCs	High chimerism No time consuming crossing into genetic disease models Chemical myeloablation No irradiation	Variability in myeloablation and reconstitution	(112)
Parabiosis	BMDM	HSC source of blood monocytes is replaced with modified/labeled HSCs	Constant influx of donor cells No myeloablation required No time consuming crossing into genetic disease models	Technically challenging Low chimerism	(39)
(b) Genetic lineage tracing models					
Ccr2$^{RFP/wt}$; Cx3cr1$^{GFP/wt}$	Monocytes (red) MG (green)	Differential labeling of Cx3cr1^{hi}:Ccr2^{neg} MG (green) and of Cx3cr1^{lo}:Ccr2^{hi} monocytes (red)	MG and monocytes contain reporter for labeling	Recruitment to the brain leads to increased Cx3cr1 levels in monocyte-derived macrophages Ccr2 expression is downregulated in monocyte-derived macrophage upon differentiation	(114, 115)
Flt3-Cre	Monocytes and HSC-derived monocyte precursors	Label/modification induced in Flt3$^+$ monocyte precursors	Useful for lineage tracing of myeloid precursors Useful complementary approach to MG restricted lineage tracing	Cre expression or transmittance restricted to male mice	(118)
Cx3cr1-Cre^{ER}	MG	Recombination is induced in all Cx3cr1$^+$ cells upon tamoxifen pulse. Long-living MG retain the label/modification. While monocytes vanish and are replenished from precursors that were generated after Cre recombination in response to the tamoxifen pulse	Long-term labeling/modification is restricted to MG	Spontaneous modification reported in one model Low recombination in mMF (40–50%)	(42)
Sall1-Cre^{ER}	MG	Label/modification induced in Sall1$^+$ MG	Sall1 expression is stable also in response to different stimuli	Targeting of non-hematopoietic cells in the liver, heart and kidney	(119)
(c) Cell-type restricted marker expression					
CD45	MG^{lo} BMDM^{hi}	MG display lower surface expression	No requirement for combination of several markers	Activated in MG upregulate CD45 BMDM in brain malignancies downregulate CD45	(112, 113, 120)
Tmem119 P2ry12 Siglech	MG^{hi} BMDM^{lo}	MG show high expression	Applicable for mouse and human	Downregulation in MG during activation BMDM in GBM show increased expression	(58, 94, 112)
Sall1	MG^{hi} BMDM^{lo}	MG show high expression	Applicable for mouse and human Stable expression level at different activation levels	Low expression found on non-leukocytes in the liver, heart, and kidney	(119)
Itga4/Cd49d	MG^{lo} BMDM^{hi}	BMDM show high expression	Applicable for mouse and human Stable expression level at different activation levels	Expression found on T cells	(112)

BMT, bone marrow transplantation; TBI, total body irradiation; HPI, head protected irradiation; BMDM, bone marrow-derived macrophage; HSC, hematopoietic stem cell; MG, microglia; BM, bone marrow; CNS, central nervous system.

The gating strategy employed by Chen et al. relied on the discrimination of macrophages and microglia by CD45 expression levels and used a mouse model that is based on Cx3cr1 and Ccr2 promoter activity for reporter gene expression (111). Hence, it is possible that the CD45^{hi}Cx3cr1^{med} population represents a mixed population of microglia that upregulate CD45 concomitant with Cx3cr1 downregulation.

Bowman et al. used a comprehensive set of different lineage tracing models to unravel the extent of macrophage recruitment to GBM and brain metastasis (112). The lineage tracing models used in this study were based on specific labeling of microglia or BMDM using the tamoxifen inducible Cx3cr1CreER-IRIS-YFP;R26-LSL-TdTom model (42) or the Flt3Cre;R26mTmG model (118) (see also Table 1). Using these complementary lineage-tracing strategies, it was demonstrated that BMDMs contribute to the TAM pool in different GBM and brain metastasis models in the absence of IR. BMDMs constituted approximately 50% of the TAM population in GBM and 25% in brain metastasis. RNA sequencing of FACS sorted myeloid populations revealed distinct clustering of all TAM populations from normal microglia and monocytes. Within the TAM cluster, further cell and tumor type-specific clusters were identified (112). The TAM cluster showed enrichment of cell cycle related genes, upregulation of complement-related factors, extracellular matrix components, proteases, lipid metabolism mediators, and clotting factors. Interestingly, the authors found several microglia-enriched genes (e.g., Tmem119, Olfml3, Lag3, Jam2, and Sparc) to be upregulated in TAM-BMDM, while other microglia genes (e.g., Sall1, P2yr12, and Mef2c) were not induced in TAM-BMDM (112). These results further indicate that macrophages acquire tissue-resident gene expression upon infiltration into foreign tissue as previously proposed by Lavin et al. (54) and Gosselin et al. (53). Analysis of the epigenetic imprinting of TAM-MG and TAM-BMDM revealed enrichment of Fos/Jun and Pu.1 binding sites in both populations. In addition to those shared motifs, it was demonstrated that TAM-BMDM showed enriched enhancer usage for Runx and Creb/bZip motifs, while TAM-MG peaks were enriched for Smad3 and Mef2a. Interestingly, based on enriched genes it appears that TAM-MG rather exert pro-inflammatory functions as evident by an upregulation of cytokines such as Tnf and Ccl4 as well as classical complement components (e.g., C4b, C2, and Cfh), a pathway that was previously shown to be associated with synaptic pruning and host defense (121). By contrast, TAM-BMDM showed gene signatures that indicate functions in wound healing, antigen presentation and immune suppression (112) (**Figure 4**). Insight into the genetic and epigenetic landscape of TAM-MG and TAM-BMDM in GBM provides evidence for complex networks of tissue and disease imprinting that at least in part can be attributed to the ontological origin of the cells and can be linked to functional differences. In the future, it will be very interesting to analyze in more detail the heterogeneity of different subpopulations at different stages of disease progression as recently done for AD mouse models and to perform functional validation of the proposed mechanisms for each subpopulation. Moreover, it is still unknown if brain tumors of different origin such as oligodendroglioma or brain metastasis induce similar genetic and epigenetic imprinting like gliomas or if the signatures are fundamentally different.

In addition to the important insight into the identity of TAM populations in GBM, Bowman et al. took advantage of the RNAseq data obtained from lineage tracing models to identify markers that discriminate macrophages and microglia in a tissue- and disease independent manner. The authors identified and validated Itga4/CD49d that is specifically repressed in microglia, as a marker to distinguish BMDMs from microglia in primary and metastatic brain cancer in mouse models as well as human samples (112). A list of lineage tracing models and markers that allow for discrimination of microglia and recruited myeloid cells and highlights advantages and limitations is provided in **Table 1**. For review see also (122).

Within the last few years, a growing number of studies performed in-depth analyses on GBM patient samples. While the primary focus of these studies was rather on tumor cell centric questions, the results obtained from GBM sequencing allow conclusions on the tumor microenvironment. Wang et al. reported increased infiltration of macrophages during disease progression with the highest extent of macrophage/microglia accumulation in mesenchymal GBM (MES) versus non-MES (68). Moreover, NF1 loss that is frequently found in MES correlated with increased infiltration of macrophages/microglia. Venteicher et al. performed single-cell RNAseq experiments on GBMs with different IDH mutational status. In line with results from mouse models of GBM, it was reported that also in human GBM, the balance between microglia to macrophages shifts toward a higher representation of macrophage programs over microglia signatures (65). This effect was noted by downregulation of microglia core genes, such as CX3CR1, P2RY12, P2RY13, and SELPLG, concomitant with upregulation of macrophage-like signatures including increased expression of CD163, IFITM2, IFITM3, TAGLN2, F13A1, and TGFB1. Microglia and macrophages displayed signatures that reflected an inflammatory program consisting of cytokines (IL1, IL8, TNF), chemokines (CCL3, CCL4), NfkB-related genes as well as immediate early genes. In line with previous reports, expression analysis suggests that the GBM environment alters expression profiles of macrophages, thus reducing their transcriptional difference from microglia. Interestingly, the authors also identified a range of factors that correlated with increased macrophage infiltration. Of those 24 identified genes, three were components of the complement system (C1A, C1S, C4A) (65). Similarly, Darmanis et al. correlated gene expression within in the immune cell cluster (containing >95% macrophages/microglia and approximately 4.5% dendritic cells) with macrophage and microglia core genes to classify the cells into macrophages and microglia (67). Consistent with data obtained in GBM mouse models, it was shown that cells with macrophage-like signatures were found rather within tumor lesions, while microglia were localized at the tumor edge (67, 111). Interestingly, as previously proposed by Bowman et al., analysis of gene signatures revealed that more pro-inflammatory markers (e.g., CCL2, CCL4, TNF, IL6R, and IL1A/B) were expressed in the tumor periphery, whereas anti-inflammatory marker (e.g., IL1RN and TGFBI) were enriched in the tumor core (67, 112).

Taken together, while controversy remains on the extent of peripheral recruitment to GBM at different stages of tumor progression, there is accumulating evidence that the ontological

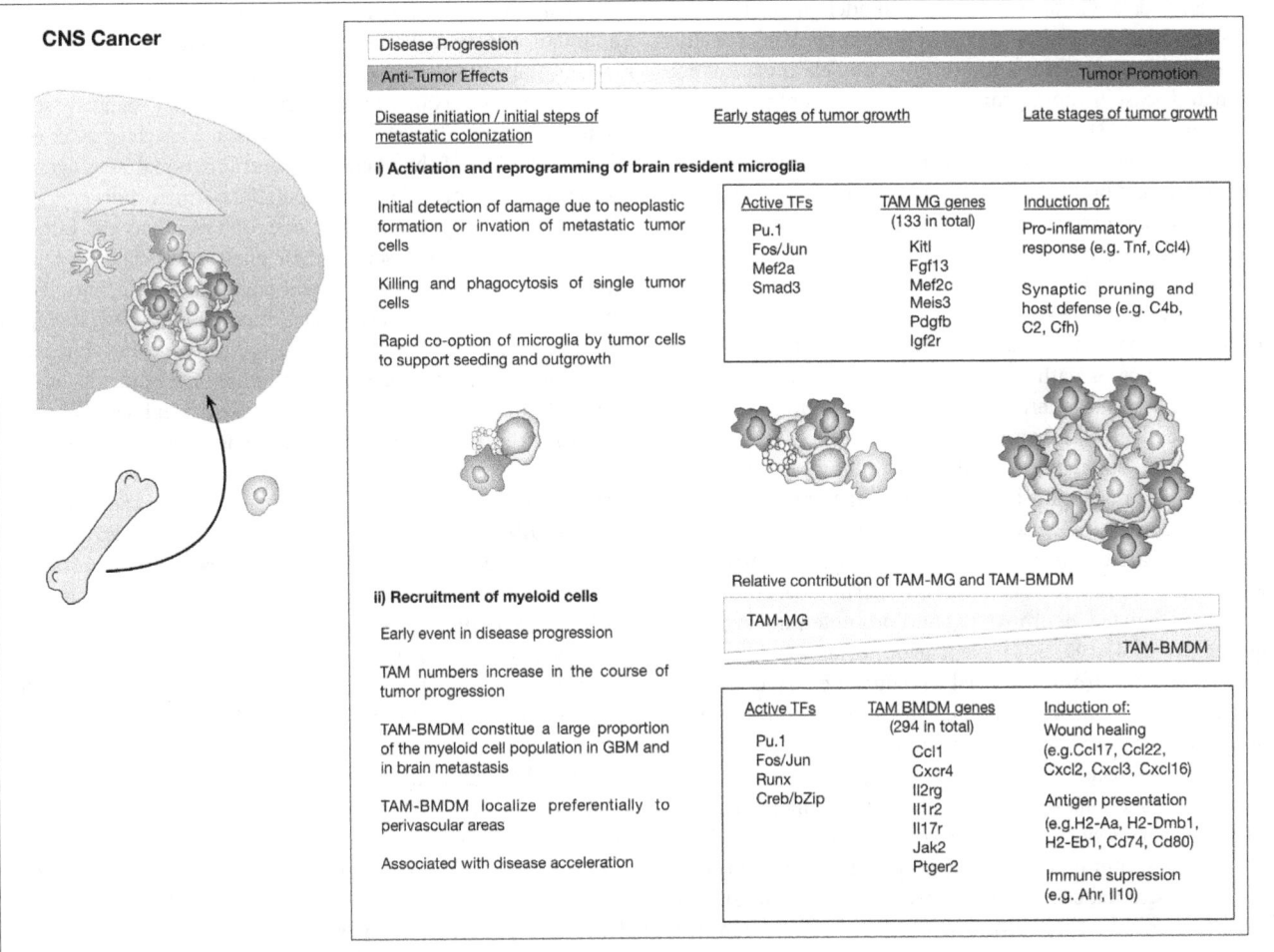

FIGURE 4 | Microglia and macrophage activation states and effector functions at different stages of disease progression in brain cancer. (i) Initial stages during neoplastic transformation in primary brain cancers or tumor cell seeding in brain metastasis are detected by microglia. As part of their role in host defense, microglia induce apoptosis in cancer cells. Tumor cells that escape microglia-mediated killing rapidly co-opt them and exploit their function to foster tumor growth. (ii) Brain tumor formation is associated with pronounced recruitment of macrophages from the periphery that starts at early stages and leads to disease acceleration. Transcriptomic analysis identified gene signatures that define tumor-associated microglia (TAM-MG) and macrophages (TAM-BMDM). The respective signatures indicate that TAM-MG rather induce pro-inflammatory responses and exert host defense functions while TAM-BMDMs were associated with wound healing, antigen presentation, and immune suppression.

origin of myeloid cells in brain cancers impacts the nature of genetic programs that are induced. Tumor-specific education of myeloid cells is expected to determine their effector functions during disease progression.

Lessons Learnt From Neurodegenerative Disorders and Brain Cancers

A series of recent studies utilized in depth transcriptomic analysis to molecularly define myeloid subpopulations at different stages of disease progression. However, a systematic analysis with the aim to interrelate the findings from different datasets has been missing. To close this gap of knowledge, Friedman et al. performed a comprehensive meta-analysis of purified mouse CNS myeloid cell profiles from different conditions across multiple studies including ischemic, infectious, inflammatory, neoplastic, demyelinating, and neurodegenerative conditions (123). Importantly, it was noted that in all comparisons microglia/brain-associated

myeloid cell enriched genes that distinguish them from myeloid cells/macrophages from the periphery were downregulated. It appears that in response to any perturbation, including normal aging, the genes that separate microglia from other macrophages and are, thus, likely involved in microglia function, show reduced expression. Friedman et al. identified several modules that are shared across pathological conditions, including the neurodegeneration-associated module with high similarities to DAM (92, 123). In addition to the DAM-like population, a unique microglia subset that expressed an interferon-related gene module was identified. This module showed increased representation with progressive β-amyloid pathology. This is in line with the distinct clusters observed in the CKp25 model that showed enrichment of interferon-responding genes and likely reflect subpopulations depending on exposure to type 1 or type 2 interferons (93, 123). Moreover, the authors compared data sets from mouse models with human samples to evaluate the translational capacity of the experimental data. Compared to mouse models, a greater range of cell-type variability was observed in the comparison of healthy and diseased human tissues, indicating that CNS inflammation in human disease is more pronounced than in commonly used mouse models (123).

Taken together, data obtained from studies on neurodegenerative disorders and brain cancers suggest that there is considerable overlap in certain tissue-specific gene signatures that are induced within the brain environment in response to disturbance of tissue homeostasis. This is apparent in the phenomenon that gene signatures in microglia and macrophages become more similar once both cell types are present in the brain (123). Moreover, a number of genes or pathways were found to be similarly dysregulated in neurodegenerative disorders and brain cancers and there is evidence of dichotomous roles of myeloid subpopulations/activation states under multiple pathological conditions. At early stages, myeloid cells are considered as protective cells that preserve tissue integrity by scavenging debris or eliminating intruders as part of their host defense mechanism. Advanced disease stages are rather associated with detrimental effects that cause tissue damage, neuronal loss and promote tumor growth (**Figures 3** and **4**) (99, 124). Neoplastic formation leads to a more rapid switch from beneficial to damaging effects and the extent of recruitment from the periphery appears to be more pronounced in brain cancers compared to neurodegeneration. It is increasingly recognized that tumor cells rapidly co-opt stromal cells and functionally reprogram their environment to generate a cancer permissive niche to foster tumor growth (124, 125). In this process, tumor cells are known to exploit housekeeping functions of stromal cells, such as host defense or wound healing mechanisms for their own benefit (126, 127). By contrast, recruitment of peripheral myeloid cells in neurodegenerative diseases such as AD appears to occur at later stages of disease progression and to a lesser extent. Protective programs of microglia have been shown to be preserved for an extended period during disease progression and to contribute in limiting disease propagation. However, at late disease stages, microglia function is no longer sufficient to prevent detrimental pathological events (99). Moreover, there is increasing evidence that dysregulation of certain pathways, e.g., the Trem2–Apoe pathway, represent a switch from protective to damaging effects (96).

As discussed in more detail in the next paragraph, detailed mechanistic insight is critical to develop therapeutic strategies that are targeted against aberrant functions of individual subpopulations at defined stages of disease progression as otherwise physiologically essential functions or protective programs might be blocked which consequently results in minimal therapeutic efficacy and adverse effects.

PERSPECTIVES FOR THERAPIES TARGETING MICROGLIA/MACROPHAGES IN NEURODEGENERATIVE DISEASE AND BRAIN CANCER

Macrophages/microglia-targeted therapies are emerging in the field of neurodegenerative disorders and cancer (128, 129). The

rationale for environment-targeted therapies is based on high abundance of stromal cells in different pathologies as well as their critical impact on disease outcome. Moreover, it was reasoned that the risk for acquired resistance is lower in genetically stable non-malignant cells, compared to genetically instable cancer cells. Recent analysis of the phenotypic and functional heterogeneity of macrophage/microglia subpopulations in neurodegeneration and cancer clearly indicate that macrophage/microglia-targeted therapies have to be based on their disease-associated specificities to achieve high therapeutic efficacy without inducing adverse effects. A variety of macrophage/microglia-targeted strategies have been tested in pre-clinical settings using genetic and pharmacological approaches. Several inhibitors have already entered phase 2 clinical trials (130). Most of those strategies aim at either blocking the recruitment of macrophages or depleting them (131). Given the importance of CSF1R downstream signaling for the differentiation and survival of macrophages and microglia, many studies tested the efficacy of blocking the ligands (CSF1 and IL34) or the receptor. However, most of these studies reported no or low efficacy when used as monotherapy, while combination of CSF1R inhibition with standard of care (e.g., chemotherapy or irradiation) led to synergistic anti-tumor effects in, e.g., glioma and breast cancer (132, 133). By contrast, Pyonteck et al. demonstrated that monotherapy with the CSF1R inhibitor BLZ945 resulted in improved survival and tumor regression in a model for proneural GBM. CSF1R inhibition in this model did not result in depletion of TAMs. Instead, TAMs showed reduced expression of several M2-like markers. The authors concluded that CSF1R-induced depolarization of TAMs might be more efficient than depletion of TAMs (101, 131). The same group recently demonstrated that long-term CSF1R inhibitor treatment led to acquired resistance driven by a compensatory IGF1–IGF1R signaling loop between macrophages and tumor cells, resulting in enhanced glioma cell survival and invasion (134). CSF1R inhibitors are currently in clinical trials to test their efficacy in GBM patients. The clinical trial using the CSF1R inhibitor PLX3397 in recurrent GBM (NCT01349036) was recently completed. PLX3397 was well tolerated but showed no efficacy in the recruited patient cohort (130). Additional studies that test CSF1R inhibitors in combination with standard of care or immune therapy are currently ongoing [e.g., BLZ945 with PRD001 anti-programed cell death-1 (PD1) in solid tumors including recurrent GBM (NCT02829723) and PLX3397 with temozolomide and radiotherapy in newly diagnosed GBM (NCT01790503)].

Colony stimulating factor 1 receptor inhibition was also tested in models for neurodegenerative disease to limit damaging neuroinflammation at disease end stage. Elmore et al. depleted virtually the entire microglia pool using the CSF1R/c-KIT inhibitor PLX3397 with no impairment of behavior and cognition. After withdrawal of the inhibitor, microglia rapidly repopulated the brain, returning to normal numbers within two weeks. Replenishment of microglia after CSF1R inhibition occurred from nestin$^+$ progenitor cells that induced expression of microglia-associated genes such as Iba1, Cx3cr1, Tmem119, Siglech, Pu.1, and Trem2 (135). Repopulating microglia were shown to be functional and responsive to inflammatory challenge similar to resident microglia (136). Hence, CSF1R-mediated microglia depletion might provide a powerful tool to resolve tissue destructive inflammation. Using the selective CSF1R inhibitor PLX5562, it was demonstrated that treatment with lower doses (leading to 30% depletion) strongly reduced microglia accumulation at amyloid plaques in the 3xTg-AD model. While plaque burden was not reduced, treatment led to improved cognition (137). Interestingly, CSF1 signaling can be regulated by Trem2, which suggests that effects of Trem2 on microglia could in part be mediated by the CSF1 signaling cascade (82). Likewise, Trem2 deficiency resulted in reduced accumulation of microglia around amyloid plaques. Krasemann et al. demonstrated that the Trem2–Apoe pathway induces a switch from a homeostatic to neurodegenerative phenotype in microglia. It was, therefore, proposed that modulation of the neurodegenerative phenotype through targeting of the Trem2–Apoe pathway might allow restoring homeostatic microglia and treat neurodegenerative disorders (96).

The recent advances in our understanding of niche-, stage-, and activation state-dependent roles of microglia and macrophages in different brain malignancies certainly open new opportunities for therapeutic intervention. Stage- or population-specific gene signature might also serve as new prognostic biomarkers to identify high-risk patients based on inflammatory fingerprints. Thorough functional validation of candidate genes that are associated with dysregulated microglia/macrophage function is needed to identify druggable targets for therapies aiming at reverting disease-promoting into protective effects or to maintain beneficial house-keeping functions as proposed for the Trem2–Apoe pathway.

CONCLUDING REMARKS

The history of microglia in brain malignancies started almost a century ago with their initial description by Pio del Rio-Hortega. Their ontological origin and biological function in health and disease has been controversially discussed ever since. While in-depth analysis down to the single-cell level provided critical insight into the heterogeneity of microglia and their blood-borne counterparts, we are just at the beginning to understand how different subpopulations or activation states regulate CNS homeostasis at steady state and how aberrant functions affect disease progression. Further investigation of the mechanisms that drive microglia and macrophage dysregulation will hopefully provide scientific rationale for the development of novel targeted therapies that provide better treatment options for patients to improve prognosis and quality of life.

AUTHOR CONTRIBUTIONS

The author confirms being the sole contributor of this work and approved it for publication.

ACKNOWLEDGMENTS

The author thanks Ursula Dietrich, Hind Medyouf, Katja Niesel, Anna Salamero-Boix, and Michael Schulz for critical reading of the manuscript and fruitful discussion on the topic. LS is supported by the Max-Eder Junior Group Leader Program (German Cancer Aid, Deutsche Krebshilfe).

REFERENCES

Louveau A. [Cerebral lymphatic drainage: implication in the brain immune privilege]. Med Sci (Paris) (2015) 31(11):953–6. doi:10.1051/medsci/20153111005

Reemst K, Noctor SC, Lucassen PJ, Hol EM. The indispensable roles of microglia and astrocytes during brain development. Front Hum Neurosci (2016) 10:566. doi:10.3389/fnhum.2016.00566

Aloisi F. The role of microglia and astrocytes in CNS immune surveillance and immunopathology. Adv Exp Med Biol (1999) 468:123–33. doi:10.1007/ 978-1-4615-4685-6_10

Nimmerjahn A, Kirchhoff F, Helmchen F. Resting microglial cells are highly dynamic surveillants of brain parenchyma in vivo. Science (2005) 308(5726):1314–8. doi:10.1126/science.1110647

Aloisi F. Immune function of microglia. Glia (2001) 36(2):165–79. doi:10.1002/glia.1106

Hanisch UK, Kettenmann H. Microglia: active sensor and versatile effector cells in the normal and pathologic brain. Nat Neurosci (2007) 10(11):1387–94. doi:10.1038/nn1997

Paolicelli RC, Bolasco G, Pagani F, Maggi L, Scianni M, Panzanelli P, et al. Synaptic pruning by microglia is necessary for normal brain deve- lopment. Science (2011) 333(6048):1456–8. doi:10.1126/science.1202529

Sierra A, Encinas JM, Deudero JJ, Chancey JH, Enikolopov G, Overstreet- Wadiche LS, et al. Microglia shape adult hippocampal neurogenesis through apoptosis-coupled phagocytosis. Cell Stem Cell (2010) 7(4):483–95. doi:10.1016/j.stem.2010.08.014

Wlodarczyk A, Holtman IR, Krueger M, Yogev N, Bruttger J, Khorooshi R, et al. A novel microglial subset plays a key role in myelinogenesis in develop- ing brain. EMBO J (2017) 36(22):3292–308. doi:10.15252/embj.201696056

Goldmann T, Wieghofer P, Jordao MJ, Prutek F, Hagemeyer N, Frenzel K, et al. Origin, fate and dynamics of macrophages at central nervous system interfaces. Nat Immunol (2016) 17(7):797–805. doi:10.1038/ni.3423

Korin B, Ben-Shaanan TL, Schiller M, Dubovik T, Azulay-Debby H, Boshnak NT, et al. High-dimensional, single-cell characterization of the brain's immune compartment. Nat Neurosci (2017) 20(9):1300–9. doi:10.1038/nn.4610

Chinnery HR, Ruitenberg MJ, McMenamin PG. Novel characterization of monocyte-derived cell populations in the meninges and choroid plexus and their rates of replenishment in bone marrow chimeric mice. J Neuropathol Exp Neurol (2010) 69(9):896–909. doi:10.1097/NEN.0b013e3181edbc1a

Meeker RB, Williams K, Killebrew DA, Hudson LC. Cell trafficking through the choroid plexus. Cell Adh Migr (2012) 6(5):390–6. doi:10.4161/ cam.21054

Herz J, Filiano AJ, Smith A, Yogev N, Kipnis J. Myeloid cells in the central nervous system. Immunity (2017) 46(6):943–56. doi:10.1016/j. immuni.2017.06.007

Sierra A, de Castro F, Del Rio-Hortega J, Rafael Iglesias-Rozas J, Garrosa M, Kettenmann H. The "Big-Bang" for modern glial biology: translation and comments on Pio del Rio-Hortega 1919 series of papers on microglia. Glia (2016) 64(11):1801–40. doi:10.1002/glia.23046

Kitamura T, Miyake T, Fujita S. Genesis of resting microglia in the gray matter of mouse hippocampus. J Comp Neurol (1984) 226(3):421–33. doi:10.1002/cne.902260310

de Groot CJ, Huppes W, Sminia T, Kraal G, Dijkstra CD. Determination of the origin and nature of brain macrophages and microglial cells in mouse central nervous system, using non-radioactive in situ hybridization and immunoperoxidase techniques. Glia (1992) 6(4):301–9. doi:10.1002/ glia.440060408

Matsumoto Y, Fujiwara M. Absence of donor-type major histocompatibility complex class I antigen-bearing microglia in the rat central nervous system of radiation bone marrow chimeras. J Neuroimmunol (1987) 17(1):71–82. doi:10.1016/0165-5728(87)90032-4

Hickey WF, Kimura H. Perivascular microglial cells of the CNS are bone marrow-derived and present antigen in vivo. Science (1988) 239(4837): 290–2. doi:10.1126/science.3276004

Lassmann H, Schmied M, Vass K, Hickey WF. Bone marrow derived elements and resident microglia in brain inflammation. Glia (1993) 7(1): 19–24. doi:10.1002/glia.440070106

McKercher SR, Torbett BE, Anderson KL, Henkel GW, Vestal DJ, Baribault H, et al. Targeted disruption of the PU.1 gene results in multiple hematopoietic abnormalities. EMBO J (1996) 15(20):5647–58.

Beers DR, Henkel JS, Xiao Q, Zhao W, Wang J, Yen AA, et al. Wild-type microglia extend survival in PU.1 knockout mice with familial amyo- trophic lateral sclerosis. *Proc Natl Acad Sci U S A* (2006) 103(43):16021-6. doi:10.1073/pnas.0607423103

Bertrand JY, Jalil A, Klaine M, Jung S, Cumano A, Godin I. Three pathways to mature macrophages in the early mouse yolk sac. *Blood* (2005) 106(9): 3004–11. doi:10.1182/blood-2005-02-0461

McGrath KE, Koniski AD, Malik J, Palis J. Circulation is established in a stepwise pattern in the mammalian embryo. *Blood* (2003) 101(5):1669–76. doi:10.1182/blood-2002-08-2531

Naito M, Umeda S, Yamamoto T, Moriyama H, Umezu H, Hasegawa G, et al. Development, differentiation, and phenotypic heterogeneity of murine tissue macrophages. *J Leukoc Biol* (1996) 59(2):133–8. doi:10.1002/jlb.59.2.133

Schulz C, Gomez Perdiguero E, Chorro L, Szabo-Rogers H, Cagnard N, Kierdorf K, et al. A lineage of myeloid cells independent of Myb and hema- topoietic stem cells. *Science* (2012) 336(6077):86–90. doi:10.1126/science. 1219179

Perdiguero EG, Geissmann F. The development and maintenance of resi- dent macrophages. *Nat Immunol* (2016) 17(1):2–8. doi:10.1038/ni.3341

Ginhoux F, Greter M, Leboeuf M, Nandi S, See P, Gokhan S, et al. Fate map- ping analysis reveals that adult microglia derive from primitive macrophages. *Science* (2010) 330(6005):841–5. doi:10.1126/science.1194637

Kierdorf K, Erny D, Goldmann T, Sander V, Schulz C, Perdiguero EG, et al. Microglia emerge from erythromyeloid precursors via Pu.1- and Irf8-dependent pathways. *Nat Neurosci* (2013) 16(3):273–80. doi:10.1038/ nn.3318

Matcovitch-Natan O, Winter DR, Giladi A, Vargas Aguilar S, Spinrad A, Sarrazin S, et al. Microglia development follows a stepwise program to regulate brain homeostasis. *Science* (2016) 353(6301):aad8670. doi:10.1126/ science.aad8670

Dai XM, Ryan GR, Hapel AJ, Dominguez MG, Russell RG, Kapp S, et al. Targeted disruption of the mouse colony-stimulating factor 1 receptor gene results in osteopetrosis, mononuclear phagocyte deficiency, increased primitive progenitor cell frequencies, and reproductive defects. *Blood* (2002) 99(1):111–20. doi:10.1182/blood.V99.1.111

Yoshida H, Hayashi S, Kunisada T, Ogawa M, Nishikawa S, Okamura H, et al. The murine mutation osteopetrosis is in the coding region of the macrophage colony stimulating factor gene. *Nature* (1990) 345(6274):442–4. doi:10.1038/345442a0

Lin H, Lee E, Hestir K, Leo C, Huang M, Bosch E, et al. Discovery of a cytokine and its receptor by functional screening of the extracellular pro- teome. *Science* (2008) 320(5877):807–11. doi:10.1126/science.1154370

Nandi S, Gokhan S, Dai XM, Wei S, Enikolopov G, Lin H, et al. The CSF-1 receptor ligands IL-34 and CSF-1 exhibit distinct developmental brain expression patterns and regulate neural progenitor cell maintenance and maturation. *Dev Biol* (2012) 367(2):100–13. doi:10.1016/j.ydbio.2012.03.026

Davies LC, Jenkins SJ, Allen JE, Taylor PR. Tissue-resident macrophages. *Nat Immunol* (2013) 14(10):986–95. doi:10.1038/ni.2705

Kierdorf K, Katzmarski N, Haas CA, Prinz M. Bone marrow cell recruitment to the brain in the absence of irradiation or parabiosis bias. *PLoS One* (2013) 8(3):e58544. doi:10.1371/journal.pone.0058544

Mildner A, Schmidt H, Nitsche M, Merkler D, Hanisch UK, Mack M, et al. Microglia in the adult brain arise from Ly-6ChiCCR2+ monocytes only under defined host conditions. *Nat Neurosci* (2007) 10(12):1544–53. doi:10.1038/nn2015

Ajami B, Bennett JL, Krieger C, Tetzlaff W, Rossi FM. Local self-renewal can sustain CNS microglia maintenance and function throughout adult life. *Nat Neurosci* (2007) 10(12):1538–43. doi:10.1038/nn2014

Ajami B, Bennett JL, Krieger C, McNagny KM, Rossi FM. Infiltrating mono- cytes trigger EAE progression, but do not contribute to the resident microglia pool. *Nat Neurosci* (2011) 14(9):1142–9. doi:10.1038/nn.2887

Varvel NH, Grathwohl SA, Baumann F, Liebig C, Bosch A, Brawek B, et al. Microglial repopulation model reveals a robust homeostatic process for replacing CNS myeloid cells. *Proc Natl Acad Sci U S A* (2012) 109(44):18150–5. doi:10.1073/pnas.1210150109

Heppner FL, Greter M, Marino D, Falsig J, Raivich G, Hovelmeyer N, et al. Experimental autoimmune encephalomyelitis repressed by microglial paralysis. *Nat Med* (2005) 11(2):146–52. doi:10.1038/nm1177

Parkhurst CN, Yang G, Ninan I, Savas JN, Yates JRIII, Lafaille JJ, etal. Microglia promote learning-dependent synapse formation through brain-derived neu- rotrophic factor. *Cell* (2013) 155(7):1596–609. doi:10.1016/j.cell.2013.11.030

Bruttger J, Karram K, Wortge S, Regen T, Marini F, Hoppmann N, et al. Genetic cell ablation reveals clusters of local self-renewing microglia in the mammalian central nervous system. *Immunity* (2015) 43(1):92–106. doi:10.1016/j.immuni.2015.06.012

Snippert HJ, van der Flier LG, Sato T, van Es JH, van den Born M, Kroon-Veenboer C, et al. Intestinal crypt homeostasis results from neutral competition between

symmetrically dividing Lgr5 stem cells. *Cell* (2010) 143(1):134–44. doi:10.1016/j.cell.2010.09.016

Tay TL, Mai D, Dautzenberg J, Fernandez-Klett F, Lin G, Sagar, et al. A new fate mapping system reveals context-dependent random or clonal expansion of microglia. *Nat Neurosci* (2017) 20(6):793–803. doi:10.1038/ nn.4547

Gautier EL, Shay T, Miller J, Greter M, Jakubzick C, Ivanov S, et al. Gene- expression profiles and transcriptional regulatory pathways that underlie the identity and diversity of mouse tissue macrophages. *Nat Immunol* (2012) 13(11):1118–28. doi:10.1038/ni.2419

Mass E, Ballesteros I, Farlik M, Halbritter F, Gunther P, Crozet L, et al. Specification of tissue-resident macrophages during organogenesis. *Science* (2016) 353:6304. doi:10.1126/science.aaf4238

Haldar M, Kohyama M, So AY, Kc W, Wu X, Briseno CG, et al. Heme-mediated SPI-C induction promotes monocyte differentiation into iron-recycling macrophages. *Cell* (2014) 156(6):1223–34. doi:10.1016/j.cell.2014.01.069

Kohyama M, Ise W, Edelson BT, Wilker PR, Hildner K, Mejia C, et al. Role for Spi-C in the development of red pulp macrophages and splenic iron homeostasis. *Nature* (2009) 457(7227):318–21. doi:10.1038/nature07472

Rosas M, Davies LC, Giles PJ, Liao CT, Kharfan B, Stone TC, et al. The tran- scription factor Gata6 links tissue macrophage phenotype and proliferative renewal. *Science* (2014) 344(6184):645–8. doi:10.1126/science.1251414

Blecher-Gonen R, Barnett-Itzhaki Z, Jaitin D, Amann-Zalcenstein D, Lara-Astiaso D, Amit I. High-throughput chromatin immunoprecipita- tion for genome-wide mapping of in vivo protein-DNA interactions and epigenomic states. *Nat Protoc* (2013) 8(3):539–54. doi:10.1038/nprot. 2013.023

Buenrostro JD, Giresi PG, Zaba LC, Chang HY, Greenleaf WJ. Transposition of native chromatin for fast and sensitive epigenomic profiling of open chromatin, DNA-binding proteins and nucleosome position. *Nat Methods* (2013) 10(12):1213–8. doi:10.1038/nmeth.2688

Gosselin D, Link VM, Romanoski CE, Fonseca GJ, Eichenfield DZ, Spann NJ, et al. Environment drives selection and function of enhancers controlling tissue- specifi macrophage identities. *Cell* (2014) 159(6):1327–40. doi:10.1016/j.cell.2014.11.023

Lavin Y, Winter D, Blecher-Gonen R, David E, Keren-Shaul H, Merad M, et al. Tissue-resident macrophage enhancer landscapes are shaped by the local microenvironment. *Cell* (2014) 159(6):1312–26. doi:10.1016/j.cell. 2014.11.018

Ghisletti S, Barozzi I, Mietton F, Polletti S, De Santa F, Venturini E, et al. Identification and characterization of enhancers controlling the inflam- matory gene expression program in macrophages. *Immunity* (2010) 32(3): 317–28. doi:10.1016/j.immuni.2010.02.008

Heinz S, Benner C, Spann N, Bertolino E, Lin YC, Laslo P, et al. Simple com- binations of lineage-determining transcription factors prime cis-regulatory elements required for macrophage and B cell identities. *Mol Cell* (2010) 38(4):576–89. doi:10.1016/j.molcel.2010.05.004

Gosselin D, Glass CK. Epigenomics of macrophages. *Immunol Rev* (2014) 262(1):96–112. doi:10.1111/imr.12213

Butovsky O, Jedrychowski MP, Moore CS, Cialic R, Lanser AJ, Gabriely G, et al. Identification of a unique TGF-beta-dependent molecular and func- tional signature in microglia. *Nat Neurosci* (2014) 17(1):131–43. doi:10.1038/ nn.3599

Greter M, Lelios I, Pelczar P, Hoeffel G, Price J, Leboeuf M, et al. Stroma- derived interleukin-34 controls the development and maintenance of langerhans cells and the maintenance of microglia. *Immunity* (2012) 37(6):1050–60. doi:10.1016/j.immuni.2012.11.001

Okabe Y, Medzhitov R. Tissue-specific signals control reversible program of localization and functional polarization of macrophages. *Cell* (2014) 157(4):832–44. doi:10.1016/j.cell.2014.04.016

Prinz M, Priller J. Tickets to the brain: role of CCR2 and CX3CR1 in myeloid cell entry in the CNS. *J Neuroimmunol* (2010) 224(1–2):80–4. doi:10.1016/j.jneuroim.2010.05.015

Gieryng A, Pszczolkowska D, Bocian K, Dabrowski M, Rajan WD, Kloss M, et al. Immune microenvironment of experimental rat C6 gliomas resembles human glioblastomas. *Sci Rep* (2017) 7(1):17556. doi:10.1038/ s41598-017-17752-w

Lambertsen KL, Deierborg T, Gregersen R, Clausen BH, Wirenfeldt M, Nielsen HH, et al. Differences in origin of reactive microglia in bone mar- row chimeric mouse and rat after transient global ischemia. *J Neuropathol Exp Neurol* (2011) 70(6):481–94. doi:10.1097/NEN.0b013e31821db3aa

Gieryng A, Pszczolkowska D, Walentynowicz KA, Rajan WD, Kaminska B. Immune microenvironment of gliomas. *Lab Invest* (2017) 97(5):498–518. doi:10.1038/labinvest.2017.19

Venteicher AS, Tirosh I, Hebert C, Yizhak K, Neftel C, Filbin MG, et al. Decoupling genetics, lineages, and microenvironment in IDH-mutant gliomas by single-cell RNA-seq. *Science* (2017) 355(6332):1–11. doi:10.1126/ science.aai8478

Gosselin D, Skola D, Coufal NG, Holtman IR, Schlachetzki JCM, Sajti E, et al. An environment-dependent transcriptional network specifies human microglia identity. *Science* (2017) 356(6344):1–11. doi:10.1126/science. aal3222

Darmanis S, Sloan SA, Croote D, Mignardi M, Chernikova S, Samghababi P, et al. Single-cell RNA-seq analysis of infiltrating neoplastic cells at the migrating front of human glioblastoma. *Cell Rep* (2017) 21(5):1399–410. doi:10.1016/j. celrep.2017.10.030

Wang Q, Hu B, Hu X, Kim H, Squatrito M, Scarpace L, et al. Tumor evolution of glioma-intrinsic gene expression subtypes associates with immunological changes in the microenvironment. *Cancer Cell* (2017) 32(1):42–56.e6. doi:10.1016/j.ccell.2017.06.003

O'Callaghan JP, Sriram K, Miller DB. Defining "neuroinflammation". *Ann N Y Acad Sci* (2008) 1139:318–30. doi:10.1196/annals.1432.032

Puentes F, Malaspina A, van Noort JM, Amor S. Non-neuronal cells in ALS: role of glial, immune cells and blood-CNS barriers. *Brain Pathol* (2016) 26(2):248–57. doi:10.1111/bpa.12352

Ellwardt E, Zipp F. Molecular mechanisms linking neuroinflammation and neurodegeneration in MS. *Exp Neurol* (2014) 262(Pt A):8–17. doi:10.1016/j. expneurol.2014.02.006

Hemmer B, Kerschensteiner M, Korn T. Role of the innate and adaptive immune responses in the course of multiple sclerosis. *Lancet Neurol* (2015) 14(4):406–19. doi:10.1016/S1474-4422(14)70305-9

Dendrou CA, Fugger L, Friese MA. Immunopathology of multiple sclerosis. *Nat Rev Immunol* (2015) 15(9):545–58. doi:10.1038/nri3871

Prinz M, Priller J. The role of peripheral immune cells in the CNS in steady state and disease. *Nat Neurosci* (2017) 20(2):136–44. doi:10.1038/nn.4475

Alzheimer A, Stelzmann RA, Schnitzlein HN, Murtagh FR. An English translation of Alzheimer's 1907 paper, "Uber eine eigenartige Erkankung der Hirnrinde". *Clin Anat* (1995) 8(6):429–31. doi:10.1002/ca.980080612

Hickman RA, Faustin A, Wisniewski T. Alzheimer disease and its growing epidemic: risk factors, biomarkers, and the urgent need for therapeutics. *Neurol Clin* (2016) 34(4):941–53. doi:10.1016/j.ncl.2016.06.009

Jay TR, Miller CM, Cheng PJ, Graham LC, Bemiller S, Broihier ML, et al. TREM2 deficiency eliminates TREM2+ inflammatory macrophages and ameliorates pathology in Alzheimer's disease mouse models. *J Exp Med* (2015) 212(3):287–95. doi:10.1084/jem.20142322

Simard AR, Soulet D, Gowing G, Julien JP, Rivest S. Bone marrow-derived microglia play a critical role in restricting senile plaque formation in Alzheimer's disease. *Neuron* (2006) 49(4):489–502. doi:10.1016/j.neuron.2006.01.022

Wang Y, Ulland TK, Ulrich JD, Song W, Tzaferis JA, Hole JT, et al. TREM2- mediated early microglial response limits diffusion and toxicity of amyloid plaques. *J Exp Med* (2016) 213(5):667–75. doi:10.1084/jem.20151948

Heppner FL, Ransohoff RM, Becher B. Immune attack: the role of inflamma- tion in Alzheimer disease. *Nat Rev Neurosci* (2015) 16(6):358–72. doi:10.1038/ nrn3880

Tejera D, Heneka MT. Microglia in Alzheimer's disease: the good, the bad and the ugly. *Curr Alzheimer Res* (2016) 13(4):370–80. doi:10.2174/1567205 013666151116125012

Wang Y, Cella M, Mallinson K, Ulrich JD, Young KL, Robinette ML, et al. TREM2 lipid sensing sustains the microglial response in an Alzheimer's disease model. *Cell* (2015) 160(6):1061–71. doi:10.1016/j.cell.2015.01.049

Yamasaki R, Lu H, Butovsky O, Ohno N, Rietsch AM, Cialic R, et al. Differential roles of microglia and monocytes in the inflamed central nervous system. *J Exp Med* (2014) 211(8):1533–49. doi:10.1084/jem.20132477

Mildner A, Schlevogt B, Kierdorf K, Bottcher C, Erny D, Kummer MP, et al. Distinct and non-redundant roles of microglia and myeloid subsets in mouse models of Alzheimer's disease. *J Neurosci* (2011) 31(31):11159–71. doi:10.1523/ JNEUROSCI.6209-10.2011

Elowitz MB, Levine AJ, Siggia ED, Swain PS. Stochastic gene expression in a single cell. *Science* (2002) 297(5584):1183–6. doi:10.1126/science.1070919

Geissmann F, Gordon S, Hume DA, Mowat AM, Randolph GJ. Unravelling mononuclear phagocyte heterogeneity. *Nat Rev Immunol* (2010) 10(6): 453–60. doi:10.1038/nri2784

Hume DA. Differentiation and heterogeneity in the mononuclear phagocyte system. *Mucosal Immunol* (2008) 1(6):432–41. doi:10.1038/mi.2008.36

Zeisel A, Munoz-Manchado AB, Codeluppi S, Lonnerberg P, La Manno G, Jureus A, et al. Brain structure. Cell types in the mouse cortex and hippo- campus revealed by single-cell RNA-seq. *Science* (2015) 347(6226):1138–42. doi:10.1126/science. aaa1934

Macaulay IC, Ponting CP, Voet T. Single-cell multiomics: multiple measure- ments from single cells. *Trends Genet* (2017) 33(2):155–68. doi:10.1016/j. tig.2016.12.003

Stubbington MJT, Rozenblatt-Rosen O, Regev A, Teichmann SA. Single- cell transcriptomics to explore the immune system in health and disease. *Science* (2017) 358(6359):58–63. doi:10.1126/science.aan6828

Darmanis S, Sloan SA, Zhang Y, Enge M, Caneda C, Shuer LM, et al. A survey of human brain transcriptome diversity at the single cell level. *Proc Natl Acad Sci U S A* (2015) 112(23):7285–90. doi:10.1073/pnas.1507125112

Keren-Shaul H, Spinrad A, Weiner A, Matcovitch-Natan O, Dvir-Szternfeld R, Ulland TK, et al. A unique microglia type associated with restricting develop- ment of Alzheimer's disease. *Cell* (2017) 169(7):1276–90.e17. doi:10.1016/j. cell.2017.05.018

Mathys H, Adaikkan C, Gao F, Young JZ, Manet E, Hemberg M, et al. Temporal tracking of microglia activation in neurodegeneration at single-cell resolution. *Cell Rep* (2017) 21(2):366–80. doi:10.1016/j.celrep.2017.09.039

Satoh J, Kino Y, Asahina N, Takitani M, Miyoshi J, Ishida T, et al. TMEM119 marks a subset of microglia in the human brain. *Neuropathology* (2016) 36(1):39–49. doi:10.1111/neup.12235

Mrdjen D, Pavlovic A, Hartmann FJ, Schreiner B, Utz SG, Leung BP, et al. High-dimensional single-cell mapping of central nervous system immune cells reveals distinct myeloid subsets in health, aging, and disease. *Immunity* (2018) 48(2):380–95.e6. doi:10.1016/j.immuni.2018.01.011

Krasemann S, Madore C, Cialic R, Baufeld C, Calcagno N, El Fatimy R, et al. The TREM2-APOE pathway drives the transcriptional phenotype of dysfunctional microglia in neurodegenerative diseases. *Immunity* (2017) 47(3):566–81.e9. doi:10.1016/j.immuni.2017.08.008

Paradowska-Gorycka A, Jurkowska M. Structure, expression pattern and biological activity of molecular complex TREM-2/DAP12. *Hum Immunol* (2013) 74(6):730–7. doi:10.1016/j.humimm.2013.02.003

Jay TR, Hirsch AM, Broihier ML, Miller CM, Neilson LE, Ransohoff RM, et al. Disease progression-dependent effects of TREM2 deficiency in a mouse model of Alzheimer's disease. *J Neurosci* (2017) 37(3):637–47. doi:10.1523/ JNEUROSCI.2110-16.2016

Hansen DV, Hanson JE, Sheng M. Microglia in Alzheimer's disease. *J Cell Biol* (2017) 217(2):459–72. doi:10.1083/jcb.201709069

Gao H, Danzi MC, Choi CS, Taherian M, Dalby-Hansen C, Ellman DG, et al. Opposing functions of microglial and macrophagic TNFR2 in the patho- genesis of experimental autoimmune encephalomyelitis. *Cell Rep* (2017) 18(1):198–212. doi:10.1016/j.celrep.2016.11.083

Pyonteck SM, Akkari L, Schuhmacher AJ, Bowman RL, Sevenich L, Quail DF, et al. CSF-1R inhibition alters macrophage polarization and blocks glioma progression. *Nat Med* (2013) 19(10):1264–72. doi:10.1038/nm.3337

Sevenich L, Bowman RL, Mason SD, Quail DF, Rapaport F, Elie BT, et al. Analysis of tumour- and stroma-supplied proteolytic networks reveals a brain-metastasis-promoting role for cathepsin S. *Nat Cell Biol* (2014) 16(9):876–88. doi:10.1038/ncb3011

Lin EY, Pollard JW. Macrophages: modulators of breast cancer progression. *Novartis Found Symp* (2004) 256:158–68.

Joyce JA, Pollard JW. Microenvironmental regulation of metastasis. *Nat Rev Cancer* (2009) 9(4):239–52. doi:10.1038/nrc2618

Quail DF, Joyce JA. The microenvironmental landscape of brain tumors. *Cancer Cell* (2017) 31(3):326–41. doi:10.1016/j.ccell.2017.02.009

Bingle L, Brown NJ, Lewis CE. The role of tumour-associated macrophages in tumour progression: implications for new anticancer therapies. *J Pathol* (2002) 196(3):254–65. doi:10.1002/path.1027

Hussain SF, Yang D, Suki D, Aldape K, Grimm E, Heimberger AB. The role of human glioma-infiltrating microglia/macrophages in mediating anti- tumor immune responses. *Neuro Oncol* (2006) 8(3):261–79. doi:10.1215/ 15228517-2006-008

Franklin RA, Liao W, Sarkar A, Kim MV, Bivona MR, Liu K, et al. The cellular and molecular origin of tumor-associated macrophages. *Science* (2014) 344(6186):921–5. doi:10.1126/science.1252510

Xu H, Zhu J, Smith S, Foldi J, Zhao B, Chung AY, et al. Notch-RBP-J signaling regulates the transcription factor IRF8 to promote inflammatory macrophage polarization. *Nat Immunol* (2012) 13(7):642–50. doi:10.1038/ni.2304

Zhu Y, Herndon JM, Sojka DK, Kim KW, Knolhoff BL, Zuo C, et al. Tissue-resident macrophages in pancreatic ductal adenocarcinoma originate from embryonic hematopoiesis and promote tumor progression. *Immunity* (2017) 47(2):323–38. e6. doi:10.1016/j.immuni.2017.07.014

Chen Z, Feng X, Herting CJ, Garcia VA, Nie K, Pong WW, et al. Cellular and molecular identity of tumor-associated macrophages in glioblastoma. *Cancer Res* (2017) 77(9):2266–78. doi:10.1158/0008-5472.CAN-16-2310

Bowman RL, Klemm F, Akkari L, Pyonteck SM, Sevenich L, Quail DF, et al. Macrophage ontogeny underlies differences in tumor-specific education in brain malignancies. *Cell Rep* (2016) 17(9):2445–59. doi:10.1016/j. celrep.2016.10.052

Muller A, Brandenburg S, Turkowski K, Muller S, Vajkoczy P. Resident microglia, and not peripheral macrophages, are the main source of brain tumor mononuclear cells. *Int J Cancer* (2015) 137(2):278–88. doi:10.1002/ ijc.29379

Jung S, Aliberti J, Graemmel P, Sunshine MJ, Kreutzberg GW, Sher A, et al. Analysis of fractalkine receptor CX(3)CR1 function by targeted deletion and green fluorescent protein reporter gene insertion. *Mol Cell Biol* (2000) 20(11):4106–14. doi:10.1128/MCB.20.11.4106-4114.2000

Cook DN, Chen SC, Sullivan LM, Manfra DJ, Wiekowski MT, Prosser DM, et al. Generation and analysis of mice lacking the chemokine fractalkine. *Mol Cell Biol* (2001) 21(9):3159–65. doi:10.1128/MCB.21.9.3159-3165.2001

Sedgwick JD, Ford AL, Foulcher E, Airriess R. Central nervous system microglial cell activation and proliferation follows direct interaction with tissue-infiltrating T cell blasts. *J Immunol* (1998) 160(11):5320–30.

Ford AL, Goodsall AL, Hickey WF, Sedgwick JD. Normal adult ramified microglia separated from other central nervous system macrophages by fl w cytometric sorting. Phenotypic differences defined and direct ex vivo anti- gen presentation to myelin basic protein-reactive CD4+ T cells compared. *J Immunol* (1995) 154(9):4309–21.

Boyer SW, Schroeder AV, Smith-Berdan S, Forsberg EC. All hemato- poietic cells develop from hematopoietic stem cells through Flk2/Flt3- positive progenitor cells. *Cell Stem Cell* (2011) 9(1):64–73. doi:10.1016/j. stem.2011.04.021

Buttgereit A, Lelios I, Yu X, Vrohlings M, Krakoski NR, Gautier EL, et al. Sall1 is a transcriptional regulator defining microglia identity and function. *Nat Immunol* (2016) 17(12):1397–406. doi:10.1038/ni.3585

O'Koren EG, Mathew R, Saban DR. Fate mapping reveals that microglia and recruited monocyte-derived macrophages are defi tively distin- guishable by phenotype in the retina. *Sci Rep* (2016) 6:20636. doi:10.1038/ srep20636

Stephan AH, Barres BA, Stevens B. The complement system: an unexpected role in synaptic pruning during development and disease. *Annu Rev Neurosci* (2012) 35:369–89. doi:10.1146/annurev-neuro-061010-113810

Koeniger T, Kuerten S. Splitting the "Unsplittable": dissecting resident and infiltrating macrophages in experimental autoimmune encephalomyelitis. *Int J Mol Sci* (2017) 18(10):1–17. doi:10.3390/ijms18102072

Friedman BA, Srinivasan K, Ayalon G, Meilandt WJ, Lin H, Huntley MA, et al. Diverse brain myeloid expression profiles reveal distinct microglial activation states and aspects of Alzheimer's disease not evident in mouse models. *Cell Rep* (2018) 22(3):832–47. doi:10.1016/j.celrep.2017.12.066

Hambardzumyan D, Gutmann DH, Kettenmann H. The role of microglia and macrophages in glioma maintenance and progression. *Nat Neurosci* (2016) 19(1):20–7. doi:10.1038/nn.4185

Quail DF, Joyce JA. Microenvironmental regulation of tumor progression and metastasis. *Nat Med* (2013) 19(11):1423–37. doi:10.1038/nm.3394

Chuang HN, van Rossum D, Sieger D, Siam L, Klemm F, Bleckmann A, et al. Carcinoma cells misuse the host tissue damage response to invade the brain. *Glia* (2013) 61(8):1331–46. doi:10.1002/glia.22518

Dvorak HF. Tumors: wounds that do not heal-redux. *Cancer Immunol Res* (2015) 3(1):1–11. doi:10.1158/2326-6066.CIR-14-0209

Brown JM, Recht L, Strober S. The promise of targeting macrophages in cancer therapy. *Clin Cancer Res* (2017) 23(13):3241–50. doi:10.1158/1078- 0432.CCR-16-3122

Salter MW, Stevens B. Microglia emerge as central players in brain disease. *Nat Med* (2017) 23(9):1018–27. doi:10.1038/nm.4397

Butowski N, Colman H, De Groot JF, Omuro AM, Nayak L, Wen PY, et al. Orally administered colony stimulating factor 1 receptor inhibitor PLX3397 in recurrent glioblastoma: an Ivy Foundation Early Phase Clinical Trials Consortium phase II study. *Neuro Oncol* (2016) 18(4):557–64. doi:10.1093/ neuonc/nov245

Quail DF, Joyce JA. Molecular pathways: deciphering mechanisms of resistance to macrophage-targeted therapies. *Clin Cancer Res* (2017) 23(4):876–84. doi:10.1158/1078-0432.CCR-16-0133

Zhu Y, Knolhoff BL, Meyer MA, Nywening TM, West BL, Luo J, et al. CSF1/ CSF1R blockade reprograms tumor-infiltrating macrophages and improves response to T-cell checkpoint immunotherapy in pancreatic cancer models. *Cancer Res* (2014) 74(18):5057–69. doi:10.1158/0008-5472.CAN-13-3723

Stafford JH, Hirai T, Deng L, Chernikova SB, Urata K, West BL, et al. Colony stimulating factor 1 receptor inhibition delays recurrence of glioblastoma after radiation by altering myeloid cell recruitment and polarization. *Neuro Oncol* (2016) 18(6):797–806. doi:10.1093/neuonc/nov272

Quail DF, Bowman RL, Akkari L, Quick ML, Schuhmacher AJ, Huse JT, et al. The tumor microenvironment underlies acquired resistance to CSF-1R inhibition in gliomas. *Science* (2016) 352(6288):aad3018. doi:10.1126/ science.aad3018

Elmore MR, Najafi AR, Koike MA, Dagher NN, Spangenberg EE, Rice RA, et al. Colony-stimulating factor 1 receptor signaling is necessary for microg- lia viability, unmasking a microglia progenitor cell in the adult brain. *Neuron* (2014) 82(2):380–97. doi:10.1016/j.neuron.2014.02.040

Elmore MR, Lee RJ, West BL, Green KN. Characterizing newly repopulated microglia in the adult mouse: impacts on animal behavior, cell morphology, and neuroinflammation. *PLoS One* (2015) 10(4):e0122912. doi:10.1371/ journal. pone.0122912

Dagher NN, Najafi AR, Kayala KM, Elmore MR, White TE, Medeiros R, et al. Colony-stimulating factor 1 receptor inhibition prevents microg- lial plaque association and improves cognition in 3xTg-AD mice. *J Neuroinflammation* (2015) 12:139. doi:10.1186/s12974-015-0366-9

4

A Novel Method to Image Macropinocytosis *in Vivo*

Lunhao Chen [1,2], Daxiao Cheng [1], Jiachen Chu [1,3], Ting Zhang [1], Zhuoer Dong [4], Huifang Lou [1], Liya Zhu [1] and Yijun Liu [1*]

[1] Department of Neurobiology, Key Laboratory of Medical Neurobiology, Ministry of Health of China, Zhejiang Provincial Key Laboratory of Neurobiology, Zhejiang University School of Medicine, Hangzhou, China, [2] Department of Orthopedic Surgery, The First Affiliated Hospital, Zhejiang University School of Medicine, Hangzhou, China, [3] Department of Physiology, Johns Hopkins University School of Medicine, Baltimore, MD, United States, [4] Middle School Attached to Northwestern Polytechnical University, Xi'an, China

Here we described an experimental protocol for *in vivo* imaging of macropinocytosis and subsequent intracellular events. By microinjection, we delivered fluorescence dextrans together with or without ATPγS into transparent *Drosophila* melanogaster embryos. Using a confocal microscope for live imaging, we monitored the generation of dextran-positive macropinosomes and subsequent intracellular events. Our protocol provides a continent and reliable way for investigating macropinocytosis and its underlying mechanisms, especially when combined with genetic strategies.

Keywords: macropinocytosis, live imaging, *Drosophila*, embryo, hemocyte, *in vivo*

Correspondence:
Yijun Liu
yjliu@zju.edu.cn

INTRODUCTION

In eukaryotic cells, macropinocytosis is the most efficient way to internalize extracellular fluid through plasma membrane-formed large vacuoles called macropinosomes (Racoosin and Swanson, 1993; Swanson and Watts, 1995; Lim and Gleeson, 2011). As an ancient cellular behavior, macropinocytosis is essential for many physiological and pathological processes, such as nutrients uptake, pathogen capture, antigen presentation, and tumorigenesis (Kerr and Teasdale, 2009; Diken et al., 2011; Liu and Roche, 2015; Bloomfield and Kay, 2016). Sharing similar intracellular mechanism, macropinocytosis is thought to be largely homologous to phagocytosis, neuronal bulk endocytosis and other actin-driven endocytosis (Bloomfield and Kay, 2016).

Macropinocytosis provides a non-selective route to internalize extracellular fluids. In cancer cells, macropinocytosis is utilized for nutrient uptake to support metabolic needs and promote tumor growth (Commisso et al., 2013). Several infectious pathogens, such as bacteria, virus and protozoa, opportunistically hijack macropinocytosis to invade host cells and evade immune responses (Haraga et al., 2008; Gobeil et al., 2013). Observation of macropinocytosis will provide insight into the underlying regulatory molecular mechanisms and enable the physiological control of macropinocytosis for drug delivery in anti-cancer or -infection therapies. However, most observations of macropinocytosis were obtained from *in vitro* experiments or unicellular organisms (Chubb et al., 2000; Chen et al., 2015), e.g., *Dictyostelium* amoebae, instead of naturalistic models that do not fully reflect the complexity of *in vivo* situations, limiting their application. Therefore, considerable gaps remain in the knowledge of the relevance of macropinocytosis, especially the lack of optical imaging approaches, in living organisms. It is essential to develop consistent and reliable methods for *in vivo* macropinocytosis studies.

Most *Drosophila* melanogaster (fruit fly) genes are evolutionarily conserved with human and other mammals (Reiter and Bier, 2002). With its short life cycle and genetic amenability, *Drosophila* provides attractive model systems for various researches (Brand and Perrimon, 1993). After removal of chorions, *Drosophila* embryos become transparent, but still tolerance to subsequent operations for live imaging, rendering this model feasible for *in vivo* cell behavioral and cell biological studies.

In the present study, we described a protocol for *in vivo* studies of macropinocytosis. By microinjection, fluorescence-labeled dextrans were delivered into *Drosophila* embryos. Engulfed by *Drosophila* hemocytes, which resemble mammalian macrophages, fluorescent dextrans were internalized with associated membrane, resulting in formation and subsequent transportation of macropinosomes. For microscopic methods, macropinocytosis was fluorescently visualized and monitored in live embryos. This method provides a novel way for observation of the organization and subsequent processing of macropinosomes *in vivo*, and an ideal model for revealing the underlying mechanisms of macropinocytosis.

MATERIALS AND METHODS

Drosophila Stocks

A stable line *srp-Gal4;UAS-GFP* was used to visualize hemocytes with green fluorescent protein (GFP) in embryos. F1 embryos were crossed from *Srp-Gal4* and *UAS-tau-GFP* for microtubule labeling in the hemocytes. All crosses were raised on standard *Drosophila* medium at 25°C with 12:12 h light/dark cycle. The *Drosophila* line *srp-gal4; UAS-GFP line* is a kindly gift from Prof. Henry Sun.

Regents and Equipment
Injection Solutions

Hank's Balanced Salt Solution (HBSS, Invitrogen, Carlsbad, CA, USA) was used for dilution of fluorescent dextrans and adenosine 5=-O-(3-thio) triphosphate (ATPγS, Sigma-Aldrich, St. Louis, MO, USA) to the final concentration of 5 mg/ml and 1 mM, respectively. Cascade blue labeled 3-kDa fluorescent dextran (CB3S), tetramethylrhodamine (TRITC) labeled 3-, 10-, 40-, and 70-kDa fluorescent dextrans (TRD3S, TRD10S, TRD40S, and TRD70S) were all purchased from Invitrogen.

Heptane Glue

To stick and stabilize embryos for subsequent injection, heptane glue was prepared as previously described (Brust-Mascher and Scholey, 2009). In brief, one pack of double sticky tapes were unrolled and dissolved with 50 ml heptane (Ourche, China). Seal the bottle and mild shake the solution for at least 12 h until it is clear and sticky.

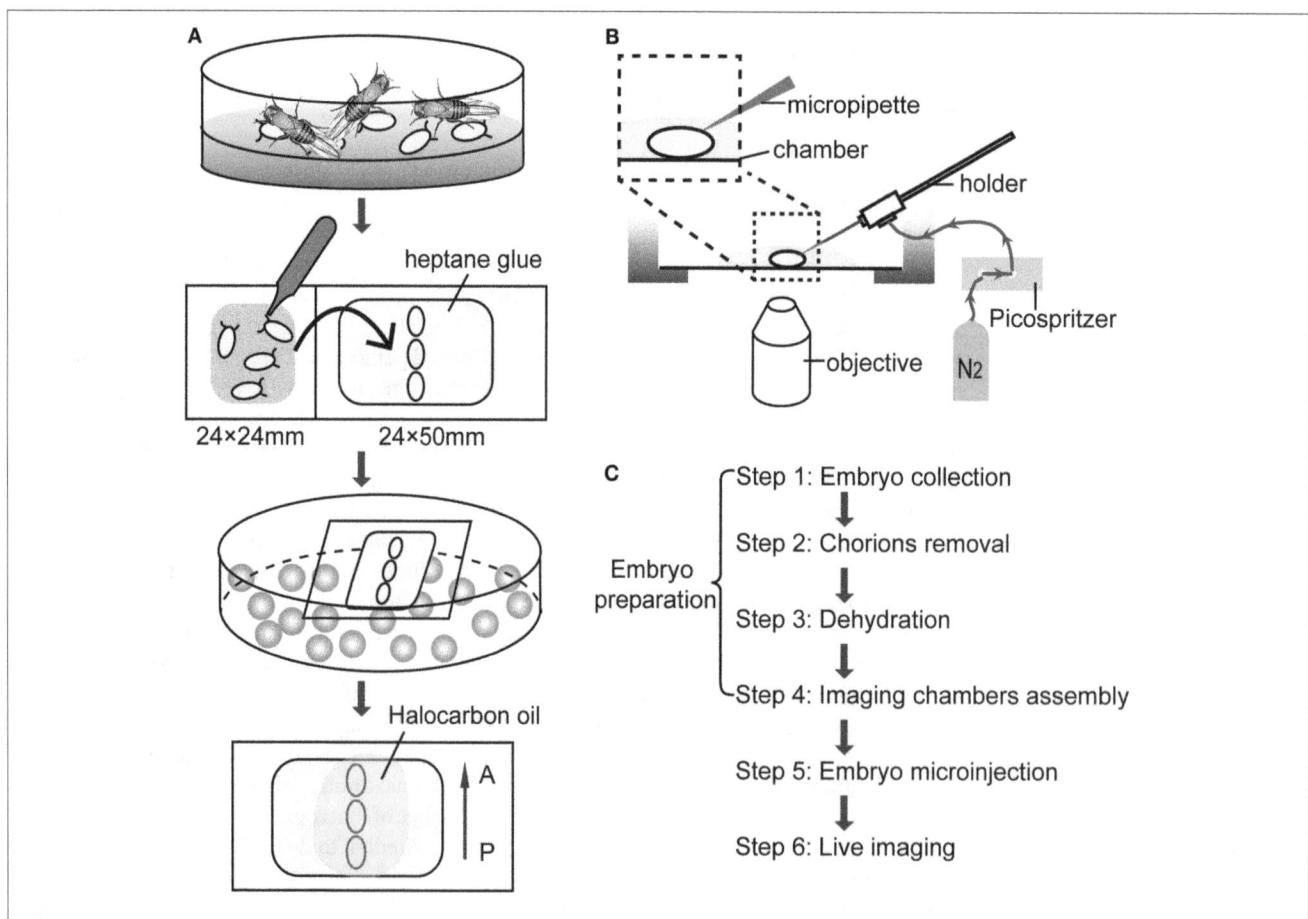

FIGURE 1 | Microinjection procedures for *in vivo* imaging. **(A)** The schematic diagram of embryos preparation. A, anterior; P, posterior. **(B)** The diagram of the microinjection system. **(C)** The flow diagram for live imaging of macropinocytosis in embryonic hemocytes.

Juice Agar Plates

Combine 2 g agar with 100 ml fruit juice, add ddH₂O to a final volume of 250 ml. Boil in microwave and pour the solution into 60 mm diameter Petri dishes and cool down at room temperature for 1 h. Scatter some dry yeast on the surface before use.

Micropipettes

Micropipettes for microinjection of dextrans are pre-pulled from borosilicate glass tubes (outer diameter: 1.0 mm, inner diameter: 0.5 mm, BF100-50-15, Sutter Instrument, Novato, CA, USA) by a micropipette puller (P-97, Sutter Instrument) to form a tip of ~5 μm in diameter. A microforge (MF-830, Narishige, Japan) was used for quality control.

Glass Coverslips

24*24 mm and 55*24 mm coverslips with thickness 0.13–0.16 mm were used (Stars, China).

Microinjection Preparation

The microinjection system was adapted from micropipette assay system for microglia migration as described previously (Wu et al., 2014). In short, the nitrogen cylinder was connected with Picospritzer (Picospritzer III, Parker Hannifin, Cleveland, OH, USA) and set the output pressure to 0.1 MPa. Connect the "OUTPUT" signal of the electronic stimulator with the micropipette holder and attach the holder to the micromanipulator (MP-225, Sutter Instrument). Set the pulse "DURATION" at 50–100 ms. About 1-2 μl injection solution with or without ATPγS was filled into the micropipette by a 1 ml syringe and make sure without any air bubbles. By local micropipette ejection (Lohof et al., 1992), injection solution is pulse-ejected into embryos controlled by pressing the "MANUAL" button.

Embryos Preparation (Figure 1A)

Step 1. Embryos Collection

100-200 adult *Drosophila* were transferred into an embryo collection cage (Brust-Mascher and Scholey, 2009) and adapt for 1 day before collection. When collection starts, change juice plates and collect newborn embryos per hr. Collected embryos were incubated for at least 5 h until hemocytes matured and GFP expression (Tepass et al., 1994; Miller et al., 2002).

Step 2. Chorions removal

Under a stereoscopic microscope (SZ51, Olympus, Japan), chorions were carefully removed from embryos by rolling them on the double sticky tape by fine tweezers (0208-5-PO, Dumont, Swiss). After removal of chorions, transparent embryos without chorions were paralleled arranged and attached on a coverslip with heptane glue.

Step 3. Dehydration

Put embryos with the coverslip into a dryer which bottom filled with allochroic silicates (Sinopharm, China). Embryos were dehydrated for ~5 min to prevent leakage of body fluids.

Step 4. Imaging chambers assembly

Transfer dehydrated embryos to an imaging chamber with a cover glass bottom. A droplet (about 20 μl) of Halocarbon oil 700 (H8898, Sigma-Aldrich) was added to each embryo providing appropriate humidity and enough oxygen.

TABLE 1 | Troubleshooting table.

Problem	Possible reason	Solution
Embryos are carried away during inserting.	Embryos are glued not enough.	Prepare thicker heptane glue to stable embryos.
	The tip is not sharp enough.	Adjust the tip size of micropipettes during preparation.
Body fluids flow out during injection.	Not enough Dehydration.	Prolong dehydration to 6-7 min.
	The flow from the micropipette tip is too high.	Use the micropipette with an appropriate tip size.
		Reduce the output pressure or the pulse duration.
Dextran-positive macropinosomes cannot be seen after 30 min of injection.	The micropipette tip is sealed or its tip is too small.	Replace the micropipette with a newly-made one. Use the micropipette with an appropriate tip size.
	Air bubbles are trapped in the micropipette tip.	Gently flick the micropipette to discharge bubbles or reload the injection solution.
	Large dextrans are filtered during diffusion.	Use small dextrans. 3 kDa-dextran is recommended.
	Embryos are not healthy enough.	Use another healthy embryo.
		Operate embryos as quickly as possible.
High background in the extracellular space	The micropipette tip is too large.	Adjust the tip size of micropipettes.
	The output pressure is too high or pulse duration is too long.	Reduce the output pressure or the pulse duration.

Embryo Microinjection

Assembly the injection equipment and imaging chambers as shown in the **Figure 1B**. Under a confocal microscope (FV1200, Olympus) with a 60x/NA 1.2 water objective, move the micropipette tip to the abdominal level of embryos by the micromanipulator. Carefully move the embryo against the micropipette tip and make sure the tip sticking into the embryo

at the center of the optic field. Wait for 30 sec and press the "MANUAL" button to operate a single pulse of 10-ms duration. The total injection volume is about 20–30 nl.

Live Imaging

After the dextran injection, 3D Time-lapse imaging was captured using 60x water dipping objectives, stacks of images were

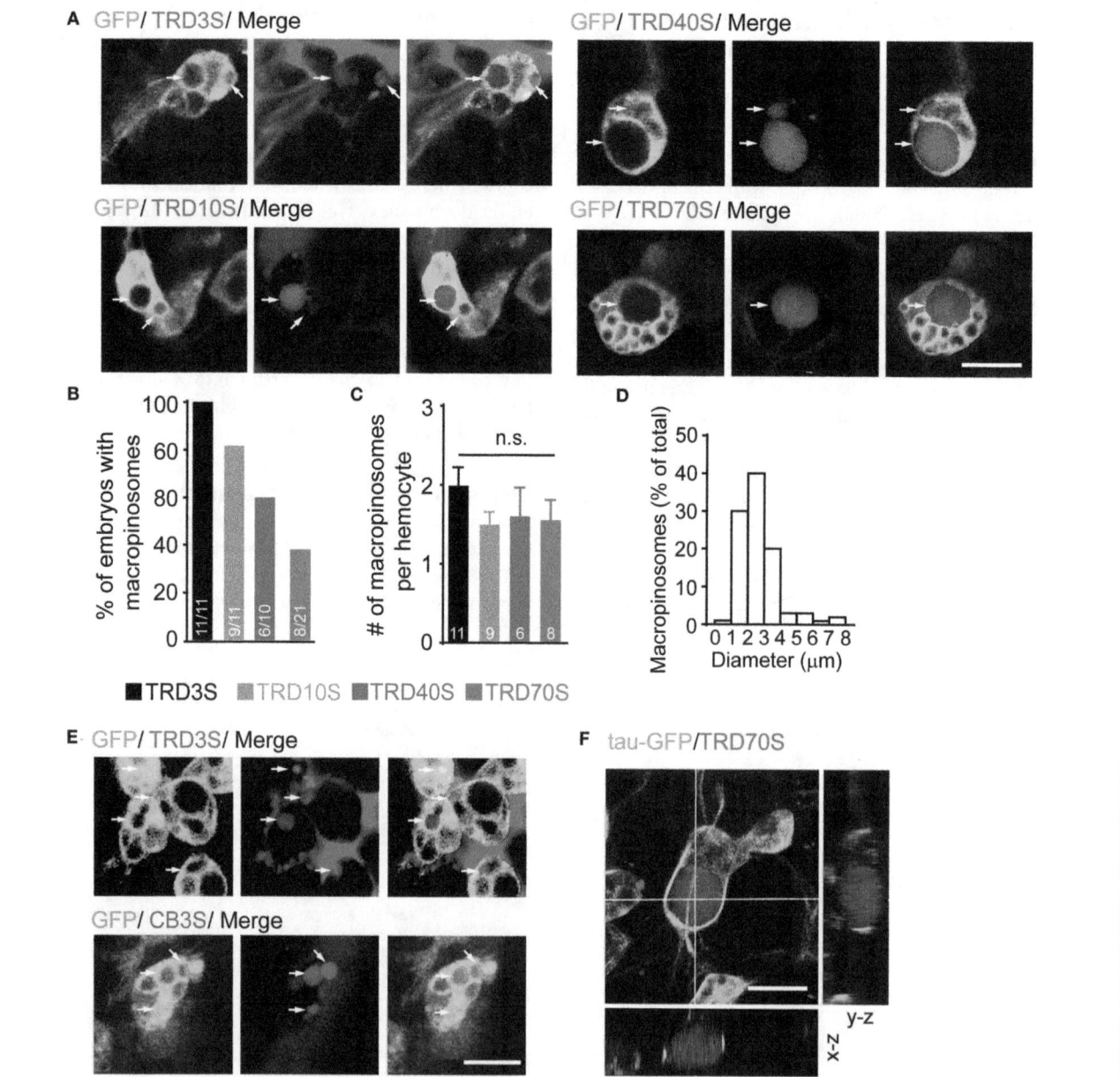

FIGURE 2 | ATPγS induced macropinosomes in *Drosophila* embryonic hemocytes. **(A)** Hemocytes (green) uptook fluorescent dextrans ranged from 3-kDa (TRD3S) to 70-kDa (TRD70S) dextrans and generated macropinosomes (arrows). **(B)** Success rates were defined as the percentage of embryos where hemocytes with macropinosomes 1.5 h after completion of different-sized dextrans injected. **(C)** Numbers (#) of macropinosomes in each hemocyte. *P*-values were calculated using one-way ANOVA among groups; n.s.: non-significant. **(D)** The size-distribution pattern of macropinosomes containing TRD3S (*n* = 100 macropinosomes, from 11 embryos). **(E)** Both TRITC labeled (red, TRD3S) and cascade blue labeled (blue, CB3S) dextrans were feasible to be uptaken by GFP-positive hemocytes (green), and formed macropinosomes (arrows). **(F)** In-depth 3D reconstruction analysis of ATPγS induced macropinosomes using a spatial deconvolution. Note that TRD70S dextran (red) labeled macropinosomes (red) and their surrounding GFP-positive microtubule-structures (green). Images were displayed in x-y (top), x-z (bottom), and y-z (right) projections. Scale bars, 10 μm.

acquired with a step size of $1\,\mu m$ for a depth of $10\,\mu m$ below embryo surface. Time-lapse movies were then generated between 3D stacks for 90 min without an interval. The time-lapse stack images were reconstructed and analyzed using Imaris software (Bitplane AG) and Image J (National Institute of Health, USA) software. The flow chart from embryo collection to live imaging was illustrated in **Figure 1C**.

Statistical Analysis

Statistical analysis was performed with STATA software (Version 13.0, Stata Corp, USA). Data are presented as means $\pm$ standard errors of the means (SEM). Statistical comparisons were assessed using student t-test or one-way ANOVA among three groups or above. Differences were considered to be significant at a P level of <0.05.

Troubleshooting

Troubleshooting advices can be found in **Table 1**.

RESULTS

Using our embryo microinjection method, we first sought to determine whether embryonic hemocytes were capable for evoking macropinocytosis. After induction by ATPγS, a non-hydrolysable ATP analog, macropinocytosis was observed using a confocal microscope. Five standard rules are followed to identify formed macropinosomes; (1) dextran-positive; (2) approximately pellet and ellipse-shaped; (3) surrounded by GFP-positive cytoplasm;(4) larger than $0.2\,\mu m$ in diameter; (5) fluorescence intensities of macropinosomes are comparable with or higher than that of the extracellular space.

In our observation, responding to injected ATPγS, GFP-positive hemocytes efficiently engulfed large volumes of extracellular fluids containing fluorescent-labeled dextran with diverse molecular weights and different fluorophores (**Figures 2A,E**). However, macropinosomes were seen in almost 100% of embryos after 3-kDa dextran injection, whereas only 81.8, 60.0, and 38.1% of embryos generated macropinosomes after injection of 10, 40, and 70-kDa dextran, respectively (**Figure 2B**). To clarify whether hemocytes have equal ability to uptake dextrans with different sizes, we measured the number of macropinosomes in each hemocyte. Our result shows that about 2 macropinosomes were generated in each hemocyte and there were no statistically significant differences among hemocytes uptake 3-kDa to 70-kDa dextrans (**Figure 2C**). In addition, our result showed that dextran-positive pinosomes were heterogeneous in sizes, ranging from 0.2 to $10\,\mu m$, and approximately 90% of the macropinosomes were 1 to $4\,\mu m$ in diameter (**Figure 2D**). Combined with genetic strategies, e.g., GAL4/UAS system, this model is appropriate for molecular studies, especially in endocytosis and its subsequent events. For instance, using hemocyte-specific promoter *srp-Gal4* to drive GFP fused tubulin-associated protein tau (tau-GFP), microtubule-filaments could be visualized and further analyzed in formation and trafficking of nascent macropinosomes (**Figure 2F**).

To clarify whether ATPγS was necessary for hemocytes to induce macropinosomes in our system. We injected 3 or 70-kD dextrans without ATPγS into embryonic hemocyte and observed spontaneous macropinocytosis in GFP positive hemocytes. Compared with ATPγS induced macropinosomes, spontaneous macropinosomes were much smaller (**Figure 3**), raising the difficulty for *in vivo* observation. Therefore, we recommend ATPγS induction to promote macropinocytosis in this model.

To further test whether this method is suitable for monitoring the cellular and subcellular events of macropinocytosis, we injected fluorescence-labeled dextrans together with ATPγS into the embryos with GFP-labeled hemocytes. As shown in **Figure 4A**, GFP-hemocyte extracellular dextrans were internalized along with cell surface ruffling and generated TRD70S-positive macropinosomes. Using low molecular weight TRITC-dextrans, TRD3S, we monitored the intracellular events of ATPγS-induced macropinosomes. During a 15 min observation period, dextran-containing macropinosomes transported in the cell body of a migrating hemocytes, indicating the method is reliable for cellular events and cell behavioral recordings (**Figure 4B**).

DISCUSSION

Adenosine triphosphate, or ATP, is the principal molecule for intracellular energy transfer in cells. Extracellular ATP is also an essential messenger for several physiologic and pathological processes (Lim and Gleeson, 2011; Li et al., 2013; Cisneros-Mejorado et al., 2015). Sensed by purinergic receptors, extracellular ATP activates intracellular signaling pathway and induces membrane ruffling and macropinocytosis (Grimmer et al., 2002; Li et al., 2013).

Here, we present an ATP analog microinjection-based system for *in vivo* observation of macropinocytosis and its subsequent cellular events. In our observation, ATPγS are

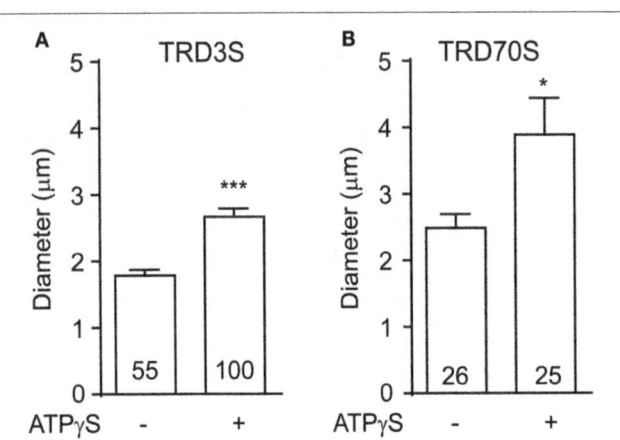

FIGURE 3 | ATPγS stimulation increased the vesicle size of macropinosomes. The diameter of macropinosomes were analyzed at 1.5 h after TRD3S **(A)** or TRD70S **(B)** injected. P-values of significance (indicated with asterisks, $^*p < 0.05$, $^{***}p < 0.001$) were calculated using student t-test.

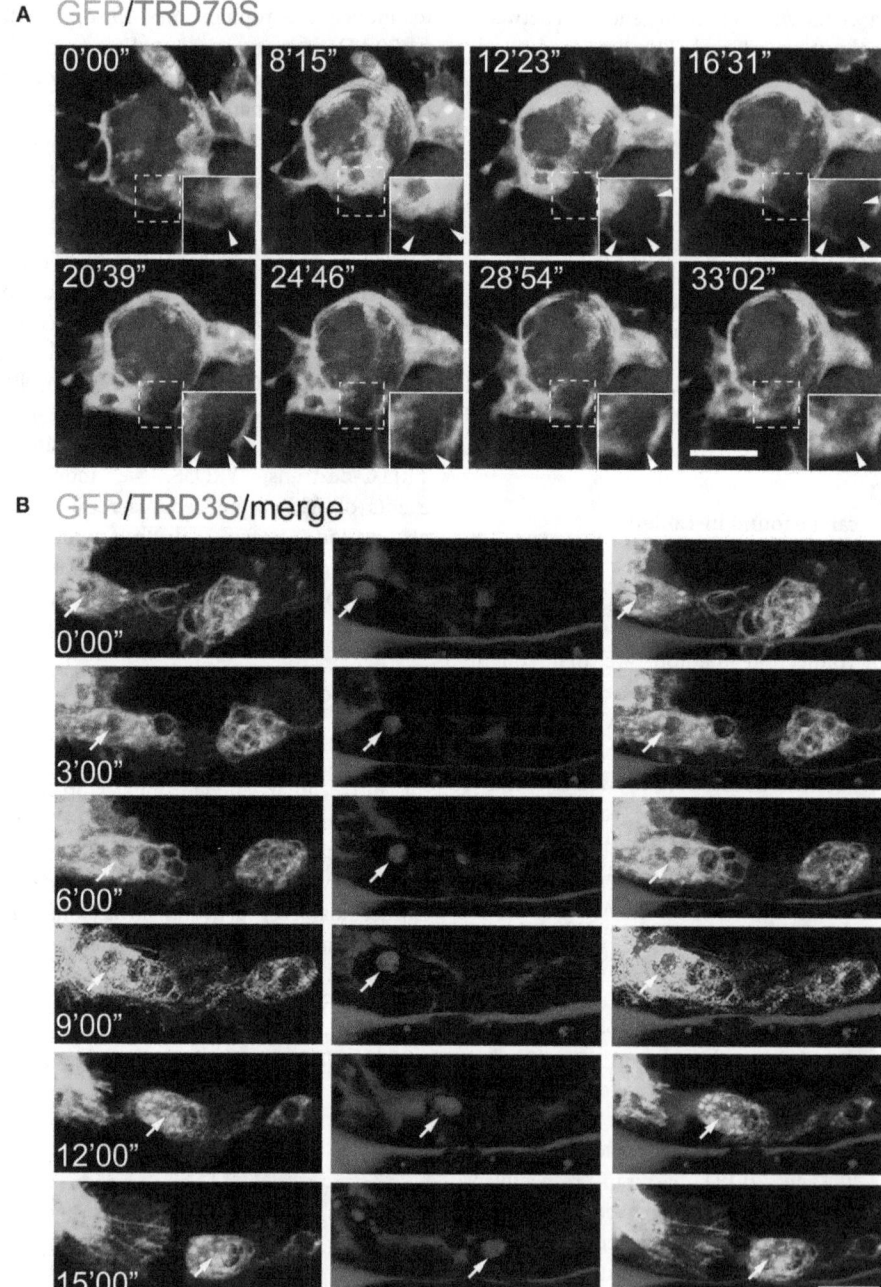

FIGURE 4 | Recording of cellular and subcellular behaviors using the ATPγS microinjection-based imaging system. **(A)** Dynamic changes of macropinosomes, from formation to fusion. Note that macropinosomes (red) were formed from the surface membrane ruffles and internalized after enclosure (arrowheads). The newly formed macropinosomes were indicated by dashed boxes and corresponding enlarged parts were presented at right corners. **(B)** TRD3S macropinosomes (red) transported within migrating hemocytes (green). The arrow indicates the location of a trafficking vesicle. Scale bars, 10 μm.

capable to induce larger macropinosomes engulfed by *Drosophila* embryonic hemocytes using different molecular weights of dextrans (**Figures 2, 3**). In addition, small dextrans are more efficient for labeling macropinosomes. Moreover, this method is feasible to investigate cellular events of macropinosomes, such as generation, fusion, and trafficking (**Figure 4**).

Although macropinocytosis is a nonselective process in cell culture systems, natural structures in living tissues may filter molecules and cause different diffusion properties in different size dextrans. Therefore, we tried to find out which size of dextran would be suitable for our system. In our experiment, dextrans ranged from 3-kDa to 70-kDa were all capable of labeling macropinosomes. However, small dextran (TRD3S) labeled macropinosomes could be seen in almost 100% of injected embryos, whereas large dextran (TRD70S) could only successfully label macropinosomes in about 40% of embryos under same conditions (**Figure 2B**). These data suggest that small molecules are more efficient for labeling macropinosomes in our system. In addition, although induced by different-sized dextrans, each hemocyte generated approximate numbers of macropinosomes. Taken together, application of larger dextran will reduce the success rate of induction without changing the uptake ability of hemocytes. It also suggests a potential filter effect of extracellular matrix in live tissues.

There are some difference of macropinocytosis between *in vivo* and *in vitro* systems. In previously studies, cultured macrophages and microglia exhibit impressive capabilities in internalization of extracellular fluids by macropinosomes (Racoosin and Swanson, 1993; Chen et al., 2015; Canton et al., 2016; Fu et al., 2016), which could generate within 1 min followed by centripetally migration and rapid shrink (Racoosin and Swanson, 1993; Lee and Knecht, 2002). It is a relatively short window to observe each phase. By contrast, *in vivo* macropinocytosis takes longer to generate macropinosomes in about 30 min (**Figure 3A**), providing enough time for observation. In addition, primary macrophages and the Drosophila hemocyte S2 cell line generate macropinosomes with high density (Gupta et al., 2009; Canton et al., 2016), raising the difficulty for observation of each vesicle. In contrast, no more than 5 macropinosomes were observed in our *in vivo* model. The sparse labeling provides a convenient approach to distinguish and monitor macropinosomes.

Combined with genetic tools and other strategies, this method is suitable to investigate molecular functions in macropinocytosis. For instance, using *srp-GAL4;UAS-tau-GFP*

transgenic *Drosophila*, microtubules were visualized around ATPγS induced macropinosomes (**Figure 2F**). Together with previous *in vitro* studies (Gilberti and Knecht, 2015), our results provide essential *in vivo* cues supporting that microtubule-associated structures may regulate macropinosomes formation and subsequent processes.

To expend applications, the method could be used to: (1) reveal the underlying mechanism in macropinocytosis; (2) uncover the way by which macrophages/microglia used for pathogens internalization; (3) develop new approaches for drug delivery via macropinocytosis. Taken together, this method provides novel sights for *in vivo* investigation of macropinocytosis and associated processes.

AUTHOR CONTRIBUTIONS

LC and YL designed the research. LC, DC, JC, TZ, ZD, and YL performed the experiments together; LC, JC, and YL wrote the paper; LZ and HL helped with animal raising and experimental preparation; LC with JC analyzed the data; YL supervised the entire study. All authors discussed the results and commented on the manuscript.

ACKNOWLEDGMENTS

We thank Prof. Henry Sun and Prof. Yongqing Zhang for providing transgenic fly stocks. We are grateful to the Core Facilities of Zhejiang University Institute of Neuroscience for technical assistance. This work was supported by grants from the Major State Basic Research Program of China (2016YFA0501000), the National Natural Science Foundation of China (31501128, 31490590, and 31490592).

REFERENCES

Bloomfield, G., and Kay, R. R. (2016). Uses and abuses of macropinocytosis. *J. Cell Sci.* 129, 2697–2705. doi: 10.1242/jcs.176149

Brand, A. H., and Perrimon, N. (1993). Targeted gene expression as a means of altering cell fates and generating dominant phenotypes. *Development* 118, 401–415.

Brust-Mascher, I., and Scholey, J. M. (2009). Microinjection techniques for studying mitosis in the *Drosophila melanogaster* syncytial embryo. *J. Vis. Exp.* 31:1382. doi: 10.3791/1382

Canton, J., Schlam, D., Breuer, C., Gutschow, M., Glogauer, M., and Grinstein, S. (2016). Calcium-sensing receptors signal constitutive macropinocytosis and facilitate the uptake of NOD2 ligands in macrophages. *Nat. Commun.* 7:11284. doi: 10.1038/ncomms11284

Chen, C., Li, H. Q., Liu, Y. J., Guo, Z. F., Wu, H. J., Li, X., et al. (2015). A novel size-based sorting mechanism of pinocytic luminal cargoes in microglia. *J. Neurosci.* 35, 2674–2688. doi: 10.1523/JNEUROSCI.4389-14.2015

Chubb, J. R., Wilkins, A., Thomas, G. M., and Insall, R. H. (2000). The Dictyostelium RasS protein is required for macropinocytosis, phagocytosis and the control of cell movement. *J. Cell Sci.* 113, 709–719.

Cisneros-Mejorado, A., Perez-Samartin, A., Gottlieb, M., and Matute, C. (2015). ATP signaling in brain: release, excitotoxicity and potential therapeutic targets. *Cell. Mol. Neurobiol.* 35, 1–6. doi: 10.1007/s10571-014-0092-3

Commisso, C., Davidson, S. M., Soydaner-Azeloglu, R. G., Parker, S. J., Kamphorst, J. J., Hackett, S., et al. (2013). Macropinocytosis of protein is an amino acid supply route in Ras-transformed cells. *Nature* 497, 633–637. doi: 10.1038/nature12138

Diken, M., Kreiter, S., Selmi, A., Britten, C. M., Huber, C., Tureci, O., et al. (2011). Selective uptake of naked vaccine RNA by dendritic cells is driven by macropinocytosis and abrogated upon DC maturation. *Gene Ther.* 18, 702–708. doi: 10.1038/gt.2011.17

Fu, P., Tang, R., Yu, Z., Li, C., Chen, X., Xie, M., et al. (2016). Rho-associated kinase inhibitors promote microglial uptake via the ERK signaling pathway. *Neurosci. Bull.* 32, 83–91. doi: 10.1007/s12264-016-0013-1

Gilberti, R. M., and Knecht, D. A. (2015). Macrophages phagocytose nonopsonized silica particles using a unique microtubule-dependent pathway. *Mol. Biol. Cell* 26, 518–529. doi: 10.1091/mbc.E14-08-1301

Gobeil, L. A., Lodge, R., and Tremblay, M. J. (2013). Macropinocytosis-Like HIV-1 internalization in macrophages is CCR5 dependent and leads to efficient but delayed degradation in endosomal compartments. *J. Virol.* 87, 735–745. doi: 10.1128/Jvi.01802-12

Grimmer, S., van Deurs, B., and Sandvig, K. (2002). Membrane ruffling and macropinocytosis in A431 cells require cholesterol. *J. Cell Sci.* 115(Pt 14), 2953–2962. doi: 10.1016/j.yexcr.2007.02.012

Gupta, G. D., Swetha, M. G., Kumari, S., Lakshminarayan, R., Dey, G., and Mayor, S. (2009). Analysis of endocytic pathways in Drosophila cells reveals a conserved role for GBF1 in internalization via GEECs. *PLoS ONE* 4:e6768. doi: 10.1371/journal.pone.0006768

Haraga, A., Ohlson, M. B., and Miller, S. I. (2008). Salmonellae interplay with host cells. *Nat. Rev. Microbiol.* 6, 53–66. doi: 10.1038/nrmicro1788

Kerr, M. C., and Teasdale, R. D. (2009). Defining macropinocytosis. *Traffic* 10, 364–371. doi: 10.1111/j.1600-0854.2009.00878.x

Lee, E., and Knecht, D. A. (2002). Visualization of actin dynamics during macropinocytosis and exocytosis. *Traffic* 3, 186–192. doi: 10.1034/j.1600-0854.2002.030304.x

Li, H. Q., Chen, C., Dou, Y., Wu, H. J., Liu, Y. J., Lou, H. F., et al. (2013). P2Y4 receptor-mediated pinocytosis contributes to amyloid beta-induced self-uptake by microglia. *Mol. Cell. Biol.* 33, 4282–4293. doi: 10.1128/MCB.00544-13

Lim, J. P., and Gleeson, P. A. (2011). Macropinocytosis: an endocytic pathway for internalising large gulps. *Immunol. Cell Biol.* 89, 836–843. doi: 10.1038/icb.2011.20

Liu, Z., and Roche, P. A. (2015). Macropinocytosis in phagocytes: regulation of MHC class-II-restricted antigen presentation in dendritic cells. *Front. Physiol.* 6:1. doi: 10.3389/fphys.2015.00001

Lohof, A. M., Quillan, M., Dan, Y., and Poo, M. M. (1992). Asymmetric modulation of cytosolic cAMP activity induces growth cone turning. *J. Neurosci.* 12, 1253–1261. doi: 10.1523/jneurosci.12-04-012 53.1992

Miller, J. M., Oligino, T., Pazdera, M., Lopez, A. J., and Hoshizaki, D. K. (2002). Identification of fat-cell enhancer regions in *Drosophila melanogaster. Insect Mol. Biol.* 11, 67–77. doi: 10.1139/g95-065

Racoosin, E. L., and Swanson, J. A. (1993). Macropinosome maturation and fusion with tubular lysosomes in macrophages. *J. Cell Biol.* 121, 1011–1020. doi: 10.1083/jcb.121.5.1011

Reiter, L. T., and Bier, E. (2002). Using Drosophila melanogaster to uncover human disease gene function and potential drug target proteins. *Exp. Opin. Ther. Targets* 6, 387–399. doi: 10.1517/14728222.6.3.387

Swanson, J. A., and Watts, C. (1995). Macropinocytosis. *Trends Cell Biol.* 5, 424–428.

Tepass, U., Fessler, L. I., Aziz, A., and Hartenstein, V. (1994). Embryonic origin of hemocytes and their relationship to cell death in *Drosophila. Development* 120, 1829–1837.

Wu, H. J., Liu, Y. J., Li, H. Q., Chen, C., Dou, Y., Lou, H. F., et al. (2014). Analysis of microglial migration by a micropipette assay. *Nat. Protoc.* 9, 491–500. doi: 10.1038/nprot.2014.015

5

Dcf1 Deficiency Attenuates the Role of Activated Microglia During Neuroinflammation

Jiao Wang[1], Jie Li[1], Qian Wang[1], Yanyan Kong[2], Fangfang Zhou[1], Qian Li[1], Weihao Li[1], Yangyang Sun[1], Yanli Wang[3], Yihui Guan[2], Minghong Wu[4]* and Tieqiao Wen[1]*

[1] Laboratory of Molecular Neural Biology, School of Life Sciences, Shanghai University, Shanghai, China, [2] Positron Emission Computed Tomography Center, Huashan Hospital, Fudan University, Shanghai, China, [3] Institute of Nanochemistry and Nanobiology, Shanghai University, Shanghai, China, [4] Shanghai Applied Radiation Institute, School of Environmental and Chemical Engineering, Shanghai University, Shanghai, China

*Correspondence:
Minghong Wu
mhwu@shu.edu.cn
Tieqiao Wen
wtq@t.shu.edu.cn

Microglia serve as the principal immune cells and play crucial roles in the central nervous system, responding to neuroinflammation via migration and the execution of phagocytosis. Dendritic cell-derived factor 1 (Dcf1) is known to play an important role in neural stem cell differentiation, glioma apoptosis, dendritic spine formation, and Alzheimer's disease (AD), nevertheless, the involvement of the *Dcf1* gene in the brain immune response has not yet been reported. In the present paper, the RNA-sequencing and function enrichment analysis suggested that the majority of the down-regulated genes in $Dcf1^{-/-}$ (Dcf1-KO) mice are immune-related. *In vivo* experiments showed that *Dcf1* deletion produced profound effects on microglial function, increased the expression of microglial activation markers, such as ionized calcium binding adaptor molecule 1 (Iba1), Cluster of Differentiation 68 (CD68) and translocator protein (TSPO), as well as certain proinflammatory cytokines (Cxcl1, Ccl7, and IL17D), but decreased the migratory and phagocytic abilities of microglial cells, and reduced the expression levels of some other proinflammatory cytokines (Cox-2, IL-1β, IL-6, TNF-α, and Csf1) in the mouse hippocampus. Furthermore, *in vitro* experiments revealed that in the absence of lipopolysaccharide (LPS), the majority of microglia were ramified and existed in a resting state, with only approximately 10% of cells exhibiting an amoeboid-like morphology, indicative of an activated state. LPS treatment dramatically increased the ratio of activated to resting cells, and *Dcf1* downregulation further increased this ratio. These data indicated that *Dcf1* deletion mediates neuroinflammation and induces dysfunction of activated microglia, preventing migration and the execution of phagocytosis. These findings support further investigation into the biological mechanisms underlying microglia-related neuroinflammatory diseases, and the role of *Dcf1* in the immune response.

Keywords: *Dcf1*, microglia, neuroinflammation, cytokines, migration, phagocytosis

Abbreviations: *Ccl7*, chemokine (C-C motif) ligand 7; CD68, Cluster of Differentiation 68; *Cox-2*, cyclooxygenase-2; *Csf1*, colony stimulating factor 1; *Cxcl1*, chemokine (C-X-C motif) ligand 1; DAPI, 4′,6-diamidino-2-phenylindole; *Dcf1*, dendritic cell-derived factor 1; Dcf1-KO, $Dcf1^{-/-}$; ^{18}F-DPA-714, N,N-Diethyl-2-(2-(4-(2-[^{18}F]fluoroethoxy) phenyl)-5,7-dimethylpyrazolo [1,5-a]pyrimidin-3-yl) acetamide; Iba1, ionized calcium-binding adapter molecule 1; *IL-1β*, interleukin-1β; *IL-6*, interleukin-6; *IL17D*, interleukin 17D; LPS, lipopolysaccharide; PBS, phosphate-buffered saline; PET, positron emission tomography; qPCR, real-time quantitative PCR; RT, room temperature; *Tnfsf11*, tumor necrosis factor (ligand) superfamily, member 11; TNF-α, tumor necrosis factor alpha; TSPO, translocator protein; WT, wile-type.

INTRODUCTION

Neuroinflammation is widely regarded as a chronic innate immune response in the brain and a potentially pathogenic factor in a number of neurodegenerative diseases such as Alzheimer's disease (AD), as well as traumatic brain injury (Panicker et al., 2015; Andreasson et al., 2016; Fernandez-Calle et al., 2017). Recently, due to its key signaling steps in the initiation of immune activation, greater attention has been paid to the potential of neuroinflammation as a therapeutic target (Panicker et al., 2015).

Microglia, which comprise approximately 20% of all glial cells, are the principal immune cells in the central nervous system and play a critical role in host defense against invading microorganisms and neoplastic cells (Rio-Hortega, 1932; Gonzalez-Scarano and Baltuch, 1999; von Bernhardi et al., 2015). In the normal adult brain, microglia display a remarkable branched, ramified morphological phenotype and are dispersed throughout the entire brain (Amor and Woodroofe, 2014). Upon injury, microglia undergo transformation to an amoeboid-like morphology, migrate to the site of injury, and execute phagocytosis (Andreasson et al., 2016). Microglia can also be activated by pathogen-associated molecules. Moreover, microglia also play a role in the regulation of activity-triggered synaptic plasticity and the remodeling of neural circuits, and further contribute to learning and memory (Koeglsperger et al., 2013; Parkhurst et al., 2013; Sofroniew, 2015). In the AD mouse brain, microglia have been shown to be clustered at the sites of Aβ plaques, with an activated, amoeboid-like morphology (Eisenberg and Jucker, 2012). Despite microglia being tightly packed and ubiquitously positioned in the tissue of young mice, coverage is impaired in old mice, and particularly more severely in 9-month-old $APP_{Sw,Ind}$ Tg mice, leaving tissue devoid of microglial processes (Baron et al., 2014). It has been suggested that inflammation may be involved in the pathogenesis of AD (Miklossy, 2008). In the aged brain, microglia extend ramified processes into the surrounding tissue (Mosher and Wysscoray, 2014). A recent study using two-photon microscopy in the living brain of murine models of AD to examine microglial behavior, reported data showing that microglia in the aged brain were less motile and had fewer processes (Meyer-Luehmann et al., 2008), which supports the notion that aging is accompanied by impaired microglial function (Streit et al., 2008). However, despite recent progress, the understanding of the cellular and molecular mechanisms that mediate microglial activation is still far from comprehensive.

Dendritic cell-derived factor 1 (Dcf1) is a membrane protein that plays an important role in neural stem cell differentiation, glioma apoptosis, dendritic spine formation, and social interaction, as well as amyloid precursor protein metabolism (Wen et al., 2002; Wang et al., 2008; Ullrich et al., 2010; Xie et al., 2014; Liu et al., 2017a,b). Downregulation of the *Dcf1* gene facilitates differentiation of neural stem cells into astrocytes (Wang et al., 2008) and deletion of *Dcf1* leads to dendritic spine dysplasia in the mouse hippocampus (Liu et al., 2017a). Therefore, *Dcf1* is an important regulator of neural development.

It is known that certain neural development-regulating molecules also play important roles in the regulation of the immune response in the brain (Garay and McAllister, 2010). To explore the function of *Dcf1* in the neural immune system, we investigated the effect of *Dcf1* deletion on the activation of microglia and expression of proinflammatory cytokines under different conditions *in vitro* and *in vivo*. We found that *Dcf1* deletion produced profound effects on microglial function, increased the expression of microglial activation markers such as TSPO, Iba1, and CD68 as well as some proinflammatory cytokines, but decreased the migration and phagocytosis abilities of microglial cells and the expression levels of other proinflammatory cytokines.

MATERIALS AND METHODS

Positron Emission Tomography (PET)

PET experiments were performed using a Siemens Inveon PET/CT system (Siemens Medical Solutions, Knoxville, United States) and conducted by the Huashan Hospital of China, according to the standard protocols and procedures (Kong et al., 2016). ^{18}F-DPA-714 was given via the catheter system intravenously in a slow bolus. Isoflurane is an inhaled anesthetic that is mobilized through the respiratory tract and into the body of mice under the influence of oxygen. Dynamic PET was performed for 60 min on isoflurane-anesthetized male nude mice after intravenous injection of ^{18}F-DPA-714. The experiments were carried out in compliance with national laws for the conduct of animal experimentation and were approved by the Animal Ethics Committee of Shanghai University.

Immunohistochemical Staining

Brain samples from WT and *Dcf1*-KO mice (C57BL/6 male mice, 2–3 months-old) were cut by frozen sectioning. Slices were rinsed 3 times with PBS and permeabilized with 0.1% Triton X-100 in PBS for 40 min. The slices were subsequently blocked in 5% bovine serum albumin (Invitrogen, United States) in PBS at RT for 2 h, followed by incubation with a goat anti-Iba1 monoclonal primary antibody (1:500, Abcam, United States) at 4°C overnight. The following day, the slices were washed 3 times with PBS, incubated sequentially with a donkey anti-goat IgG secondary antibody Alexa 488 (1:1000, Abcam, United States) at RT for 2 h and the nuclear stain DAPI (Invitrogen, United States) at RT for 10 min, and finally washed 3 times with PBS. Fluorescence intensity was detected using a Zeiss LSM710 fluorescence microscope. All animals were treated in accordance with the guidelines of the Society for Neuroscience Ethics Committee on Animal Research. The study design was approved by the Animal Ethics Committee of Shanghai University.

Cell Culture

BV2 cells, a mouse microglia cell line, were cultured in Dulbecco's Modified Eagle Medium (Invitrogen, United States) supplemented with 10% fetal bovine serum (Invitrogen, United States) and 1% penicillin/streptomycin (Invitrogen, United States), and maintained at 37°C in a 95% humidified atmosphere with 5% CO_2. At approximately 90% confluence, the cells were detached with 0.1% trypsin-EDTA (Invitrogen, United States), seeded onto appropriate plates with fresh medium, and incubated overnight.

Transfection

BV2 cells were seeded onto 24-well plates at a density of 1×10^5 cells/well and cultured overnight at 37°C in an atmosphere of 5% CO_2. The following day, cells were transfected with the psiRNA-hH1neo plasmid or the psiRNA-*Dcf1* plasmid using *Lipofectamine*™ 2000 (Invitrogen, United States), according to the manufacturer's protocol.

Observation of BV2 Microglia Cell Morphology

BV2 cells were cultured on 24-well plates and transfected as described above. 24 h post-transfection, cells were stimulated with 1000 ng/ml LPS for 12 h, followed by collection of bright-field images using a Nikon microscope. The cells were then rinsed with PBS, fixed in 4% paraformaldehyde in PBS at RT for 10 min, and permeabilized with 0.1% Triton X-100 in PBS for 10 min. The cells were subsequently blocked in 2% bovine serum albumin in PBS at RT for 1 h, followed by incubation with a goat anti-Iba1 monoclonal primary antibody (1:500, Abcam, United States) at 4°C overnight. The following day, the cells were washed 3 times with PBS, and incubated sequentially with a donkey anti-goat IgG secondary antibody Alexa 488 (1:1000, Abcam, United States) at RT for 2 h, the cytoskeleton red fluorescent probe ActinRed (1:50, KeyGEN BioTECH, China) at RT for 20 min, and DAPI at RT for 5 min, and finally washed 3 times with PBS. Fluorescence intensity was detected using a Zeiss LSM710 fluorescence microscope.

Total RNA Extraction, cDNA Synthesis, and Real-Time Quantitative PCR (qPCR)

BV2 cells were cultured on 24-well plates and transfected as described above. 24 h post-transfection, cells were treated with 1000 ng/ml LPS for 12 h. Subsequently, the total RNA was extracted using a total RNA extraction kit (Promega, United States), according to the manufacturer's protocol. The total RNA in the WT and *Dcf1*-KO hippocampal tissue was extracted in the same manner. The concentration of RNA was determined by measuring the absorbance at 260 nm, and 2 μg RNA was used for cDNA synthesis using an RT master mix (TaKaRa, Japan). QPCR amplification was performed in at least triplicate using a mixture of Top Green qPCR super mix (Transgen, China), cDNA samples, and designated primers (**Table 1**). The relative gene expression was calculated by comparing the CT value of the gene of interest with that of *Gapdh*, the internal control.

Western Blotting

The total protein in the WT and *Dcf1*-KO hippocampal tissue was extracted using cell lysis buffer (Beyotime, China), according to the manufacturer's protocol. For protein extraction from BV2 cells, transfected cells cultured on 24-well plates were stimulated

TABLE 1 | List of primers used for qPCR.

Gene name	Primer sequence (5′–3′)
Dcf1	Upstream: CGCTGCTGCTGTTGACTATG
	Downstream: GTAGGTGTGCAAGGGGTAGG
Ccl7	Upstream: GCTGCTTTCAGCATCCAAGTG
	Downstream: CCAGGGACACCGACTACTG
Cxcl1	Upstream: CTGGGATTCACCTCAAGAACATC
	Downstream: CAGGGTCAAGGCAAGCCTC
Csf1	Upstream: ATGAGCAGGAGTATTGCCAAGG
	Downstream: TCCATTCCCAATCATGTGGCTA
IL17D	Upstream: AGCACACCCGTCTTCTCTC
	Downstream: GCTGGAGTTCGCACTGTCC
Tnfsf11	Upstream: CAGCATCGCTCTGTTCCTGTA
	Downstream: CTGCGTTTTCATGGAGTCTCA
Cox-2	Upstream: CAGTTTATGTTGTCTGTCCAGAGTTTC
	Downstream: CCAGCACTTCACCCATCAGTT
IL-6	Upstream: AACGATGATGCACTTGCAGA
	Downstream: CTCTGAAGGACTCTGGCTTTG
IL-1β	Upstream: CTTCCTTGTGCAAGTGTCTG
	Downstream: CAGGTCATTCTCATCACTGTC
TNF-α	Upstream: AAATTCGAGTGACAAGCCTGTAG
	Downstream: GAGAACCTGGGAGTAGACAAGGT
Gapdh	Upstream: TCACCACCATGGAGAAGGC
	Downstream: GCTAAGCAGTTGGTGGTGCA

with 1000 ng/ml LPS for 12 h. Following the treatment, cells were washed twice with ice-cold PBS and the total protein was extracted using cell lysis buffer (Beyotime, China), according to the manufacturer's protocol. Protein samples were separated by sodium dodecyl sulfate polyacrylamide gel electrophoresis and electroblotted onto nitrocellulose membranes. The membranes were blocked with 5% bovine serum albumin in PBS at RT for 1 h and then incubated with the following primary antibodies at 4°C overnight: goat anti-Iba1 (1:1000, Abcam, United States), rabbit anti-IL17D (1:500, Abcam, United States), rabbit anti-CD68 (1:500, Abways, China), rabbit anti-Cox-2 (1:500, Wanleibio, China), rabbit anti-IL-6 (1:500, Wanleibio, China), rabbit anti-IL-1β (1:500, Wanleibio, China), and rabbit anti-TNF-α (1:500, Wanleibio, China). The following day, the membranes were incubated with a mouse anti-GAPDH (1:1000, Abcam, United States) at RT for 1 h, followed by an infrared dye 700-conjugated goat anti-mouse IgG (1:10000, Zemed, United States) and either an infrared dye 800-conjugated goat anti-rabbit IgG (1:10000, Zemed, United States) or an infrared dye 700-conjugated donkey anti-goat IgG secondary antibody (1:10000, Zemed, United States) at RT for a further 1 h. Visualization and quantification was carried out using LI-COR Odyssey scanner and software (LI-COR Biosciences). The relative protein expression level was normalized to Gapdh of the same lane, and data were obtained from four independent immunoblots.

Cell Migration Assay

BV2 cells were seeded at a density of 1×10^5 cells per well on a 24-well plate and cultured for 24 h at 37°C with 5% CO_2, followed by

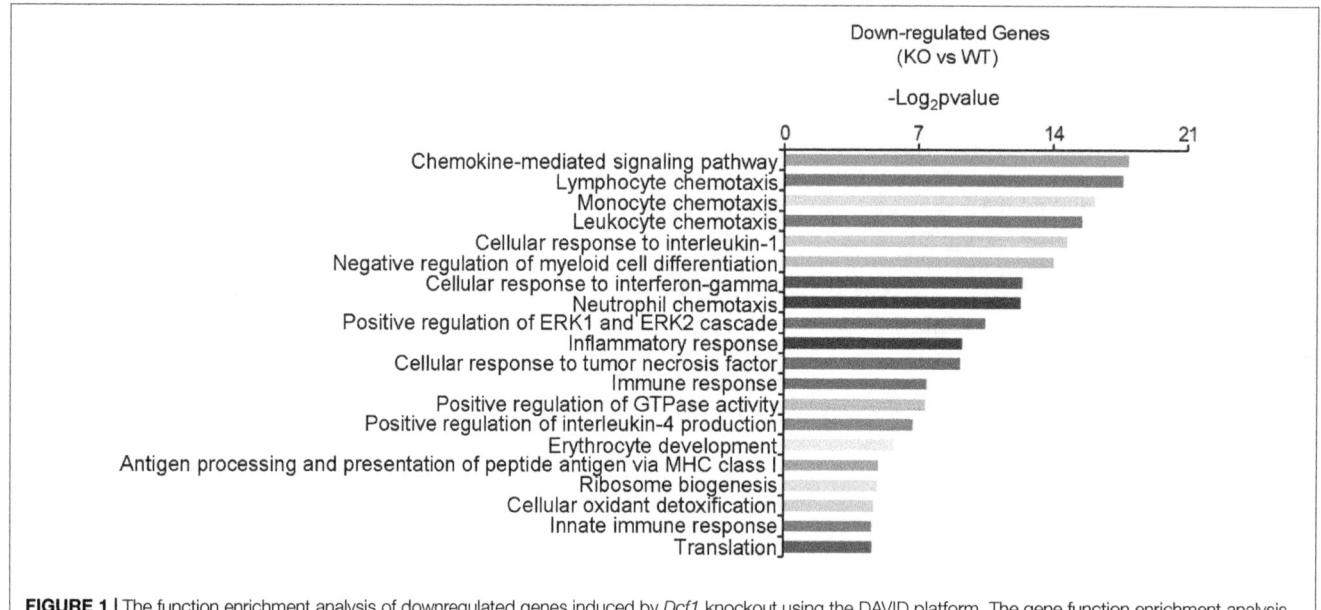

FIGURE 1 | The function enrichment analysis of downregulated genes induced by *Dcf1* knockout using the DAVID platform. The gene function enrichment analysis of downregulated genes from RNA-sequencing results in WT and KO mice. The DAVID platform was used for analysis. The results showed that the majority of the downregulated genes were immune-related.

transfection as described above. 36 h post-transfection, a wound healing assay was used to evaluate alterations in the migration rate. Briefly, lineation was carried out at the central region of cell growth in each well using a P-20 pipette tip, and the cells were observed every 12 h for 48 h using a Nikon Ti-S fluorescence microscope. The results were analyzed using the Image Pro Plus software[1].

Cell Phagocytosis Assay

BV2 cells were seeded at a density of 1×10^5 cells per well on a 24-well plate and cultured for 24 h at 37°C with 5% CO_2, followed by transfection as described above. 36 h post-transfection, fresh culture medium containing 5 µl grapheme quantum dots (2–3 µm) was added to each well and incubated at 37°C for 5 min, followed by a PBS wash, fixation with 4% paraformaldehyde for 30 min and permeabilization with 0.1% Triton X-100 in PBS for 10 min at RT. Cells were subsequently incubated with ActinRed at a dilution of 1:50 for 30 min at RT and washed three times with PBS. The phagocytic activity of the cells was evaluated by confocal microscopy.

Statistical Analysis

All data were analyzed using the Graphpad Prism software and were presented as the mean ± SEM. The mRNA and protein expression levels of WT and *Dcf1*-KO mice were analyzed using a *t*-test. The microscope images were analyzed using the Image Pro Plus software. The changes in cell morphology, mRNA and protein expression levels, and the migratory and phagocytic capacities of BV2 cells were analyzed using a one-way Analysis of Variance. Significance was set to $p < 0.05$.

[1]http://dx.doi.org/10.17504/protocols.io.iascaee

RESULTS

Dcf1 Deletion Downregulates the Expression of Immune-Related Genes in the Hippocampus

In order to gain insight into the molecular activities with which *Dcf1* may be involved in the nervous system, we examined and compared the mRNA levels in the hippocampus of both WT and *Dcf1*-KO mice by RNA sequencing and function enrichment analysis using DAVID (The Database for Annotation, Visualization, and Integrated Discovery). We found that the majority of downregulated genes in *Dcf1*-KO mice were immune-related (**Figure 1**). Since microglia are the major components of the immune system in the brain, we hypothesized that *Dcf1* may regulate microglial function. To test this hypothesis, we assessed the effects of *Dcf1* deletion on microglial activation and the production of cytokines in microglial cells using *Dcf1*-KO mice. In addition, the effects of *Dcf1* downregulation by RNAi on the LPS-induced changes in morphology, migratory and phagocytic capacity, and the expression levels of proinflammatory cytokines, in cultured BV2 cells were evaluated as described below.

Dcf1 Deletion Induces Microglial Activation *in Vivo*

The activation of microglia is characterized by an increase in the expression level of TSPO (translocator protein) and Iba1 (ionized calcium binding adaptor molecule 1). Recently, a technique was developed to monitor the expression level of TSPO by PET imaging of a radiolabeled TSPO-binding tracer, ^{18}F-DPA-714 (Auvity et al., 2017; Rizzo et al., 2017; Saba et al., 2017). Using this PET imaging technique, we found that TSPO expression was significantly increased in certain brain regions of *Dcf1*-KO mice, including the hippocampus, as compared with that of WT mice (**Figures 2A–C**), indicating that *Dcf1* deletion upregulated the ratio of activated microglia to resting microglia. Consistently, the expression level of Iba1 was also significantly increased in *Dcf1*-KO mice, as reflected by immunostaining and Western blotting analysis, which both show that the mean density in KO mice is 1.5 times that in WT mice (**Figures 2D–F**). Furthermore, Western blotting analysis shows that the expression level of CD68 was significantly increased in *Dcf1*-KO mice (**Figure 2G**).

Dcf1 Deletion Induces Abnormal Expression of Proinflammatory Cytokines *in Vivo*

Cytokines have been reported to promote neuronal differentiation and remodeling in the brain (Jeon and Kim, 2016). Many reports have shown that proinflammatory cytokines were dramatically increased in activated microglial cells (Hwang et al., 2014). In the central nervous system, central cytokines such as IL-6, TNF-α, and IL-1β are secreted from microglia, and are considered to be involved in neuronal development and neuroplasticity (Moynagh, 2005; Jeon and Kim, 2016). Thus, to investigate the molecular differences in proinflammatory cytokines between WT and *Dcf1*-KO mouse hippocampal tissue, we examined both the mRNA and protein expression levels of Cox-2, IL-1β, Tnfsf11, Cxcl1, Ccl7, IL-6, IL17D, TNF-α, and Csf1. As illustrated in **Figure 3A**, the mRNA levels of *Ccl7* and *IL17D* dramatically increased by 2-fold in KO as compared with WT mice, in addition to *Tnfsf11* and *Cxcl1*, although the latter two were not significantly increased. Moreover, the mRNA levels of *Cox-2, IL-1β, IL-6, TNF-α,* and *Csf1* were significantly decreased by approximately 50% in the *Dcf1*-KO mice. Western blotting was used to verify these changes. As can be seen in **Figures 3B,C**, the protein expression levels of Cox-2, IL-1β, IL-6, and TNF-α were consistently significantly reduced in the *Dcf1*-KO mice as compared with the WT mice, and IL17D was significantly increased by approximately 40%.

Downregulation of *Dcf1* Alters the LPS-Induced Morphological Change in Cultured BV2 Microglial Cells

To better understand the role that *Dcf1* plays in the changes in the microglial morphology induced by inflammatory stimulation, we investigated the effect of *Dcf1* downregulation on the morphological changes of cultured BV2 cells caused by LPS treatment. LPS is an outer membrane component of *Gram-negative* bacteria and a strong stimulator of microglial cells (Qin et al., 2004). As shown in Supplementary Figure S1 and Supplementary Table S13, psiRNA-*Dcf1* plasmid was used to knock down the *Dcf1* gene, and qPCR analysis of the *Dcf1* mRNA level revealed a significant decrease in the LPS + psiRNA-*Dcf1* group, with no significant differences seen among the other groups. Cell morphology was examined using immunostaining against ActinRed and Iba1. As shown in **Figure 4**, in the absence of LPS, the vast majority of cells were ramified and existed in a resting state, with approximately 10% of cells exhibiting an amoeboid-like morphology, indicative of the activated state. LPS

treatment dramatically increased the ratio of activated to resting cells, with *Dcf1* downregulation further increasing this ratio. This result is consistent with the elevated activation of microglia cells seen in *Dcf1*-KO mice.

Downregulation of *Dcf1* Affects the Expression of Cytokines *in Vitro*

To examine whether the silencing of *Dcf1* affected the inflammatory response at the molecular level, qPCR was performed in order to quantify the mRNA expression of the nine cytokines examined above *in vivo* studies. The relative abundance of each mRNA was expressed relative to *Gapdh*. As illustrated in **Figure 5A**, the mRNA levels of *Cox-2, IL-1β, Cxcl1, Ccl7, IL-6, TNF-α,* and *Csf1* were significantly increased in the LPS + vehicle group as compared with the group in the absence of LPS (blank). However, downregulation of *Dcf1* in the LPS + psiRNA-*Dcf1* group significantly decreased the levels of *Cox-2, IL-1β, IL-6, TNF-α,* and *Csf1*. In contrast, *Cxcl1* and *Ccl7* were dramatically increased, and IL17D was also elevated, although not significantly, as compared with the LPS + vehicle group. Western blotting was performed to further confirm whether downregulation of *Dcf1* affected the expression of these cytokines. As shown in **Figures 5B,C**, cells treated with LPS had increased expression of proinflammatory cytokines such as Cox-2, IL-1β, IL-6, and TNF-α, however, these four proinflammatory factors were significantly reduced upon *Dcf1* downregulation as compared with the LPS + vehicle group. These data demonstrated that LPS-stimulated BV2 microglial cells had increased the expression of proinflammatory cytokines, whereas downregulation of the *Dcf1* dramatically decreased the expression of most factors that were detected in LPS-stimulated BV2 cells, with the exception of Cxcl1, Ccl7, and IL17D. These results suggested that downregulation of *Dcf1* suppressed the expression of the majority of proinflammatory factors in activated BV2 microglial cells, supporting the previous results (**Figure 3**) showing differences in the expression of these genes between the hippocampal tissue from WT and *Dcf1*-KO mice.

Downregulation of *Dcf1* Decreases the Migratory Ability of Microglia

Microglial cells respond to neuroinflammation with the processes of migration (Andreasson et al., 2016) and phagocytosis (Dheen et al., 2007). It has been shown that once activated, microglia migrate toward injured areas, and that this process is controlled by the presence of cytokines and chemokines (Noda and Suzumura, 2012). It can be speculated therefore, that *Dcf1* may affect microglial migration; and thus, a wound-healing assay was employed to study the effect of *Dcf1* downregulation on the migratory ability of BV2 cells. The representative images show the scratched areas of BV2 cells in different groups from 0 to 48 h (**Figure 6A**). Upon deletion of *Dcf1*, the migration rate of BV2 cells was increased by approximately 2-fold as compared with the LPS + vehicle group at 48 h. Statistical analysis reveals that *Dcf1* downregulation significantly decreased the migratory ability of BV2 cells (**Figure 6B**), suggesting that *Dcf1* may be involved in LPS-stimulated microglial cell migration.

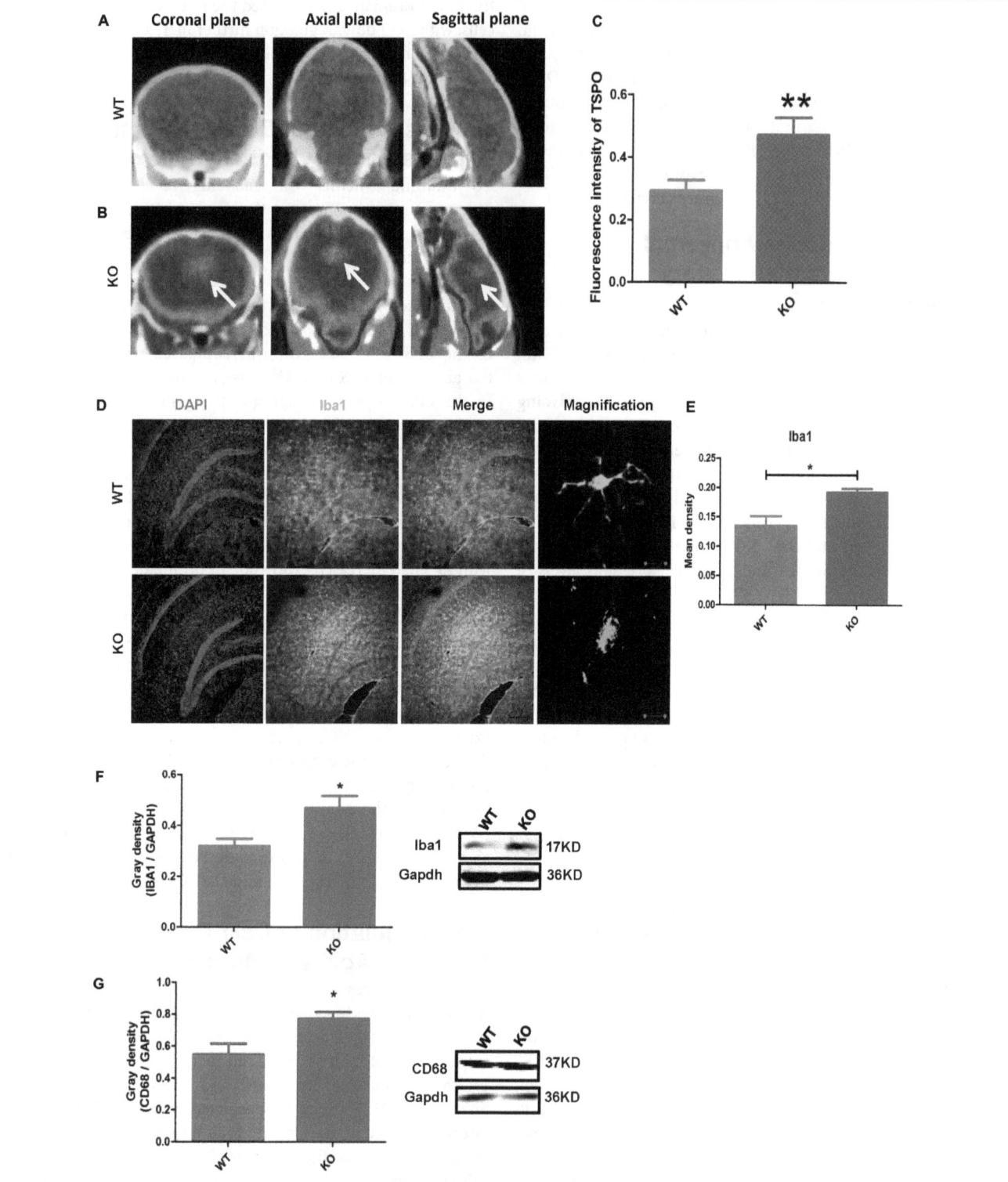

FIGURE 2 | *Dcf1* deletion induces activation of microglial cells *in vivo*. ¹⁸F-DPA-714 (green) was used to trace TSPO (a biomarker of microglia) by PET to observe the activity of microglia *in vivo*. **(A,B)** Brain observation by PET in WT **(A)** and *Dcf1*-KO **(B)** mice. White arrowhead denotes activated microglia. **(C)** Quantitation of the TSPO in WT and *Dcf1*-KO mouse brain (Supplementary Table S1). **(D)** Immunohistochemical observation of microglia from WT and *Dcf1*-KO mouse brain sections. Microglial cells were detected by the Iba1 biomarker (green), and the nuclei were counterstained with DAPI (blue). Scale bars represent 200 μm. Higher magnification of confocal images were shown in right panel. Scale bars, 10 μm. **(E)** Quantitation of the mean density of Iba1 staining in WT and *Dcf1*-KO mouse brain sections (Supplementary Table S2) (mean ± SEM). $n = 4$. *$p < 0.05$. Protein expression of Iba1 **(F)** and CD68 **(G)** in WT and *Dcf1*-KO mouse brain tissue (Supplementary Tables S3, S4). Quantification of protein expression levels normalized to Gapdh. Data are expressed as the mean ± SEM. $n = 3$. *$p < 0.05$; **$p < 0.01$.

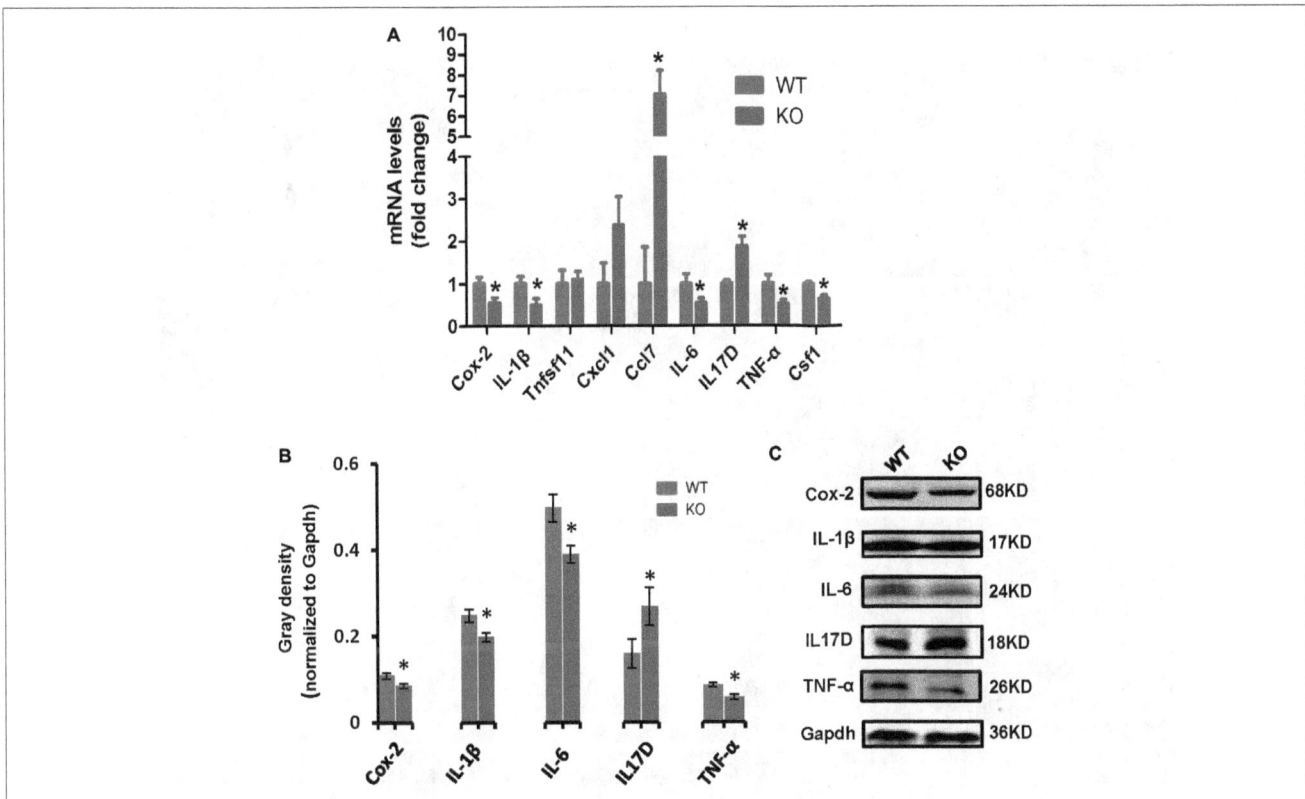

FIGURE 3 | Expression of proinflammatory cytokines in WT and *Dcf1*-KO mice. **(A)** The mRNA levels of *Cox-2, IL-1β, Tnfsf11, Cxcl1, Ccl7, IL-6, IL17D, TNF-α,* and *Csf1* were assessed by qPCR in WT and KO mice. The relative abundance of each mRNA was expressed relative to *Gapdh*. Data are expressed as the mean ± SEM. *n* = 3. *$p < 0.05$ (Supplementary Table S5). **(B)** Protein expression of Cox-2, IL-1β, IL-6, IL17D, and TNF-α were assessed by Western blotting in WT and KO mice. Quantification of protein expression levels normalized to Gapdh. Datas are expressed as the mean ± SEM. *n* = 3. *$p < 0.05$ (Supplementary Table S6). **(C)** Sample Western blotting shown for *Cox-2, IL-1β, IL-6, IL17D, TNF-α, and Gapdh.*

Downregulation of *D*cf1 in LPS-Activated Microglia Results in a Diminished Phagocytic Capacity

To explore the effect of *Dcf1* downregulation on the phagocytic ability of LPS-activated microglia, a cell phagocytosis assay was performed. Following transfection, cells were incubated with medium containing green fluorescent grapheme quantum dots, and subjected to confocal microscopy. The cell skeleton was labeled with ActinRed (red) to form a composite image with which to count phagocytic cells (**Figure 7A**). The ratio of phagocytic cells to total cell number was calculated to evaluate the phagocytic capacity (**Figure 7B**). Image analysis shows that compared with the blank, the phagocytic capacity of BV2 cells following treatment with LPS was increased by approximately 10%. Moreover, downregulation of *Dcf1* led to a significant decrease of approximately 50% in microglial phagocytic ability as compared with the LPS + vehicle group.

DISCUSSION

Here, we provide evidence of neuroinflammatory responses induced by *Dcf1* deficiency (**Figure 8**). The RNA-sequencing

and function enrichment analysis show that the majority of the downregulated genes in *Dcf1*-KO mice were immune-related (**Figure 1**), suggesting that *Dcf1* may play a role in brain immunity. Previous work has identified that microglial cells are responsible for surveillance immunity in the CNS and are activated in response to inflammation (Streit, 2002). TSPO is consistently raised in activated microglia of the CNS. In the present study, we found that *Dcf1* deficiency induced brain immunity and activation of microglial cells, as reflected by the upregulation of TSPO, IbaI, and CD68 *in vivo* (**Figure 2** and Supplementary Figure S2). *In vitro*, downregulation of *Dcf1* increased the morphological transformation of BV2 microglial cells to an amoeboid-like structure, which indicated an activated state (**Figure 4**). Following brain injury, microglial cells rapidly respond by activating the proinflammatory process, releasing inflammatory mediators and resolving the inflammatory response (Carniglia et al., 2017). The neuroinflammatory cytokines, Cxcl1, Ccl7, and IL17D were increased in *Dcf1*-KO mice (**Figure 3** and Supplementary Figure S3). Interestingly, a defect in *Dcf1* reduced the expression of other proinflammatory factors including Cox-2, IL-1β, IL-6, TNF-α, and Csf1 (**Figure 3**), implying that *Dcf1* influences multiple inflammatory responses. *In vitro*, downregulation of *Dcf1* decreased the expression of the majority detected cytokines, with the exception of Cxcl1,

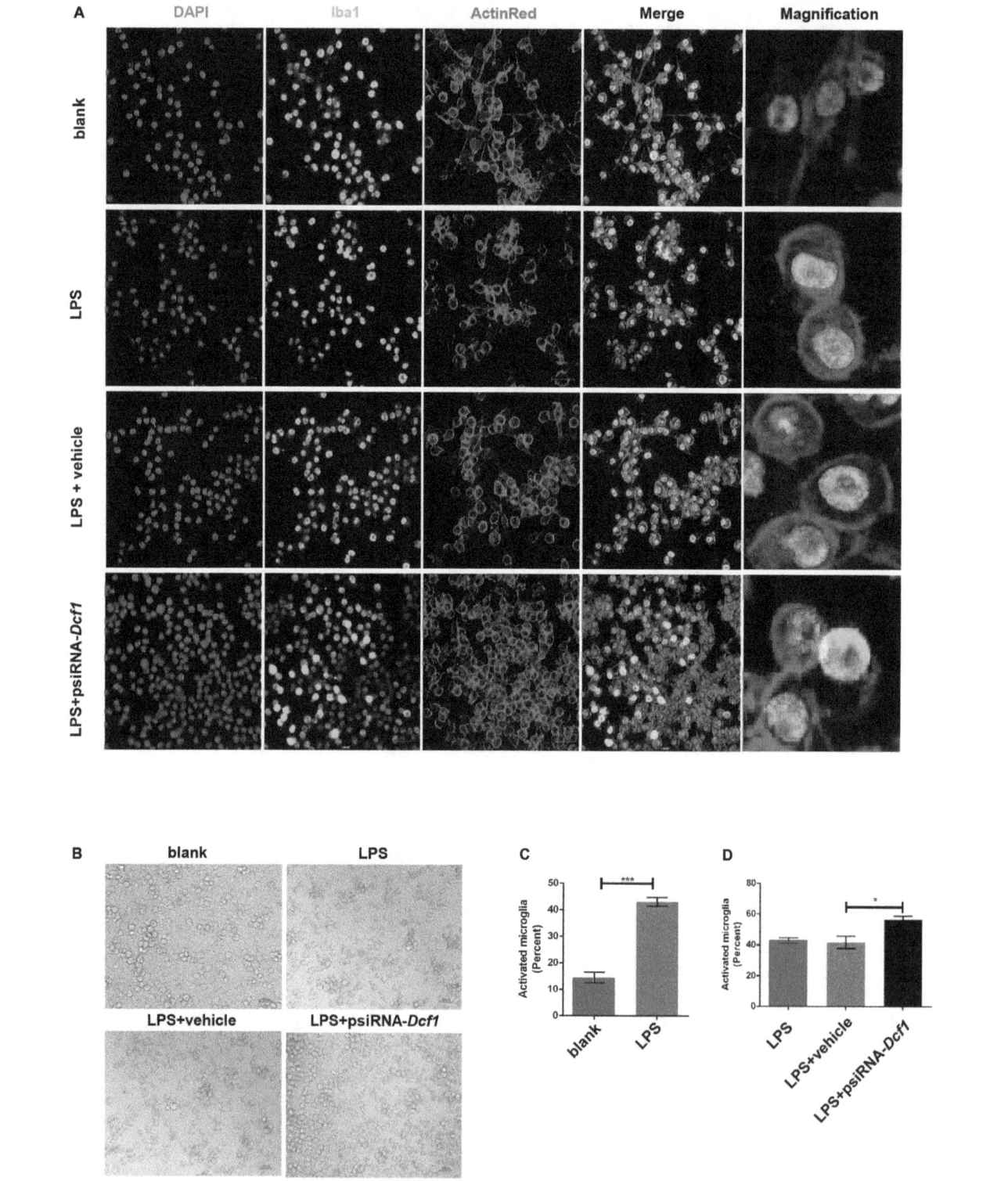

FIGURE 4 | Morphology of LPS-stimulated BV2 microglial cells *in vitro*. BV_2 microglial cells were transfected with the psiRNA-hH1neo plasmid or the psiRNA-*Dcf1* plasmid. 24 h post-transfection, BV2 microglia were stimulated with LPS (1000 ng/ml) and incubated for 12 h. **(A)** Immunofluorescence observation of the morphology in LPS-stimulated BV2 microglia *in vitro*. BV2 microglial cells were detected by the Iba1 marker (green), the cell skeleton by ActinRed (red), and the nuclei by DAPI (blue). Scale bars represent 50 μm. Higher magnification of confocal images were shown in right panel. Scale bars, 10 μm. **(B)** Bright field images of BV2 microglia. Scale bars represent 100 μm. **(C)** Comparison of the percentage of activated BV2 microglia stimulated with LPS (Supplementary Table S7). **(D)** Comparison of the percentage of LPS-activated BV2 microglia transfected with psiRNA-*Dcf1*. Data are expressed as the mean ± SEM. $n = 6$. $*p < 0.05$; $***p < 0.001$ (Supplementary Table S8).

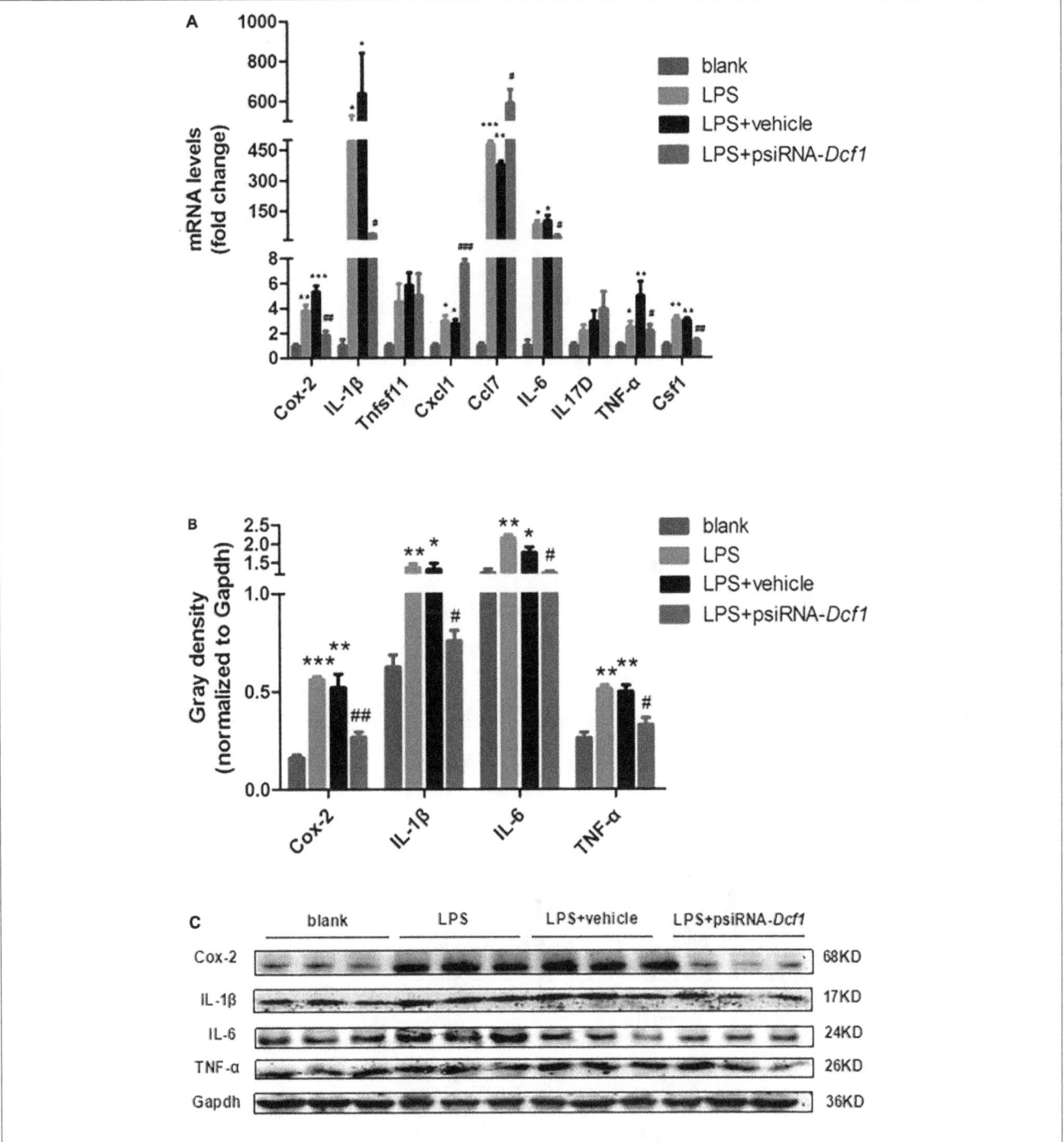

FIGURE 5 | Effects of *Dcf1* downregulation on proinflammatory cytokines expression in LPS-stimulated BV2 microglia. BV$_2$ microglial cells were transfected with the psiRNA-hH1neo plasmid or the psiRNA-*Dcf1* plasmid. 24 h post-transfection, BV2 microglia were stimulated with LPS (1000 ng/ml) and incubated for 12 h. **(A)** The expression of *Cox-2, IL-1β, Tnfsf11, Cxcl1, Ccl7, IL-6, IL17D, TNF-α*, and *Csf1* were assessed by qPCR. The relative abundance of each mRNA was expressed relative to *Gapdh* (Supplementary Table S9). **(B)** Protein expression of Cox-2, IL-1β, IL-6, and TNF-α were assessed by Western blotting. Quantification of protein expression levels normalized to Gapdh. Data are expressed as the mean ± SEM. $n = 3$. *$p < 0.05$; **$p < 0.01$; ***$p < 0.001$ vs. blank. #$p < 0.05$; ##$p < 0.01$; ###$p < 0.001$ vs. LPS + vehicle (Supplementary Table S10). **(C)** Sample Western blots shown for *Cox-2, IL-1β, IL-6, TNF-α, and Gapdh.*

Ccl7, and IL17D (**Figure 5** and Supplementary Figure S4). Moreover, *Dcf1* knockdown diminished migratory (**Figure 6**) and phagocytic (**Figure 7**) abilities of BV2 cells, indicating that *Dcf1* deletion induced microglial dysfunction. Therefore, the deficiency of *Dcf1* induced an abnormal activation of microglial cells and disturbed the release of neuroinflammatory cytokines, which may destroy the immune homeostasis in the brain.

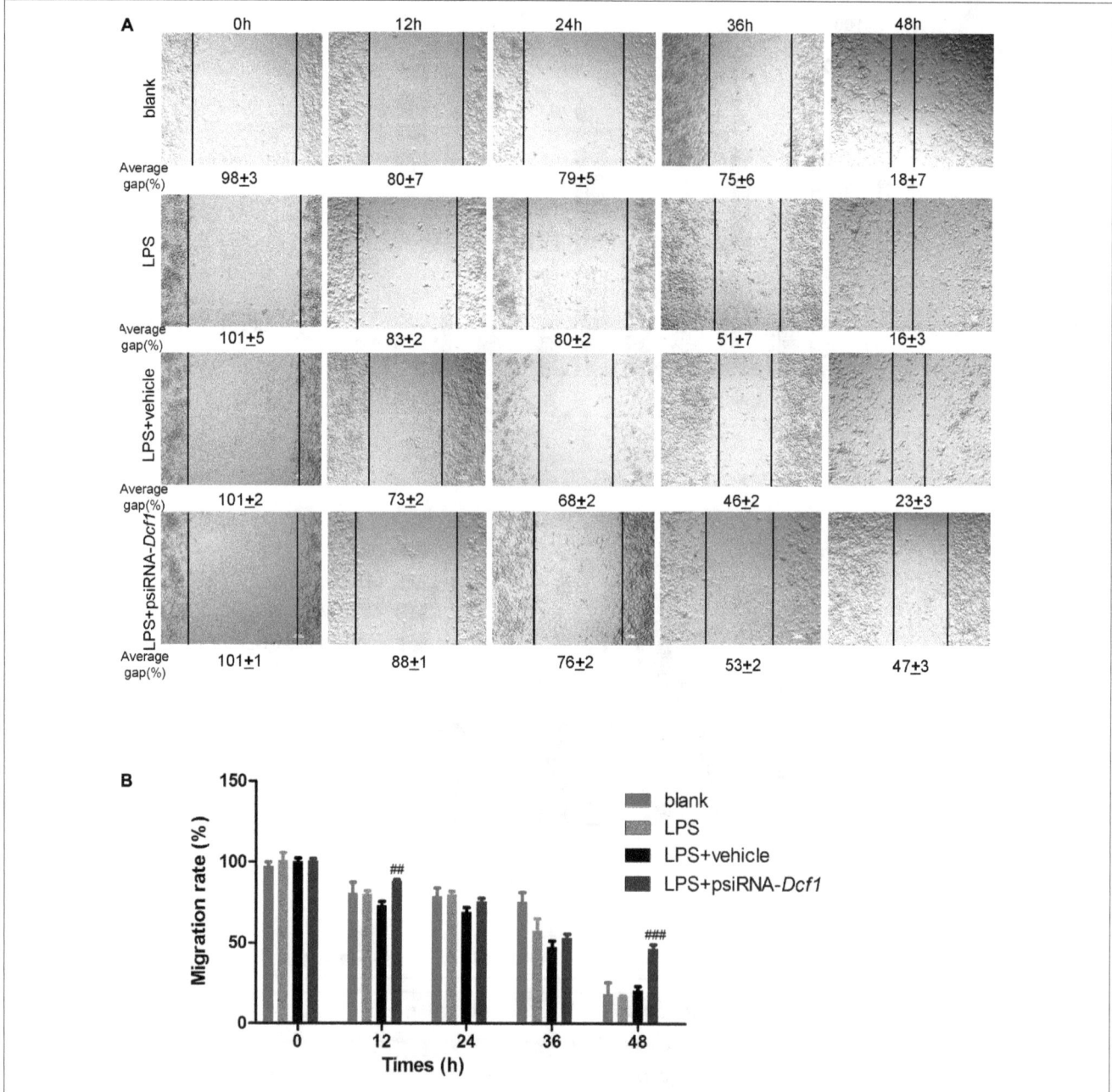

FIGURE 6 | *Dcf1* deletion decreases the migratory capacity of BV2 microglial cells. BV2 microglial cells were transfected with the psiRNA-hH1neo plasmid or the psiRNA-*Dcf1* plasmid. 24 h post-transfection, BV2 microglia were stimulated with LPS (1000 ng/ml) and incubated for 12 h. **(A)** Representative images of the scratched areas in each condition at different time points were photographed. The average gap (AG, %) was used to quantify the relative migration of the cells. Scale bars represent 200 μm. **(B)** Statistical analysis of the BV2 microglial migration rate. Data are expressed as the mean ± SEM. $n = 8$. $^{\#\#}p < 0.01$; $^{\#\#\#}p < 0.001$ vs. LPS + vehicle (Supplementary Table S11).

In order to assess the effects of *Dcf1* on microglia activation, PET technology was used to detect and monitor neuroinflammation *in vivo* (Auvity et al., 2017). TSPO was used as a biomarker for brain inflammation (Kong et al., 2016), since it is poorly expressed in the brain under normal physiological conditions, but is upregulated in activated microglial cells in response to inflammation or brain injury. Moreover, ^{18}F-DPA-714, a novel TSPO radiotracer, has been used to detect and

monitor neuroinflammation in various central system diseases (Wang et al., 2014; Lavisse et al., 2015; Kong et al., 2016). An increase in TSPO reflects increased microglial activation, which is a key event in the neuroinflammatory response (Dickens et al., 2014; Kong et al., 2016). **Figure 2** showed a PET image in which the green TSPO radiotracer was significantly increased and aggregated in *Dcf1*-KO mice, indicating activation of microglial cells. Moreover, morphological changes in mitochondria and

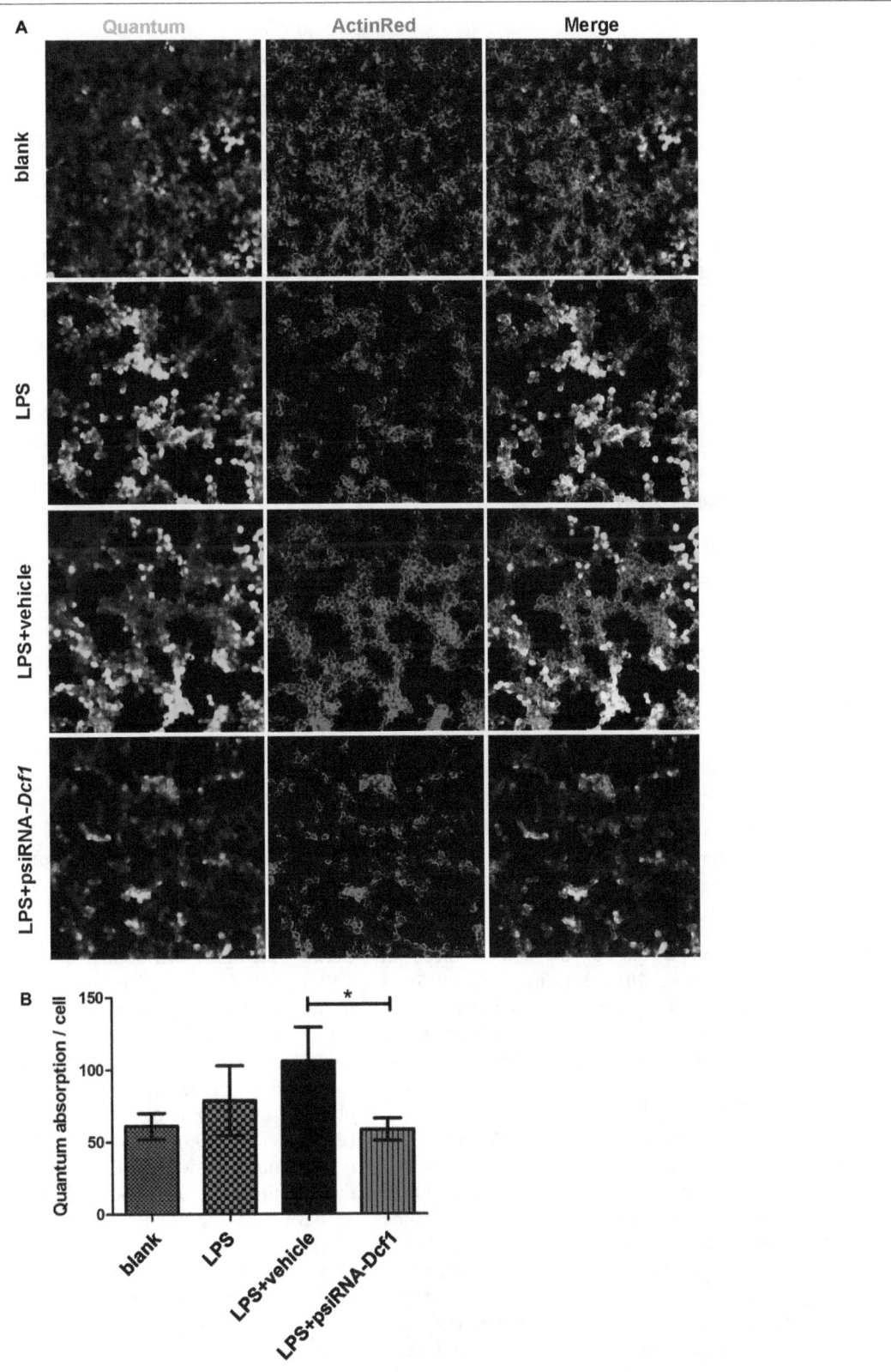

FIGURE 7 | *Dcf1* deletion suppresses the phagocytic ability of BV2 microglia cells. BV2 microglial cells were transfected with the psiRNA-hH1neo plasmid or the psiRNA-*Dcf1* plasmid. 24 h post-transfection, BV2 microglia were stimulated with LPS (1000 ng/ml) and incubated for 12 h. **(A)** Image showing the phagocytic ability of BV2 microglia cells. Quanta were spontaneous green, and the cell skeleton was detected by ActinRed (red). Scale bars represent 20 μm. **(B)** Analysis of the average quantum absorption in each cell to assess the phagocytic activity of BV2 microglia. Data are expressed as the mean ± SEM. $n = 4$. *$p < 0.05$ (Supplementary Table S12).

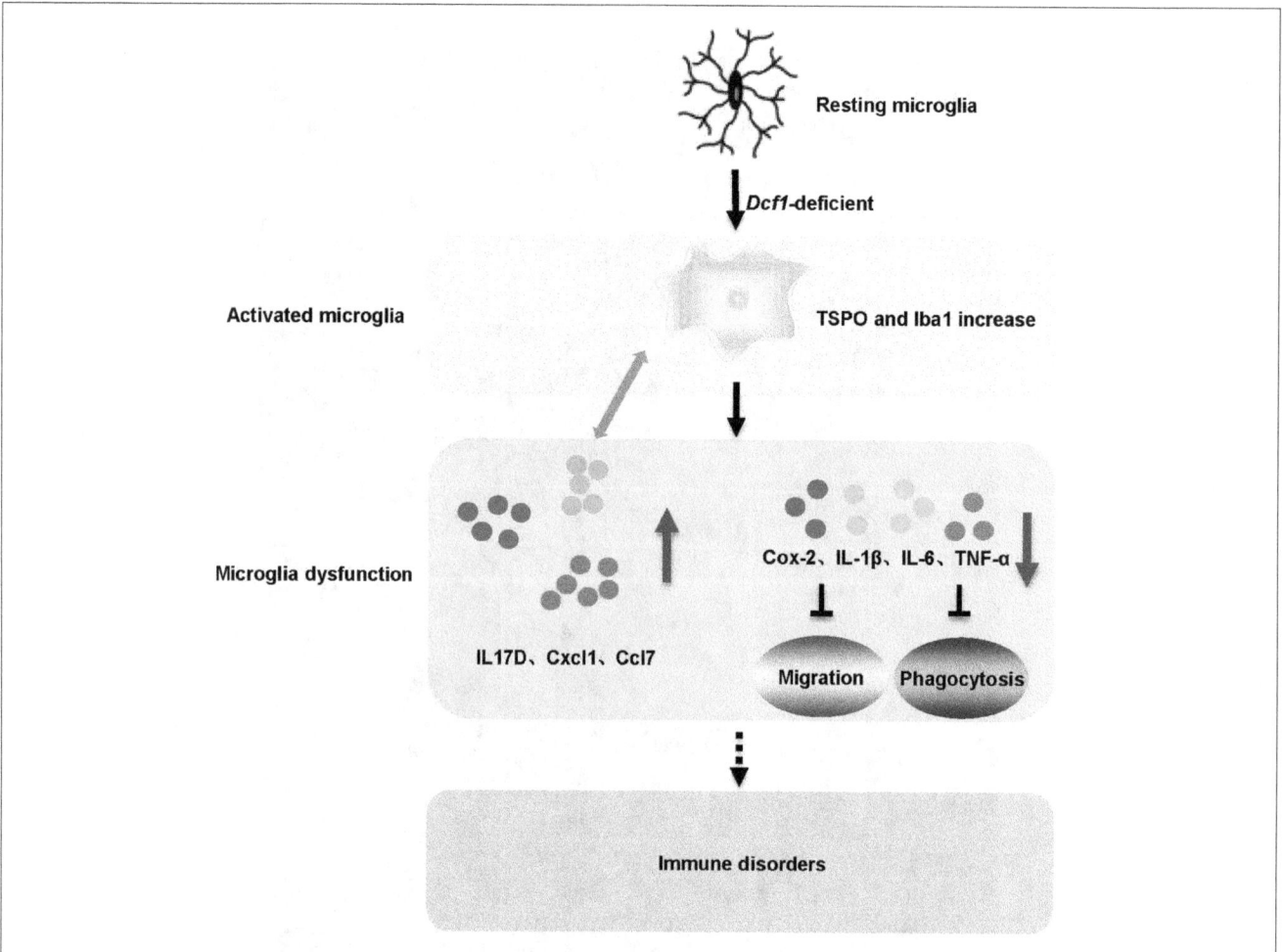

FIGURE 8 | Schematic diagram of microglial activation and dysfunction induced by *Dcf1* deficiency. *Dcf1*-deficient activated microglia, induced subsequent aberrant proinflammatory cytokines release and microglial dysfunction, which blocked the migratory and phagocytic abilities of activated microglia.

the resulting dysfunction is one of the critical steps in neuroinflammation (Xie et al., 2014; Martins et al., 2015). It has been reported that TSPO is primarily located in the outer mitochondrial membrane, and our previous data have shown that *Dcf1* is also localized within mitochondria (Xie et al., 2014), suggesting that *Dcf1* may interact with TSPO in the mitochondria of microglial cells and influence their activation.

The neuroinflammatory cytokines Cxcl1, Ccl7, and IL17D were increased during local inflammatory reaction, which can expand the immune recruitment (Boissiere-Michot et al., 2014). Cxcl1 is a chemokine produced by glial cells that attracts immune cells to the brain (Wohleb et al., 2013; Miller and Raison, 2016), and Ccl7 has been reported to exert potent proinflammatory actions through chemotaxis of monocyte-derived macrophages and other inflammatory leukocytes in the central neural system (Stuart et al., 2015). The nature of these factors and their participation in neuroinflammatory responses (Johnson et al., 2011; Jung et al., 2015) are consistent with our results of increased *Cxcl1* and *Ccl7* (**Figure 3**) accompanied by the activation of microglia (**Figure 2**). This process also included an increase in IL17D (**Figure 3**), implying an important function of IL17D

during microglial activation. Moreover, Cox-2, IL-1β, IL-6, TNF-α, and Csf1 are pleiotropic cytokines that are involved in various immune responses (Liew, 2003; Kishimoto, 2006; De et al., 2014; Leppkes et al., 2014), the levels of which were decreased in both *Dcf1*-KO mice (**Figure 3**) and LPS-stimulated BV2 cells (**Figure 5**). The same phenomena were also observed in BV2 microglial cells activated by Prostaglandin E2 (Petrova et al., 1999), where the mRNA levels of *IL-6* and *TNF-α* were modestly decreased. Reduced IL-1β did indeed inhibit the secretion of IL-6, TNF-α, and Cox-2; and high levels of IL-1β have been suggested to potentially induce the production of IL-6, TNF-α, and Cox-2 (Erta et al., 2012; Seibert and Masferrer, 1994). This is consistent with our results showing that lower expression levels of IL-1β may inhibit the secretion of IL-6, TNF-α, and Cox-2. The roles of central cytokines in the brain are not yet fully understood (Jeon and Kim, 2016), resulting in the paradox phenomena seen among different research studies (Felger and Lotrich, 2013; Farooq et al., 2017). Moreover, certain signaling pathways in dendritic cells, and immune cell types, have been shown to prevent the production of cytokines such as IL-6 (Hansen and Caspi, 2010), suggesting that *Dcf1* may also regulate the secretion of cytokines in the brain.

Microglia were activated *in vivo* detecting by ^{18}F-DPA-714 in the *Dcf1*-KO brain (**Figure 2**), and silencing of *Dcf1* significantly promoted the LPS-induced changes in microglial morphology *in vitro* (**Figure 4**), suggesting that downregulation of *Dcf1* increased the activation of BV2 microglia cells. Iba1 is a microglia-specific protein (Ohsawa et al., 2004), and as such, in order to confirm the expression level of Iba1 in *Dcf1*-KO mice, immunohistochemical staining and Western blotting were performed. We found that the ratio of activated to resting microglial cells were dramatically increased, and the protein expression of Iba1 and CD68 was consistently elevated in *Dcf1*-KO mice (**Figures 2F,G**). It has previously been reported that the expression of Iba1 is regulated by cytokines and interferons (Imai and Kohsaka, 2002), and our data showed that the levels of the cytokines *Cxcl1*, *Ccl7*, and *IL17D*, were increased (**Figure 3A**), which may have induced a higher expression of Iba1. In addition, the levels of *Cox-2*, *IL-6*, *IL-1β*, *TNF-α*, and *Csf1* were all reduced (**Figure 3A**), implying that deletion of *Dcf1* interfered with the production of various inflammatory factors.

When neuronal damage occurs, microglia adopt an activated state and exert diversified functions including migration, phagocytosis, and the production of various cytokines and chemokines (Noda and Suzumura, 2012). We found that in the presence of LPS, downregulation of *Dcf1* decreased the migratory capacity of BV2 microglial cells compared with the LPS + vehicle group (**Figure 6**). Cell migration during the immune response appears to be modulated by two fundamental processes: cell adhesion systems located at the site of inflammation, and chemotactic signals elicited through cytokines and chemokines (Simson and Foster, 2000). It has also been reported that the release of inflammatory cytokines and chemokines initiates the inflammatory response and leads to the migration of microglia toward sites of injury (Zhou et al., 2014). In the present study, the decreased mRNA levels of the majority detected proinflammatory cytokines (**Figure 5**) in conjunction with the RNA sequencing results (**Figure 1**), in which the majority of downregulated genes were related to neuroimmune responses, showed a defect in the first step of microglial migration process. Therefore, the microglial migratory ability was reduced upon downregulation of *Dcf1* (**Figure 6**), which was accompanied by the downregulation of proinflammatory cytokines.

Our results showed that *Dcf1* deletion activated microglial cells in the brain *in vivo* (**Figures 2A,B**), and that downregulation of *Dcf1* led to a significant decrease in the phagocytic capacity of BV2 microglia *in vitro* (**Figure 7**). In contrast, enhanced microglia-activated phagocytosis has been proposed to be required for the removal of injured neurons, axons, and myelin sheaths (Fu et al., 2014). It has also been reported that cytokines was involved in the regulation of phagocytic capacity (Koenigsknecht-Talboo and Landreth, 2005; Tamura et al., 2017), and the activation of microglia induced cytokine secretion (such as IL-6 and TNF-α) and phagocytosis (Heo et al., 2015). However, our results showed that the

downregulation of *Dcf1* inhibited the expression of the majority detected proinflammatory cytokines including IL-6 and TNF-α (**Figure 5**). Considering that the production of proinflammatory cytokines is typically accompanied by an increased phagocytic ability, it is reasonable to suggest that the decreased secretion of different proinflammatory cytokines impeded the phagocytic ability of microglia in the present study. Moreover, it has been reported that activated microglia migrate toward injured areas, and subsequently phagocytose foreign substances or unwanted self-debris (Noda and Suzumura, 2012). Furthermore, phagocytes such as microglia follow the cytokine gradient to the infected area (Strohmeyer et al., 2005). Our results indicated that the migratory ability of microglia was impaired by the downregulation of *Dcf1* (**Figure 6**), and that the diminished migratory ability maybe the first inhibitory step in the blockade of phagocytic capacity (**Figure 7**).

Interestingly, we were not the first to notice the paradox phenomena between activated microglia and the reduced secretion of proinflammatory cytokines, since it has been reported that deviations from microglial homeostasis induce diseases (Yirmiya et al., 2015). Our present results showed that *Dcf1* deletion decreased the expression of proinflammatory cytokines such as Cox-2, IL-1β, IL-6, Csf1and TNF-α (**Figure 3**), thus, we suggested that the downregulation of these proinflammatory cytokines (**Figure 3**) impaired the normal function of activated microglia, including the migratory (**Figure 6**) and phagocytic (**Figure 7**) abilities. Moreover, *Dcf1* deletion also increased the expression of other proinflammatory cytokines such as IL17D and Ccl7 (**Figure 3**), which may be secreted by the activated microglia. From these results, we can also speculate that the absence of *Dcf1* may induce the abnormal function of the immune system, causing an aberrant secretion of proinflammatory factors and block the normal immune response involving activated microglia. This demonstrates that the intimate details of abnormal inflammatory responses remain unclear, and that further research is required to determine the biological mechanisms induced by *Dcf1* deletion.

CONCLUSION

In conclusion, our data indicates that *Dcf1*-deficient microglia induced aberrant proinflammatory cytokines release and subsequent microglial dysfunction, which blocked the migratory and phagocytic abilities of activated microglia. Taken together, these observations provide novel insight into the role of *Dcf1* in activated microglial cells during the neuroimmune response, and further lay the foundation for the elucidation of the mechanism underlying neuroinflammatory-related diseases.

AUTHOR CONTRIBUTIONS

JW, JL, and QW designed the experiments. JL, QW, and FZ conducted most of the experiments, with assistance from YK,

QL, WL, YS, YW, and YG. JL, QW, and YK collected data and contributed to the statistical analysis. JW, JL, and QW analyzed the data and wrote the manuscript. JW, MW, and TW obtained funding and revised the manuscript. All authors read and approved the final manuscript.

FUNDING

This work was sponsored by the National Natural Science Foundation of China (Grant Number: 31500827), Young Eastern Scholar (Grant Number: QD2015033), the Natural Science Foundation of Shanghai (Grant Number: 14ZR1414400), the National Natural Science Foundation of China (Grant Numbers: 81471162; 81571345; and 41430644) and the Science and Technology Commission of Shanghai (Grant Number: 14JC1402400), Program for Changjiang Scholars and Innovative Research Team in University (No. IRT_17R71).

ACKNOWLEDGMENTS

We would like to thank Dr. Natalie Ward (Medical College of Wisconsin, Wauwatosa, WI, United States) for editing this manuscript and Dr. Yonghua Ji (Shanghai University, Shanghai, China) for providing the BV2 cells.

REFERENCES

Amor, S., and Woodroofe, M. N. (2014). Innate and adaptive immune responses in neurodegeneration and repair. *Immunology* 141, 287–291. doi: 10.1111/imm. 12134

Andreasson, K. I., Bachstetter, A. D., Colonna, M., Ginhoux, F., Holmes, C., Lamb, B., et al. (2016). Targeting innate immunity for neurodegenerative disorders of the central nervous system. *J. Neurochem.* 138, 653–693. doi: 10. 1111/jnc.13667

Auvity, S., Goutal, S., Theze, B., Chaves, C., Hosten, B., Kuhnast, B., et al. (2017). Evaluation of TSPO PET imaging, a marker of glial activation, to study the neuroimmune footprints of morphine exposure and withdrawal. *Drug Alcohol Depend.* 170, 43–50. doi: 10.1016/j.drugalcdep.2016.10.037

Baron, R., Babcock, A. A., Nemirovsky, A., Finsen, B., and Monsonego, A. (2014). Accelerated microglial pathology is associated with Abeta plaques in mouse models of Alzheimer's disease. *Aging Cell* 13, 584–595. doi: 10.1111/acel.12210

Boissiere-Michot, F., Lazennec, G., Frugier, H., Jarlier, M., Roca, L., Duffour, J., et al. (2014). Characterization of an adaptive immune response in microsatellite-instable colorectal cancer. *Oncoimmunology* 3:e29256. doi: 10.4161/onci.29256

Carniglia, L., Ramirez, D., Durand, D., Saba, J., Turati, J., Caruso, C., et al. (2017). Neuropeptides and microglial activation in inflammation, pain, and neurodegenerative diseases. *Mediators Inflamm.* 2017:5048616. doi: 10.1155/2017/5048616

De, I., Nikodemova, M., Steffen, M. D., Sokn, E., Maklakova, V. I., Watters, J. J., et al. (2014). CSF1 overexpression has pleiotropic effects on microglia in vivo. *Glia* 62, 1955–1967. doi: 10.1002/glia.22717

Dheen, S. T., Kaur, C., and Ling, E. A. (2007). Microglial activation and its implications in the brain diseases. *Curr. Med. Chem.* 14, 1189–1197. doi: 10. 2174/092986707780597961

Dickens, A. M., Vainio, S., Marjamaki, P., Johansson, J., Lehtiniemi, P., Rokka, J., et al. (2014). Detection of microglial activation in an acute model of neuroinflammation using PET and radiotracers 11C-(R)-PK11195 and 18F-GE-180. *J. Nucl. Med.* 55, 466–472. doi: 10.2967/jnumed.113.125625

Eisenberg, D., and Jucker, M. (2012). The amyloid state of proteins in human diseases. *Cell* 148, 1188–1203. doi: 10.1016/j.cell.2012.02.022

Erta, M., Quintana, A., and Hidalgo, J. (2012). Interleukin-6, a major cytokine in the central nervous system. *Int. J. Biol. Sci.* 8, 1254–1266. doi: 10.7150/ijbs.4679

Farooq, R. K., Asghar, K., Kanwal, S., and Zulqernain, A. (2017). Role of inflammatory cytokines in depression: focus on interleukin-1beta. *Biomed. Rep.* 6, 15–20. doi: 10.3892/br.2016.807

Felger, J. C., and Lotrich, F. E. (2013). Inflammatory cytokines in depression: neurobiological mechanisms and therapeutic implications. *Neuroscience* 246, 199–229. doi: 10.1016/j.neuroscience.2013.04.060

Fernandez-Calle, R., Vicente-Rodriguez, M., Gramage, E., Pita, J., Perez-Garcia, C., Ferrer-Alcon, M., et al. (2017). Pleiotrophin regulates microglia-mediated neuroinflammation. *J. Neuroinflammation* 14:46. doi: 10.1186/s12974-017-0823-8

Fu, R., Shen, Q., Xu, P., Luo, J. J., and Tang, Y. (2014). Phagocytosis of microglia in the central nervous system diseases. *Mol. Neurobiol.* 49, 1422–1434. doi: 10.1007/s12035-013-8620-6

Garay, P. A., and McAllister, A. K. (2010). Novel roles for immune molecules in neural development: implications for neurodevelopmental disorders. *Front. Synaptic Neurosci.* 2:136. doi: 10.3389/fnsyn.2010.00136

Gonzalez-Scarano, F., and Baltuch, G. (1999). Microglia as mediators of inflammatory and degenerative diseases. *Annu. Rev. Neurosci.* 22, 219–240. doi: 10.1146/annurev.neuro.22.1.219

Hansen, A. M., and Caspi, R. R. (2010). Glutamate joins the ranks of immunomodulators. *Nat. Med.* 16, 856–858. doi: 10.1038/nm0810-856

Heo, D. K., Lim, H. M., Nam, J. H., Lee, M. G., and Kim, J. Y. (2015). Regulation of phagocytosis and cytokine secretion by store-operated calcium entry in primary isolated murine microglia. *Cell. Signal.* 27, 177–186. doi: 10.1016/j.cellsig.2014. 11.003

Hwang, I. K., Choi, J. H., Nam, S. M., Park, O. K., Yoo, D. Y., Kim, W., et al. (2014). Activation of microglia and induction of pro-inflammatory cytokines in the hippocampus of type 2 diabetic rats. *Neurol. Res.* 36, 824–832. doi: 10.1179/1743132814Y.0000000330

Imai, Y., and Kohsaka, S. (2002). Intracellular signaling in M-CSF-induced microglia activation: role of Iba1. *Glia* 40, 164–174. doi: 10.1002/glia.10149

Jeon, S. W., and Kim, Y. K. (2016). Neuroinflammation and cytokine abnormality in major depression: cause or consequence in that illness? *World J. Psychiatry* 6, 283–293. doi: 10.5498/wjp.v6.i3.283

Johnson, E. A., Dao, T. L., Guignet, M. A., Geddes, C. E., Koemeter-Cox, A. I., and Kan, R. K. (2011). Increased expression of the chemokines CXCL1 and MIP-1alpha by resident brain cells precedes neutrophil infiltration in the brain following prolonged soman-induced status epilepticus in rats. *J. Neuroinflammation* 8:41. doi: 10.1186/1742-2094-8-41

Jung, K. H., Das, A., Chai, J. C., Kim, S. H., Morya, N., Park, K. S., et al. (2015). RNA sequencing reveals distinct mechanisms underlying BET inhibitor JQ1-mediated modulation of the LPS-induced activation of BV-2 microglial cells. *J. Neuroinflammation* 12:36. doi: 10.1186/s12974-015-0260-5

Kishimoto, T. (2006). Interleukin-6: discovery of a pleiotropic cytokine. *Arthritis Res. Ther.* 8(Suppl. 2):S2. doi: 10.1186/ar1916

Koeglsperger, T., Li, S., Brenneis, C., Saulnier, J. L., Mayo, L., Carrier, Y., et al. (2013). Impaired glutamate recycling and GluN2B-mediated neuronal calcium overload in mice lacking TGF-beta1 in the CNS. *Glia* 61, 985–1002. doi: 10. 1002/glia.22490

Koenigsknecht-Talboo, J., and Landreth, G. E. (2005). Microglial phagocytosis induced by fibrillar beta-amyloid and IgGs are differentially regulated by proinflammatory cytokines. *J. Neurosci.* 25, 8240–8249. doi: 10.1523/JNEUROSCI.1808-05.2005

Kong, X., Luo, S., Wu, J. R., Wu, S., De Cecco, C. N., Schoepf, U. J., et al. (2016). (18)F-DPA-714 PET imaging for detecting neuroinflammation in rats with chronic hepatic encephalopathy. *Theranostics* 6, 1220–1231. doi: 10.7150/thno. 15362

Lavisse, S., Inoue, K., Jan, C., Peyronneau, M. A., Petit, F., Goutal, S., et al. (2015). [18F]DPA-714 PET imaging of translocator protein TSPO (18 kDa) in the normal and excitotoxically-lesioned nonhuman primate brain. *Eur. J. Nucl. Med. Mol. Imaging* 42, 478–494. doi: 10.1007/s00259-014-2962-9

Leppkes, M., Roulis, M., Neurath, M. F., Kollias, G., and Becker, C. (2014). Pleiotropic functions of TNF-alpha in the regulation of the intestinal epithelial response to inflammation. *Int. Immunol.* 26, 509–515. doi: 10.1093/intimm/dxu051

Liew, F. Y. (2003). The role of innate cytokines in inflammatory response. *Immunol. Lett.* 85, 131–134. doi: 10.1016/S0165-2478(02)00238-9

Liu, Q., Feng, R., Chen, Y., Luo, G., Yan, H., Chen, L., et al. (2017a). Dcf1 triggers dendritic spine formation and facilitates memory acquisition. *Mol. Neurobiol.* 55, 763–775. doi: 10.1007/s12035-016-0349-6

Liu, Q., Shi, J., Lin, R., and Wen, T. (2017b). Dopamine and dopamine receptor D1 associated with decreased social interaction. *Behav. Brain Res.* 324, 51–57. doi: 10.1016/j.bbr.2017.01.045

Martins, C., Hulkova, H., Dridi, L., Dormoy-Raclet, V., Grigoryeva, L., Choi, Y., et al. (2015). Neuroinflammation, mitochondrial defects and neurodegeneration in mucopolysaccharidosis III type C mouse model. *Brain* 138, 336–355. doi: 10.1093/brain/awu355

Meyer-Luehmann, M., Spires-Jones, T. L., Prada, C., Garcia-Alloza, M., de Calignon, A., Rozkalne, A., et al. (2008). Rapid appearance and local toxicity of amyloid-beta plaques in a mouse model of Alzheimer's disease. *Nature* 451, 720–724. doi: 10.1038/nature06616

Miklossy, J. (2008). Chronic inflammation and amyloidogenesis in Alzheimer's disease – role of spirochetes. *J. Alzheimers Dis.* 13, 381–391. doi: 10.3233/JAD-2008-13404

Miller, A. H., and Raison, C. L. (2016). The role of inflammation in depression: from evolutionary imperative to modern treatment target. *Nat. Rev. Immunol.* 16, 22–34. doi: 10.1038/nri.2015.5

Mosher, K. I., and Wyss-coray, T. (2014). Microglial dysfunction in brain aging and Alzheimer's disease. *Biochem. Pharmacol.* 88, 594–604. doi: 10.1016/j.bcp.2014.01.008

Moynagh, P. N. (2005). The interleukin-1 signalling pathway in astrocytes: a key contributor to inflammation in the brain. *J. Anat.* 207, 265–269. doi: 10.1111/j.1469-7580.2005.00445.x

Noda, M., and Suzumura, A. (2012). Sweepers in the CNS: microglial migration and phagocytosis in the Alzheimer disease pathogenesis. *Int. J. Alzheimers Dis.* 2012:891087. doi: 10.1155/2012/891087

Ohsawa, K., Imai, Y., Sasaki, Y., and Kohsaka, S. (2004). Microglia/macrophage-specific protein Iba1 binds to fimbrin and enhances its actin-bundling activity. *J. Neurochem.* 88, 844–856. doi: 10.1046/j.1471-4159.2003.02213.x

Panicker, N., Saminathan, H., Jin, H., Neal, M., Harischandra, D. S., Gordon, R., et al. (2015). Fyn kinase regulates microglial neuroinflammatory responses in cell culture and animal models of Parkinson's disease. *J. Neurosci.* 35, 10058–10077. doi: 10.1523/JNEUROSCI.0302-15.2015

Parkhurst, C. N., Yang, G., Ninan, I., Savas, J. N., Yates, J. R. III, Lafaille, J. J., et al. (2013). Microglia promote learning-dependent synapse formation through brain-derived neurotrophic factor. *Cell* 155, 1596–1609. doi: 10.1016/j.cell.2013.11.030

Petrova, T. V., Akama, K. T., and Van Eldik, L. J. (1999). Selective modulation of BV-2 microglial activation by prostaglandin E(2). Differential effects on endotoxin-stimulated cytokine induction. *J. Biol. Chem.* 274, 28823–28827. doi: 10.1074/jbc.274.40.28823

Qin, L., Liu, Y., Wang, T., Wei, S. J., Block, M. L., Wilson, B., et al. (2004). NADPH oxidase mediates lipopolysaccharide-induced neurotoxicity and proinflammatory gene expression in activated microglia. *J. Biol. Chem.* 279, 1415–1421. doi: 10.1074/jbc.M307657200

Rio-Hortega, P. D. (1932). "Microglia," in *Cytology and Cellular Pathology of the Nervous System*, Vol. 2, ed. W. Penfield (New York, NY Hoeber).

Rizzo, G., Veronese, M., Tonietto, M., Bodini, B., Stankoff, B., Wimberley, C., et al. (2017). Generalization of endothelial modelling of TSPO PET imaging:

considerations on tracer affinities. *J. Cereb. Blood Flow Metab.* doi: 10.1177/0271678X17742004 [Epub ahead of print]. doi: 10.1177/0271678X17742004

Saba, W., Goutal, S., Auvity, S., Kuhnast, B., Coulon, C., Kouyoumdjian, V., et al. (2017). Imaging the neuroimmune response to alcohol exposure in adolescent baboons: a TSPO PET study using ^{18}F-DPA-714. *Addict. Biol.* doi: 10.1111/adb.12548 [Epub ahead of print]. doi: 10.1111/adb.12548

Seibert, K., and Masferrer, J. L. (1994). Role of inducible cyclooxygenase (COX-2) in inflammation. *Receptor* 4, 17–23.

Simson, L., and Foster, P. S. (2000). Chemokine and cytokine cooperativity: eosinophil migration in the asthmatic response. *Immunol. Cell Biol.* 78, 415–422. doi: 10.1046/j.1440-1711.2000.00922.x

Sofroniew, M. V. (2015). Astrocyte barriers to neurotoxic inflammation. *Nat. Rev. Neurosci.* 16, 249–263. doi: 10.1038/nrn3898

Streit, W. J. (2002). Microglia as neuroprotective, immunocompetent cells of the CNS. *Glia* 40, 133–139. doi: 10.1002/glia.10154

Streit, W. J., Miller, K. R., Lopes, K. O., and Njie, E. (2008). Microglial degeneration in the aging brain–bad news for neurons? *Front. Biosci.* 13:3423–3438.

Strohmeyer, R., Kovelowski, C. J., Mastroeni, D., Leonard, B., Grover, A., and Rogers, J. (2005). Microglial responses to amyloid beta peptide opsonization and indomethacin treatment. *J. Neuroinflammation* 2:18.

Stuart, M. J., Singhal, G., and Baune, B. T. (2015). Systematic review of the neurobiological relevance of chemokines to psychiatric disorders. *Front. Cell. Neurosci.* 9:357. doi: 10.3389/fncel.2015.00357

Tamura, T., Aoyama, M., Ukai, S., Kakita, H., Sobue, K., and Asai, K. (2017). Neuroprotective erythropoietin attenuates microglial activation, including morphological changes, phagocytosis, and cytokine production. *Brain Res.* 1662, 65–74. doi: 10.1016/j.brainres.2017.02.023

Ullrich, S., Munch, A., Neumann, S., Kremmer, E., Tatzelt, J., and Lichtenthaler, S. F. (2010). The novel membrane protein TMEM59 modulates complex glycosylation, cell surface expression, and secretion of the amyloid precursor protein. *J. Biol. Chem.* 285, 20664–20674. doi: 10.1074/jbc.M109.055608

von Bernhardi, R., Eugenin-von Bernhardi, L., and Eugenin, J. (2015). Microglial cell dysregulation in brain aging and neurodegeneration. *Front. Aging Neurosci.* 7:124. doi: 10.3389/fnagi.2015.00124

Wang, L., Wang, J., Wu, Y., Wu, J., Pang, S., Pan, R., et al. (2008). A novel function of dcf1 during the differentiation of neural stem cells in vitro. *Cell. Mol. Neurobiol.* 28, 887–894. doi: 10.1007/s10571-008-9266-1

Wang, Y., Yue, X., Kiesewetter, D. O., Wang, Z., Lu, J., Niu, G., et al. (2014). [(18)F]DPA-714 PET imaging of AMD3100 treatment in a mouse model of stroke. *Mol. Pharm.* 11, 3463–3470. doi: 10.1021/mp500234d

Wen, T., Gu, P., and Chen, F. (2002). Discovery of two novel functional genes from differentiation of neural stem cells in the striatum of the fetal rat. *Neurosci. Lett.* 329, 101–105. doi: 10.1016/S0304-3940(02)00585-2

Wohleb, E. S., Powell, N. D., Godbout, J. P., and Sheridan, J. F. (2013). Stress-induced recruitment of bone marrow-derived monocytes to the brain promotes anxiety-like behavior. *J. Neurosci.* 33, 13820–13833. doi: 10.1523/JNEUROSCI.1671-13.2013

Xie, Y., Li, Q., Yang, Q., Yang, M., Zhang, Z., Zhu, L., et al. (2014). Overexpression of DCF1 inhibits glioma through destruction of mitochondria and activation of apoptosis pathway. *Sci. Rep.* 4:3702. doi: 10.1038/srep03702

Yirmiya, R., Rimmerman, N., and Reshef, R. (2015). Depression as a microglial disease. *Trends Neurosci.* 38, 637–658. doi: 10.1016/j.tins.2015.08.001

Zhou, X., He, X., and Ren, Y. (2014). Function of microglia and macrophages in secondary damage after spinal cord injury. *Neural Regen. Res.* 9, 1787–1795. doi: 10.4103/1673-5374.143423

MeCP2 Deficiency in Neuroglia: New Progress in the Pathogenesis of Rett Syndrome

Xu-Rui Jin[1,2], Xing-Shu Chen[1]* and Lan Xiao[1]*

[1] Department of Histology and Embryology, Faculty of Basic Medicine, Collaborative Program for Brain Research, Third Military Medical University, Chongqing, China, [2] The Cadet Brigade of Clinic Medicine, Third Military Medical University, Chongqing, China

*Correspondence:
Xing-Shu Chen
xingshuchen2011@163.com
Lan Xiao
xiaolan35@hotmail.com

Rett syndrome (RTT) is an X-linked neurodevelopmental disease predominantly caused by mutations of the methyl-CpG-binding protein 2 (MeCP2) gene. Generally, RTT has been attributed to neuron-centric dysfunction. However, increasing evidence has shown that glial abnormalities are also involved in the pathogenesis of RTT. Mice that are MeCP2-null specifically in glial cells showed similar behavioral and/or neuronal abnormalities as those found in MeCP2-null mice, a mouse model of RTT. MeCP2 deficiency in astrocytes impacts the expression of glial intermediate filament proteins such as fibrillary acidic protein (GFAP) and S100 and induces neuron toxicity by disturbing glutamate metabolism or enhancing microtubule instability. MeCP2 deficiency in oligodendrocytes (OLs) results in down-regulation of myelin gene expression and impacts myelination. While MeCP2-deficient microglia cells fail in response to environmental stimuli, release excessive glutamate, and aggravate impairment of the neuronal circuit. In this review, we mainly focus on the progress in determining the role of MeCP2 in glial cells involved in RTT, which may provide further insight into a therapeutic intervention for RTT.

Keywords: Rett syndrome (RTT), MeCP2, astrocyte, oligodendrocyte, microglia

INTRODUCTION

Rett Syndrome (RTT) is an X-linked autism spectrum disorder that affects 1 in every 10,000–15,000 newborns in the United States (Chahrour and Zoghbi, 2007). It is specially characterized by a period of seemingly normal development that lasts for 6–18 months after birth. Subsequently, microcephaly and stereotypic hand wringing start to appear (Nomura, 2005) and more special symptoms appear as the age increases, such as a loss of motor coordination, ataxia, gait apraxia, seizures, poor sleep efficiency or parkinsonian features (Roze et al., 2007). In an overwhelming majority (more than 95%) of RTT patients, the syndrome is caused by mutations in a gene called methyl-CpG-binding protein 2 (MeCP2), a transcriptional corepressor that can bind to methylated CpG islands and complex with Sin3 homolog A (Sin3A) and histone deacetylases (HDACs) to regulate gene expression (Lyst and Bird, 2015). Moreover, different mutations are associated with disease severity, and there are approximately 30 types of mutations that can cause RTT. Patients with the R270X mutation and frame shift deletions in a $(CCACC)_n$-rich region present with the most typical symptoms (Bienvenu et al., 2000).

Generally, the symptoms of RTT are attributed to neuronal dysfunction. Substantial evidence shows that neuronal, morphological and functional abnormalities are involved in RTT pathogenesis. For instance, MeCP2-deficient neurons have an abnormal morphology (e.g., fewer dendritic spines and reduced arborization) (Zhou et al., 2006; Smrt et al., 2007;

Palmer et al., 2008). The number of synapses is decreased in hippocampi of MeCP2-null mice and conversely, the change was elevated in MeCP2-overexpressing mice (Johnston et al., 2001; Chao et al., 2007; Banerjee et al., 2012). Moreover, the re-expression of *MECP2* in MeCP2-null neurons effectively rescued behavioral abnormalities in mice (Luikenhuis et al., 2004).

However, increasing evidence has shown that white matter damage and/or glial cell (i.e., astrocyte, oligodendrocyte and microglia) dysfunction induced by a change in the DNA methylation state is also involved in the pathogenesis of RTT (Ballas et al., 2009; Maezawa and Jin, 2010; Okabe et al., 2012; Durand et al., 2013; Nguyen et al., 2013). Recently, it was reported that MeCP2-null astrocytes are incapable of supporting the normal development of co-cultured wild-type (WT) neurons (Williams et al., 2014). MeCP2-null microglia and astrocytes have been reported to be toxic to neurons through non-cell autonomous mechanisms, including a slower rate of glutamate (Glu) clearance and release of excessive Glu, as well as glial connexin (Maezawa et al., 2009; Maezawa and Jin, 2010). This latter finding proposes a viewpoint that RTT is not simply a disease of neurons alone, but a complex disease in which glial cells might play a vital role in the pathological process. Hence, we will attempt to summarize the progress of glial abnormalities involved in the pathogenesis of RTT in this review.

MeCP2 AND RTT

MeCP2 is one of the members of the methyl-CpG-binding domain protein (MBD) family, which is functionally involved in chromatin remodeling or transcriptional regulation. There are two crucial domains in MeCP2:, one is MBD and another is the transcriptional repression domain (TRD), which can recruit different protein partners, such as HDACs and Sin3A, to form a transcriptional repression complex and regulate target gene expression (Du et al., 2015) (**Figure 1**). The *MeECP2* gene consists of four exons (exon 1–4) and three introns (intron 1–3) and is located on the X chromosome. The transcriptional level of MeCP2 exon 1 (E1) is much higher than other exons in the brain, and mutations in MeCP2 E1 are sufficient to cause RTT (Fichou et al., 2009). Furthermore, the MeCP2 isoform has a time-specific expression pattern during brain development. MeCP2 E1 in the mouse hippocampus was detected as early as at E14, whereas MeCP2 E2 was detected at E18 (Olson et al., 2014). Generally, *MECP2* was believed to bind to methylated CpG islands; however, a recent study showed that MeCP2 can bind to non-CG methylated DNA and influence the transcription of disease-relevant genes in the adult mouse brain (Chen et al., 2015; Luo and Ecker, 2015) (**Figure 1**). Those results provide insight into the molecular mechanism of MECP2 in the delayed onset of RTT.

Generally, almost 95% of RTT patients carry mutations in the MeCP2 gene, and recent findings demonstrated that two additional genes, cyclin-dependent kinase like 5 (CDKL5) (Evans et al., 2005) and fork head box G1 (FOXG1), can also be involved in the pathogenesis of this syndrome (Mencarelli et al., 2010). Furthermore, CDKL5 has been shown to have the ability to promote the release of MeCP2 from DNA by phosphorylating MeCP2 (Mari et al., 2005; Bertani et al., 2006), while a direct functional relationship between these two molecules in RTT is controversial.

MeCP2 DEFICIENCY AND NEURONAL DYSFUNCTION

The function of MeCP2 in brain is multifarious, including modulation of neurogenesis, synaptic development and maintenance of neural circuits (Chahrour and Zoghbi, 2007; Banerjee et al., 2012; Lyst and Bird, 2015). It was demonstrated that MeCP2 is essential for neurogenesis in Xenopus embryos, and deficiency of MeCP2 resulted in a decreased number of neuronal precursors (Stancheva et al., 2003). *In vitro*, MeCP2 mutant mesenchymal stem cells presented impaired neural differentiation and increased the rate of senescence (Squillaro et al., 2012). MeCP2-deficient neurons have decreased numbers of axons and dendrites (Nguyen et al., 2012), and the neurite complexity was reduced in cultured MeCP2-null embryonic primary cortical neurons (Vacca et al., 2016), indicating that MeCP2 plays a crucial role in modulating neuronal differentiation and terminal maturation. Recent studies have revealed that MeCP2 is also involved in neuronal cell fate specification and migration (Feldman et al., 2016). It was found that neural progenitor cells (NPCs) lacking MeCP2 exhibited delayed corticogenesis with respect to abnormal migration of NPCs from the subventricular and ventricular zones into the cortical plate (Bedogni et al., 2016).

Some of the abnormal social behaviors of patients with RTT, such as anxiety and autistic features, are thought to be caused by MeCP2 deficiency in the neurons of certain special brain areas, including the forebrain, hypothalamus and basolateral amygdala (Armstrong et al., 1998; Gemelli et al., 2006; Fyffe et al., 2008; Chao et al., 2010; Williams et al., 2014). *MECP2* conditional knockout in glutamatergic neurons, but not in inhibitory neurons, leads to more serious RTT-like symptoms in mice (Meng et al., 2016). It has been shown that the balance between Glu excitatory synapses and GABAergic (gamma-amino butyric acid) inhibitory synapses is disrupted in RTT (Nelson and Valakh, 2015). MeCP2 knockdown reduces the excitatory synapse number and attenuates synaptic scale-up by reducing the expression of metabotropic glutamate receptor 2 (GluR2), and there is a major down-regulation of GABAergic inhibitory synapses in MeCP2 knockout mice (He et al., 2014; Kang et al., 2014). Thus, correcting the abnormal MeCP2 level by adding the gene back or over-expressing is a valid method to study the function of MeCP2 and treatment for RTT. For instance, dendritic abnormalities and behavioral changes can be ameliorated by reactivation of MeCP2 expression in MeCP2-null mice (Armstrong et al., 1998; Stearns et al., 2007; Robinson et al., 2012). Recently, MECP2 gene therapy by intracisternal injection of transgenic adeno-associated virus 9 (AAV9/hMECP2) has been shown to extend survival of MeCP2-deficient (Mecp$^{2-/y}$) mice without apparent toxicity (Matagne et al., 2017; Sinnett et al., 2017). Moreover, as RTT

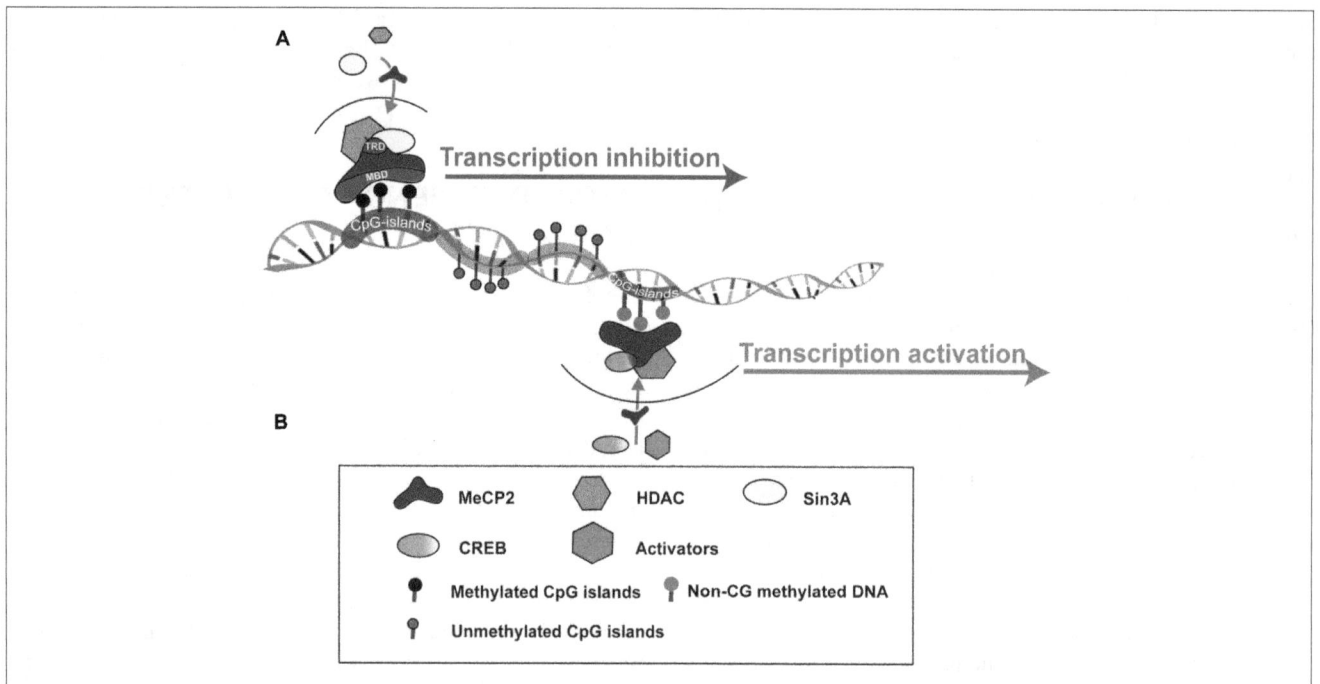

FIGURE 1 | Schematic to show how MeCP2 regulates target gene expression. **(A)** MeCP2 recruits a transcriptional corepressor complex containing Sin3A and histone deacetylase (HDAC) to methylated CpG islands and results in target gene transcription inhibition. TRD, transcriptional repression domain; MBD, methyl-CpG-binding domain. **(B)** MeCP2 is able to active gene transcription by recruiting CREB and other transcriptional factors to non-CG methylted DNA regions.

is always caused by heterozygous mutations in an X-linked gene and there exists a prevalent wild-type MeCP2 allele on the inactive X chromosome, reactivation of the inactive X chromosome-linked wild-type allele may represent another way to rescue MeCP2 deficiency. Recently, several trans-acting X-chromosome inactivation factors (XCIFs) have been identified and the inhibitor of two XCIFs (PDPK1, AURKA) has been demonstrated to exhibit reactive expression of the WT MeCP2 allele on the inactive X chromosome, which can be considered as a potential therapy for RTT (Bhatnagar et al., 2014).

In addition to RTT, evidence has shown that MeCP2 regulates the expression of its downstream genes, such as brain-derived neurotrophic factor (BDNF) (Chen et al., 2003) and ubiquitin-protein ligase E3A (UBE3a), the latter being involved in Angelman syndrome (Makedonski et al., 2005). In addition, MeCP2 also regulates the expression of distal-less homeobox 5/6 (Dlx5/6) genes, which are necessary for spinal skeletal development and are related to definite symptoms of RTT patients, such as scoliosis and microcephaly (Nakashima et al., 2010).

MeCP2 DEFICIENCY AND GLIAL CELL DYSFUNCTION

Glial cells are non-excitable cells that are functional support neurons and maintain the stability of neuronal structure and function. In the central nervous system (CNS), glial cells mainly include astrocytes, OLs and microglia. Previous studies have revealed that MeCP2 is present in a majority of neurons but

is absent from glial cells (Shahbazian et al., 2002). However, increasing evidence suggests that the abnormality of MeCP2 also plays an important role in white matter damage and/or glial dysfunction in RTT (Ballas et al., 2009; Maezawa and Jin, 2010; Okabe et al., 2012; Durand et al., 2013; Nguyen et al., 2013). Mice with MeCP2 loss specifically in glial cells presented Rett-like symptoms due to neuronal toxicity via a non-cell autonomous mechanism, and the restoration of MeCP2 in glial cells can rescue some of these defects (Lioy et al., 2011; Nguyen et al., 2013; Cronk et al., 2015). Evidence from magnetic resonance spectroscopy (MRS) indicated an increased glia-to-neuron ratio in the white matter of RTT patients along with ongoing axonal damage and glial abnormalities (Khong et al., 2002). The contribution of MeCP2 deficiency to dysfunction of different types of glial cells varies in RTT pathogenesis.

MeCP2 IN ASTROCYTES

Astrocytes are the most abundant of all glial cell types and are well known for supporting neurons and maintaining brain function. Although the expression level of MeCP2 in astrocytes is lower, it is important for astrocyte differentiation and function. After *MECP2* was specifically knocked out in neural stem cells (NSCs), these NSCs tend to differentiate into more astrocytes (Andoh-Noda et al., 2015). While the growth rate of MeCP2-deficient astrocytes is significantly slower and more interleukin (IL)-1β and IL-6 are released in response to immune stimulation *in vitro*, no obvious morphological change was found in those cells (De Filippis et al., 2012). MeCP2-deficient astrocytes that were differentiated from induced pluripotent stem cell (iPSC) lines from RTT patients have negative regulatory effects on neuronal morphology and function (Williams et al., 2014). Moreover, subsequent studies found that MeCP2-null astrocytes have certain abnormalities in target gene regulation and are toxic to neurons mainly due to abnormal Glu metabolism. Similarly, it was shown that a state of MeCP2 deficiency can spread in the brain of MeCP2$^{-/+}$ mice via a non-cell autonomous mechanism (Maezawa et al., 2009). This suggests that abnormal astrocytes are able to deteriorate the function of neurons, which speeds up the process of RTT. It was found that preferential re-expression of Mecp2 in astrocytes dramatically improved RTT symptoms and the lifetimes of MeCP2-deficient mice (Kifayathullah et al., 2010; Lioy et al., 2011; Zachariah et al., 2012).

Target Gene Dysregulation in MeCP2-Null Astrocytes

The expression of astroglial marker genes GFAP and S100β was significantly higher in MeCP2-null astrocytes than that in WT astrocytes (Forbes-Lorman et al., 2014). It was noted that MECP2 small interference RNA (siRNA) increased the expression of GFAP in the female amygdala (Forbes-Lorman et al., 2014). These observations suggest that GFAP and S100β genes are suppressed by MeCP2. Although the mechanism by which MeCP2 regulates these two genes is unclear, evidence has shown that MeCP2 E1 can couple with the Sin3A/HDAC complex, which can bind to the GFAP promoter and regulate GFAP transcription. On embryonic day 11.5 (E11.5), the promoter of the GFAP gene is highly methylated, which may facilitate assembly of MeCP2-Sin3A/HDAC complexes and suppress GFAP gene transcription. On E14, the promoter of the GFAP gene is demethylated, which may lead to the disintegration of this complex and results in GFAP gene expression. In the early stages of astrocyte differentiation, MeCP2 also has a vital effect by binding to the promoter of the S100β gene, while this binding is gradually reduced in the later stages, as demethylation of a specific CpG site occurs (Cheng et al., 2011; Forbes-Lorman et al., 2014).

In addition to astroglial marker genes, many other target genes such as solute carrier family member 38, member 1 (Slc38a1), neuronal regeneration-related protein (Nrep), and nuclear receptor subfamily 2, group F, member 2 (Nr2f2) are also regulated by MeCP2. Slc38a1 is a rate-limiting transporter of glutamine (Gln) across the plasma membrane, Nrep is a transcriptional factor that is involved in glial mobility and neoplasia, and Nr2f2 is a transcription factor that is necessary for glial differentiation from NSCs (Mackenzie and Erickson, 2004; Yasui et al., 2013; Delepine et al., 2015). In addition, Nr2f2 can regulate some target genes including those encoding chromogranin B (Chgb), chemokine (C-C motif) ligand 2 (CCL2), and lipocalin 2 (LCN2). Chgb is a highly efficient system that is directly involved in monoamine accumulation. In MeCP2-deficient mice, the expression of Nr2f2 is up-regulated, which may down-regulate Chgb and cause excess monoamine accumulation in the

extracellular fluid and impair neurons. This finding suggests that influence of MeCP2 deficiency can be aggravated through a downstream target gene cascade in astrocytes (Kloukina-Pantazidou et al., 2013; Delepine et al., 2015). Interestingly, in response to LPS stimulation, MeCP2-deficient astrocytes released fewer cytokines such as IL-1β and IL-6 (Maezawa et al., 2009).

Microtubule Instability in MeCP2-Null Astrocytes

Cellular morphology, division, migration and intracellular transportation of vesicles are controlled by microtubules (MTs), which assemble from α- and β-tubulin dimers. The acetylation modification of tubulin is deemed to be a characteristic of stable MTs (Palazzo et al., 2003). In Mecp2$^{308/y}$ and Mecp2 p. Arg294* iPSC-derived astrocytes, the level of acetylated tubulin is reduced, and histone deacetylase 6 (HDAC6) is overexpressed (Delepine et al., 2016). A similar result was also reported in MeCP2-deficient fibroblasts and MeCP2-null neurons (Gold et al., 2015). MT growth speed of Mecp2$^{308/y}$ astrocytes was higher, and MT polymerization was significantly increased. Moreover, MT-dependent lysosome vesicle mobility was obviously increased in both Mecp2$^{308/y}$ and Mecp2 p.Arg294* iPSC-derived astrocytes; however, the percentage of highly directional vesicles was reduced (Delepine et al., 2016). A special type of directional vesicle cellular transport can regulate Glu uptake, which suggests an abnormal Glu uptake rate may be associated with MT-dependent vesicle mobility in RTT (Li et al., 2015). The aforementioned phenomenon was related to the decreased expression of stathmin 2 (STMN2), which could inhibit MT polymerization (Nectoux et al., 2012). Moreover, expression of TUBA1B, which encodes the ubiquitous α-tubulin, is down-regulated in brain tissue of patients with RTT (Abuhatzira et al., 2009). These findings suggest a role of MeCP2 as an activator for MT-associated genes. Molecularly, a recent study revealed a potential mechanism by which sumoylated MeCP2 releases CREB from the repressor complex and increases the transcription of CREB-regulated genes such as BDNF (Tai et al., 2016). Interestingly, sumoylation of MeCP2 was found to be decreased in RTT, and further study is warranted to examine the sumoylation status of MeCP2 in glial cells of RTT. In addition, epothilone D, a brain-penetrating MT-stabilizing natural product, can rescue MT growth velocity, and epothilone D corrects the abnormal behavioral symptoms of Mecp2$^{308/y}$ mice (Delepine et al., 2016). Treatment via targeting MTs seems to be a new approach for RTT therapy. Notably, when co-cultured with MeCP2-deficient fibroblasts, MT stability of WT human fibroblasts is reduced (Delepine et al., 2013). All of this evidence suggests that MT impairment seems to be a common phenomenon of MeCP2 deficiency.

Abnormal Glutamate Metabolism in MeCP2-Deficient Astrocytes

Glutamate is an important signaling molecule in the CNS. At a high extracellular concentration, it is a potent cytotoxin that can induce both neuronal and glial death through excitotoxicity or oxidative stress. Extracellular glutamate concentration is maintained at an appropriate level predominantly by active transport mediated by excitatory amino acid transporters (EAATs) of astrocytes (Lehmann et al., 2009). Normally, when astrocytes are incubated with high levels of extracellular Glu, EAAT1/EAAT2 expression is rapidly decreased. However, when the down-regulation of EAAT1/2 was impaired, EAAT1/EAAT2 expression had no obvious changes in MeCP2-null astrocytes (Okabe et al., 2012). Regarding the underlying mechanism, it was found that HDAC I and II served as repressors for EAAT2 promoter activity (Karki et al., 2014), thus, MeCP2 deficiency may fail to recruit HDACs to inhibit EAAT gene expression upon environment stimulation. Notably, the Glu clearance rate of MeCP2-null astrocytes was lower than that of WT astrocytes *in vitro* (Okabe et al., 2012). Moreover, the glutamine synthetase (GS) protein was significantly higher in MeCP2-null astrocytes than that in WT astrocytes (Okabe et al., 2012). All of this evidence seems to support the concept that both Glu clearance and production are abnormal in MeCP2-deficient astrocytes, and that these abnormalities may contribute to the pathological process of RTT (**Figure 2**).

Spread of MeCP2 Deficiency in Astrocytes

An interesting phenomenon is that MeCP2 expression in $Mecp2^{-/+}$ mice is much lower at 7 months than that at 1 month, when proliferation rates of astrocytes are higher.

In vitro, MeCP2 expression was significantly decreased in WT astrocytes that were co-cultured with MeCP2-deficient astrocytes for a long period, indicating that MeCP2 deficiency can spread among astrocytes progressively and eventually lead to a dysfunctional brain (Maezawa et al., 2009). Recent studies have shown that MeCP2 deficiency can spread among astrocytes via connexin, especially connexin 43 (Cx43)-mediated gap junction. When Cx43 was knocked down with siRNA, the spread of MeCP2 deficiency was significantly reduced (Maezawa et al., 2009). This finding suggests that astroglial Cx may be a new target for preventing the phenotype of RTT.

Abnormal Astrocytes May Be the Main Contributor to the Disordered Breathing Pattern

RTT symptoms always include a severely disordered breathing pattern and reduced CO_2 sensitivity. In the respiratory control center, highly chemosensitive astrocytes can respond to the physiological decrease in pH with a vigorous increase in intracellular Ca^{2+} and release of ATP (Gourine et al., 2010). MeCP2 deficiency in astrocytes induced an obvious depression of the ventilatory responses to an increased level of CO_2, which is similar to global MeCP2-null mice (Zhang et al., 2011). Additionally, although neurons constitute the respiratory control network and determine the ventilatory response to CO_2, MeCP2 mutation selectively in neurons leads to an absolute lower depression of the CO_2 response than MeCP2 loss in astrocytes (Fyffe et al., 2008; Garg et al., 2015).

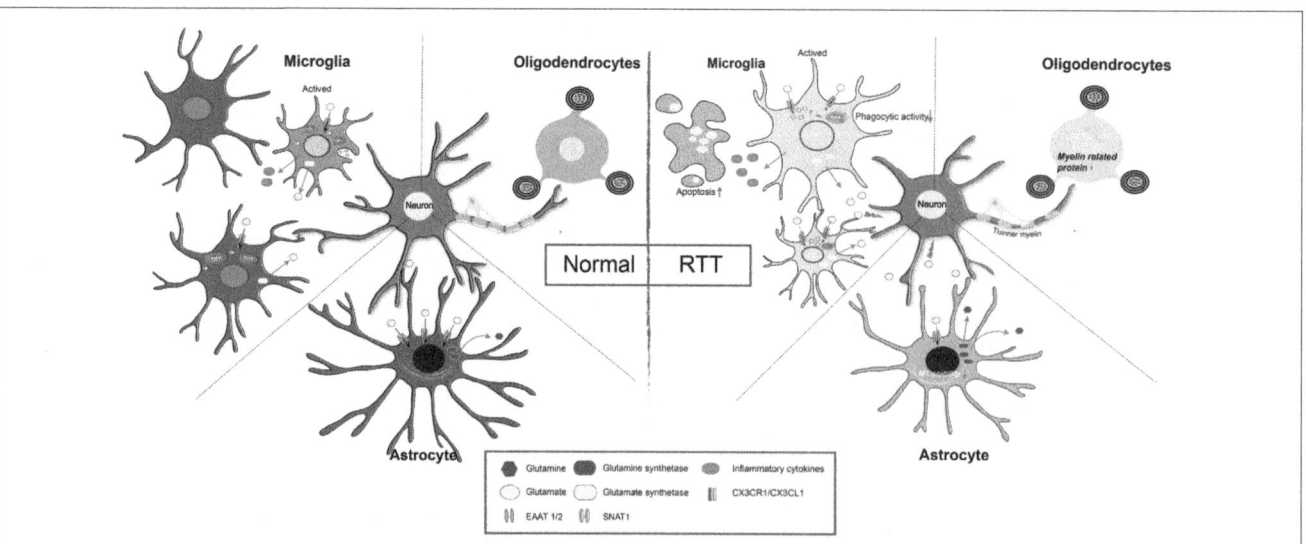

FIGURE 2 | Glia and neuron interaction under pathologic condition of RTT. Normal: Extracellular glutamate (Glu) can be uptake by excitatory amino acid transporter 1 and 2 (EAAT1/2) in astrocyte and sodium-coupled neutral amino acid transporter 1 (SNAT1) in microglial cell, and is utilized by glutamine synthetase to reduce its toxicity to neuron. Both astrocytes and microglia can be active upon stimulation, and the active microglia has capacity of phagocytosis. Furthermore, CX3CR1 (a chemokine receptor in microglial cell) and its neuronal ligand CX3CL1 mediates neuron-microglia interaction. Myelinating oligodendrocyte is important to maintain axonal integrity. RTT: MeCP2 deficiency in astrocytes leads to a decreased expression of EAAT1/2, and increased expression of glutamine synthetase, which results in a high concentration of extracellular Glu, and thus impairs neuronal activity. Additionally, those astrocytes present decreases microtubule stability. In microglia, MeCP2 deficiency results in decreased phagocytic activity and overloads of Glu due to upregulation of SNAT1. The MeCP2-deficient in oligodendrocyte results in thinner myelin due to down-regulation of myelin-related genes.

After re-expression of MeCP2 in astrocytes, the respiratory phenotype is rescued (Garg et al., 2015), suggesting that the disordered breathing pattern may be caused by abnormal astrocytes in RTT patients. However, recent evidence suggests that RTT-like breathing abnormality was only developed in 20% of $Mecp2^{+/-}$ mice (Johnson et al., 2015). Imaging of calcium signaling in ventral medullary astrocytes reveals that these phenomena may be related to a reduction in the ability of MeCP2-deficient astrocytes to sense $P_{CO_2}/[H^+]$ and that the CO_2-induced $[Ca^{2+}]_i$ response is impaired. Additionally, the ATP response of MeCP2-deficient astrocytes is not affected. Normally, ATP propagates astrocytic Ca^{2+} excitation, depolarizes chemoreceptor neurons, and induces adaptive increases in breathing. This evidence suggests that the most apparent deficiency of the MeCP2-null astrocyte is the ability to sense physiological increases in CO_2, while the downstream signaling pathway can still be activated by ATP (Bissonnette and Knopp, 2008; Zhang et al., 2011; Turovsky et al., 2015).

MeCP2 IN MICROGLIA

Mecp2 Regulates Microglial Activation

Microglia are considered to be resident myeloid-derived cells distributed throughout the brain and account for over 10% of CNS cells (Kawabori and Yenari, 2015). Microglial activation can occur in response to stimuli such as glucocorticoids, hypoxia, and inflammation (Kiernan et al., 2016). In Mecp2-null mice, microglial cells are relatively small but become larger in response to environmental stimuli compared to those of wild-type mice, and these microglial cells are lost with disease progression. Moreover, the expression of glucocorticoid-induced transcriptional signature genes and a subset of hypoxia-inducible genes were up-regulated in Mecp2-deficient microglia, suggesting the regulatory role of MeCP2 in microglia activation (Cronk et al., 2015) (**Figure 2**). From another point of view, a gene array study confirmed that the function of differentially expressed genes, as MeCP2 deficiency is strongly related to regulating the activation states of microglia (Zhao et al., 2017). In addition, studies have shown that bone marrow transplantation increases the number of microglia in the brain (Derecki et al., 2012). MeCP2-null mice that were transplanted with WT bone marrow partly restored behavior and functional abnormalities (Derecki et al., 2012). However, another study reached the opposite conclusion that WT marrow transplantation did not positively influence MeCP2-null mice (Wang et al., 2015). The beneficial effect declined when the phagocytic activity of these transplanted cells was inhibited and the ability to clear apoptotic targets was reduced. This result indicated that the phagocytic activity of microglia is impaired in MeCP2-null microglia and that a non-cell autonomous mechanism is involved in spreading the MeCP2-deficient state (Maezawa and Jin, 2010). In contrast to microglia, the MeCP2 knockout of myeloid-derived cells led to the release of more tumor necrosis factor-α (TNF-α) and inflammatory cytokines, such as IL-6, TNF-α, and IL-3 (O'Driscoll et al., 2015).

Mutant MECP2 in Microglia Impairs the Neuronal Circuit

In the developing CNS, microglia have been demonstrated to play an important role in shaping neuronal circuit structure by microglia-synapse interactions or removing excess synapses, and an etiology of disrupted synaptic function can be detected in mouse models of RTT (Johnston et al., 2001; Noutel et al., 2011). Recent research offered a new viewpoint of the effect of MECP2-defective microglia on the neuronal circuit. It was reported that in MECP2-null mice, microglia contribute to RTT pathogenesis by excessively engulfing and thereby eliminating, which is concomitant with synaptic loss at the end stages of the disease (Schafer et al., 2016). Intriguingly, gain or loss of Mecp2 expression specifically in microglia cannot induce the abovementioned phenomena (Schafer et al., 2016), indicating that abnormalities in microglia may accelerate the pathological process by impairing the neuronal circuit as a secondary effect, not the primary cause, of RTT. In addition, the result of another research study also implicitly demonstrated this point that restoring the expression of Mecp2 in myeloid cells by bone marrow transplantation made no sense in the Mecp2-null mouse (Wang et al., 2015).

A recent study showed that ablating CX3CR1, a chemokine receptor of microglia mediating neuron–microglia interaction by pairing with its neuronal ligand CX3CL1, can rescue the negative effect of MeCP2-deficient microglia on neurons. After blocking the interaction, the survival of MECP2-null mice was remarkably improved and behavioral abnormalities were partly restored. These results indicated that blocking intercellular interaction in brains of RTT patients might be a novel therapeutic approach for RTT (Horiuchi et al., 2016).

MeCP2-Deficient Microglia Leads to Neurotoxicity

Activation of glutamate receptors and the expression level of Glu transporters are important for maintaining the plasticity of glutamatergic synapses (Groc et al., 2006; Yuan and Bellone, 2013). MeCP2-deficient microglia released a fivefold higher level of Glu, which was associated with neurotoxic or synaptotoxic activity, as the neuronal abnormality can be partially restored by blocking the NMDA receptor (Maezawa and Jin, 2010). In addition, normal neurons co-cultured with MeCP2 mutant microglia exhibited functional and morphological disorder such as thinner and shorter dendrites and a reduced number of excitatory synapses, because the microtubule was disrupted by a decrease in MAP2 and acetylated tubulin (Maezawa and Jin, 2010; Cronk et al., 2015; Jin et al., 2015). Moreover, the level of the connexin channel protein Cx32, which can release Glu, was significantly up-regulated in MeCP2-null microglia, and the blockade of Cx32 can partially ameliorate the excessive release of Glu (Maezawa and Jin, 2010).

From the perspective of bioenergetics, MeCP2 loss in microglia led to mitochondrial dysfunction by impairing Glu homeostasis (Jin et al., 2015). In MeCP2-deficient microglia, the overexpression of the Glu transporter sodium-coupled neutral amino acid transporter 1 (SNAT1) promoted microglial uptake

TABLE 1 | A summary of the abnormal features in glial cells induced by MeCP2 abnormalities.

MECP2 abnormalities	Features			PMID
	Astrocyte	**Oligodendrocyte**	**Microglia**	
Mecp2 [tm1.1Bird]	Slower growth rate, reduced cytokine release			19386901
	Increased MT growth velocity, decreased MT stability			23351786
	Decreased CO_2 sensitivity			PMC4532534
	Increased glutamate production			22532851
	Down-regulation of SNAT1, glutamate overproduction, less ATP production			25673846
	Up-regulation of glucocorticoid-induced transcription and hypoxia-inducible gene expression			25902482
	Be activated but subsequently depleted with disease progression			25902482
	Toxic to neurons by increasing glutamate production and release, up-regulation of Cx32 expression			20392956
Mecp2[308/y]	Atrophy			22157810
	Down-regulation of Nr2f2 expression			26208914
	Down-regulation of Stathmin expression, increased MT growth velocity, decreased MT stability			22252744, 26604147
	Up-regulation of MAG expression in corpus callosum, down-regulation of PLP expression in cerebellum			20697302
	Down-regulation of CNP expression in subcortical white matter and hippocampus			22334035
Mecp2p.Arg294* Mecp2[loxJ/y]/NG2Cre	Increased MT growth velocity, decreased MT stability and vesicular transport			26604147
	Down-regulation of PLP and MBP expression			24285883
	Thinner myelin sheet			
	More active and developed severe hind limb clasping phenotypes			
Mecp2 siRNA	Up-regulation of GFAP, S100β and BDNF expression			24269336, 19386901
	Up-regulation of MBP, PLP, MOG and MOBP expression			26140854

of Glu; the excessive Glu was translated into the mitochondria, coupled with the rising mitochondrial reactive oxygen and damaged the mitochondria (Jin et al., 2015). As a consequence, adenosine triphosphate (ATP) production is significantly reduced in MeCP2-deficient microglia, and this ATP reduction leads to a greater potential for the apoptosis of abnormal microglia (Jin et al., 2015) (**Figure 2**).

MeCP2 IN OLIGODENDROGLIA

Oligodendrocytes are the myelinating cells in the brain that maintain nerve impulse conduction and provide nutrition for axons (recently reviewed in Bergles and Richardson, 2016). There are several markers that distinguish the oligodendroglial cells at different stages. OL progenitor cells (OPCs) are positive for platelet-derived growth factor receptor α (PDGFR-α) or neural antigen 2 (NG2); immature OLs are O4-positive, and mature OLs express myelin-associated glycoprotein (MAG), proteolipid protein (PLP), CNP and myelin basic protein (MBP) (He and Lu, 2013).

Some studies have found the expression of MeCP2 in oligodendroglial lineage cells (KhorshidAhmad et al., 2016), which may be associated with oxidative damage of white matter that occurred in the early stages of RTT (Durand et al., 2013). In MeCP2-null mice, although the OL morphology is unchanged, the expression of myelin-related genes is partially changed in some specific regions. For example, in the frontal cortex of MeCP2-null mice, no changes in global expression levels of the MBP, MAG and PLP were found, but MAG expression

was significantly increased in the corpus callosum, while PLP expression was significantly decreased in the cerebellum (Vora et al., 2010). Furthermore, in *Mecp2*[308/y]-mutant mice, GFAP, Cx43, Cx45, Cx40 and Cx32 were unchanged, but the CNP expression level was decreased (Wu et al., 2012). Additionally, cultured OLs with MeCP2 knocked down exhibited an increase in myelin genes, including MBP, PLP, myelin oligodendrocyte glycoprotein (MOG), myelin-associated oligodendrocyte basic protein (MOBP) and Ying Yang 1 (YY1) (KhorshidAhmad et al., 2016), indicating MeCP2 is a negative regulator for myelin protein expression (Sharma et al., 2015). As a co-repressor of MeCP2, HDACs have also been found to be functionally involved in oligodendroglial development (Huang et al., 2015). Moreover, when MeCP2 was specifically knocked out in NG2[+] OPCs, the mice displayed more active behaviors, with severe hind limb-clasping phenotypes (Nguyen et al., 2013). The PLP and MBP expression levels were lower in those mice, and thinner myelin was present. After restoring the expression of MeCP2 in the OPCs, the ethological abnormality was significantly corrected in both female and male mice (Nguyen et al., 2013) (**Figure 2**). It is noteworthy that MeCP2 re-expression in OPCs, compared with that in astrocytes or microglia cells, shows more potent ability to prolong the lifespan of MeCP2[stop/y] mice (Nguyen et al., 2013). However, the expression of myelin protein (MBP) was mildly or not rescued (CNP, MOG, or PLP) after re-expression of MeCP2 in oligodendrocytes in otherwise MeCP2-null mice, suggesting that a likely non-cell autonomous mechanism regulated the expression of OL-related genes (Nguyen et al., 2013). This mechanism is maybe related to the excessive Glu released by other glial cell types, such as astrocytes or microglia, because excessive Glu might partly restore the development of OLs through activating *n*-methyl-D-aspartic acid (NMDA) receptors in MeCP2-null brains (Lundgaard et al., 2013). In addition, white matter damage in RTT patients has been related to impaired nuclear factor kappa-B (NF-κB) signaling: there were additional copies of the inhibitor of the kappa light polypeptide gene, kinase gamma (IKBKG) enhancer in B-cells of patients with Xq28 duplications, and three of the five presented white matter anomalies in brain MRI (Philippe et al., 2013).

PERSPECTIVES

In this review, we summarized glial abnormalities induced by MeCP2 deficiency and how these cells are toxic to neurons and disturb the neural circuit (**Table 1**). The mice with MeCP2-null specifically in glial cells showed behavioral and neural synaptic abnormalities that are similar to those in MeCP2-null mice, and re-expression of MeCP2 in glial cells can restore those abnormalities and prolong the life span of MeCP2-null mice. Additionally, we discussed the role of MeCP2 in glial cells to gain a better understanding of the pathology process of RTT and ultimately a new glial-target strategy for RTT treatment. For

instance, Glu metabolism abnormalities may be a predominant component of the RTT pathology process, especially in glial cells. Some drugs such as riluzole, a glutamatergic modulator that has been approved for treating amyotrophic lateral sclerosis (ALS) by primarily rectifying extensive Glu levels, may also be utilized to treat RTT. Moreover, the regulatory mechanism of MeCP2 in the oligodendroglial development involved in RTT pathogenesis is understudied.

AUTHOR CONTRIBUTIONS

X-RJ and X-SC wrote the manuscript, LX designed and revised the manuscript.

ACKNOWLEDGMENTS

This work was supported by grants from the National Natural Science Foundation of China (31100771), Natural Science Foundation Project of CQ CSTC (2016jcyjA0600, CSTCKJCXLJRC07), and Postdoctoral Foundation of CQ CSTC (2011018).

REFERENCES

Abuhatzira, L., Shemer, R., and Razin, A. (2009). MeCP2 involvement in the regulation of neuronal alpha-tubulin production. *Hum. Mol. Genet.* 18, 1415–1423. doi: 10.1093/hmg/ddp048

Andoh-Noda, T., Akamatsu, W., Miyake, K., Matsumoto, T., Yamaguchi, R., Sanosaka, T., et al. (2015). Differentiation of multipotent neural stem cells derived from Rett syndrome patients is biased toward the astrocytic lineage. *Mol. Brain* 8:31. doi: 10.1186/s13041-015-0121-2

Armstrong, D. D., Dunn, K., and Antalffy, B. (1998). Decreased dendritic branching in frontal, motor and limbic cortex in Rett syndrome compared with trisomy 21. *J. Neuropathol. Exp. Neurol.* 57, 1013–1017. doi: 10.1097/00005072-199811000-00003

Ballas, N., Lioy, D. T., Grunseich, C., and Mandel, G. (2009). Non-cell autonomous influence of MeCP2-deficient glia on neuronal dendritic morphology. *Nat. Neurosci.* 12, 311–317. doi: 10.1038/nn.2275

Banerjee, A., Castro, J., and Sur, M. (2012). Rett syndrome: genes, synapses, circuits, and therapeutics. *Front. Psychiatry* 3:34. doi: 10.3389/fpsyt.2012.00034

Bedogni, F., Cobolli Gigli, C., Pozzi, D., Rossi, R. L., Scaramuzza, L., Rossetti, G., et al. (2016). Defects during Mecp2 null embryonic cortex development precede the onset of overt neurological symptoms. *Cereb. Cortex* 26, 2517–2529. doi: 10.1093/cercor/bhv078

Bergles, D. E., and Richardson, W. D. (2016). Oligodendrocyte development and plasticity. *Cold Spring Harb. Perspect. Biol.* 8:a020453. doi: 10.1101/cshperspect.a020453

Bertani, I., Rusconi, L., Bolognese, F., Forlani, G., Conca, B., De Monte, L., et al. (2006). Functional consequences of mutations in CDKL5, an X-linked gene involved in infantile spasms and mental retardation. *J. Biol. Chem.* 281, 32048–32056. doi: 10.1074/jbc.M606325200

Bhatnagar, S., Zhu, X., Ou, J., Lin, L., Chamberlain, L., Zhu, L. J., et al. (2014). Genetic and pharmacological reactivation of the mammalian inactive X chromosome. *Proc. Natl. Acad. Sci. U.S.A.* 111, 12591–12598. doi: 10.1073/pnas.1413620111

Bienvenu, T., Carrie, A., De Roux, N., Vinet, M. C., Jonveaux, P., Couvert, P., et al. (2000). MECP2 mutations account for most cases of typical forms of Rett syndrome. *Hum. Mol. Genet.* 9, 1377–1384. doi: 10.1093/hmg/9.9.1377

Bissonnette, J. M., and Knopp, S. J. (2008). Effect of inspired oxygen on periodic breathing in methy-CpG-binding protein 2 (Mecp2) deficient mice. *J. Appl. Physiol.* 104, 198–204. doi: 10.1152/japplphysiol.00843.2007

Chahrour, M., and Zoghbi, H. Y. (2007). The story of Rett syndrome: from clinic to neurobiology. *Neuron* 56, 422–437. doi: 10.1016/j.neuron.2007.10.001

Chao, H. T., Chen, H., Samaco, R. C., Xue, M., Chahrour, M., Yoo, J., et al. (2010). Dysfunction in GABA signalling mediates autism-like stereotypies and Rett syndrome phenotypes. *Nature* 468, 263–269. doi: 10.1038/nature09582

Chao, H. T., Zoghbi, H. Y., and Rosenmund, C. (2007). MeCP2 controls excitatory synaptic strength by regulating glutamatergic synapse number. *Neuron* 56, 58–65. doi: 10.1016/j.neuron.2007.08.018

Chen, L., Chen, K., Lavery, L. A., Baker, S. A., Shaw, C. A., Li, W., et al. (2015). MeCP2 binds to non-CG methylated DNA as neurons mature, influencing transcription and the timing of onset for Rett syndrome. *Proc. Natl. Acad. Sci. U.S.A.* 112, 5509–5514. doi: 10.1073/pnas.1505909112

Chen, W. G., Chang, Q., Lin, Y., Meissner, A., West, A. E., Griffith, E. C., et al. (2003). Derepression of BDNF transcription involves calcium-dependent phosphorylation of MeCP2. *Science* 302, 885–889. doi: 10.1126/science.1086446

Cheng, P. Y., Lin, Y. P., Chen, Y. L., Lee, Y. C., Tai, C. C., Wang, Y. T., et al. (2011). Interplay between SIN3A and STAT3 mediates chromatin conformational changes and GFAP expression during cellular differentiation. *PLOS ONE* 6:e22018. doi: 10.1371/journal.pone.0022018

Cronk, J. C., Derecki, N. C., Ji, E., Xu, Y., Lampano, A. E., Smirnov, I., et al. (2015). Methyl-CpG binding protein 2 regulates microglia and macrophage gene expression in response to inflammatory stimuli. *Immunity* 42, 679–691. doi: 10.1016/j.immuni.2015.03.013

De Filippis, B., Fabbri, A., Simone, D., Canese, R., Ricceri, L., Malchiodi-Albedi, F., et al. (2012). Modulation of RhoGTPases improves the behavioral phenotype and reverses astrocytic deficits in a mouse model of Rett syndrome. *Neuropsychopharmacology* 37, 1152–1163. doi: 10.1038/npp.2011.301

Delepine, C., Meziane, H., Nectoux, J., Opitz, M., Smith, A. B., Ballatore, C., et al. (2016). Altered microtubule dynamics and vesicular transport in mouse and human MeCP2-deficient astrocytes. *Hum. Mol. Genet.* 25, 146–157. doi: 10.1093/hmg/ddv464

Delepine, C., Nectoux, J., Bahi-Buisson, N., Chelly, J., and Bienvenu, T. (2013). MeCP2 deficiency is associated with impaired microtubule stability. *FEBS Lett.* 587, 245–253. doi: 10.1016/j.febslet.2012.11.033

Delepine, C., Nectoux, J., Letourneur, F., Baud, V., Chelly, J., Billuart, P., et al. (2015). Astrocyte transcriptome from the Mecp2(308)-truncated mouse model of rett syndrome. *Neuromol. Med.* 17, 353–363. doi: 10.1007/s12017-015-8363-9

Derecki, N. C., Cronk, J. C., Lu, Z., Xu, E., Abbott, S. B., Guyenet, P. G., et al. (2012). Wild-type microglia arrest pathology in a mouse model of Rett syndrome. *Nature* 484, 105–109. doi: 10.1038/nature10907

Du, Q., Luu, P. L., Stirzaker, C., and Clark, S. J. (2015). Methyl-CpG-binding domain proteins: readers of the epigenome. *Epigenomics* 7, 1051–1073. doi: 10.2217/epi.15.39

Durand, T., De Felice, C., Signorini, C., Oger, C., Bultel-Ponce, V., Guy, A., et al. (2013). F(2)-Dihomo-isoprostanes and brain white matter damage in stage 1 Rett syndrome. *Biochimie* 95, 86–90. doi: 10.1016/j.biochi.2012.09.017

Evans, J. C., Archer, H. L., Colley, J. P., Ravn, K., Nielsen, J. B., Kerr, A., et al. (2005). Early onset seizures and Rett-like features associated with mutations in CDKL5. *Eur. J. Hum. Genet.* 13, 1113–1120. doi: 10.1038/sj.ejhg.5201451

Feldman, D., Banerjee, A., and Sur, M. (2016). Developmental dynamics of Rett syndrome. *Neural Plast.* 2016:6154080. doi: 10.1155/2016/6154080

Fichou, Y., Nectoux, J., Bahi-Buisson, N., Rosas-Vargas, H., Girard, B., Chelly, J., et al. (2009). The first missense mutation causing Rett syndrome specifically affecting the MeCP2_e1 isoform. *Neurogenetics* 10, 127–133. doi: 10.1007/s10048-008-0161-1

Forbes-Lorman, R. M., Kurian, J. R., and Auger, A. P. (2014). MeCP2 regulates GFAP expression within the developing brain. *Brain Res.* 1543, 151–158. doi: 10.1016/j.brainres.2013.11.011

Fyffe, S. L., Neul, J. L., Samaco, R. C., Chao, H. T., Ben-Shachar, S., Moretti, P., et al. (2008). Deletion of Mecp2 in Sim1-expressing neurons reveals a critical role for MeCP2 in feeding behavior, aggression, and the response to stress. *Neuron* 59, 947–958. doi: 10.1016/j.neuron.2008.07.030

Garg, S. K., Lioy, D. T., Knopp, S. J., and Bissonnette, J. M. (2015). Conditional depletion of methyl-CpG-binding protein 2 in astrocytes depresses the hypercapnic ventilatory response in mice. *J. Appl. Physiol.* 119, 670–676. doi: 10.1152/japplphysiol.00411.2015

Gemelli, T., Berton, O., Nelson, E. D., Perrotti, L. I., Jaenisch, R., and Monteggia, L. M. (2006). Postnatal loss of methyl-CpG binding protein 2 in the forebrain is sufficient to mediate behavioral aspects of Rett syndrome in mice. *Biol. Psychiatry* 59, 468–476. doi: 10.1016/j.biopsych.2005.07.025

Gold, W. A., Lacina, T. A., Cantrill, L. C., and Christodoulou, J. (2015). MeCP2 deficiency is associated with reduced levels of tubulin acetylation and can be restored using HDAC6 inhibitors. *J. Mol. Med.* 93, 63–72. doi: 10.1007/s00109-014-1202-x

Gourine, A. V., Kasymov, V., Marina, N., Tang, F., Figueiredo, M. F., Lane, S., et al. (2010). Astrocytes control breathing through pH-dependent release of ATP. *Science* 329, 571–575. doi: 10.1126/science.1190721

Groc, L., Gustafsson, B., and Hanse, E. (2006). AMPA signalling in nascent glutamatergic synapses: there and not there! *Trends Neurosci.* 29, 132–139. doi: 10.1016/j.tins.2006.01.005

He, L., and Lu, Q. R. (2013). Coordinated control of oligodendrocyte development by extrinsic and intrinsic signaling cues. *Neurosci. Bull.* 29, 129–143. doi: 10.1007/s12264-013-1318-y

He, L. J., Liu, N., Cheng, T. L., Chen, X. J., Li, Y. D., Shu, Y. S., et al. (2014). Conditional deletion of Mecp2 in parvalbumin-expressing GABAergic cells results in the absence of critical period plasticity. *Nat. Commun.* 5:5036. doi: 10.1038/ncomms6036

Horiuchi, M., Smith, L., Maezawa, I., and Jin, L. W. (2016). CX3CR1 ablation ameliorates motor and respiratory dysfunctions and improves survival of a Rett syndrome mouse model. *Brain Behav. Immun.* 60, 106–116. doi: 10.1016/j.bbi.2016.02.014

Huang, N., Niu, J., Feng, Y., and Xiao, L. (2015). Oligodendroglial development: new roles for chromatin accessibility. *Neuroscientist* 21, 579–588. doi: 10.1177/1073858414565467

Jin, L. W., Horiuchi, M., Wulff, H., Liu, X. B., Cortopassi, G. A., Erickson, J. D., et al. (2015). Dysregulation of glutamine transporter SNAT1 in Rett syndrome microglia: a mechanism for mitochondrial dysfunction and neurotoxicity. *J. Neurosci.* 35, 2516–2529. doi: 10.1523/JNEUROSCI.2778-14.2015

Johnson, C. M., Cui, N., Zhong, W., Oginsky, M. F., and Jiang, C. (2015). Breathing abnormalities in a female mouse model of Rett syndrome. *J. Physiol. Sci.* 65, 451–459. doi: 10.1007/s12576-015-0384-5

Johnston, M. V., Jeon, O. H., Pevsner, J., Blue, M. E., and Naidu, S. (2001). Neurobiology of Rett syndrome: a genetic disorder of synapse development. *Brain Dev.* 23(Suppl. 1), S206–S213. doi: 10.1016/S0387-7604(01)00351-5

Kang, S. K., Kim, S. T., Johnston, M. V., and Kadam, S. D. (2014). Temporal- and location-specific alterations of the GABA recycling system in Mecp2 KO mouse brains. *J. Cent. Nerv. Syst. Dis.* 6, 21–28. doi: 10.4137/JCNSD.S14012

Karki, P., Webb, A., Smith, K., Johnson, J Jr, Lee, K., Son, D. S., et al. (2014). Yin Yang 1 is a repressor of glutamate transporter EAAT2, and it mediates manganese-induced decrease of EAAT2 expression in astrocytes. *Mol. Cell. Biol.* 34, 1280–1289. doi: 10.1128/MCB.01176-13

Kawabori, M., and Yenari, M. A. (2015). The role of the microglia in acute CNS injury. *Metab. Brain Dis.* 30, 381–392. doi: 10.1007/s11011-014-9531-6

Khong, P. L., Lam, C. W., Ooi, C. G., Ko, C. H., and Wong, V. C. (2002). Magnetic resonance spectroscopy and analysis of MECP2 in Rett syndrome. *Pediatr. Neurol.* 26, 205–209. doi: 10.1016/S0887-8994(01)00385-X

KhorshidAhmad, T., Acosta, C., Cortes, C., Lakowski, T. M., Gangadaran, S., and Namaka, M. (2016). Transcriptional regulation of brain-derived neurotrophic factor (BDNF) by methyl CpG binding protein 2 (MeCP2): a novel mechanism for re-myelination and/or myelin repair involved in the treatment of multiple sclerosis (MS). *Mol. Neurobiol.* 53, 1092–1107. doi: 10.1007/s12035-014-9074-1

Kiernan, E. A., Smith, S. M., Mitchell, G. S., and Watters, J. J. (2016). Mechanisms of microglial activation in models of inflammation and hypoxia: implications for chronic intermittent hypoxia. *J. Physiol.* 594, 1563–1577. doi: 10.1113/JP271502

Kifayathullah, L. A., Arunachalam, J. P., Bodda, C., Agbemenyah, H. Y., Laccone, F. A., and Mannan, A. U. (2010). MeCP2 mutant protein is expressed in astrocytes as well as in neurons and localizes in the nucleus. *Cytogenet. Genome. Res.* 129, 290–297. doi: 10.1159/000315906

Kloukina-Pantazidou, I., Chrysanthou-Piterou, M., Havaki, S., and Issidorides, M. R. (2013). Chromogranin A and vesicular monoamine transporter 2 immunolocalization in protein bodies of human locus coeruleus neurons. *Ultrastruct. Pathol.* 37, 102–109. doi: 10.3109/01913123.2012.750410

Lehmann, C., Bette, S., and Engele, J. (2009). High extracellular glutamate modulates expression of glutamate transporters and glutamine synthetase in cultured astrocytes. *Brain Res.* 1297, 1–8. doi: 10.1016/j.brainres.2009.08.070

Li, D., Herault, K., Zylbersztejn, K., Lauterbach, M. A., Guillon, M., Oheim, M., et al. (2015). Astrocyte VAMP3 vesicles undergo Ca2+ -independent cycling and modulate glutamate transporter trafficking. *J. Physiol.* 593, 2807–2832. doi: 10.1113/JP270362

Lioy, D. T., Garg, S. K., Monaghan, C. E., Raber, J., Foust, K. D., Kaspar, B. K., et al. (2011). A role for glia in the progression of Rett's syndrome. *Nature* 475, 497–500. doi: 10.1038/nature10214

Luikenhuis, S., Giacometti, E., Beard, C. F., and Jaenisch, R. (2004). Expression of MeCP2 in postmitotic neurons rescues Rett syndrome in mice. *Proc. Natl. Acad. Sci. U.S.A.* 101, 6033–6038. doi: 10.1073/pnas.0401626101

Lundgaard, I., Luzhynskaya, A., Stockley, J. H., Wang, Z., Evans, K. A., Swire, M., et al. (2013). Neuregulin and BDNF induce a switch to NMDA receptor-dependent myelination by oligodendrocytes. *PLOS Biol.* 11:e1001743. doi: 10.1371/journal.pbio.1001743

Luo, C., and Ecker, J. R. (2015). Epigenetics. Exceptional epigenetics in the brain. *Science* 348, 1094–1095. doi: 10.1126/science.aac5832

Lyst, M. J., and Bird, A. (2015). Rett syndrome: a complex disorder with simple roots. *Nat. Rev. Genet.* 16, 261–275. doi: 10.1038/nrg3897

Mackenzie, B., and Erickson, J. D. (2004). Sodium-coupled neutral amino acid (System N/A) transporters of the SLC38 gene family. *Pflugers. Arch.* 447, 784–795. doi: 10.1007/s00424-003-1117-9

Maezawa, I., and Jin, L. W. (2010). Rett syndrome microglia damage dendrites and synapses by the elevated release of glutamate. *J. Neurosci.* 30, 5346–5356. doi: 10.1523/JNEUROSCI.5966-09.2010

Maezawa, I., Swanberg, S., Harvey, D., Lasalle, J. M., and Jin, L. W. (2009). Rett syndrome astrocytes are abnormal and spread MeCP2 deficiency through gap junctions. *J. Neurosci.* 29, 5051–5061. doi: 10.1523/JNEUROSCI.0324-09.2009

Makedonski, K., Abuhatzira, L., Kaufman, Y., Razin, A., and Shemer, R. (2005). MeCP2 deficiency in Rett syndrome causes epigenetic aberrations at the PWS/AS imprinting center that affects UBE3A expression. *Hum. Mol. Genet.* 14, 1049–1058. doi: 10.1093/hmg/ddi097

Mari, F., Azimonti, S., Bertani, I., Bolognese, F., Colombo, E., Caselli, R., et al. (2005). CDKL5 belongs to the same molecular pathway of MeCP2 and it is responsible for the early-onset seizure variant of Rett syndrome. *Hum. Mol. Genet.* 14, 1935–1946. doi: 10.1093/hmg/ddi198

Matagne, V., Ehinger, Y., Saidi, L., Borges-Correia, A., Barkats, M., Bartoli, M., et al. (2017). A codon-optimized Mecp2 transgene corrects breathing deficits and improves survival in a mouse model of Rett syndrome. Neurobiol. Dis. 99, 1–11. doi: 10.1016/j.nbd.2016.12.009

Mencarelli, M. A., Spanhol-Rosseto, A., Artuso, R., Rondinella, D., De Filippis, R., Bahi-Buisson, N., et al. (2010). Novel FOXG1 mutations associated with the congenital variant of Rett syndrome. J. Med. Genet. 47, 49–53. doi: 10.1136/jmg.2009.067884

Meng, X., Wang, W., Lu, H., He, L. J., Chen, W., Chao, E. S., et al. (2016). Manipulations of MeCP2 in glutamatergic neurons highlight their contributions to Rett and other neurological disorders. Elife 5:e14199. doi: 10.7554/eLife.14199

Nakashima, N., Yamagata, T., Mori, M., Kuwajima, M., Suwa, K., and Momoi, M. Y. (2010). Expression analysis and mutation detection of DLX5 and DLX6 in autism. Brain Dev. 32, 98–104. doi: 10.1016/j.braindev.2008.12.021

Nectoux, J., Florian, C., Delepine, C., Bahi-Buisson, N., Khelfaoui, M., Reibel, S., et al. (2012). Altered microtubule dynamics in Mecp2-deficient astrocytes. J. Neurosci. Res. 90, 990–998. doi: 10.1002/jnr.23001

Nelson, S. B., and Valakh, V. (2015). Excitatory/inhibitory balance and circuit homeostasis in autism spectrum disorders. Neuron 87, 684–698. doi: 10.1016/j.neuron.2015.07.033

Nguyen, M. V., Du, F., Felice, C. A., Shan, X., Nigam, A., Mandel, G., et al. (2012). MeCP2 is critical for maintaining mature neuronal networks and global brain anatomy during late stages of postnatal brain development and in the mature adult brain. J. Neurosci. 32, 10021–10034. doi: 10.1523/JNEUROSCI.1316-12.2012

Nguyen, M. V., Felice, C. A., Du, F., Covey, M. V., Robinson, J. K., Mandel, G., et al. (2013). Oligodendrocyte lineage cells contribute unique features to Rett syndrome neuropathology. J. Neurosci. 33, 18764–18774. doi: 10.1523/JNEUROSCI.2657-13.2013

Nomura, Y. (2005). Early behavior characteristics and sleep disturbance in Rett syndrome. Brain Dev. 27(Suppl. 1), S35–S42. doi: 10.1016/j.braindev.2005.03.017

Noutel, J., Hong, Y. K., Leu, B., Kang, E., and Chen, C. (2011). Experience-dependent retinogeniculate synapse remodeling is abnormal in MeCP2-deficient mice. Neuron 70, 35–42. doi: 10.1016/j.neuron.2011.03.001

O'Driscoll, C. M., Lima, M. P., Kaufmann, W. E., and Bressler, J. P. (2015). Methyl CpG binding protein 2 deficiency enhances expression of inflammatory cytokines by sustaining NF-kappaB signaling in myeloid derived cells. J. Neuroimmunol. 283, 23–29. doi: 10.1016/j.jneuroim.2015.04.005

Okabe, Y., Takahashi, T., Mitsumasu, C., Kosai, K., Tanaka, E., and Matsuishi, T. (2012). Alterations of gene expression and glutamate clearance in astrocytes derived from an MeCP2-null mouse model of Rett syndrome. PLOS ONE 7:e35354. doi: 10.1371/journal.pone.0035354

Olson, C. O., Zachariah, R. M., Ezeonwuka, C. D., Liyanage, V. R., and Rastegar, M. (2014). Brain region-specific expression of MeCP2 isoforms correlates with DNA methylation within Mecp2 regulatory elements. PLOS ONE 9:e90645. doi: 10.1371/journal.pone.0090645

Palazzo, A., Ackerman, B., and Gundersen, G. G. (2003). Cell biology: Tubulin acetylation and cell motility. Nature 421:230. doi: 10.1038/421230a

Palmer, A., Qayumi, J., and Ronnett, G. (2008). MeCP2 mutation causes distinguishable phases of acute and chronic defects in synaptogenesis and maintenance, respectively. Mol. Cell. Neurosci. 37, 794–807. doi: 10.1016/j.mcn.2008.01.005

Philippe, O., Rio, M., Malan, V., Van Esch, H., Baujat, G., Bahi-Buisson, N., et al. (2013). NF-kappaB signalling requirement for brain myelin formation is shown by genotype/MRI phenotype correlations in patients with Xq28 duplications. Eur. J. Hum. Genet. 21, 195–199. doi: 10.1038/ejhg.2012.140

Robinson, L., Guy, J., Mckay, L., Brockett, E., Spike, R. C., Selfridge, J., et al. (2012). Morphological and functional reversal of phenotypes in a mouse model of Rett syndrome. Brain 135, 2699–2710. doi: 10.1093/brain/aws096

Roze, E., Cochen, V., Sangla, S., Bienvenu, T., Roubergue, A., Leu-Semenescu, S., et al. (2007). Rett syndrome: an overlooked diagnosis in women with stereotypic hand movements, psychomotor retardation, Parkinsonism, and dystonia? Mov. Disord. 22, 387–389. doi: 10.1002/mds.21276

Schafer, D. P., Heller, C. T., Gunner, G., Heller, M., Gordon, C., Hammond, T., et al. (2016). Microglia contribute to circuit defects in Mecp2 null mice independent

of microglia-specific loss of Mecp2 expression. Elife 5:e15224. doi: 10.7554/eLife.15224

Shahbazian, M. D., Antalffy, B., Armstrong, D. L., and Zoghbi, H. Y. (2002). Insight into Rett syndrome: MeCP2 levels display tissue- and cell-specific differences and correlate with neuronal maturation. Hum. Mol. Genet. 11, 115–124. doi: 10.1093/hmg/11.2.115

Sharma, K., Singh, J., Pillai, P. P., and Frost, E. E. (2015). Involvement of MeCP2 in regulation of myelin-related gene expression in cultured rat oligodendrocytes. J. Mol. Neurosci. 57, 176–184. doi: 10.1007/s12031-015-0597-3

Sinnett, S. E., Hector, R. D., Gadalla, K. K. E., Heindel, C., Chen, D., Zaric, V., et al. (2017). Improved MECP2 gene therapy extends the survival of MeCP2-null mice without apparent toxicity after intracisternal delivery. Mol. Ther. Methods Clin. Dev. 5, 106–115. doi: 10.1016/j.omtm.2017.04.006

Smrt, R. D., Eaves-Egenes, J., Barkho, B. Z., Santistevan, N. J., Zhao, C., Aimone, J. B., et al. (2007). Mecp2 deficiency leads to delayed maturation and altered gene expression in hippocampal neurons. Neurobiol. Dis. 27, 77–89. doi: 10.1016/j.nbd.2007.04.005

Squillaro, T., Alessio, N., Cipollaro, M., Melone, M. A., Hayek, G., Renieri, A., et al. (2012). Reduced expression of MECP2 affects cell commitment and maintenance in neurons by triggering senescence: new perspective for Rett syndrome. Mol. Biol. Cell 23, 1435–1445. doi: 10.1091/mbc.E11-09-0784

Stancheva, I., Collins, A. L., Van Den Veyver, I. B., Zoghbi, H., and Meehan, R. R. (2003). A mutant form of MeCP2 protein associated with human Rett syndrome cannot be displaced from methylated DNA by notch in Xenopus embryos. Mol. Cell. 12, 425–435. doi: 10.1016/S1097-2765(03)00276-4

Stearns, N. A., Schaevitz, L. R., Bowling, H., Nag, N., Berger, U. V., and Berger-Sweeney, J. (2007). Behavioral and anatomical abnormalities in Mecp2 mutant mice: a model for Rett syndrome. Neuroscience 146, 907–921. doi: 10.1016/j.neuroscience.2007.02.009

Tai, D. J., Liu, Y. C., Hsu, W. L., Ma, Y. L., Cheng, S. J., Liu, S. Y., et al. (2016). MeCP2 SUMOylation rescues Mecp2-mutant-induced behavioural deficits in a mouse model of Rett syndrome. Nat. Commun. 7:10552. doi: 10.1038/ncomms10552

Turovsky, E., Karagiannis, A., Abdala, A. P., and Gourine, A. V. (2015). Impaired CO2 sensitivity of astrocytes in a mouse model of Rett syndrome. J. Physiol. 593, 3159–3168. doi: 10.1113/JP270369

Vacca, M., Tripathi, K. P., Speranza, L., Aiese Cigliano, R., Scalabrì, F., Marracino, F., et al. (2016). Effects of Mecp2 loss of function in embryonic cortical neurons: a bioinformatics strategy to sort out non-neuronal cells variability from transcriptome profiling. BMC Bioinformat. 17(Suppl. 2):14. doi: 10.1186/s12859-015-0859-7

Vora, P., Mina, R., Namaka, M., and Frost, E. E. (2010). A novel transcriptional regulator of myelin gene expression: implications for neurodevelopmental disorders. Neuroreport 21, 917–921. doi: 10.1097/WNR.0b013e32833da500

Wang, J., Wegener, J. E., Huang, T. W., Sripathy, S., De Jesus-Cortes, H., Xu, P., et al. (2015). Wild-type microglia do not reverse pathology in mouse models of Rett syndrome. Nature 521, E1–E4. doi: 10.1038/nature14671

Williams, E. C., Zhong, X., Mohamed, A., Li, R., Liu, Y., Dong, Q., et al. (2014). Mutant astrocytes differentiated from Rett syndrome patients-specific iPSCs have adverse effects on wild-type neurons. Hum. Mol. Genet. 23, 2968–2980. doi: 10.1093/hmg/ddu008

Wu, W., Gu, W., Xu, X., Shang, S., and Zhao, Z. (2012). Downregulation of CNPase in a MeCP2 deficient mouse model of Rett syndrome. Neurol. Res. 34, 107–113. doi: 10.1179/016164111X13214359296301

Yasui, D. H., Xu, H., Dunaway, K. W., Lasalle, J. M., Jin, L. W., and Maezawa, I. (2013). MeCP2 modulates gene expression pathways in astrocytes. Mol. Autism 4:3. doi: 10.1186/2040-2392-4-3

Yuan, T., and Bellone, C. (2013). Glutamatergic receptors at developing synapses: the role of GluN3A-containing NMDA receptors and GluA2-lacking AMPA receptors. Eur. J. Pharmacol. 719, 107–111. doi: 10.1016/j.ejphar.2013.04.056

Zachariah, R. M., Olson, C. O., Ezeonwuka, C., and Rastegar, M. (2012). Novel MeCP2 isoform-specific antibody reveals the endogenous MeCP2E1 expression in murine brain, primary neurons and astrocytes. *PLOS ONE* 7:e49763. doi: 10.1371/journal.pone.0049763

Zhang, X., Su, J., Cui, N., Gai, H., Wu, Z., and Jiang, C. (2011). The disruption of central CO2 chemosensitivity in a mouse model of Rett syndrome. *Am. J. Physiol. Cell Physiol.* 301, C729–C738. doi: 10.1152/ajpcell.00334. 2010

Zhao, D., Mokhtari, R., Pedrosa, E., Birnbaum, R., Zheng, D., and Lachman, H. M. (2017). Transcriptome analysis of microglia in a mouse model of Rett syndrome: differential expression of genes associated with microglia/macrophage activation and cellular stress. *Mol. Autism* 8:17. doi: 10.1186/s13229-017-0134-z

Zhou, Z., Hong, E. J., Cohen, S., Zhao, W. N., Ho, H. Y., Schmidt, L., et al. (2006). Brain-specific phosphorylation of MeCP2 regulates activity-dependent Bdnf transcription, dendritic growth, and spine maturation. *Neuron* 52, 255–269. doi: 10.1016/j.neuron.2006.09.037

7

Role of Glial Immunity in Lifespan Determination: A *Drosophila* Perspective

Ilias Kounatidis[1]* and Stanislava Chtarbanova[2]*

[1] *Cell Biology, Development, and Genetics Laboratory, Department of Biochemistry, University of Oxford, Oxford, United Kingdom,* [2] *Department of Biological Sciences, University of Alabama, Tuscaloosa, AL, United States*

Correspondence:
Ilias Kounatidis
ilias.kounatidis@bioch.ox.ac.uk;
Stanislava Chtarbanova
schtarbanova@ua.edu

Increasing body of evidence indicates that proper glial function plays an important role in neuroprotection and in organismal physiology throughout lifespan. Work done in the model organism *Drosophila melanogaster* has revealed important aspects of glial cell biology in the contexts of longevity and neurodegeneration. In this mini review, we summarize recent findings from work done in the fruit fly *Drosophila* about the role of glia in maintaining a healthy status during animal's life and discuss the involvement of glial innate immune pathways in lifespan and neurodegeneration. Overactive nuclear factor kappa B (NF-κB) pathways and defective phagocytosis appear to be major contributors to lifespan shortening and neuropathology. Glial NF-κB silencing on the other hand, extends lifespan possibly through an immune–neuroendocrine axis. Given the evolutionary conservation of NF-κB innate immune signaling and of macrophage ontogeny across fruit flies, rodents, and humans, the above observations in glia could potentially support efforts for therapeutic interventions targeting to ameliorate age-related pathologies.

Keywords: *Drosophila*, innate immunity, glia, lifespan, neurodegeneration, phagocytosis

INTRODUCTION

Organismal aging is a complex phenomenon resulting in the progressive decline of physiological functions and increased susceptibility to death (1). Both, genetic and environmental factors are believed to contribute to the aging process and lifespan (1–3). Work in several model organisms including the invertebrates *Drosophila melanogaster* and *Caenorhabditis elegans* have identified genes and cellular pathways conserved through evolution that affect longevity such as the insulin-like pathway or the target of rapamycin (TOR) pathway (4, 5). Thus, these model organisms proved to be of valuable use for studying the molecular mechanisms that underlie aging.

Drosophila, the common fruit fly, is an excellent versatile model organism to investigate the interplay between innate immune function and brain physiology among the effects of this interaction to host lifespan. There is a high degree of evolutionary conservation of the molecular mechanisms of innate immunity between flies and mammals. For instance, the two *Drosophila* nuclear factor kappa B (NF-κB)-based pathways, namely Toll and Immune deficiency (IMD) share similarities with the mammalian Toll-like receptor pathways and tumor necrosis factor receptor 1 pathways, respectively (6–10). In the context of bacterial and fungal infections, activation of these pathways leads to the translocation of NF-κB factors (Relish for IMD and Dif and Dorsal for Toll pathway, respectively) from the cytoplasm into the nucleus of the cell allowing transcription and synthesis of potent antimicrobial peptides (AMPs) (10). Phagocytosis is another powerful mechanism to eliminate cellular debris or infection that has been conserved during evolution (11, 12). In mammals, phagocytosis is mediated by cell surface receptors, which bind bacteria or apoptotic bodies either directly or via opsonins (13). In flies, several phagocytic recognition receptors have been identified on hemocytes (the fly macrophage-like cells), among which is the EGF-like repeat-containing protein Draper (12). Draper has also been implicated in the removal of apoptotic neurons during *Drosophila* nervous system development (14) and metamorphosis (15) as well as in phagocytosis of axonal debris after axonal injury (16–18). Flies have also significantly contributed to advances in studies of neurodegeneration such as the identification of novel neuroprotective genes and provided information about conserved processes required for maintaining the structural integrity of the central nervous system (CNS) (19). Moreover, several human neurodegenerative diseases such as Alzheimer's, Parkinson's, and Huntington's disease have been effectively modeled in *Drosophila* yielding insights into the molecular base of these disorders (20).

The chronic inflammatory status that accompanies human aging, also known as inflammaging, is considered a significant risk factor for many chronic pathologies including cancer, cardiovascular and neurodegenerative disorders (21). In the context of aging, increased levels of pro-inflammatory cytokines such as TNF-alpha and Interleukine (IL)-6 are found upregulated in brain tissue (22). With age, mammalian microglia, which are the brain immune cells exhibit primed profile characterized by increased activation and enhanced secretion of pro-inflammatory cytokines (23, 24). Decline in microglial function, migration and chemotaxis are also observed with age (24). For instance, microglia's engulfment capacity of amyloid-beta (Aβ) (25) or alpha-synuclein (α-Syn) (26) oligomers, whose accumulation is characteristic for Alzheimer's and Parkinson's disease, respectively, are compromised in aged animals. Moreover, activated microglia and neuroinflammatory profiles are observed in most neurodegenerative disorders including Huntington's (27), Alzheimer's (28, 29), and Parkinson's (30–32) diseases and are believed to underlie the onset, severity, and progression of these disorders (24). Similar to mammalian models, both chronic innate immune activation (4, 33) as well as decline in phagocytic activity of glia (18) are observed in the aging *Drosophila* brain. It is thus apparent that glial immunity is linked to both, healthy aging and age-dependent neurodegeneration. In the mammalian brain, under normal physiological conditions, microglia provide the first line of defense against brain injury and infection. These cells are able to sense pathogens *via* pathogen recognition receptors, activate innate immune signaling pathways, phagocytose microorganisms, and clear cellular debris (34). Microglia also have the capacity to secrete neurotrophic factors and anti-inflammatory molecules, therefore, playing a protective role in these contexts. On the other hand, the neurodegenerative process itself can trigger inflammation (34–36), leading to detrimental effects on the brain. It is, therefore, important to understand the mechanisms by which, changes in the same signaling pathway (e.g., NF-kB) lead to two distinct phenotypes, namely healthy aging associated with neuroprotection and neurodegeneration.

Glial cells are essential players in CNS development and in maintaining homeostasis in this tissue (37). Glial cells provide trophic support to neurons, regulate ionic homeostasis in the brain, and serve as immune cells that are armed to respond to injury or infection (37). Increasing body of evidence indicates that dysfunction of diverse cellular processes specifically in glial cells may have profound impact on animal's survival and, therefore, affect life expectancy. We review here recent discoveries of the role played by glial cells in animal's lifespan, as well as how glial innate immune pathways relate to organismal longevity in the model organism *Drosophila melanogaster*.

GLIAL TYPES AND THEIR CONTRIBUTION TO HEALTHY AGING

Glial cells play major roles in nervous system development, synapse formation, plasticity, and brain homeostasis (38, 39). Five major morphologically distinct classes of glial cells with diverse functions can be appreciated in the brain of adult *Drosophila* (38–40) and additional glial subtypes in its visual system (40, 41). Among brain glia, perineurial and subperineurial glia form the blood–brain barrier (BBB) to isolate the brain from the potassium-rich hemolymph (insect blood) assuring optimal neuronal function (42–44). To meet the high-energy demands of neurons, glia supply neurons with metabolites through an evolutionary conserved process known as metabolic coupling (45). *Drosophila* BBB glia support neurons by providing them metabolic factors derived from the break down of the sugar trehalose (45). Glia-, but not neuron-specific silencing of the genes encoding the glycolytic enzymes *Trehalase* and *Pyruvate kinase* substantially shortens flies' lifespan and leads to neurodegeneration (45). Interestingly, glia-specific knock down of another gene encoding an enzyme involved in glycolysis, *Aldolase*, leads to neurodegeneration and shortened lifespan (46). Together, these studies indicate that glial glycolysis plays an important role in healthy aging and neuroprotection. Along the same lines, mutations in the gene encoding the glia-enriched monocarboxylate transporter *Chaski*, lead to shortened lifespan, synaptic dysfunction, and locomotor impairment pointing to an important role for glial transport of metabolites during the animal's lifespan (47). This work is of particular importance as it is becoming increasingly evident that metabolic changes, among which lower brain glucose metabolism, accompanies aging and Alzheimer's disease in humans (48). Metabolomics analysis of mouse brain samples reveals compromised energy state in the aging brain, possibly affecting glial cells that supply glycolytic substrates to neurons (49).

Ensheathing glia, which express the engulfment receptor Draper are the main subtype of adult *Drosophila* glial cells that phagocytose axonal debris following nerve injury (50). Cortex glia play an important trophic role for neurons in the adult brain (38) and are important for neuronal excitability as downregulation of a potassium-dependent sodium/calcium exchanger in this glial subtype leads to seizures in the adult (51). Astrocytes, which share morphological and functional properties with mammalian astrocytes (38) are major contributors to the maintenance of neurotransmitter homeostasis and are involved in regulating circadian rhythmicity in the adult (52). In the *Drosophila*, adult visual system several glial subtypes among, which epithelial glia play a role in synaptic transmission and the processing of visual information (41). It was recently reported that lipid droplet accumulation in these glia due to mitochondrial dysfunction in neurons contributes to neurodegeneration (53). The recently discovered Semper glia, which share functional features with mammalian visual system glial cells such as Müller glia, astrocytes, and oligodendrocytes provide support to photoreceptors and prevent light-induced retinal degeneration (40).

Another, unique, microglia-like glial subtype has been recently discovered in *Drosophila*. These cells, called MANF immunoreactive cells (MiCs) are transiently present in the metamorphosing pupal brain, but not in the adult stage, and upon certain conditions in glia, including (*i*) genetic silencing of dmMANF (Mesencephalic astrocyte-derived neurotrophic factor), (*ii*) induction of autophagy via overexpression of *Atg-1* or the dominant-negative form of Target of rapamycin (*TOR^{TED}*), and (*iii*) challenge of innate immunity by ectopic expression of IMD-pathway receptors *PGRP-LC* or *PGRP-LE* (54). Although MiCs have features reminiscent of mammalian microglia, they were not observed in brains of 10-day-old fly mutants that exhibit neurodegeneration such as *ATM^8* (discussed later in the text) and *swiss cheese* or of *Drosophila* α-Syn model of Parkinson's disease. It appears that MiCs are immunoreactive; they express the NF-κB transcription factor Relish in their nucleus as well as the phagocytic receptor Draper on their surface (54). Interestingly, glia-specific silencing of *dmMANF* also leads to neurodegeneration and shorter lifespan in the adult suggesting a homeostatic role for this gene in glia (55). How exactly MiCs contribute to this phenotype is not known. More studies are needed in order to fully characterize this intriguing glial cell population and how exactly they relate to brain immune function and healthy lifespan. Glial subtypes and their function in the adult are presented in **Table 1**.

ROLE OF GLIAL IMMUNITY IN NEURODEGENERATION AND SHORTENED LIFESPAN

Prolonged activation of inflammatory responses often translates into harmful consequences ultimately leading to reduction of animal's lifespan. In the context of brain aging, persistent IMD pathway activation, as well as defective glial clearance function are associated with neurodegenerative phenotypes. Several studies in *Drosophila* highlight the role of glial innate immune responses (phagocytosis as well as NF-κB activation) in promoting neurodegeneration and shortening lifespan. One of the first reports correlating glial activation of the NF-κB ortholog Relish with neurodegenerative phenotypes and lifespan shortening is a study done in a fly model of Ataxia-telangiectasia (A-T) (61). This work shows that glial cells in the fly brain are responsible for increased innate immune activation when ATM kinase activity is reduced. AMPs, which are direct NF-κB transcriptional targets, are up regulated exclusively in glial cells in *ATM^8* mutants leading to Caspase-3 activation in neighboring neurons suggesting that neurodegeneration could be driven by increase in glial immunity (61). Flies in which *ATM* is specifically silenced in glial cells exhibit shortened lifespan, premature defects in locomotor activity, and spongiform brain pathology in conjunction with activation of

TABLE 1 | Glial subtypes and their location and functions in the adult.

Glial subtype	Function in adult	Location	Reference
Cortex glia	– Trophic support to neurons – Regulation of seizure susceptibility	– Brain cortex – Wrap neuronal cell bodies and processes	Kremer et al. (39) Stork et al. (56) Melom et al. (51)
Astrocyte-like glia	– Maintenance of neurotransmitter homeostasis – Circadian rhythm regulation	– Brain neuropil	Kremer et al. (39) Rival et al. (57) Stork et al. (58) Suh et al. (59) Ng et al. (52)
Ensheathing glia	– Phagocytosis of debris after injury – Regulation of olfactory circuit plasticity	– Brain neuropil – Associated with axon tracts	Kremer et al. (39) Doherty et al. (50) Kazama et al. (60)
Perineurial glia	– Blood–brain barrier (BBB) formation and chemoisolaion – Sugar import into the CNS	– Brain surface	Kremer et al. (39) Featherstone (44) Hindle et al. (43) Miller et al. (46) Volkenhoff et al. (45)
Subperineurial glia	– BBB formation and chemoisolaion	– Brain surface	Kremer et al. (39) Featherstone (44) Hindle et al. (43)
MANF immunoreactive cells	– Microglia-like cells	– Pupal brain neuropil	Stratoulias et al. (54)
Adult visual system glia	– Role in synaptic transmission – Prevent light-induced retinal degeneration	– Optic lobe	Chotard et al. (41) Charlton-Perkins et al. (40)

Caspase-3 indicative for neurodegeneration (61). Results from a subsequent study done by the same group show that the degree of activation of the innate immune response correlates with the severity of neurodegeneration and lifespan duration in *ATM^8* mutants and that glial overexpression of a constitutively active form of Relish (Rel-D) leads to neurodegeneration (62). The fact that innate immune activation in brain tissue contributes to neuropathology is supported by findings from other groups that implicate both, activation of the IMD and Toll pathways in *Drosophila* models of light-induced retinal degeneration and Alzheimer's disease, respectively (63, 64). Pan-neuronal activation of constitutively active Relish results in increased lethality at eclosion pointing to a toxic effect of prolonged immunity in nerve cells (63); however, glial-specific effects of innate immune activation on lifespan and neurodegeneration in these models remains to be determined.

Another piece of evidence for direct implication of the NF-κB-dependent innate immune response in neurodegeneration and longevity comes from the finding that mutation in *defense repressor 1* (*Dnr1*), a negative regulator of the IMD pathway acting at the level of the caspase Dredd, leads to progressive neurodegeneration and reduced lifespan in a Relish-dependent manner (65).

In the same study, the authors report that bacterially induced progressive neurodegeneration and resulting locomotor defects are suppressed when *Relish* is specifically knocked down in glial cells. This work goes further by demonstrating that glia-specific overexpression of several AMPs also leads to progressive neurodegeneration. Intriguingly, the overexpression of single AMP-coding genes in glia is sufficient to cause neuropathology and this effect is direct because glial knock down of *Relish* does not suppress *Defensin-* nor *Drosomycin-*induced neurodegeneration (65). A subsequent study demonstrates that glia-specific overexpression of individual AMPs results in impaired locomotor activity and shortened lifespan providing additional evidence for the effect of these NF-κB target genes on fitness and longevity (33).

The fly IMD pathway is tightly regulated at almost every step of the signaling cascade from the surface to the nucleus of the cell (8, 66). In addition to *Dnr-1*, mutants for other intracellular safeguards of the pathway such as *Pirk, Trabid,* and *Transglutaminase* (*Tg*) that act at the level of the adaptor protein Imd, the kinase TAK1, and the transcription factor Relish, respectively, also exhibit

shortened lifespan and neurodegeneration along with brain-specific upregulation of AMPs (33) (**Figure 1**). Glial silencing of *Relish* in *Trabid* mutants suppresses the age-dependent locomotor impairments and neurodegeneration in these flies and also restores lifespan to almost wild-type levels (33). Altogether, these studies attribute a role for glial NF-κB activation and downstream effectors such as the AMPs in lifespan and neurodegeneration.

Equally important to the overactive NF-κB/Relish branch of the immune response, are the alterations in glial phagocytosis, which are also associated with enhanced neurodegeneration and reduction in lifespan. A recent report showed that while protein levels of the engulfment receptor Draper are reduced in an age-dependent manner, glia's efficiency in removing cellular debris, such as the ones deriving from degenerating neurons, declines (18). In the context of healthy aging, reduced Draper levels follow an age-associated regression of glial phosphoinositide-3-kinase signaling that mediates TOR-dependent translation of *draper* mRNA (**Figure 1**) while in situations of neuronal injuries a STAT92E-dependent transcriptional upregulation of *draper* has

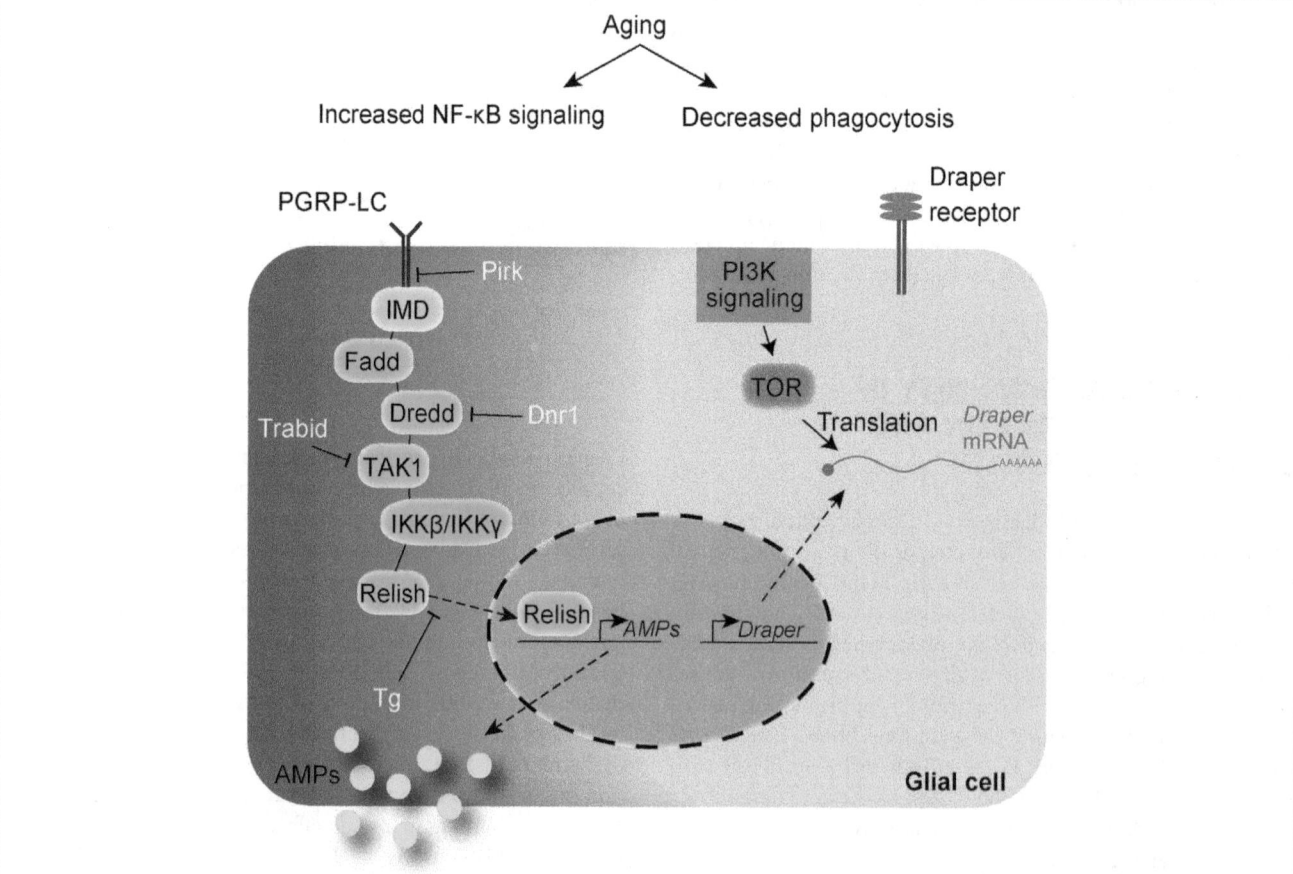

FIGURE 1 | Age-dependent changes in innate immune pathways in *Drosophila* glial cells: immune deficiency (IMD) pathway (on the left) shows age-dependent activation resulting in increased levels of antimicrobial peptides (AMPs) in middle and old-aged adults in absence of microbial challenge. Mutations in genes encoding specific IMD negative regulators namely *Dnr1*, *trabid*, *Transglutaminase (Tg,)* and *pirk* release the pathway allowing activation of Relish and subsequent transcription of downstream genes including those encoding AMPs. Aging also affects expression of the phagocytic receptor Draper (on the right) leading to inefficient phagocytic capacity of glial cells. Draper expression levels in the healthy brain are regulated *via* phosphoinositide-3-kinase signaling activity that mediates TOR-dependent translation of *draper* mRNA in glial cells. Age related decline in the activity of this signaling cascade leads to reduction in protein levels of Draper in glia.

been described (18, 67). Moreover, flies mutant for Draper exhibit short lifespan (68) and age-dependent neurodegeneration (69), pointing to a neuroprotective role for the phagocytic branch of the innate immune response. Flies, in which *Draper* is silenced specifically in glial cells, also exhibit age-dependent neurodegeneration (69). Persistent apoptotic neurons throughout the lifespan of *Draper* mutants that are not efficiently processed by glial cells due to defects in phagosome maturation appear to be the main reason for the observed phenotypes (69). However, how exactly corpse processing defects cause neurodegeneration remains to be determined.

ROLE OF GLIAL IMMUNITY IN LIFESPAN EXTENSION

Kounatidis and colleagues (33) describe an age-dependent shift in IMD-related AMP transcription in *Drosophila* including tissues like the brain of adult flies that is accompanied by neurological impairments such as decreased locomotor performance and increased neurodegeneration in 50-day-old wild-type flies (33). In the same study, the suppression of age-dependent progressive immunity by silencing three components of the IMD pathway, namely *Imd, Dredd,* and *Relish* in glial cells results in increased transcription of the *adipokinetic hormone* (*AKH*)-coding gene concomitant with high nutrient levels later in life and an extension of active lifespan (33). AKH is the fly ortholog of the mammalian gonadotropin-releasing hormone (70), which in mice controls an NF-κB-dependent immune-neuroendocrine axis that is involved in organismal aging (71). A remarkable deceleration of the aging process is recorded in mice upon hypothalamic blocking of NF-κB and the upstream kinase IKK-β (inhibitor of nuclear factor kappa-B kinase subunit beta), followed by an increased median longevity by nearly 20% (71). Similarly, in flies, glial-specific NF-κB immune signaling suppression results in 61% increase in median longevity compared to controls, which is also accompanied by increased locomotor activity in older age (33).

Recent studies in rodents that employ antiaging drugs link lifespan extension with reduction of age-associated overproduction of the pro-inflammatory cytokine TNF-alpha by microglia in both hypothalamus and hippocampus (72). Collectively, these studies implicate glial NF-κB signaling in lifespan determination and point to a role for the immune-neuroendocrine axis in this process.

CONCLUSION/FUTURE DIRECTIONS

It is becoming increasingly evident that glial cells play an important role in neuroprotection and in organismal physiology throughout lifespan. In the recent years, studies in the model organism *Drosophila*

have revealed numerous aspects of glial contribution toward both, healthy aging and the development and progression of age-related pathologies of the nervous system. Dysregulation of glial innate immune reactions such as improper NF-κB signaling or impaired Draper-based phagocytosis results in early onset neurodegeneration and lifespan shortening. Thus, both branches of the innate immune response seem to contribute in host neuroprotection and longevity. Additional work is needed to investigate whether these two pieces of the innate immune response possess synergistic properties and identify possible cellular factors that regulate both the inflammatory and phagocytic pathways in glial cells.

Injection of Alzheimer's disease-related Aβ oligomers (AβOs) into the brains of mice and macaques results in activated pro-inflammatory IKKβ/NF-κB signaling in the hypothalamus and subsequent induction of peripheral glucose intolerance (73). The majority of dementia-related diseases share an inflammation-based branch. Given the evolutionary conservation of innate immune signaling in flies, mice, and humans, strategies like challenging the inflammatory effect of NF-κB pathways could be proved an effective strategy in both "healthy aging" status and in cases of predisposition to age-related neurological diseases.

Additional avenues for future research will involve the studies of the exact mechanisms by which glial effectors downstream of NF-κB such as the AMPs induce neurotoxicity and shorten lifespan. Amyloid-β peptides, which are involved in Alzheimer's disease pathology have antimicrobial properties (74) and can protect the mouse brain from infection with pathogenic bacteria (75). AMPs can exert bactericidal effects by inserting into and disrupting bacterial membranes (76, 77). Interestingly, modeling studies have suggested that Aβ peptides can also insert into lipid bilayers and potentially form pores in cellular membranes, and thus can damage cells and lead to neurodegeneration (78). Flies can offer an excellent experimental system to address this question and provide important new insights into the mechanisms of AMP toxicity.

Interpolations that boost glial engulfment activity can delay age-dependent processes like delayed clearance of damaged neurons and cellular debris (18). Furthermore, it has been shown that enhanced glial engulfment reverses Aβ accumulation as well as associated behavioral phenotypes in a *Drosophila* AD model (68).

Therefore, glial immune signaling will potentially provide a new cohort of molecular foci for therapeutic interventions in cases of common incurable neurodegenerative disorders in the aging population.

AUTHOR CONTRIBUTIONS

IK and SC wrote the manuscript and designed the figures.

REFERENCES

Lopez-Otin C, Blasco MA, Partridge L, Serrano M, Kroemer G. The hallmarks of aging. *Cell* (2013) 153:1194–217. doi:10.1016/j.cell.2013.05.039

Rodriguez-Rodero S, Fernandez-Morera JL, Menendez-Torre E, Calvanese V, Fernandez AF, Fraga MF. Aging genetics and aging. *Aging Dis* (2011) 2:186–95.

Kirkwood TB. Time of our lives. What controls the length of life? *EMBO Rep* (2005) 6(Spec No):S4–8. doi:10.1038/sj.embor.7400419

He Y, Jasper H. Studying aging in *Drosophila*. *Methods* (2014) 68:129–33. doi:10.1016/j.ymeth.2014.04.008

Wood WB. Aging of *C. elegans*: mosaics and mechanisms. *Cell* (1998) 95:147–50. doi:10.1016/S0092-8674(00)81744-4

Chtarbanova S, Imler JL. Microbial sensing by Toll receptors: a historical perspective. *Arterioscler Thromb Vasc Biol* (2011) 31:1734–8. doi:10.1161/ATVBAHA.108.179523

Kleino A, Silverman N. The *Drosophila* IMD pathway in the activation of the humoral immune response. *Dev Comp Immunol* (2014) 42:25–35. doi:10.1016/j.dci.2013.05.014

Myllymaki H, Valanne S, Ramet M. The *Drosophila* IMD signaling pathway. *J Immunol* (2014) 192:3455–62. doi:10.4049/jimmunol.1303309

Valanne S, Wang JH, Ramet M. The *Drosophila* Toll signaling pathway. *J Immunol* (2011) 186:649–56. doi:10.4049/jimmunol.1002302

Kounatidis I, Ligoxygakis P. *Drosophila* as a model system to unravel the layers of innate immunity to infection. *Open Biol* (2012) 2:120075. doi:10.1098/rsob.120075

Wang LH, Kounatidis I, Ligoxygakis P. *Drosophila* as a model to study the role of blood cells in inflammation, innate immunity and cancer. *Front Cell Infect Microbiol* (2014) 3:113. doi:10.3389/fcimb.2013.00113

Ulvila J, Vanha-Aho LM, Ramet M. *Drosophila* phagocytosis – still many unknowns under the surface. *APMIS* (2011) 119:651–62. doi:10.1111/ j.1600-0463.2011.02792.x

Gordon S. Phagocytosis: an immunobiologic process. *Immunity* (2016) 44: 463–75. doi:10.1016/j.immuni.2016.02.026

Kurant E, Axelrod S, Leaman D, Gaul U. Six-microns-under acts upstream of Draper in the glial phagocytosis of apoptotic neurons. *Cell* (2008) 133:498–509. doi:10.1016/j.cell.2008.02.052

Hilu-Dadia R, Hakim-Mishnaevski K, Levy-Adam F, Kurant E. Draper- mediated JNK signaling is required for glial phagocytosis of apoptotic neurons during *Drosophila* metamorphosis. *Glia* (2018) 66(7):1520–32. doi:10.1002/ glia.23322

MacDonald JM, Beach MG, Porpiglia E, Sheehan AE, Watts RJ, Freeman MR. The *Drosophila* cell corpse engulfment receptor Draper mediates glial clear- ance of severed axons. *Neuron* (2006) 50:869–81. doi:10.1016/j.neuron.2006. 04.028

Awasaki T, Tatsumi R, Takahashi K, Arai K, Nakanishi Y, Ueda R, et al. Essential role of the apoptotic cell engulfment genes draper and ced-6 in programmed axon pruning during *Drosophila* metamorphosis. *Neuron* (2006) 50:855–67. doi:10.1016/j.neuron.2006.04.027

Purice MD, Speese SD, Logan MA. Delayed glial clearance of degenerating axons in aged *Drosophila* is due to reduced PI3K/Draper activity. *Nat Commun* (2016) 7:12871. doi:10.1038/ncomms12871

Lessing D, Bonini NM. Maintaining the brain: insight into human neuro-degeneration from *Drosophila melanogaster* mutants. *Nat Rev Genet* (2009) 10:359–70. doi:10.1038/nrg2563

McGurk L, Berson A, Bonini NM. *Drosophila* as an in vivo model for human neurodegenerative disease. *Genetics* (2015) 201:377–402. doi:10.1534/genetics.115.179457

Franceschi C, Campisi J. Chronic inflammation (inflammaging) and its potential contribution to age-associated diseases. *J Gerontol A Biol Sci Med Sci* (2014) 69(Suppl 1):S4–9. doi:10.1093/gerona/glu057

Michaud M, Balardy L, Moulis G, Gaudin C, Peyrot C, Vellas B, et al. Proinflammatory cytokines, aging, and age-related diseases. *J Am Med Dir Assoc* (2013) 14:877–82. doi:10.1016/j.jamda.2013.05.009

Rawji KS, Mishra MK, Michaels NJ, Rivest S, Stys PK, Yong VW. Immunosenescence of microglia and macrophages: impact on the ageing central nervous system. *Brain* (2016) 139:653–61. doi:10.1093/brain/ awv395

Spittau B. Aging microglia-phenotypes, functions and implications for age-related neurodegenerative diseases. *Front Aging Neurosci* (2017) 9:194. doi:10.3389/fnagi.2017.00194

Floden AM, Combs CK. Microglia demonstrate age-dependent interaction with amyloid-beta fibrils. *J Alzheimers Dis* (2011) 25:279–93. doi:10.3233/ JAD-2011-101014

Bliederhaeuser C, Grozdanov V, Speidel A, Zondler L, Ruf WP, Bayer H, et al. Age-dependent defects of alpha-synuclein oligomer uptake in microglia and monocytes. *Acta Neuropathol* (2016) 131:379–91. doi:10.1007/s00401- 015-1504-2

Crotti A, Benner C, Kerman BE, Gosselin D, Lagier-Tourenne C, Zuccato C, et al. Mutant Huntingtin promotes autonomous microglia activation via myeloid lineage-determining factors. *Nat Neurosci* (2014) 17:513–21. doi:10.1038/nn.3668

Lopategui Cabezas I, Herrera Batista A, Penton Rol G. The role of glial cells in Alzheimer disease: potential therapeutic implications. *Neurologia* (2014) 29:305–9. doi:10.1016/j.nrl.2012.10.006

Jones SV, Kounatidis I. Nuclear factor-kappa B and Alzheimer disease, unifying genetic and environmental risk factors from cell to humans. *Front Immunol* (2017) 8:1805. doi:10.3389/fimmu.2017.01805

Nolan YM, Sullivan AM, Toulouse A. Parkinson's disease in the nuclear age of neuroinflammation. *Trends Mol Med* (2013) 19:187–96. doi:10.1016/j.molmed.2012.12.003

More SV, Kumar H, Kim IS, Song SY, Choi DK. Cellular and molecular mediators of neuroinflammation in the pathogenesis of Parkinson's disease. *Mediators Inflamm* (2013) 2013:952375. doi:10.1155/2013/952375

Taylor JM, Main BS, Crack PJ. Neuroinflammation and oxidative stress: co-conspirators in the pathology of Parkinson's disease. *Neurochem Int* (2013) 62:803–19. doi:10.1016/j.neuint.2012.12.016

Kounatidis I, Chtarbanova S, Cao Y, Hayne M, Jayanth D, Ganetzky B, et al. NF-kappaB immunity in the brain determines fly lifespan in healthy aging and age-related neurodegeneration. *Cell Rep* (2017) 19:836–48. doi:10.1016/j.celrep.2017.04.007

Le W, Wu J, Tang Y. Protective microglia and their regulation in Parkinson's disease. *Front Mol Neurosci* (2016) 9:89. doi:10.3389/fnmol.2016.00089

Wyss-Coray T, Mucke L. Inflammation in neurodegenerative disease – a double-edged sword. *Neuron* (2002) 35:419–32. doi:10.1016/S0896-6273(02) 00794-8

Wilcock DM, Munireddy SK, Rosenthal A, Ugen KE, Gordon MN, Morgan D. Microglial activation facilitates Abeta plaque removal following intracra-nial anti-Abeta antibody administration. *Neurobiol Dis* (2004) 15:11–20. doi:10.1016/j.nbd.2003.09.015

Barres BA. The mystery and magic of glia: a perspective on their roles in health and disease. *Neuron* (2008) 60:430–40. doi:10.1016/j.neuron.2008.10.013

Freeman MR. *Drosophila* central nervous system glia. *Cold Spring Harb Perspect Biol* (2015) 7:a020552. doi:10.1101/cshperspect.a020552

Kremer MC, Jung C, Batelli S, Rubin GM, Gaul U. The glia of the adult *Drosophila* nervous system. *Glia* (2017) 65:606–38. doi:10.1002/glia.23115

Charlton-Perkins MA, Sendler ED, Buschbeck EK, Cook TA. Multifunctional glial support by Semper cells in the *Drosophila* retina. *PLoS Genet* (2017) 13:e1006782. doi:10.1371/journal.pgen.1006782

Chotard C, Salecker I. Glial cell development and function in the *Drosophila* visual system. *Neuron Glia Biol* (2007) 3:17–25. doi:10.1017/ S1740925X07000592

Limmer S, Weiler A, Volkenhoff A, Babatz F, Klambt C. The *Drosophila* blood-brain barrier: development and function of a glial endothelium. *Front Neurosci* (2014) 8:365. doi:10.3389/fnins.2014.00365

Hindle SJ, Bainton RJ. Barrier mechanisms in the *Drosophila* blood-brain barrier. *Front Neurosci* (2014) 8:414. doi:10.3389/fnins.2014.00414

Featherstone DE. Glial solute carrier transporters in *Drosophila* and mice. *Glia* (2011) 59:1351–63. doi:10.1002/glia.21085

Volkenhoff A, Weiler A, Letzel M, Stehling M, Klambt C, Schirmeier S. Glial glycolysis is essential for neuronal survival in *Drosophila*. *Cell Metab* (2015) 22:437–47. doi:10.1016/j.cmet.2015.07.006

Miller D, Hannon C, Ganetzky B. A mutation in *Drosophila* aldolase causes temperature-sensitive paralysis, shortened lifespan, and neurodegeneration. *J Neurogenet* (2012) 26:317–27. doi:10.3109/01677063.2012.706346

Delgado MG, Oliva C, Lopez E, Ibacache A, Galaz A, Delgado R, et al. Chaski, a novel *Drosophila* lactate/pyruvate transporter required in glia cells for survival under nutritional stress. *Sci Rep* (2018) 8:1186. doi:10.1038/ s41598-018-19595-5

Cunnane S, Nugent S, Roy M, Courchesne-Loyer A, Croteau E, Tremblay S, et al. Brain fuel metabolism, aging, and Alzheimer's disease. *Nutrition* (2011) 27:3–20. doi:10.1016/j.nut.2010.07.021

Ivanisevic J, Stauch KL, Petrascheck M, Benton HP, Epstein AA, Fang M, et al. Metabolic drift in the aging brain. *Aging (Albany NY)* (2016) 8:1000–20. doi:10.18632/aging.100961

Doherty J, Logan MA, Tasdemir OE, Freeman MR. Ensheathing glia function as phagocytes in the adult *Drosophila* brain. *J Neurosci* (2009) 29:4768–81. doi:10.1523/JNEUROSCI.5951-08.2009

Melom JE, Littleton JT. Mutation of a NCKX eliminates glial microdomain calcium oscillations and enhances seizure susceptibility. *J Neurosci* (2013) 33:1169–78. doi:10.1523/JNEUROSCI.3920-12.2013

Ng FS, Tangredi MM, Jackson FR. Glial cells physiologically modulate clock neurons and circadian behavior in a calcium-dependent manner. *Curr Biol* (2011) 21:625–34. doi:10.1016/j.cub.2011.03.027

Liu L, Zhang K, Sandoval H, Yamamoto S, Jaiswal M, Sanz E, et al. Glial lipid droplets and ROS induced by mitochondrial defects promote neurodegenera-tion. *Cell* (2015) 160:177–90. doi:10.1016/j.cell.2014.12.019

Stratoulias V, Heino TI. MANF silencing, immunity induction or autophagy trigger an unusual cell type in metamorphosing *Drosophila* brain. *Cell Mol Life Sci* (2015) 72:1989–2004. doi:10.1007/s00018-014-1789-7

Walkowicz L, Kijak E, Krzeptowski W, Gorska-Andrzejak J, Stratoulias V, Woznicka O, et al. Downregulation of DmMANF in glial cells results in neurodegeneration and affects sleep and lifespan in *Drosophila melanogaster*. *Front Neurosci* (2017) 11:610. doi:10.3389/fnins.2017.00610

Stork T, Bernardos R, Freeman MR. Analysis of glial cell development and function in *Drosophila*. *Cold Spring Harb Protoc* (2012) 2012:1–17. doi:10.1101/pdb.top067587

Rival T, Soustelle L, Strambi C, Besson MT, Iche M, Birman S. Decreasing glutamate buffering capacity triggers oxidative stress and neuropil degen- eration in the *Drosophila* brain. *Curr Biol* (2004) 14:599–605. doi:10.1016/j.cub.2004.03.039

Stork T, Sheehan A, Tasdemir-Yilmaz OE, Freeman MR. Neuron-glia inter- actions through the heartless FGF receptor signaling pathway mediate mor- phogenesis of *Drosophila* astrocytes. *Neuron* (2014) 83:388–403. doi:10.1016/j.neuron.2014.06.026

Suh J, Jackson FR. *Drosophila* ebony activity is required in glia for the circadian regulation of locomotor activity. *Neuron* (2007) 55:435–47. doi:10.1016/j.neuron.2007.06.038

Kazama H, Yaksi E, Wilson RI. Cell death triggers olfactory circuit plasticity via glial signaling in *Drosophila*. *J Neurosci* (2011) 31:7619–30. doi:10.1523/JNEUROSCI.5984-10.2011

Petersen AJ, Rimkus SA, Wassarman DA. ATM kinase inhibition in glial cells acti- vates the innate immune response and causes neurodegeneration in *Drosophila*. *Proc Natl Acad Sci U S A* (2012) 109:E656–64. doi:10.1073/pnas.1110470109

Petersen AJ, Katzenberger RJ, Wassarman DA. The innate immune response transcription factor relish is necessary for neurodegeneration in a *Drosophila* model of ataxia-telangiectasia. *Genetics* (2013) 194:133–42. doi:10.1534/genetics.113.150854

Chinchore Y, Gerber GF, Dolph PJ. Alternative pathway of cell death in *Drosophila* mediated by NF-kappaB transcription factor relish. *Proc Natl Acad Sci U S A* (2012) 109:E605–12. doi:10.1073/pnas.1110666109

Tan L, Schedl P, Song HJ, Garza D, Konsolaki M. The Toll–>NFkappaB signal- ing pathway mediates the neuropathological effects of the human Alzheimer's Abeta42 polypeptide in *Drosophila*. *PLoS One* (2008) 3:e3966. doi:10.1371/journal.pone.0003966

Cao Y, Chtarbanova S, Petersen AJ, Ganetzky B. Dnr1 mutations cause neurode- generation in *Drosophila* by activating the innate immune response in the brain. *Proc Natl Acad Sci U S A* (2013) 110:E1752–60. doi:10.1073/pnas.1306220110

Aggarwal K, Silverman N. Positive and negative regulation of the *Drosophila* immune response. *BMB Rep* (2008) 41:267–77. doi:10.5483/BMBRep.2008.41.4.267

Doherty J, Sheehan AE, Bradshaw R, Fox AN, Lu TY, Freeman MR. PI3K signaling

and Stat92E converge to modulate glial responsiveness to axonal injury. *PLoS Biol* (2014) 12:e1001985. doi:10.1371/journal.pbio.1001985

Ray A, Speese SD, Logan MA. Glial draper rescues abeta toxicity in a *Drosophila* model of Alzheimer's disease. *J Neurosci* (2017) 37(49):11881–93. doi:10.1523/JNEUROSCI.0862-17.2017

Etchegaray JI, Elguero EJ, Tran JA, Sinatra V, Feany MB, McCall K. Defective phagocytic corpse processing results in neurodegeneration and can be rescued by TORC1 activation. *J Neurosci* (2016) 36:3170–83. doi:10.1523/JNEUROSCI.1912-15.2016

Zandawala M, Tian S, Elphick MR. The evolution and nomenclature of GnRH- type and corazonin-type neuropeptide signaling systems. *Gen Comp Endocrinol* (2017). doi:10.1016/j.ygcen.2017.06.007

Zhang G, Li J, Purkayastha S, Tang Y, Zhang H, Yin Y, et al. Hypothalamic programming of systemic ageing involving IKK-beta, NF-kappaB and GnRH. *Nature* (2013) 497:211–6. doi:10.1038/nature12143

Sadagurski M, Cady G, Miller RA. Anti-aging drugs reduce hypothalamic inflammation in asex-specifi manner. *Aging Cell* (2017) 16:652–60. doi:10.1111/acel.12590

Clarke JR, Lyra ESNM, Figueiredo CP, Frozza RL, Ledo JH, Beckman D, et al. Alzheimer-associated Abeta oligomers impact the central nervous system to induce peripheral metabolic deregulation. *EMBO Mol Med* (2015) 7:190–210. doi:10.15252/emmm.201404183

Soscia SJ, Kirby JE, Washicosky KJ, Tucker SM, Ingelsson M, Hyman B, et al. The Alzheimer's disease-associated amyloid beta-protein is an antimicrobial peptide. *PLoS One* (2010) 5:e9505. doi:10.1371/journal.pone.0009505

Kumar DK, Choi SH, Washicosky KJ, Eimer WA, Tucker S, Ghofrani J, et al. Amyloid-beta peptide protects against microbial infection in mouse and worm models of Alzheimer's disease. *Sci Transl Med* (2016) 8:340ra72. doi:10.1126/scitranslmed.aaf1059

Bechinger B, Gorr SU. Antimicrobial peptides: mechanisms of action and resistance. *J Dent Res* (2017) 96:254–60. doi:10.1177/0022034516679973

Li J, Koh JJ, Liu S, Lakshminarayanan R, Verma CS, Beuerman RW. Membrane active antimicrobial peptides: translating mechanistic insights to design. *Front Neurosci* (2017) 11:73. doi:10.3389/fnins.2017.00073

Mobley DL, Cox DL, Singh RR, Maddox MW, Longo ML. Modeling amy- loid beta-peptide insertion into lipid bilayers. *Biophys J* (2004) 86:3585–97. doi:10.1529/biophysj.103.032342

8

Revisit the Candidacy of Brain Cell Types as the Cell(s) of Origin for Human High-Grade Glioma

Fangjie Shao and Chong Liu*

Department of Pathology and Pathophysiology, Zhejiang University School of Medicine, Hangzhou, China

Correspondence:
Chong Liu
chongliu77@zju.edu.cn

High-grade glioma, particularly, glioblastoma, is the most aggressive cancer of the central nervous system (CNS) in adults. Due to its heterogeneous nature, glioblastoma almost inevitably relapses after surgical resection and radio-/chemotherapy, and is thus highly lethal and associated with a dismal prognosis. Identifying the cell of origin has been considered an important aspect in understanding tumor heterogeneity, thereby holding great promise in designing novel therapeutic strategies for glioblastoma. Taking advantage of genetic lineage-tracing techniques, performed mainly on genetically engineered mouse models (GEMMs), multiple cell types in the CNS have been suggested as potential cells of origin for glioblastoma, among which adult neural stem cells (NSCs) and oligodendrocyte precursor cells (OPCs) are the major candidates. However, it remains highly debated whether these cell types are equally capable of transforming in patients, given that in the human brain, some cell types divide so slowly, therefore may never have a chance to transform. With the recent advances in studying adult NSCs and OPCs, particularly from the perspective of comparative biology, we now realize that notable differences exist among mammalian species. These differences have critical impacts on shaping our understanding of the cell of origin of glioma in humans. In this perspective, we update the current progress in this field and clarify some misconceptions with inputs from important findings about the biology of adult NSCs and OPCs. We propose to re-evaluate the cellular origin candidacy of these cells, with an emphasis on comparative studies between animal models and humans.

Keywords: cell of origin, high-grade glioma, glioblastoma, adult neural stem cells (NSCs), oligodendrocyte precursor cells (OPC), genetically engineered mouse models (GEMMs), lineage tracing

INTRODUCTION

Adult gliomas are the most common cancers of the central nervous system (CNS) (Louis, 2006; Perry and Wesseling, 2016). Despite many years of efforts in both basic research and clinical practice, the prognosis of malignant gliomas, particularly the most advanced one, glioblastoma multiforme (GBM), remains dismal. This lack of progress is largely associated with high inter- and intra-tumoral heterogeneity. Tumor tissues from not only different patients, but also from the same ones, can be stratified into distinct morphopathological groups or molecular subtypes (Verhaak et al., 2010; Snuderl et al., 2011; Brennan et al., 2013; Kim J. et al., 2015; Wang et al., 2016, 2017). Such heterogeneity is generally considered as the main reason for drug resistance and high recurrence rate during treatment.

A cell of origin is the normal progenitor from which all the neoplastic cells of a given type of cancer develop (Visvader, 2011; Chaffer and Weinberg, 2015). Identification of the cell of origin can give critical insights into the principles dictating tumor heterogeneity, therefore holding great promise in understanding the cancer etiology, and facilitating the design of effective therapeutic strategies. In this *Perspective*, we review the current progress in the research of the cell of origin of glioma. Together with new findings in NSCs and OPCs from both rodents and large-brained mammals including humans, we propose to carefully re-evaluate the candidacy of several popular cell types that have been believed as the potential cells of origin of glioma in humans.

CNS CELL TYPES RELEVANT TO GLIOMA ETIOLOGY: THEIR LINEAGE RELATIONSHIP AND SOME IMPORTANT UPDATES

Knowing the properties of neural cell types and their lineage relationship will help understanding their potential contributions to the etiology of human glioma. Neural cells in the adult CNS are grossly classified as neurons, astrocytes, oligodendrocyte precursor cells (OPCs), and oligodendrocytes. In addition to these lineage-committed progenitor and mature cells, specialized stem cells, termed adult neural stem cells (NSCs) exist within restricted regions such as the subventricular zone (SVZ) next to the lateral ventricle, and the subgranular zone (SGZ) of the hippocampus (Ming and Song, 2011), in the adult brain. Both SVZ adult NSCs and OPCs have been implicated as the major candidates for glioma cell of origin, therefore, deserving a little more discussion.

Adult Neural Stem Cells (NSCs)

Adult NSCs (also termed B1 cells), which were best studied in rodents, have been generally believed to be able to persistently self-renew, and give rise to multiple neuronal and glial cell types (Alvarez-Buylla et al., 2001). Recent progresses in NSC biology, however, may suggest a quite different scenario. By using a temporal Histone 2B-EGFP marking system or barcoded retroviral labeling-based clonal analysis, two groups independently reported that postnatal B1 cells are derived from embryonic NSCs that divide during mid-fetal development and then remain quiescent until they reactivate, thus generating progeny in the postnatal brain (Fuentealba et al., 2015; Furutachi et al., 2015). Surprisingly, clonal analysis unraveled that postnatally, a single B1 cell neither divides repeatedly to produce generations of olfactory bulb (OB) neurons, nor gives rise to cortical glial cells and OB neurons simultaneously, raising an interesting possibility that adult NSCs may not systematically self-renew (Fuentealba et al., 2015) (see also **Figure 1A**). Therefore, although adult NSCs exhibit remarkable self-renewal potential and differentiation plasticity in culture (Doetsch et al., 1999; Codega et al., 2014), it remains highly debated whether, in the brain, they conform to the hardwired definition of tissue stem

cells, as seen in the case of hematopoietic or intestinal stem cells (Batlle and Clevers, 2017).

Oligodendrocyte Precursor Cells (OPCs)

Oligodendrocyte precursor cells were initially thought to function solely as transient forms of glial progenitors, to generate mature oligodendrocytes. Nevertheless, recent studies show that even though some OPCs indeed differentiate, many of them retain the ability to self-renew (Nishiyama et al., 2009; Vigano and Dimou, 2016) (see also **Figure 1A**). By using a sensitive DNA-labeling approach to mark cells undergoing proliferation, Yeung et al. (2014) showed that all OPCs in the adult mouse brain were dividing. Strikingly, Garcia-Marques et al. (2014 observed that at the clonal level, a single OPC could give rise to up to 400 cells in the adult mouse brain, therefore unequivocally demonstrating that OPCs are a *bona fide* self-renewable cell population *in vivo*. Given that OPCs make up 5–10% of all cells in the brain (Dawson et al., 2003), using the absolute number as the criteria, OPCs should be viewed as the largest proliferation pool in the mammalian brain. In addition to self-renewal, OPCs have been reported to exhibit some lineage plasticity. Despite being a matter of intensive debate, OPCs were shown to be able to differentiate into astrocytes and/or neurons *in vivo* (Rivers et al., 2008; Zhu et al., 2008, 2011; Richardson et al., 2011), and can be reprogrammed into the NSC-like status *in vitro* (Kondo and Raff, 2000), thus resembling NSCs in ways stronger than those previously considered (Richardson et al., 2011).

THE RESEARCH PROGRESS OF GLIOMA CELLULAR ORIGINS

NSCs as the Cell of Origin: Evidence and Concerns

Adult NSCs have been widely viewed as the most possible cell of origin for high-grade glioma, given their prominent property to self-renew, and the remarkable plasticity to differentiate into multiple neural cell types (Doetsch et al., 1999; Alvarez-Buylla et al., 2001; Stiles and Rowitch, 2008). In addition, cancer stem cells (CSCs) isolated from human GBMs share many markers normally expressed by NSCs (such as Nestin, GFAP, CD133, and Sox2), and are able to form renewable NSC-like spheres in culture (Singh et al., 2004; Bao et al., 2006). Furthermore, mouse and human NSCs can be transformed *in vitro*; they gain the capacity to develop into gliomas after implantation into host mice (Bachoo et al., 2002; Duan et al., 2015). Importantly, delivery of DNA or viral vehicles into the embryonic, neonatal, or adult SVZ (the brain structure where NSCs reside) to introduce over-expression of oncogenes and/or knockout/knockdown of tumor suppressor genes could efficiently generate high-grade glioma in mice (Alcantara Llaguno et al., 2009; Marumoto et al., 2009; Breunig et al., 2015; Zuckermann et al., 2015). Intriguingly, human glioblastomas were frequently diagnosed next to the SVZ, further supporting the possibility that they originated from NSCs (Barami et al., 2009). More direct evidence was obtained from

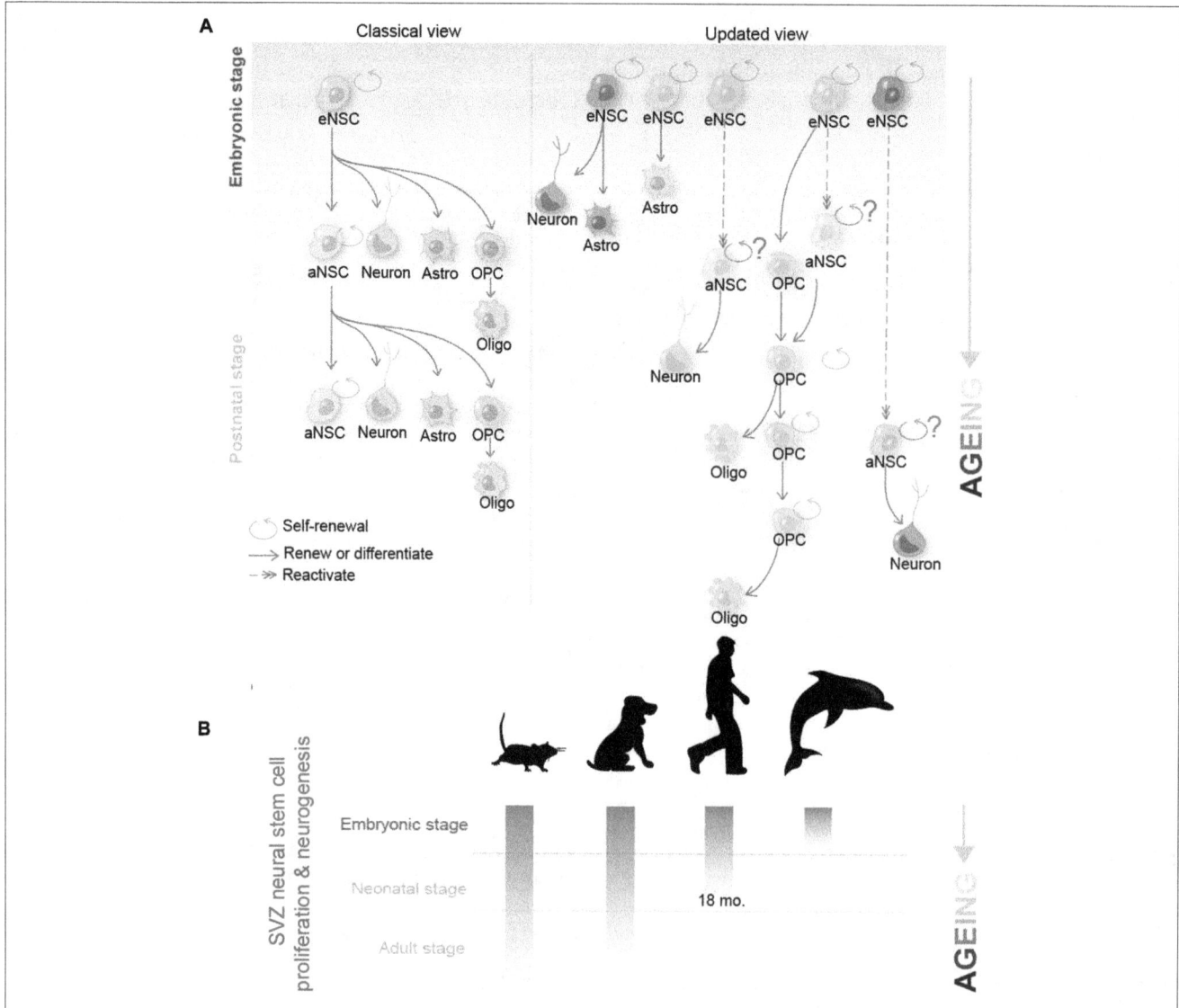

FIGURE 1 | Recent progress in the biology of neural stem cells (NSCs) and oligodendrocyte precursor cells (OPCs) provides new insights into their candidacy as the cell of origin for human glioblastoma. **(A)** Classical (left) and the updated (right) view of the NSC behavior in the brain. In the classical view, it was believed that a single NSC can repeatedly self-renew for many generations and give rise to new NSCs; at the same time, it possesses the potential to differentiate into neurons, astrocyte and OPCs. OPCs can further differentiate into mature oligodendrocytes. Recent studies using mouse models suggest that adult B1 cells (adult form of NSCs) are derived from eNSCs that actively proliferate at ~E14.5. These embryonic NSCs remain quiescent until they are reactivated at the adult stage. Though as a whole population, adult NSCs continuously proliferate and give rise to olfactory bulb (OB) neurons and glial cells, they are extremely heterogeneous at the single cell level. Clonal analysis revealed that a single adult NSC can either give rise to OB neurons or glial cells (such as astrocytes and oligodendrocytes), but rarely to both cell types. Furthermore, many, if not all, adult NSCs cannot bud off OB neurons and simultaneously self-renew, raising the question of whether adult NSCs conform to the hardwired definition of self-renewable tissue stem cells (Fuentealba et al., 2015; Furutachi et al., 2015). On the other hand, despite OPCs originally being derived from NSCs during early development, and adult NSCs contributing to the OPC pool to some extent, in the normal brain, most adult OPCs are generated from the locally resident OPCs. Clonal analysis further revealed that adult OPCs can self-renew continuously (Garcia-Marques et al., 2014; Yeung et al., 2014). **(B)** Neurogenic and proliferative activities of SVZ NSCs in rodent, dog, human, and dolphin. Please note that neurogenic activity disappears in human at ~18 months, and is completely absent in dolphin postnatally; however, both species can suffer from GBM at adulthood. Kindly refer to the main text for further details.

the lineage-tracing experiment by using genetically engineered mouse models (GEMMs). Taking advantage of NSC-specific genetic tools such as hGFAP-Cre, Nestin-Cre, or Nestin-Cre[ER], Parada and his colleagues showed that mouse NSCs are capable of transforming into high-grade gliomas after the loss of *Trp53*, *NF1* and/or *PTEN* (Zhu et al., 2005; Chen et al., 2012; Alcantara Llaguno et al., 2015).

While these multiple lines of evidence demonstrate that NSCs are capable of transforming into malignancy, several important issues should be understood. Firstly, as already mentioned, recent findings about NSC biology challenges the concept that a single SVZ aNSC can repeatedly self-renew, therefore greatly decreasing the possibility for an aNSC to accumulate mutations, as previously assumed. Secondly, the stem cell feature of CSCs need not necessarily be inherited from tissue stem cells; it can also be regained through the de-differentiation of lineage-committed progenitors or mature cells (Batlle and Clevers, 2017). Thirdly, many claimed that NSC cellular markers are not specific to NSCs. For example, the most widely used NSC marker Nestin, an intermediate filament protein expressed in radial glia and adult B1 cells, is prominently expressed in reactive astrocytes (Ernst and Christie, 2006). Although partial overlaps between brain tumor locations and the NSC niche is a good argument to support the fact that gliomas originate from adult NSCs in patients, a recent work revealed that the SVZ may merely function as a niche toward which glioma cells prefer to migrate (Qin et al., 2017).

An additional dimension of complexity comes from the nature of NSCs *per se*. As NSCs can readily differentiate into fate-committed precursors such as OPCs or mature astrocytes, it is unclear whether NSCs, after acquiring initial mutations, directly transform, or they must proceed through the status of lineage-committed cell types prior to the final transformation. In fact, by using a single-cell resolution genetic mouse model termed mosaic analysis with double markers (MADM), we have shown that introducing *p53* and *NF1* mutations into NSCs did not evidently change the proliferation rate of pre-cancerous adult NSCs, but drastically promoted the over-expansion of descendant OPCs, arguing against a direct transformation of NSCs, at least in the context of this mutation combination (Liu et al., 2011).

OPCs as the Cell of Origin: Evidence and Some Updates

Oligodendrocyte precursor cells have been proposed as an important cell of origin for glioma since they were first identified. As already mentioned, OPCs represent the largest proliferation pool in the brain, and exhibit remarkable self-renewal capacity both *in vitro* and *in vivo*, and are therefore suitable, as cells of origin, to accumulate genetic mutations. In fact, NG2, one of the most commonly used OPC cell marker, was initially isolated from a rat glioma model (Stallcup, 1981). In addition to NG2, we and others showed that many cellular markers typically expressed in OPCs, such as Olig2, PDGFRa, and O4, were also expressed in most, if not all, human malignant gliomas (Shoshan et al., 1999; Ligon et al., 2004, 2007; Rebetz et al., 2008; Ledur et al., 2016; Shao et al., 2017). Furthermore, over-expression of the oncogenic form of EGFR (EGFRvIII) under the promoter S100b, a non-stem cell marker (Raponi et al., 2007), induced gliomas recapitulating the pathological features of human oligodendroglioma (Weiss et al., 2003; Persson et al., 2010). Moreover, overexpression of PDGF-BB alone, or when combined with p53 and Pten deactivation, was shown to be able to effectively transform rat and mouse OPCs into lower-grade oligodendrogliomas or high-grade gliomas (Assanah et al., 2006; Lindberg et al., 2009; Lei et al., 2011; Lu et al., 2016). More direct evidence to support the OPC-origin of high-grade gliomas comes from fate-mapping experiments. By using OPC-specific NG2-Cre or NG2-Cre^ERT transgenic mouse lines, we and others have provided convincing evidence that OPCs, after acquiring *Trp53* and

NF1 mutations, can be directly transformed into malignant gliomas resembling the proneural subtype of GBM, whenever the mutations were introduced in early or adult stage (Liu et al., 2011; Galvao et al., 2014; Alcantara Llaguno et al., 2015).

The data from our group show that OPC-like tumor cells are universally present in all human malignant gliomas, and share remarkable similarities in many aspects with their counterparts found in mouse genetic models, in which OPCs are the defined cells of origin (Ledur et al., 2016; Shao et al., 2017). These lines of evidence collectively lead to a reasonable assumption that OPCs are important glioma cells of origin in patients.

Mature Astrocytes and Neurons as the Cells of Origin: An Unsettled Issue

Whether mature astrocytes and/or neurons are able to directly transform remains highly debated. Chow et al. (2011) utilized GFAP-Cre^ER to introduce Trp53, Pten and/or Rb1 mutations into astrocytes and induced high-grade astrocytomas in adult mice. Also using GFAP-Cre^ER, Vitucci et al. (2017) observed that murine astrocytes could transform into high-grade glioma mimicking human mesenchymal, proneural, and neural GBMs. By using Cre-activatable lentiviral vehicles that encoded shRNA against Trp53 and NF1, Friedmann-Morvinski et al. (2012) reported high incidence of GBMs when they transfected such lentiviral particles into the brains of hGFAP-Cre, Synapsin I-Cre or CamK2a-Cre transgenic mice. Therefore, the authors claimed that both mature astrocytes and neurons can function as the cells of origin for GBMs through dedifferentiation. Nevertheless, as most claimed astrocyte-specific markers such as GFAP are also expressed in NSCs (Chaker et al., 2016), and those for neurons like Synapsin-1 are also expressed in OPCs [(Cahoy et al., 2008; Zhang et al., 2014) and personal observations], further validation is necessary to exclude the possibility of targeting NSCs and/or OPCs when attempting to manipulate mature astrocytes or neurons. Highly specific genetic tools are warranted to clarify this fundamental issue.

HUMAN RELEVANCE: NEW DATA AND THE INSIGHT FROM COMPARATIVE STUDIES

Most of our current knowledge on glioma cell of origin was derived from the observations on animal models, mostly GEMM-based cancer models. One fundamental question we must confront is that how much of the landscape depicted thus far can be directly extrapolated to human cases. Despite the overall anatomical structures and developmental principles of the CNS being highly conserved among mammals, notable differences, particularly in the properties of adult NSCs, do exist among species. Recognizing these differences has important impacts on shaping our understandings of the glioma cell of origin in humans.

Adult NSCs May Not Be a Major Player in the Pathogenesis of Glioblastoma in Human or Other Large-Brained Animals

Unlike in rodents, where NSCs and neural progenitors proliferate continuously to form new neurons, in large-brained mammals, such as humans, SVZ neurogenesis declines drastically during postnatal life (Lipp and Bonfanti, 2016; Paredes et al., 2016), and fully disappears at around 18 months (Sanai et al., 2011), long before high-grade gliomas are diagnosed. Consistent with this observation, by measuring the turnover rate of nuclear bomb test-derived ^{14}C in genomic DNA, Bergmann et al. (2012, 2015) showed that there is virtually no postnatal neurogenesis in the human OB.

Direct evidence to support a lack of marked levels of neurogenesis or self-renewal of NSCs in the adult human SVZ comes from immunohistological studies, where proliferative cells were rarely found in the SVZ in adults (Wang et al., 2011; Dennis et al., 2016). Furthermore, the density of dividing cells in the SVZ is comparable to or even lower than that in other regions such as the corpus callosum (Shao et al. personal observations). Despite the suggestion that certain pathological conditions such as ischemic stroke may activate NSCs in the adult human brain (Jin et al., 2006; Marti-Fabregas et al., 2010), this conclusion was disproved by the ^{14}C turnover assay (Huttner et al., 2014). Regardless the potential of adult NSCs to be activated *in vivo* after injury, no definitive evidence yet shows an association of human glioma pathogenesis with any pathological lesions.

Comparative studies between species provide deeper insights into questioning the relevance of adult NSCs in glioma pathogenesis (as summarized in **Figure 1B**). Unlike humans, but quite similar to rodents, dogs possess SVZ neurogenesis that persists into adulthood (Malik et al., 2012). Therefore, one may expect a much higher incidence of gliomagenesis in dogs, if adult NSCs indeed play critical roles in initiating glioma. Contrary to this speculation, epidemiological studies suggest that the incidence of spontaneous brain tumors in dogs is remarkably similar to that in humans, i.e., approximately 20 in 100,000 per year (Dobson et al., 2002; Hicks et al., 2017). On the other hand, aquatic mammals such as dolphins, which lack a functional periventricular germinal layer postnatally and any detectable dividing cells within the SVZ (Parolisi et al., 2017), can surely suffer from glioblastoma (Diaz-Delgado et al., 2015). These findings, together with those in rodent NSCs, contradict the argument that adult NSCs play major roles in initiating gliomagenesis.

Adult OPCs May Function as an Important Cell of Origin with Strong Human Relevance

Unlike the great variations of cellular behaviors of adult NSCs, the renewal capacity of OPCs are largely conserved across species. For instance, immunohistological studies show that, although sparse, OPCs are the major cycling cells in the adult human brain (Geha et al., 2010). In line with this observation, ^{14}C data revealed that gray matter oligodendrocytes do not reach a plateau until the fourth decade of life, even after which the annual turnover remains as high as 2.5% (Yeung et al., 2014). These results in collection clearly demonstrate that OPCs undergo substantial renewal in the adult human brain.

Interestingly, the proliferation rate of OPCs are significantly elevated in epileptic patients (Geha et al., 2010). As epilepsies are frequently associated with glioma patients (Iuchi et al., 2015; Englot et al., 2016), these observations raise an intriguing possibility that aberrant neuronal activity may directly contribute to OPC self-renewals and, most likely, to oncogenic transformation. This hypothesis has been recently substantiated by showing that artificially enhancing the neuronal activity in GEMMs through optogenetic approaches can stimulate the proliferation of normal resident OPCs and engrafted human GBM cells (Gibson et al., 2014; Venkatesh et al., 2015).

Therefore, although comprehensive studies are warranted to systematically characterize the relative proliferating capacities of OPCs and NSCs/NPCs in the adult human brain *in situ*, given that OPCs retain a relatively decent level of self-renewal activity, and significantly outnumber NSCs in the adult human brain, they remain a highly probable candidate for the cell of origin of human GBMs.

THE RELATIONSHIP BETWEEN CELL(s) OF ORIGIN AND CANCER STEM CELLS (CSCs)

It should be noted that the "CSC" is a functional definition that can only be assessed by the capacity of a cancer cell to initiate new tumors. Some studies identified CSCs from the NSC-derived GBM mouse models and showed that these NSC-derived CSCs resemble normal NSCs in certain ways such as the expression of Nestin (Zheng et al., 2008; Alcantara Llaguno et al., 2009; Chen et al., 2012). However, the cells functioning as CSCs may not have to be derived and/or resemble normal NSCs. By using S100b- promoter-driven EGFRViii transgene, Persson et al. (2010) clearly showed that oligodendroglioma can be initiated from non NSCs, and the CSCs in this model can be identified and isolated based on their expression of NG2 (CSPG4), an OPC marker. We showed that CSCs derived from OPC-originated HGGs expressed NG2 as well as other OPC markers (such as PDGFRa and Olig2) and that the OPC feature is essential for the maintenance of the stemness of these CSCs (Liu et al., 2011; Ledur et al., 2016). Interestingly, OPC-originated CSCs gained the capacity to form spheres and to express Nestin. This latter observation implicates that Nestin is a marker for the stemness but not the cell identity in this particular case. In the human cases, NG2 have been used to enrich CSCs from oligodendrogliomas (Persson et al., 2010) and at least some GBMs (Persson et al., 2010; Al-Mayhani et al., 2011). Our own study showed that human primary GBM cell lines maintained under culture conditions that favor the enrichment of OPC-like tumor cells have enhanced malignancy (Ledur et al., 2016). In addition to OPCs, Schmid et al. (2016) provided the evidence that

TABLE 1 | Pathological features and molecular signatures of currently reported GEMMs for gliomas.

Putative cell of origin	Mutations	Approach	Molecular subtype	Pathology	Reference
NSC	Ras, Akt	RCAS/tv-a system	NA	GBM	Holland et al., 2000
	Ink4a, Arf, EGFR	Retrovirus	NA	High-grade gliomas	Bachoo et al., 2002
	H-Ras, AKT	Lentivirus + GFAP-Cre mice	NA	GBM	Marumoto et al., 2009
	Trp53, Nf1, and/or Pten	Adenovirus + Nestin-CreER	NA	A	Alcantara Llaguno et al., 2009
	PTEN, Trp53	Adenovirus-Cre	NA	High-grade gliomas	Jacques et al., 2010
	Ras; Erbb2; Pdgfra	Plasmid DNA + Electroporation	Proneural, Neural, Mesenchymal	AA, AO, AOA, GBM	Breunig et al., 2015
	Trp53, Pten, Nf1	CRISPR/Cas9 + Electroporation	NA	GBM	Zuckermann et al., 2015
	Trp53, NF1	hGFAP-Cre	NA	A, AA, GBM	Zhu et al., 2005
	Trp53, Pten	hGFAP-Cre	NA	Malignant gliomas	Zheng et al., 2008
	Nf1, Trp53, Pten	hGFAP-Cre	NA	Malignant gliomas	Chen et al., 2012
	K-Ras	BLBP-Cre	NA	Gliomatosis	Munoz et al., 2013
	Trp53, Nf1, and/or Pten	Nestin-CreER	NA	GBM	Alcantara Llaguno et al., 2015
OPC	PDGF	Retrovirus	NA	GBM	Assanah et al., 2006
	PDGF-B	RCAS/tv-a system	NA	O	Lindberg et al., 2009
	Pten, Trp53	Retrovirus + PDGF-IRES-Cre	Proneural	GBM	Lei et al., 2011
	TAZ, PDGFB	RCAS/N-tva system	Mesenchymal	Gliomas	Bhat et al., 2011
	NF1, PDGFA	RCAS/tv-a system	Mesenchymal	GBM	Ozawa et al., 2014
	PDGFB	RCAS/tv-a system	Mesenchymal	GBM	Ozawa et al., 2014
	Arf	RCAS/tv-a system	NA	A	Lindberg et al., 2014
	Ink4a, Arf	RCAS/tv-a system	NA	A	Lindberg et al., 2014
	Arf, PDGF-B	RCAS/tv-a system	NA	O	Lindberg et al., 2014
	Ink4a, Arf, PDGF-B	RCAS/tv-a system	NA	O	Lindberg et al., 2014
	Pten, Trp53	Retrovirus + PDGFB-IRES-Cre	Proneural	GBM	Lu et al., 2016
	Pten, Trp53, Olig2	Retrovirus + PDGFB-IRES-Cre	Classical	GBM	Lu et al., 2016
	Trp53	S100β-v-erbB	NA	O	Weiss et al., 2003
	ink/arf	S100β-v-erbB	NA	AO	Weiss et al., 2003
	Trp53	S100β-v-erbB	OPC like	O, GBM	Persson et al., 2010
	Trp53, NF1	NG2-Cre	Proneural	Malignant gliomas	Liu et al., 2011
	Trp53, NF1	NG2-CreER	Proneural	Malignant gliomas	Galvao et al., 2014
	Trp53, Nf1, and/or Pten	NG2-CreER	NA	Malignant gliomas	Alcantara Llaguno et al., 2015
Astrocyte	Ink4a, Arf, EGFR	Retrovirus	NA	High-grade gliomas	Bachoo et al., 2002
	Trp53, NF1	Lentivirus + GFAP-Cre mice	Mesenchymal	GBM	Friedmann-Morvinski et al., 2012
	Trp53, Pten	GFAP-CreER	Proneural, Neural, Mesenchymal	AA, GBM	Chow et al., 2011
	Trp53, Pten, Rb1	GFAP-CreER	Proneural, Neural, Mesenchymal	AA, AOA, GBM	Chow et al., 2011
	TgGZT$_{121}$, Kras^{G12D}, Pten	GFAP-CreER	Mesenchymal, Proneural, Neural	GBM	Vitucci et al., 2017
Neuron	Trp53, NF1	Lentivirus + Synapsin I-Cre or CamK2a-Cre mice	Mesenchymal	Malignant gliomas	Friedmann-Morvinski et al., 2012

A, astrocytoma; O, oligodendroglioma; AA, anaplastic astrocytoma; AO, anaplastic oligodendroglioma; AOA, anaplastic oligoastrocytoma; GBM, glioblastoma multiforme.

mature astrocytes could dedifferentiate into glioma CSCs upon transformation. Therefore, CSCs in gliomas can definitely be developed from the non-CSC cell types. The detailed lineage relationship between NSCs, lineage-committed progenitors, mature cells and CSCs remains to be fully elucidated in the future studies.

THE RELATIONSHIP BETWEEN CELLS OF ORIGIN, TUMOR SUBTYPES AND HETEROGENEITY

Cumulative evidence suggests that the same cell of origin can give rise to the GBMs manifesting different molecular features and that distinct types of cells of origin can evolve in parallel to give rise to tumors resembling similar molecular features (see also **Table 1**).

For instance, OPCs have been previously considered to mainly give rise to oligodendrogliomas and proneural subtype of GBMs (Weiss et al., 2003; Lei et al., 2011; Liu et al., 2011; Galvao et al., 2014). However, recent studies demonstrate that they can also serve as the cell of origin for astrocytoma (Lindberg et al., 2014) and other subtypes of GBMs, depending on the mutations initially introduced (Carro et al., 2009; Bhat et al., 2011, 2013; Lu et al., 2016). In particular, removal of Olig2 switches OPC-derived proneural subtype of GBMs into the classical subtype. Over-expression of TAZ or suppression of NF1, instead, readily induces OPC-derived GBMs into the mesenchymal subtype (Bhat et al., 2011; Ozawa et al., 2014). Similar observations were also obtained in astrocyte-originated GBMs, where the same GEMM can give rise to tumors with highly heterogeneous profiles (Chow et al., 2011; Schmid et al., 2016).

Importantly, the evolution routes of a defined cell of origin may also affect the molecular features of brain tumors. The recurrent GBMs from the same patients frequently switched their molecular features when compared to their primary tumor counterparts (Kim H. et al., 2015; Kim J. et al., 2015; Wang et al., 2016). Therefore, the molecular signature of a particular transformed tumor may not always reliably predict its cell of origin.

FUTURE PERSPECTIVES

Owing to genetic lineage tracing techniques and other advanced biological methods, tremendous progress has been made in understanding the glioma cell of origin during the past decade. Now, a consensus has been made that several important cell types, particularly NSCs and OPCs, are capable of transforming at least in GEMMs. However, many fundamental questions remain unanswered. For instance, is there a universal cell type functioning as the cell of origin for all gliomas in humans? Or alternatively, do different cell types give rise to gliomas with distinct pathological identities? Can different mutations drive the same cell of origin to follow the same or distinct routes toward the final transformation? When exactly do human gliomas form? GEMMs will surely continue to serve as the most important tools to address these fundamental questions. Nevertheless, we should be aware of the difference between GEMMs and patients. Newer methods and the concept of comparative pathology could help us identify what really initiates this devastating form of cancer in humans.

AUTHOR CONTRIBUTIONS

FS and CL wrote the manuscript. FS prepared the figure and the table.

ACKNOWLEDGMENTS

We thank Drs. Weijun Yang and Yingjie Wang for their critical comments.

REFERENCES

Alcantara Lluguno, S., Chen, J., Kwon, C.-H., Jackson, E. L., Li, Y., Burns, D. K., et al. (2009). Malignant astrocytomas originate from neural stem/progenitor cells in a somatic tumor suppressor mouse model. *Cancer Cell* 15, 45–56. doi: 10.1016/j.ccr.2008.12.006

Alcantara Lluguno, S. R., Wang, Z., Sun, D., Chen, J., Xu, J., Kim, E., et al. (2015). Adult lineage-restricted CNS progenitors specify distinct glioblastoma subtypes. *Cancer Cell* 28, 429–440. doi: 10.1016/j.cell.2015.09.007

Al-Mayhani, M. T., Grenfell, R., Narita, M., Piccirillo, S., Kenney-Herbert, E., Fawcett, J. W., et al. (2011). NG2 expression in glioblastoma identifies an actively proliferating population with an aggressive molecular signature. *Neuro Oncol.* 13, 830–845. doi: 10.1093/neuonc/nor088

Alvarez-Buylla, A., Garcia-Verdugo, J. M., and Tramontin, A. D. (2001). A unified hypothesis on the lineage of neural stem cells. *Nat. Rev. Neurosci.* 2, 287–293. doi: 10.1038/35067582

Assanah, M., Lochhead, R., Ogden, A., Bruce, J., Goldman, J., and Canoll, P. (2006). Glial progenitors in adult white matter are driven to form malignant gliomas by platelet-derived growth factor-expressing retroviruses. *J. Neurosci.* 26, 6781–6790. doi: 10.1523/Jneurosci.0514-06.2006

Bachoo, R. M., Maher, E. A., Ligon, K. L., Sharpless, N. E., Chan, S. S., You, M. J. J., et al. (2002). Epidermal growth factor receptor and Ink4a/Arf: convergent mechanisms governing terminal differentiation and transformation along the neural stem cell to astrocyte axis. *Cancer Cell* 1, 269–277. doi: 10.1016/S1535-6108(02)00046-6

Bao, S. D., Wu, Q. L., Mclendon, R. E., Hao, Y. L., Shi, Q., Hjelmeland, A. B., et al. (2006). Glioma stem cells promote radioresistance by preferential activation

of the DNA damage response. *Nature* 444, 756–760. doi: 10.1038/nature05236

Barami, K., Sloan, A. E., Rojiani, A., Schell, M. J., Staller, A., and Brem, S. (2009). Relationship of gliomas to the ventricular walls. *J. Clin. Neurosci.* 16, 195–201. doi: 10.1016/j.jocn.2008.03.006

Batlle, E., and Clevers, H. (2017). Cancer stem cells revisited. *Nat. Med.* 23, 1124–1134. doi: 10.1038/nm.4409

Bergmann, O., Liebl, J., Bernard, S., Alkass, K., Yeung, M. S., Maggie, S. Y., et al. (2012). The age of olfactory bulb neurons in humans. *Neuron* 74, 634–639. doi: 10.1016/j.neuron.2012.03.030

Bergmann, O., Spalding, K. L., and Frisen, J. (2015). Adult neurogenesis in humans. *Cold Spring Harb. Perspect. Biol.* 7:a018994. doi: 10.1101/cshperspect.a018994

Bhat, K. P. L., Balasubramaniyan, V., Vaillant, B., Ezhilarasan, R., Hummelink, K., Hollingsworth, F., et al. (2013). Mesenchymal differentiation mediated by NF-κB promotes radiation resistance in glioblastoma. *Cancer Cell* 24, 331–346. doi: 10.1016/j.ccr.2013.08.001

Bhat, K. P. L., Salazar, K. L., Balasubramaniyan, V., Wani, K., Heathcock, L., Hollingsworth, F., et al. (2011). The transcriptional coactivator TAZ regulates mesenchymal differentiation in malignant glioma. *Genes Dev.* 25, 2594–2609. doi: 10.1101/gad.176800.111

Brennan, C. W., Verhaak, R. G. W., Mckenna, A., Campos, B., Noushmehr, H., Salama, S. R., et al. (2013). The somatic genomic landscape of glioblastoma. *Cell* 155, 462–477. doi: 10.1016/j.cell.2013.09.034

Breunig, J. J., Levy, R., Antonuk, C. D., Molina, J., Dutra-Clarke, M., Park, H., et al. (2015). Ets factors regulate neural stem cell depletion and gliogenesis in Ras pathway glioma. *Cell Rep.* 12, 258–271. doi: 10.1016/j.celrep.2015.06.012

Cahoy, J. D., Emery, B., Kaushal, A., Foo, L. C., Zamanian, J. L., Christopherson, K. S., et al. (2008). A transcriptome database for astrocytes, neurons, and oligodendrocytes: a new resource for understanding brain development and function. *J. Neurosci.* 28, 264–278. doi: 10.1523/Jneurosci.4178-07.2008

Carro, M. S., Lim, W. K., Alvarez, M. J., Bollo, R. J., Zhao, X., Snyder, E. Y., et al. (2009). The transcriptional network for mesenchymal transformation of brain tumours. *Nature* 463, 318–325. doi: 10.1038/nature08712

Chaffer, C. L., and Weinberg, R. A. (2015). How does multistep tumorigenesis really proceed? *Cancer Discov.* 5, 22–24. doi: 10.1158/2159-8290.cd-14-0788

Chaker, Z., Codega, P., and Doetsch, F. (2016). A mosaic world: puzzles revealed by adult neural stem cell heterogeneity. *Wiley Interdiscip. Rev. Dev. Biol.* 5, 640–658. doi: 10.1002/wdev.248

Chen, J., Li, Y. J., Yu, T. S., Mckay, R. M., Burns, D. K., Kernie, S. G., et al. (2012). A restricted cell population propagates glioblastoma growth after chemotherapy. *Nature* 488, 522–526. doi: 10.1038/nature11287

Chow, L. M. L., Endersby, R., Zhu, X. Y., Rankin, S., Qu, C. X., Zhang, J. Y., et al. (2011). Cooperativity within and among Pten, p53, and Rb pathways induces high-grade astrocytoma in adult brain. *Cancer Cell* 19, 305–316. doi: 10.1016/j.ccr.2011.01.039

Codega, P., Silva-Vargas, V., Paul, A., Maldonado-Soto, A. R., Deleo, A. M., Pastrana, E., et al. (2014). Prospective identification and purification of quiescent adult neural stem cells from their in vivo niche. *Neuron* 82, 545–559. doi: 10.1016/j.neuron.2014.02.039

Dawson, M. R. L., Polito, A., Levine, J. M., and Reynolds, R. (2003). NG2-expressing glial progenitor cells: an abundant and widespread population of cycling cells in the adult rat CNS. *Mol. Cell. Neurosci.* 24, 476–488. doi: 10.1016/S1044-7431(03)00210-0

Dennis, C. V., Suh, L. S., Rodriguez, M. L., Kril, J. J., and Sutherland, G. T. (2016). Human adult neurogenesis across the ages: an immunohistochemical study. *Neuropathol. Appl. Neurobiol.* 42, 621–638. doi: 10.1111/nan.12337

Diaz-Delgado, J., Sacchino, S., Suarez-Bonnet, A., Sierra, E., Arbelo, M., Espinosa, A., et al. (2015). High-grade astrocytoma (glioblastoma multiforme) in an Atlantic spotted dolphin (*Stenella frontalis*). *J. Comp. Pathol.* 152, 278–282. doi: 10.1016/j.jcpa.2014.12.016

Dobson, J. M., Samuel, S., Milstein, H., Rogers, K., and Wood, J. L. N. (2002). Canine neoplasia in the UK: estimates of incidence rates from a population of insured dogs. *J. Small Anim. Pract.* 43, 240–246. doi: 10.1111/j.1748-5827.2002.tb00066.x

Doetsch, F., Caille, I., Lim, D. A., Garcia-Verdugo, J. M., and Alvarez-Buylla, A. (1999). Subventricular zone astrocytes are neural stem cells in the adult mammalian brain. *Cell* 97, 703–716. doi: 10.1016/S0092-8674(00)80783-7

Duan, S. L., Yuan, G. H., Liu, X. M., Ren, R. T., Li, J. Y., Zhang, W. Z., et al. (2015). PTEN deficiency reprogrammes human neural stem cells towards a glioblastoma stem cell-like phenotype. *Nat. Commun.* 6:10068. doi: 10.1038/ncomms10068

Englot, D. J., Chang, E. F., and Vecht, C. J. (2016). Epilepsy and brain tumors. *Handb. Clin. Neurol.* 134, 267–285. doi: 10.1016/b978-0-12-802997-8.00016-5

Ernst, C., and Christie, B. R. (2006). The putative neural stem cell marker, nestin, is expressed in heterogeneous cell types in the adult rat neocortex. *Neuroscience* 138, 183–188. doi: 10.1016/j.neuroscience.2005.10.065

Friedmann-Morvinski, D., Bushong, E. A., Ke, E., Soda, Y., Marumoto, T., Singer, O., et al. (2012). Dedifferentiation of neurons and astrocytes by oncogenes can induce gliomas in mice. *Science* 338, 1080–1084. doi: 10.1126/science.1226929

Fuentealba, L. C., Rompani, S. B., Parraguez, J. I., Obernier, K., Romero, R., Cepko, C. L., et al. (2015). Embryonic origin of postnatal neural stem cells. *Cell* 161, 1644–1655. doi: 10.1016/j.cell.2015.05.041

Furutachi, S., Miya, H., Watanabe, T., Kawai, H., Yamasaki, N., Harada, Y., et al. (2015). Slowly dividing neural progenitors are an embryonic origin of adult neural stem cells. *Nat. Neurosci.* 18, 657–665. doi: 10.1038/nn.3989

Galvao, R. P., Kasina, A., Mcneill, R. S., Harbin, J. E., Foreman, O., Verhaak, R. G. W., et al. (2014). Transformation of quiescent adult oligodendrocyte precursor cells into malignant glioma through a multistep reactivation process. *Proc. Natl. Acad. Sci. U.S.A.* 111, E4214–E4223. doi: 10.1073/pnas.1414389111

Garcia-Marques, J., Nunez-Llaves, R., and Lopez-Mascaraque, L. (2014). NG2-glia from pallial progenitors produce the largest clonal clusters of the brain: time frame of clonal generation in cortex and olfactory bulb. *J. Neurosci.* 34, 2305–2313. doi: 10.1523/Jneurosci.3060-13.2014

Geha, S., Pallud, J., Junier, M.-P., Devaux, B., Leonard, N., Chassoux, F., et al. (2010). NG2+/Olig2+ cells are the major cycle-related cell population of the adult human normal brain. *Brain Pathol.* 20, 399–411. doi: 10.1111/j.1750-3639.2009.00295.x

Gibson, E. M., Purger, D., Mount, C. W., Goldstein, A. K., Lin, G. L., Wood, L. S., et al. (2014). Neuronal activity promotes oligodendrogenesis and adaptive myelination in the mammalian brain. *Science* 344:1252304. doi: 10.1126/science.1252304

Hicks, J., Platt, S., Kent, M., and Haley, A. (2017). Canine brain tumours: a model for the human disease? *Vet. Comp. Oncol.* 15, 252–272. doi: 10.1111/vco.12152

Holland, E. C., Celestino, J., Dai, C., Schaefer, L., Sawaya, R. E., and Fuller, G. N. (2000). Combined activation of Ras and Akt in neural progenitors induces glioblastoma formation in mice. *Nat. Genet.* 25, 55–57. doi: 10.1038/75596

Huttner, H. B., Bergmann, O., Salehpour, M., Racz, A., Tatarishvili, J., Lindgren, E., et al. (2014). The age and genomic integrity of neurons after cortical stroke in humans. *Nat. Neurosci.* 17, 801–803. doi: 10.1038/nn.3706

Iuchi, T., Hasegawa, Y., Kawasaki, K., and Sakaida, T. (2015). Epilepsy in patients with gliomas: incidence and control of seizures. *J. Clin. Neurosci.* 22, 87–91. doi: 10.1016/j.jocn.2014.05.036

Jacques, T. S., Swales, A., Brzozowski, M. J., Henriquez, N. V., Linehan, J. M., Mirzadeh, Z., et al. (2010). Combinations of genetic mutations in the adult neural stem cell compartment determine brain tumour phenotypes. *EMBO J.* 29, 222–235. doi: 10.1038/emboj.2009.327

Jin, K. L., Wang, X. M., Xie, L., Mao, X. O., Zhu, W., Wang, Y., et al. (2006). Evidence for stroke-induced neurogenesis in the human brain. *Proc. Natl. Acad. Sci. U.S.A.* 103, 13198–13202. doi: 10.1073/pnas.0603512103

Kim, H., Zheng, S., Amini, S. S., Virk, S. M., Mikkelsen, T., Brat, D. J., et al. (2015). Whole-genome and multisector exome sequencing of primary and post-treatment glioblastoma reveals patterns of tumor evolution. *Genome Res.* 25, 316–327. doi: 10.1101/gr.180612.114

Kim, J., Lee, I. H., Cho, H. J., Park, C. K., Jung, Y. S., Kim, Y., et al. (2015). Spatiotemporal evolution of the primary glioblastoma genome. *Cancer Cell* 28, 318–328. doi: 10.1016/j.ccell.2015.07.013

Kondo, T., and Raff, M. (2000). Oligodendrocyte precursor cells reprogrammed to become multipotential CNS stem cells. *Science* 289, 1754–1757. doi: 10.1126/science.289.5485.1754

Ledur, P. F., Liu, C., He, H., Harris, A. R., Minussi, D. C., Zhou, H. Y., et al. (2016). Culture conditions tailored to the cell of origin are critical for maintaining native properties and tumorigenicity of glioma cells. *Neuro Oncol.* 18, 1413–1424. doi: 10.1093/neuonc/now062

Lei, L., Sonabend, A. M., Guarnieri, P., Soderquist, C., Ludwig, T., Rosenfeld, S., et al. (2011). Glioblastoma models reveal the connection between adult glial progenitors and the proneural phenotype. *PLoS One* 6:e20041. doi: 10.1371/journal.pone.0020041

Ligon, K. L., Alberta, J. A., Kho, A. T., Weiss, J., Kwaan, M. R., Nutt, C. L., et al. (2004). The oligodendroglial lineage marker OLIG2 is universally expressed in diffuse gliomas. *J. Neuropathol. Exp. Neurol.* 63, 499–509. doi: 10.1093/jnen/63.5.499

Ligon, K. L., Huillard, E., Mehta, S., Kesari, S., Liu, H. Y., Alberta, J. A., et al. (2007). Olig2-regulated lineage-restricted pathway controls replication competence in neural stem cells and malignant glioma. *Neuron* 53, 503–517. doi: 10.1016/j.neuron.2007.01.009

Lindberg, N., Jiang, Y., Xie, Y., Bolouri, H., Kastemar, M., Olofsson, T., et al. (2014). Oncogenic signaling is dominant to cell of origin and dictates astrocytic or oligodendroglial tumor development from oligodendrocyte precursor cells. *J. Neurosci.* 34, 14644–14651. doi: 10.1523/jneurosci.2977-14.2014

Lindberg, N., Kastemar, M., Olofsson, T., Smits, A., and Uhrbom, L. (2009). Oligodendrocyte progenitor cells can act as cell of origin for experimental glioma. *Oncogene* 28, 2266–2275. doi: 10.1038/onc.2009.76

Lipp, H. P., and Bonfanti, L. (2016). Adult neurogenesis in mammals: variations and confusions. *Brain Behav. Evol.* 87, 205–221. doi: 10.1159/000446905

Liu, C., Sage, J. C., Miller, M. R., Verhaak, R. G. W., Hippenmeyer, S., Vogel, H., et al. (2011). Mosaic analysis with double markers reveals tumor cell of origin in glioma. *Cell* 146, 209–221. doi: 10.1016/j.cell.2011.06.014

Louis, D. N. (2006). Molecular pathology of malignant gliomas. *Annu. Rev. Pathol. Mech. Dis.* 1, 97–117. doi: 10.1146/annurev.pathol.1.110304.100043

Lu, F. H., Chen, Y., Zhao, C. T., Wang, H. B., He, D. Y., Xu, L. L., et al. (2016). Olig2-dependent reciprocal shift in PDGF and EGF receptor signaling regulates tumor phenotype and mitotic growth in malignant glioma. *Cancer Cell* 29, 669–683. doi: 10.1016/j.ccell.2016.03.027

Malik, S. Z., Lewis, M., Isaacs, A., Haskins, M., Van Winkle, T., Vite, C. H., et al. (2012). Identification of the rostral migratory stream in the canine and feline brain. *PLoS One* 7:e36016. doi: 10.1371/journal.pone.0036016

Marti-Fabregas, J., Romaguera-Ros, M., Gomez-Pinedo, U., Martinez-Ramirez, S., Jimenez-Xarrie, E., Marín, R., et al. (2010). Proliferation in the human ipsilateral subventricular zone after ischemic stroke. *Ann. Neurosci.* 17, 134–135. doi: 10.5214/ans.0972-7531.1017308

Marumoto, T., Tashiro, A., Friedmann-Morvinski, D., Scadeng, M., Soda, Y., Gage, F. H., et al. (2009). Development of a novel mouse glioma model using lentiviral vectors. *Nat. Med.* 15, 110–116. doi: 10.1038/nm.1863

Ming, G. L., and Song, H. J. (2011). Adult neurogenesis in the mammalian brain: significant answers and significant questions. *Neuron* 70, 687–702. doi: 10.1016/j.neuron.2011.05.001

Munoz, D. M., Singh, S., Tung, T., Agnihotri, S., Nagy, A., Guha, A., et al. (2013). Differential transformation capacity of neuro-glial progenitors during development. *Proc. Natl. Acad. Sci. U.S.A.* 110, 14378–14383. doi: 10.1073/pnas.1303504110

Nishiyama, A., Komitova, M., Suzuki, R., and Zhu, X. Q. (2009). Polydendrocytes (NG2 cells): multifunctional cells with lineage plasticity. *Nat. Rev. Neurosci.* 10, 9–22. doi: 10.1038/nrn2495

Ozawa, T., Riester, M., Cheng, Y. K., Huse, J. T., Squatrito, M., Helmy, K., et al. (2014). Most human non-GCIMP glioblastoma subtypes evolve from a common proneural-like precursor glioma. *Cancer Cell* 26, 288–300. doi: 10.1016/j.ccr.2014.06.005

Paredes, M. F., Sorrells, S. F., Garcia-Verdugo, J. M., and Alvarez-Buylla, A. (2016). Brain size and limits to adult neurogenesis. *J. Comp. Neurol.* 524, 646–664. doi: 10.1002/cne.23896

Parolisi, R., Cozzi, B., and Bonfanti, L. (2017). Non-neurogenic SVZ-like niche in dolphins, mammals devoid of olfaction. *Brain Struct. Funct.* 222, 2625–2639. doi: 10.1007/s00429-016-1361-3

Perry, A., and Wesseling, P. (2016). Histologic classification of gliomas. *Handb. Clin. Neurol.* 134, 71–95. doi: 10.1016/b978-0-12-802997-8.00005-0

Persson, A. I., Petritsch, C., Swartling, F. J., Itsara, M., Sim, F. J., Auvergne, R., et al. (2010). Non-stem cell origin for oligodendroglioma. *Cancer Cell* 18, 669–682. doi: 10.1016/j.ccr.2010.10.033

Qin, E. Y., Cooper, D. D., Abbott, K. L., Lennon, J., Nagaraja, S., Mackay, A., et al. (2017). Neural precursor-derived pleiotrophin mediates subventricular zone invasion by glioma. *Cell* 170, 845.e19–859.e19. doi: 10.1016/j.cell.2017.07.016

Raponi, E., Agenes, F., Delphin, C., Assard, N., Baudier, J., Legraverend, C., et al. (2007). S100B expression defines a state in which GFAP-expressing cells lose their neural stem cell potential and acquire a more mature developmental stage. *Glia* 55, 165–177. doi: 10.1002/glia.20445

Rebetz, J., Tian, D., Persson, A., Widegren, B., Salford, L. G., Englund, E., et al. (2008). Glial progenitor-like phenotype in low-grade glioma and enhanced CD133-expression and neuronal lineage differentiation potential in high-grade glioma. *PLoS One* 3:e1936. doi: 10.1371/journal.pone.0001936

Richardson, W. D., Young, K. M., Tripathi, R. B., and Mckenzie, I. (2011). NG2-glia as multipotent neural stem cells: fact or fantasy? *Neuron* 70, 661–673. doi: 10.1016/j.neuron.2011.05.013

Rivers, L. E., Young, K. M., Rizzi, M., Jamen, F., Psachoulia, K., Wade, A., et al. (2008). PDGFRA/NG2 glia generate myelinating oligodendrocytes and piriform projection neurons in adult mice. *Nat. Neurosci.* 11, 1392–1401. doi: 10.1038/nn.2220

Sanai, N., Nguyen, T., Ihrie, R. A., Mirzadeh, Z., Tsai, H. H., Wong, M., et al. (2011). Corridors of migrating neurons in the human brain and their decline during infancy. *Nature* 478, 382–386. doi: 10.1038/nature10487

Schmid, R. S., Simon, J. M., Vitucci, M., Mcneill, R. S., Bash, R. E., Werneke, A. M., et al. (2016). Core pathway mutations induce de-differentiation of murine astrocytes into glioblastoma stem cells that are sensitive to radiation but resistant to temozolomide. *Neuro Oncol.* 18, 962–973. doi: 10.1093/neuonc/nov321

Shao, F. J., Jiang, W. H., Gao, Q. Q., Li, B. Z., Sun, C. R., Wang, Q. Y., et al. (2017). Frozen tissue preparation for high-resolution multiplex histological analyses of human brain specimens. *J. Neurooncol.* 135, 21–28. doi: 10.1007/s11060-017-2547-0

Shoshan, Y., Nishiyama, A., Chang, A. S., Mork, S., Barnett, G. H., Cowell, J. K., et al. (1999). Expression of oligodendrocyte progenitor cell antigens by gliomas: implications for the histogenesis of brain tumors. *Proc. Natl. Acad. Sci. U.S.A.* 96, 10361–10366. doi: 10.1073/pnas.96.18.10361

Singh, S. K., Hawkins, C., Clarke, I. D., Squire, J. A., Bayani, J., Hide, T., et al. (2004). Identification of human brain tumour initiating cells. *Nature* 432, 396–401. doi: 10.1038/nature03128

Snuderl, M., Fazlollahi, L., Le, L. P., Nitta, M., Zhelyazkova, B. H., Davidson, C. J., et al. (2011). Mosaic amplification of multiple receptor tyrosine kinase genes in glioblastoma. *Cancer Cell* 20, 810–817. doi: 10.1016/j.ccr.2011.11.005

Stallcup, W. B. (1981). The Ng2 antigen, a putative lineage marker - immunofluorescent localization in primary cultures of rat-brain. *Dev. Biol.* 83, 154–165. doi: 10.1016/S0012-1606(81)80018-8

Stiles, C. D., and Rowitch, D. H. (2008). Glioma stem cells: a midterm exam. *Neuron* 58, 832–846. doi: 10.1016/j.neuron.2008.05.031

Venkatesh, H. S., Johung, T. B., Caretti, V., Noll, A., Tang, Y. J., Nagaraja, S., et al. (2015). Neuronal activity promotes glioma growth through neuroligin-3 secretion. *Cell* 161, 803–816. doi: 10.1016/j.cell.2015.04.012

Verhaak, R. G. W., Hoadley, K. A., Purdom, E., Wang, V., Qi, Y., Wilkerson, M. D., et al. (2010). Integrated genomic analysis identifies clinically relevant subtypes of glioblastoma characterized by abnormalities in PDGFRA, IDH1, EGFR, and NF1. *Cancer Cell* 17, 98–110. doi: 10.1016/j.ccr.2009.12.020

Vigano, F., and Dimou, L. (2016). The heterogeneous nature of NG2-glia. *Brain Res.* 1638, 129–137. doi: 10.1016/j.brainres.2015.09.012

Visvader, J. E. (2011). Cells of origin in cancer. *Nature* 469, 314–322. doi: 10.1038/nature09781

Vitucci, M., Irvin, D. M., Mcneill, R. S., Schmid, R. S., Simon, J. M., Dhruv, H. D., et al. (2017). Genomic profiles of low-grade murine gliomas evolve during progression to glioblastoma. *Neuro Oncol.* 19, 1237–1247. doi: 10.1093/neuonc/nox050

Wang, C. M., Liu, F., Liu, Y. Y., Zhao, C. H., You, Y., Wang, L., et al. (2011). Identification and characterization of neuroblasts in the subventricular zone and rostral migratory stream of the adult human brain. *Cell Res.* 21, 1534–1550. doi: 10.1038/cr.2011.83

Wang, J. G., Cazzato, E., Ladewig, E., Frattini, V., Rosenbloom, D. I. S., Zairis, S., et al. (2016). Clonal evolution of glioblastoma under therapy. *Nat. Genet.* 48, 768–776. doi: 10.1038/ng.3590

Wang, Q. H., Hu, B. L., Hu, X., Kim, H., Squatrito, M., Scarpace, L., et al. (2017). Tumor evolution of glioma-intrinsic gene expression subtypes associates with immunological changes in the microenvironment. *Cancer Cell* 32, 42.e6–56.e6. doi: 10.1016/j.ccell.2017.06.003

Weiss, W. A., Burns, M. J., Hackett, C., Aldape, K., Hill, J. R., Kuriyama, H., et al. (2003). Genetic determinants of malignancy in a mouse model for oligodendroglioma. *Cancer Res.* 63, 1589–1595.

Yeung, M. S. Y., Zdunek, S., Bergmann, O., Bernard, S., Salehpour, M., Alkass, K., et al. (2014). Dynamics of oligodendrocyte generation and myelination in the human brain. *Cell* 159, 766–774. doi: 10.1016/j.cell.2014.10.011

Zhang, Y., Chen, K., Sloan, S. A., Bennett, M. L., Scholze, A. R., O'keeffe, S., et al. (2014). An RNA-sequencing transcriptome and splicing database of glia, neurons, and vascular cells of the cerebral cortex. *J. Neurosci.* 34, 11929–11947. doi: 10.1523/jneurosci.1860-14.2014

Zheng, H. W., Ying, H. Q., Yan, H. Y., Kimmelman, A. C., Hiller, D. J., and Chen, A. J. (2008). p53 and Pten control neural and glioma stem/progenitor cell renewal and differentiation. *Nature* 455, 1129–1133. doi: 10.1038/nature07443

Zhu, X. Q., Bergles, D. E., and Nishiyama, A. (2008). NG2 cells generate both oligodendrocytes and gray matter astrocytes. *Development* 135, 145–157. doi: 10.1242/dev.004895

Zhu, X. Q., Hill, R. A., Dietrich, D., Komitova, M., Suzuki, R., and Nishiyama, A. (2011). Age-dependent fate and lineage restriction of single NG2 cells. *Development* 138, 745–753. doi: 10.1242/dev.047951

Zhu, Y., Guignard, F., Zhao, D. W., Liu, L., Burns, D. K., Mason, R. P., et al. (2005). Early inactivation of p53 tumor suppressor gene cooperating with NF1 loss induces malignant astrocytoma. *Cancer Cell* 8, 119–130. doi: 10.1016/j.ccr.2005.07.004

9

Astrocyte Senescence and Metabolic Changes in Response to HIV Antiretroviral Therapy Drugs

Justin Cohen, Luca D'Agostino, Joel Wilson, Ferit Tuzer and Claudio Torres*

Department of Pathology and Laboratory Medicine, Drexel University College of Medicine, Philadelphia, PA, United States

*Correspondence:
Claudio Torres
claudio.torres@drexelmed.edu

With the advent of highly active antiretroviral therapy (HAART) survival rates among patients infected by HIV have increased. However, even though survival has increased HIV-associated neurocognitive disorders (HAND) still persist, suggesting that HAART-drugs may play a role in the neurocognitive impairment observed in HIV-infected patients. Given previous data demonstrating that astrocyte senescence plays a role in neurocognitive disorders such as Alzheimer's disease (AD), we examined the role of HAART on markers of senescence in primary cultures of human astrocytes (HAs). Our results indicate HAART treatment induces cell cycle arrest, senescence-associated beta-galactosidase, and the cell cycle inhibitor p21. Highly active antiretroviral therapy treatment is also associated with the induction of reactive oxygen species and upregulation of mitochondrial oxygen consumption. These changes in mitochondria correlate with increased glycolysis in HAART drug treated astrocytes. Taken together these results indicate that HAART drugs induce the senescence program in HAs, which is associated with oxidative and metabolic changes that could play a role in the development of HAND.

Keywords: cellular senescence, highly active antiretroviral therapy, HIV, astrocytes, glycolysis, HIV-associated neurocognitive disorders

INTRODUCTION

With the advent of highly active antiretroviral therapy (HAART), HIV infection has transitioned from an acute, terminal illness to a chronic but manageable condition (Bhatia et al., 2012). The HIV-infected population is consequently aging, and it had been projected that by 2015 more than 50% of the HIV-infected population in the United States would be 50 years of age and older. While this is undoubtedly a major success, aging is a significant risk factor for disease (Niccoli and Partridge, 2012) and HIV patients experience a variety of age-related complications, suggesting premature aging (Capeau, 2011). One such complication is a series of neurological problems collectively known as HIV-associated neurocognitive disorders (HAND) (Heaton et al., 2010). HAND can be categorized with increasing severity from asymptomatic neurocognitive impairment, mild neurocognitive disorder, and HIV-associated dementia. While the prevalence of HIV-associated dementia in the post-HAART era has decreased in HIV-infected patients, asymptomatic neurocognitive impairment and mild neurocognitive disorder have increased (Heaton et al., 2010). The persistence of neurological problems in HIV-infected patients remains a major public health issue and the identification of mechanisms involved may lead to potential treatments.

While beneficial in their suppression of HIV, HAART drugs have a multitude of side effects including myopathy, hepatotoxicity, hypersensitivity reactions, lipodystrophy, and insulin resistance (Feeney and Mallon, 2010). *In vitro*, there has been evidence of HAART drugs inducing ER stress (Sato et al., 2012), unfolded protein response (Zhou et al., 2005), and changes to cellular metabolism (Arend et al., 2013). These side effects suggest that cells may undergo a great deal of stress in response to HAART drugs. One possible way that cells can respond to stress is to undergo cellular senescence.

Cellular senescence is an age-related phenotype originally discovered to occur *in vitro* after extensive cell passaging, and is associated with the telomere attrition that occurs with successive rounds of DNA replication (Bodnar et al., 1998). Senescence also occurs prematurely in response to other mediators. Oncogene-induced senescence can occur via the activation of tumorigenic signals such as telomere dysfunction (Suram et al., 2012) and oncogenic RAS (Serrano et al., 1997). Stress-induced premature senescence can occur in response to cytotoxic stimuli such as proteasome inhibition and oxidative stress (Chen et al., 1995; Bitto et al., 2010). Several classes of HAART drugs including nucleoside reverse transcriptase inhibitors and protease inhibitors have been shown to cause stress-induced premature senescence (Caron et al., 2008; Lefevre et al., 2010; Hernandez-Vallejo et al., 2013; Afonso et al., 2015), suggesting that HIV patients may be experiencing cellular senescence. Evidence for cellular senescence during HIV comes from a previous study showing increased senescent $CD8^+$ T-cells isolated from HIV patients (Chou et al., 2013). Regardless of the inducer, there are several phenotypes and biomarkers generally shared among senescent cells. These include cell cycle arrest, increased senescence-associated beta-galactosidase (SA β-gal) activity, expression of the cell cycle inhibitors p16 and p21, mitochondrial dysfunction, and the secretion of pro-inflammatory cytokines and proteases known as the senescence-associated secretory phenotype (SASP) (Rodier and Campisi, 2011). The pro-inflammatory environment created by the SASP has major implications for age-related decline in tissues and may contribute the chronic inflammation observed in the central nervous system (CNS) in neurological diseases such as Parkinson's and AD (Jabbari Azad et al., 2014; Yan et al., 2014) and HAND.

Senescence in the CNS is an emergent concept and few studies have examined its role as a contributor to neurodegenerative disease. Astrocytes are the most abundant cells in the brain and are involved in a variety of functions to maintain CNS homeostasis such as CNS metabolism, blood brain barrier maintenance, and ion regulation (Stobart and Anderson, 2013). Due to their numerous functions in the CNS, disruption of their physiological functions due to senescence could be a major contributor to neurological disease. Our recent work demonstrates a decrease in astrocyte-enriched genes during senescence, indicating a loss in their differentiated function (Crowe et al., 2016). This could impact brain physiology in Alzheimer's patients where we have previously reported a significant increase in the population of senescent astrocytes (Bhat et al., 2012). In the present study, we evaluated the role

of HAART drug exposure on astrocyte senescence. Human astrocytes (HAs) treated with a clinically relevant combination of nucleotide reverse transcriptase inhibitors and protease inhibitors underwent cellular senescence with expression of p16, p21, SA β-gal, and pro-inflammatory cytokines. The process was accompanied with increased oxidative stress, mitochondrial oxygen consumption, and changes in glucose metabolism with increased glucose uptake and upregulation in glycolytic intermediates. To our knowledge, our findings are the first to demonstrate HAART drug-induced senescence in a CNS cell type, which may have implications for HAND.

MATERIALS AND METHODS

Cell Culture and Drug Treatments

Human astrocytes were cultured at 37°C, 5% CO_2 in astrocyte medium supplemented with 2% fetal bovine serum, growth supplement, and penicillin/streptomycin all obtained from ScienCell Research Laboratories (Carlsbad, CA, United States). Cells were cultured until they reached ~80% confluence before passaging. At each passage, astrocytes were trypsinized, counted, and the cumulative population doubling level (CPDL) was calculated as we have previously described (Bitto et al., 2010). Cells were treated every 2–3 days for up to a week with either 0.3% DMSO as a vehicle control or the HAART drug combinations of abacavir (ABC) 10 μM and lamivudine (3TC) 5 μM or ABC, 3TC, and ritonavir (RTV) 1 μM. For the long-term experiments, cells were treated for up to 4 weeks with either 0.2% dimethyl sulfoxide (DMSO) as a vehicle control or the combinations of ABC 3 μM, 3TC 1.9 μM, atazanavir (ATV) 50 nM, and RTV 100 nM; or tenofovir (TDF) 100 nM, emtricitabine (FTC) 1.2 μM, ATV, and RTV; or TDF, FTC, and efavirenz (EFV) 125 nM. All HAART drugs were provided by the NIH AIDS Reagent Program.

Senescence-Associated β-Galactosidase Activity Assay

Senescence-associated beta-galactosidase staining was performed as previously described (Dimri et al., 1995). Briefly, following exposure to the HAART drug combinations or DMSO, astrocytes were fixed in 2% formaldehyde/0.2% glutaraldehyde for 3 min and stained for SA β-gal activity overnight. The cells were counted and positive (blue) cells were expressed as a percentage of the total. At least 200 cells were counted.

Immunoblotting

Following indicated treatment times, HAs were lysed in radioimmunoprecipitation assay (RIPA) buffer. Western blot analysis was performed under standard conditions using 15 μg of total cell proteins. Membranes were probed for antibodies against p16 [sc-56330 (**JC8**), monoclonal; BD Biosciences, San Jose, CA, United States]; p21 [sc-756 (**H-164**), polyclonal; Santa Cruz Biotechnology, Santa Cruz, CA, United States]; phosphorylated (9211) and total p38 (9212) both polyclonal (Cell Signaling Technology, Danvers, MA, United States); phosphorylated (3033) and total p65 (3034) both polyclonal (Cell Signaling

Technology, Danvers, MA, United States); Oxphos cocktail of mitochondrial ETC complexes I [ab110242 (**20E9DH10C12**)], II [ab14714 (**21A11AE7**)], III [ab14745 (**13G12AF12BB11**)], IV [ab110258 (**12C4F12**)], and V [ab14748 (**15H4C4**)] (monoclonal; Abcam, Cambridge, MA, United States); β-actin [A00702 (**2D1D10**), monoclonal; Genscript, Piscataway, NJ, United States); and β-tubulin [sc-9104 (**H-235**), polyclonal; Santa Cruz Biotechnology, Santa Cruz, CA, United States). Clone numbers are in bold.

Total Cellular ROS, Mitochondrial ROS, Mitochondrial Membrane Potential, Mitochondrial Mass, and Glucose Uptake Assessments

Determination of total cellular reactive oxygen species (ROS), mitochondrial ROS, mitochondrial membrane potential, mitochondrial mass, and glucose uptake assessments was made using flow cytometry as previously described (Nacarelli et al., 2016). Total cellular ROS was detected by incubating cells with 10 μM 2'7'-dichlorofluorescein diacetate (DCF-DA; Sigma-Aldrich, St. Louis, MO, United States) in 1% fetal bovine serum-supplemented MEM and washing twice with Krebs Ringer phosphate glucose buffer (145 mM NaCl, 5.7 mM NaH_2PO_4, 4.86 mM KCl, 0.54 mM $CaCl_2$, 1.22 mM

$MgSO_4$, and 5.5 mM glucose) following incubation. All other compounds were washed with PBS. Mitochondrial superoxide levels were assayed by incubating the cells with 5 μM Mito-Sox Red (Molecular Probes, Waltham, MA, United States). Mitochondrial membrane potential was detected by incubating cells with 25 nM tetramethylrhodamine (TMRE) (Molecular Probes, Waltham, MA, United States). Mitochondrial mass was evaluated by incubating the cells with 100 nM Mito-tracker Green FM (Molecular Probes, Waltham, MA, United States). The glucose analog 2-NBDG uptake was detected by incubating the cells with 10 μM of 2-NBDG (Molecular Probes, Waltham, MA, United States). For the previous analyses, incubation was performed at 37°C in 5% CO_2 for 30 min except 2-NBDG which required 90 min. Cells were collected in 0.25% trypsin-EDTA with complete medium. Cells were analyzed by flow cytometry using a Guava EasyCyte Mini and the Guava Express Plus program (Guava Technologies, Hayward, CA, United States). Acquisitions involved 5000 events.

Analysis of Inflammatory Factors Secreted by Astrocytes

Following the end of the treatment period, HAs were incubated with serum-free MCDB105 media. After a 24-h incubation period, media were collected and cells were

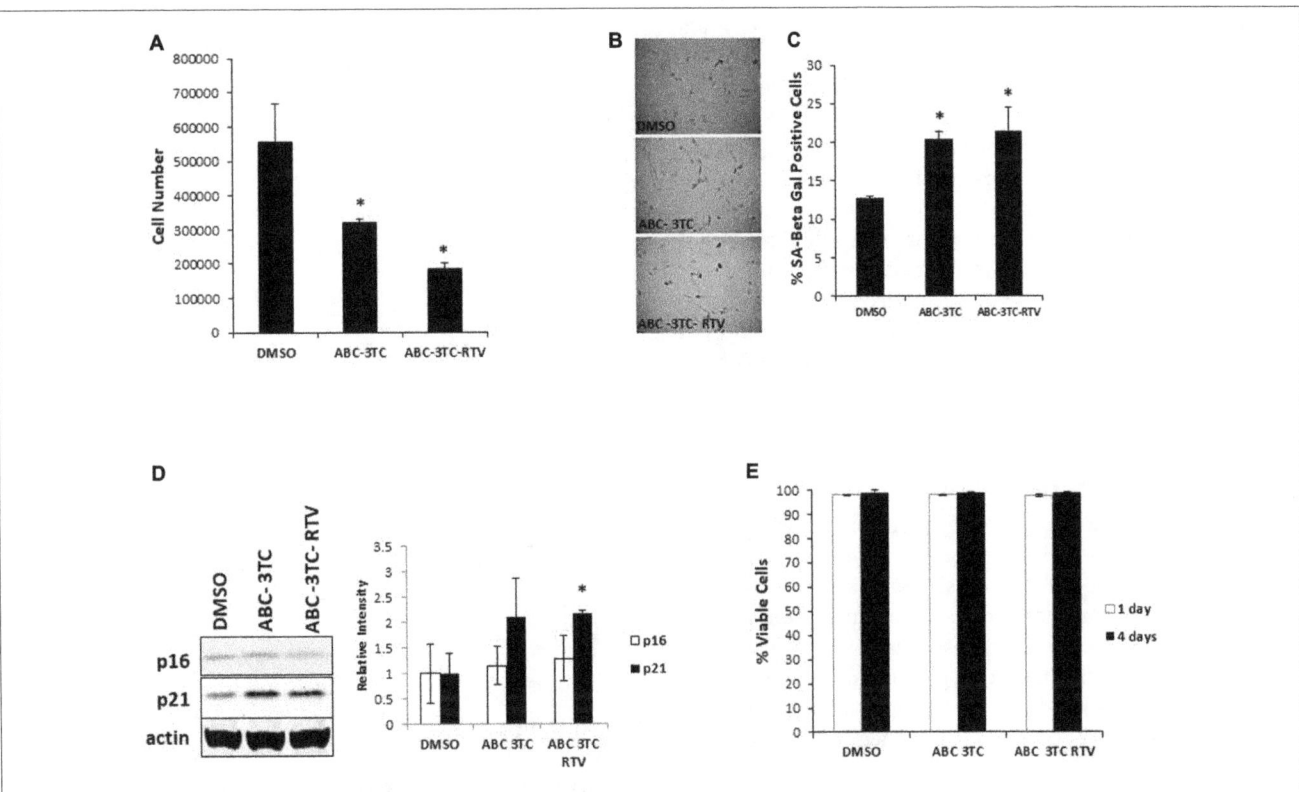

FIGURE 1 | Expression of senescence markers in HAART drug treated human astrocytes (HAs). Human astrocytes were treated with the following HAART concentrations: abacavir (ABC) 10 μM, lamivudine (3TC) 5 μM, and ritonavir (RTV) 1 μM for 7 days **(A–D)** or 4 days **(E)** in complete astrocyte media. **(A)** Cell proliferation. **(B)** Representative SA β-gal images displayed at 20× of HAs stained for SA β-gal 1 week after HAART treatment. **(C)** Quantification of B. **(D)** Left—representative Western blot illustrating protein levels of senescence markers p16 and p21. Right—quantification of the blots. **(E)** Viability, astrocytes were incubated with Guava Viacount reagent for 5 min prior to detection by flow cytometry. *p-value < 0.05, $n = 3$, error bars are SD.

trypsinized and counted to determine the cell number for normalization. Human Cytokine Array C5 (RayBiotech, Norcross, GA, United States) was used to evaluate secreted inflammatory factors in the conditioned media according to the company's protocol. The intensity of the signal on the array membranes was quantified by densitometry using ImageJ software and normalized to cell number. The HAART drug treated values were then set as relative to control values. Samples that had no change in expression due to levels being undetectable from background were set to 1. Interleukin-6 (IL-6) detection was performed via the Human IL-6 Quantikine ELISA kit (R&D Systems, Minneapolis, MN, United States) according to the product manual using conditioned media as described above. Absorbance was measured at 450 nm.

Oxygen Consumption Measurements

Oxygen consumption was determined by using a Seahorse XF24 Bioanalyzer (Seahorse Bioscience, North Billerica, MA, United States) and the XF Cell Mito Stress Test Kit as previously reported (Nacarelli et al., 2016). Cells were seeded after treatment at 25,000 cells per well. The Bioanalyzer was pre-loaded with oligomycin, carbonilcyanide p-triflouromethoxyphenylhydrazone (FCCP), and rotenone/antimycin A prior to measurement. Oxygen consumption was measured in triplicate before and after consecutive addition of oligomycin, FCCP, and rotenone/antimycin A. Respiration rates and proton leak were assessed as previously described (Hill et al., 2012) based upon oxygen consumption rate measurements. Basal respiration represents the initial oxygen consumption rate, while maximal respiration signifies oxygen consumption after FCCP addition. ATP-linked respiration denotes the oligomycin-sensitive oxygen change to basal oxygen consumption rate. Proton leak corresponds to the oligomycin-insensitive oxygen consumption rate. Non-mitochondrial sources of oxygen consumption were subtracted by normalizing to the rotenone/antimycin A-insensitive oxygen consumption rate. Acidification was based on the extracellular acidification rate. Data were normalized to cell number.

Metabolite Measurements

Metabolite measurements were performed by Human Metabolome Technologies America Inc. (Boston, MA, Unites States) using their C-Scope analysis. Samples were collected according to their protocol and sent overnight on dry ice, after which capillary electrophoresis mass spectrometry was performed. Quantifications were performed by hierarchical cluster analysis (HCA) and principal component analysis (PCA) by Human Metabolome Technologies America Inc.'s statistical software.

Statistical Analysis

Data were either compared using a two-tailed Student's t-test when two groups were involved or one-way analysis of variance (ANOVA) followed by Bonferroni correction when three groups

TABLE 1 | Senescence-associated secretory phenotype of ABC–3TC–RTV-treated human astrocytes.

Name	ABC–3TC–RTV	Name continued	ABC–3TC–RTV continued
Angiogenin	0.53	IL-8	0.75
BDNF	0.59	IP-10	0.77
EGF	0.86	Leptin	1.00
Eotaxin-1	0.81	LIF	0.98
Eotaxin-2	0.89	Light	1.05
Eotaxin-3	0.83	MCP-1	0.73
FGF-4	0.75	MCP-4	1.00
FGF-6	0.82	M-CSF	0.82
FGF-9	1.27	MDC	1.74
Fractalkine	1.03	MIF	1.31
GCP-2	1.66	MIG	1.00
G-CSF	1.00	MIP-1 beta	1.16
GDNF	1.06	NAP-2	0.93
GM-CSF	1.00	NT-3	0.60
GRO	0.73	NT-4	0.58
GRO alpha	1.00	Oncostatin M	0.84
HGF	2.64	Osteopontin	0.39
I-309	1.00	Osteoprotegerin	4.53
IGF-1	0.69	PARC	2.19
IGFBP-1	1.08	PDGF-BB	1.00
IGFBP-2	0.86	PLGF	1.16
IGFBP-3	0.56	RANTES	1.04
IGFBP-4	1.39	TARC	1.00
IL-1 alpha	1.71	TGF beta 1	1.00
IL-1 beta	1.78	TGF beta 2	0.89
IL-10	1.58	TGF beta 3	2.16
IL-12 p40/p70	1.80	Thrombopoietin	1.00
IL-13	1.00	TIMP-1	0.89
IL-16	2.34	TIMP-2	0.76
IL-3	1.17	TNF alpha	1.18
IL-4	1.00	TNF beta	0.46
IL-5	1.00	VEGF-A	4.16
IL-6	1.00	CCL23	1.08
IL-7	1.00	ENA-78	1.79

Cytokines were measured by an antibody array and values are expressed as relative to DMSO control.

were analyzed. Normality was determined using a Shapiro–Wilk Test with the caveat that low sample sizes can reduce the accuracy of normality tests. Means were derived from at least three independent experiments. Error bars on graphs reflect standard deviation (SD). Statistical significance was considered at $p < 0.05$.

RESULTS

HAART Drugs Induce Senescence Program and Inflammatory Response in Human Astrocytes

HIV-infected patients do not take individual antiretroviral drugs but rather they are put on a regimen that includes several different

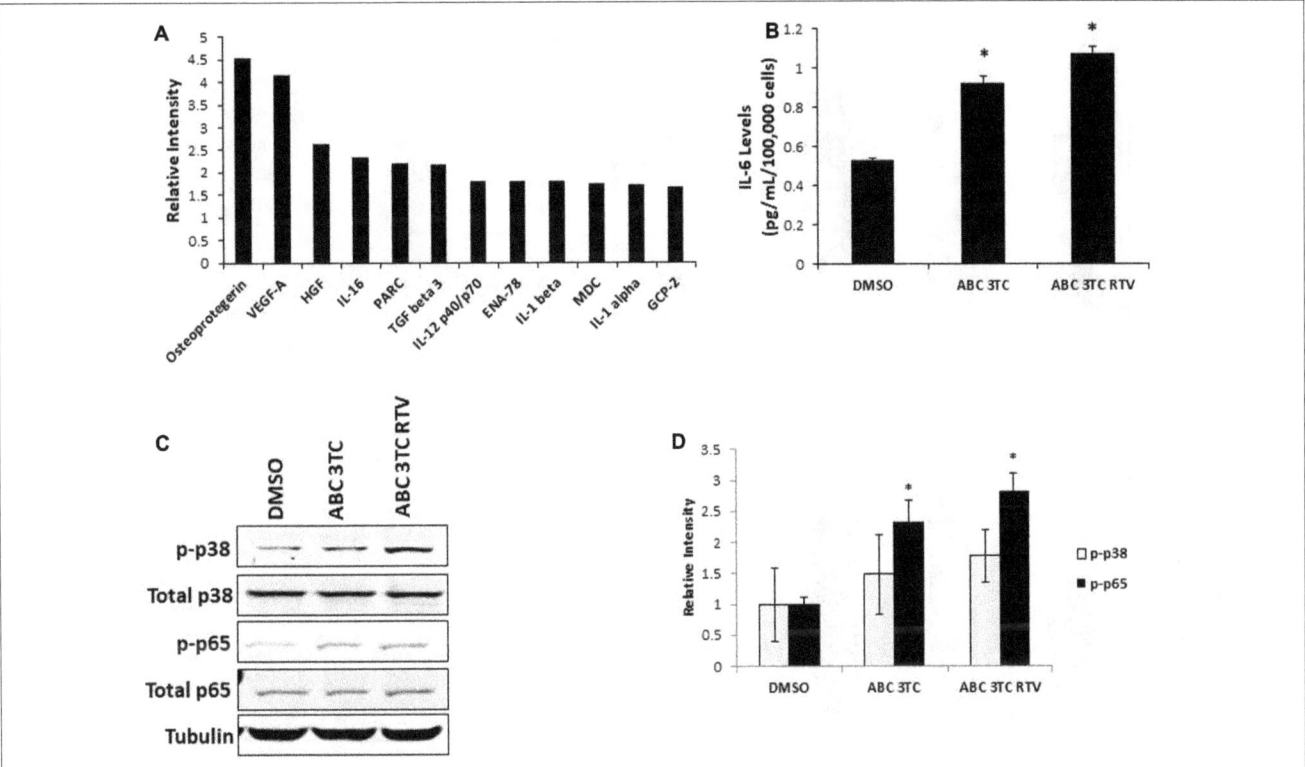

FIGURE 2 | Secretory profile of HAART drug treated HAs. Human astrocytes were treated with the following HAART concentrations: ABC 10 μM, 3TC 5 μM, and RTV 1 μM for 1 week. **(A)** Senescence-associated secretory phenotype (SASP) analysis. Astrocytes were subjected to a 24-h incubation in MCBD105 media to generate conditioned media (CM). The secretory profile was detected by incubating CM on a cytokine membrane array and normalized to cell number. Values are relative to a DMSO control. **(B)** IL-6 quantitation. CM was collected as in **(A)** and IL-6 was quantitated by ELISA. **(C)** Representative Western blot illustrating phosphorylated protein levels of inflammatory mediators p38 and p65. Total p38, total p65, and tubulin were used as a loading control. **(D)** Quantification of **(C)**. *p-value < 0.05, n = 3, error bars are SD.

drug classes. We therefore evaluated whether a clinically relevant combination of HAART drugs could affect physiology of HAs. Cells were chronically treated with the either the nucleotide reverse transcriptase inhibitors (NRTIs) ABC and 3TC alone or in combination with the protease inhibitor RTV. Determination of the effects on cell proliferation indicates that after 1 week of treatment these drugs induce a reduction in the cell number compared to vehicle control **(Figure 1A)** which paralleled an increase in SA β-gal activity **(Figures 1B,C)**. Protein levels of the senescence marker p21 but not p16 was significantly increased in the ABC–3TC–RTV combination, suggesting that the pathway may be p21-dependent **(Figure 1D)**. However, the ABC–3TC combination did not show a significant increase in either p16 or p21. In order to rule out the induction of apoptosis, cell viability was measured 1 and 4 days after treatment. No reduction in cell viability was observed compared to the vehicle **(Figure 1E)**. These results suggest that the ABC–3TC and ABC–3TC–RTV induced reduction in HA growth is the result of induction of the senescence program.

In addition, we examined the effect of several other clinically relevant HAART regimens on HA senescence at doses near what is found in the CNS (de Almeida et al., 2006). Human astrocytes were treated with the PIs ATV and RTV with an NRTI backbone of ABC and 3TC; ATV and RTV with an NRTI

backbone of TDF and FTC; or TDF and FTC with the NNRTI EFV. Studies of the effects of these drugs on the replicative capacity indicate that compared to DMSO vehicle, there was a statistically significant decrease in population doublings by 11 days of treatment, which broadened further after 35 days **(Supplementary Figure S1A)**. The effects of these drugs were observed even at lower concentrations and the increase in SA β-gal positive cells was significant starting at 1 week, which trended upward at 3 weeks **(Supplementary Figures S1B,C)**. The exception to this trend was the TDF–FTC–EFV combination, which did not increase any further than its 1 week value. Interestingly, at these lower concentrations, ABC–3TC–ATV–RTV does not induce expression of p16 and p21 after 1 week **(Supplementary Figure S1D)**, even though an increase in SA β-gal was observed. This seems to match the minimal impact on cell growth at 1 week, and suggests that changes in the activity of SA β-gal may occur before other markers of senescence.

In order to examine the effects of HAART on the SASP, we characterized the secretory pattern of ABC–3TC–RTV-treated HAs using an antibody array. A total of 68 cytokines were analyzed and their relative levels of expression compared to control untreated are shown in **Table 1**. Treatment modulated the secretion of a variety of inflammatory molecules including TGF-β3, IL-1α, and IL-1β **(Figure 2A)**. Further validation

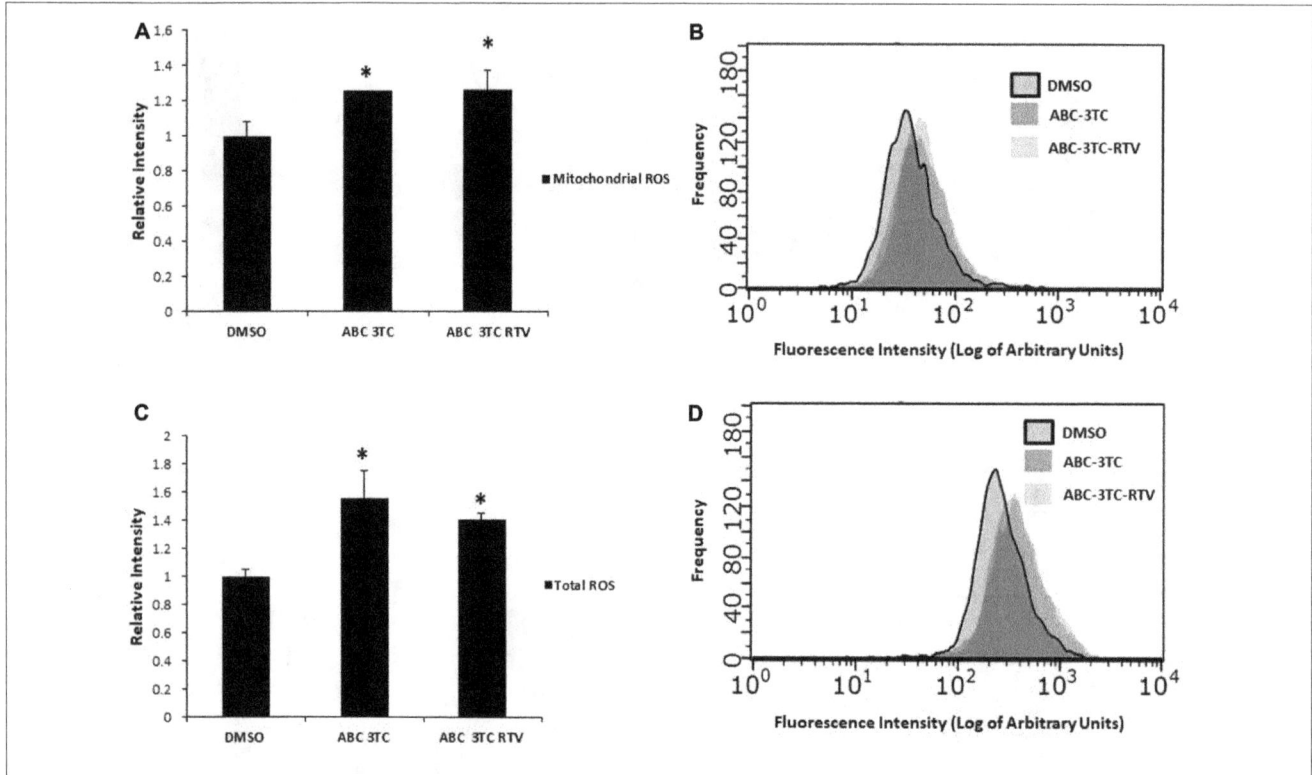

FIGURE 3 | HAART drugs induce total and mitochondrial ROS in HAs. Human astrocytes were treated for 1 week with ABC 10 μM, 3TC 5 μM, and RTV 1 μM before assaying. **(A)** Mitochondrial ROS was measured by incubation with MitoSox for 30 min followed by quantification on a flow cytometer. Bar graphs show mean intensity. **(B)** Representative histogram of data from **(A)**. **(C)** Total ROS was measured by incubation with DCF-DA for 30 min followed by quantification on a flow cytometer. Bar graphs show mean intensity. **(D)** Representative histogram of data from **(C)**. *p-value < 0.05, $n = 5$, error bars are SD.

these cytokines could be particularly interesting as they have been shown to induce senescence in neighboring cells through a process called paracrine senescence (Acosta et al., 2013). Little change was found in IL-6 using the cytokine array. This may be due to a sensitivity issue of the membrane-based analysis; therefore, we also examined IL-6 via ELISA. One week of treatment with the ABC–3TC and ABC–3TC–RTV HAART drug combination induced a significant, nearly twofold increase in IL-6 release **(Figure 2B)**. The CNS-based ABC–3TC–ATV–RTV combination was also able to induce IL-6 secretion over time in HAs, with a nearly threefold increase after 4 weeks of treatment **(Supplementary Figure S1E)**. Importantly, the pro-inflammatory transcription factor p65 (NF-κB), which has been shown to mechanistically induce senescence (Freund et al., 2011; Tilstra et al., 2012), is activated in response to both ABC–3TC and ABC–3TC–RTV treatment **(Figures 2C,D)**. Another pro-inflammatory mediator p38 while trending upward did not reach statistical significance **(Figures 2C,D)**.

HAART Drugs Induce Oxidative Stress

Accumulation of ROS can induce oxidative stress and contribute to premature senescence (Chen et al., 1995). The main source of ROS is the mitochondrial-specific superoxide anion, which can be converted to other forms of ROS to cause oxidative damage (Wang et al., 2013). We examined the effects of HAART

drug combinations on mitochondrial ROS production in HAs, and we observed that both ABC–3TC and the ABC–3TC–RTV combinations significantly increased mitochondrial ROS **(Figures 3A,B)**. Interestingly, the lower dose combination of ABC–3TC–ATV–RTV at 4 weeks of treatment was able to reach a similar level of mitochondrial ROS **(Supplementary Figure S2A)**. Due to this increase in mitochondrial ROS, we examined if there was a corresponding change in total cellular ROS. With 1 week treatment of ABC–3TC and ABC–3TC–RTV, total ROS significantly increases **(Figures 3C,D)**. A similar increase was seen after prolonged treatment with the lower dose ABC–3TC–ATV–RTV combination **(Supplementary Figure S2B)**. These results indicate that the HAART drugs induce oxidative stress in HAs.

HAART Drugs Impact Mitochondrial Respiration

The accumulation of mitochondrial ROS suggests that the HAART drugs may be affecting mitochondria. Since mitochondrial dysfunction is thought to contribute to aging and senescence (Lee and Wei, 2012) we evaluated changes in mitochondrial respiration in HAART-treated HAs by using a seahorse bioanalyzer **(Figure 4A)**. ABC–3TC–RTV treatment for 1 week increased both basal and maximal mitochondrial respiration **(Figures 4B,C)**. There was also an increase in

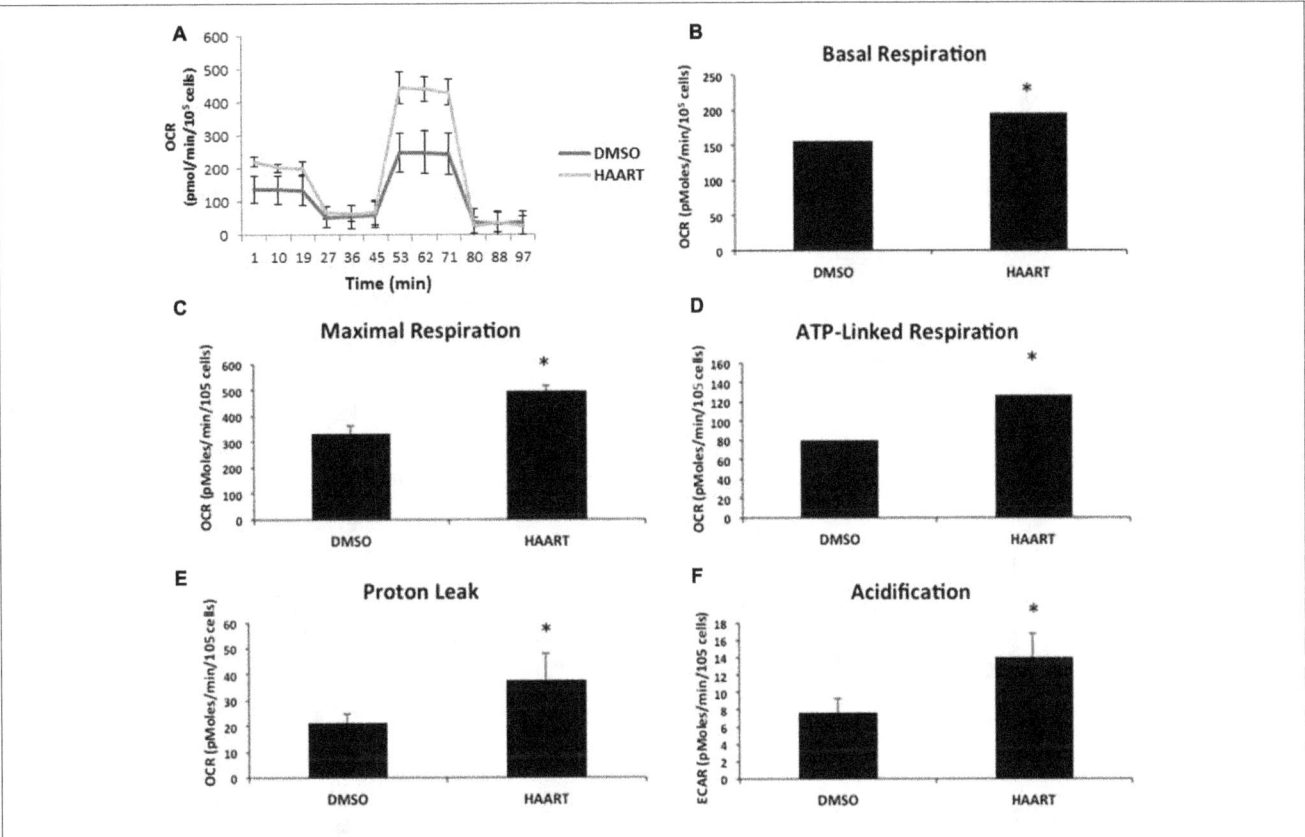

FIGURE 4 | HAART drugs induce mitochondrial respiration. Human astrocytes were treated with the following HAART drug concentrations: ABC 10 μM, 3TC 5 μM, and RTV 1 μM for 1 week before assaying. Oxygen consumption measurements were taken after HAART treatment on a Seahorse XF24 Bioanalyzer using the XF Cell Mito Stress Test Kit to acquire the oxygen consumption rate. **(A)** Representative Seahorse recording. **(B–E)** Bar graphs of the indicated oxygen consumption rate components. **(F)** Acidification measurements were acquired using the Seahorse XF24 Bioanalyzer set to measure the extracellular acidification rate. *p-value < 0.05, five independent replicates, error bars are SD.

ATP-linked respiration (**Figure 4D**). This HAART drug treatment induced an increase in mitochondrial proton leak, suggesting that it may contribute to the observed increase in mitochondrial respiration (**Figure 4E**). Measurement of the extracellular acidification rate indicates that HAART drug treatment increases acidification (**Figure 4F**). Altogether these results suggest that ABC–3TC–RTV cause over-activation of the mitochondria, which may contribute to increased mitochondrial ROS.

To determine if the increase in mitochondrial respiration was associated with other changes in the mitochondria, we first examined protein levels of the mitochondrial electron transport chain. As shown in **Figures 5A,B**, HAs treated for 1 week with ABC–3TC–RTV did not induce changes in protein levels of critical components of mitochondrial complexes. However, we observed that mitochondrial mass increased in response to the HAART drug treatment (**Figure 5C**). Since we were able to detect an increase in mitochondrial respiration and mass, we wanted to determine if this reflects a change in TMRE as a qualitative indicator of mitochondrial membrane potential. Treatment for 1 week with the HAART drugs increased TMRE fluorescence, indicating that the mitochondria may be polarized and activated (**Figure 5D**).

HAART Drugs Induce Astrocyte Glycolysis

Highly active antiretroviral therapy drug treatment severely affected astrocyte mitochondrial respiration accompanied by an increased medium acidification, suggesting an altered lactate production as a consequence of enhanced glycolysis. These results are intriguing because senescent fibroblasts were shown to have profound metabolic changes including increased glycolysis (James et al., 2015). We therefore wanted to determine if our astrocytes made senescent from HAART drug treatment have heightened glycolysis. To evaluate directly the effects of HAART on glycolysis we determined changes in glucose and glycolytic intermediates in response to the drugs. First, we examined glucose uptake using a fluorescent glucose analog, 2-NDBG. Highly active antiretroviral therapy drug treatment increased uptake of the glucose analog as measured by flow cytometry (**Figure 6A**). Glut1, the main glucose transporter in astrocytes was upregulated in response to HAART drug treatment, suggesting that the increase in 2-NDBG uptake could be due to an increase in this transporter. Glut3, which is the main glucose transporter for neurons, is not affected by HAART drug treatment (**Figures 6B,C**). In order to confirm that

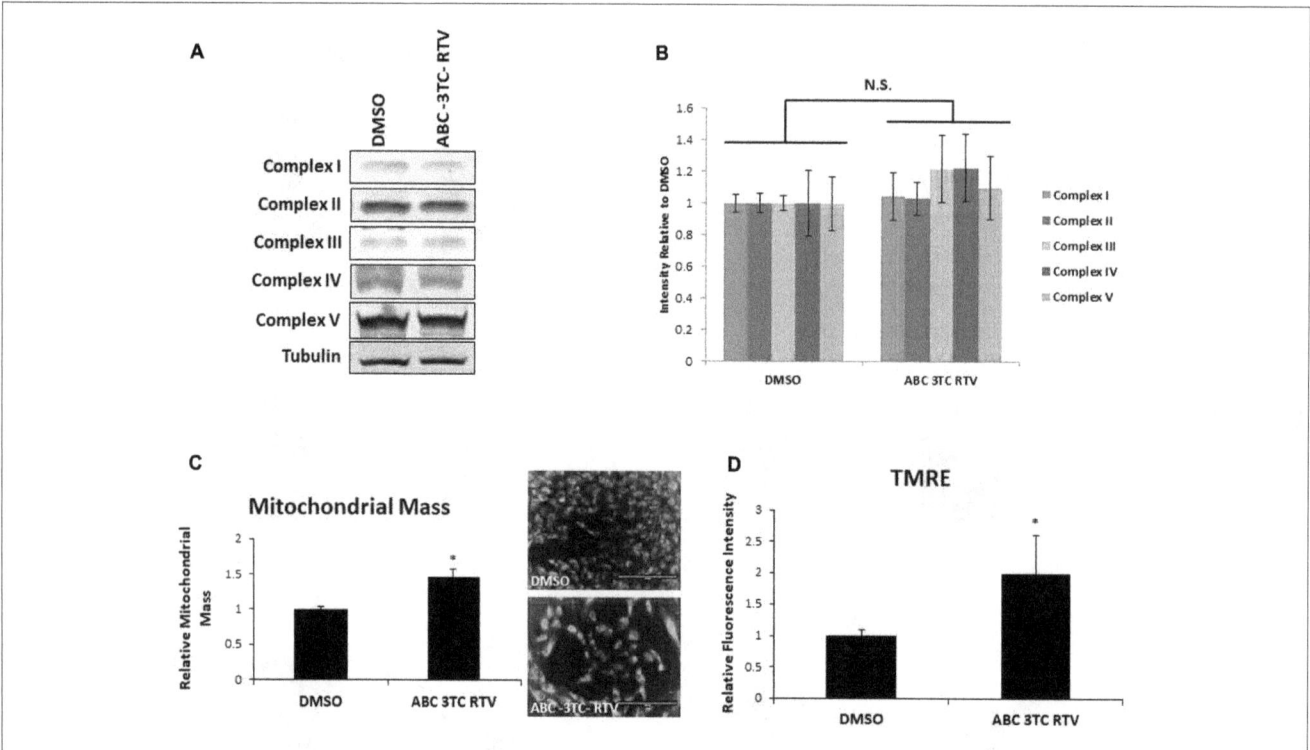

FIGURE 5 | Effect of HAART drugs on mitochondrial mass and mitochondrial membrane potential. Human astrocytes were treated with ABC 10 μM, 3TC 5 μM, and RTV 1 μM for 1 week before assaying. **(A)** Representative Western blot showing protein levels of the mitochondrial electron transport chain complexes. **(B)** Quantification of **A**. **(C)** Mitochondrial mass, cells were incubated with mitotracker for 30 min prior to quantification by flow cytometry as displayed on the left. Right is representative microscopy of fluorescence at 20× before undergoing flow cytometry. **(D)** TMRE, cells were incubated with TMRE for 30 min prior to quantification by flow cytometry. *p-value < 0.05, n = 3, error bars are SD.

there is an increase in glucose metabolism, we examined levels of metabolites associated with glycolysis. Glucose-6-phosphate (G6P), a product produced in the first step of glycolysis trends upward but does not reach statistical significance with HAART drug treatment in HAs **(Figure 6D)**. On the other hand, pyruvate, the last product of glycolysis, does show a statistically significant increase **(Figure 6E)**. Correspondingly, we observed an increase in the production of lactate **(Figure 6F)**, indicating that anaerobic glycolysis is enhanced in response to HAART. This increase in lactate production may explain the increased acidification determined by the Seahorse bioanalyzer **(Figure 4F)**. Overall, these results indicate that HAART drug treatment induces an increase in glucose metabolism in HAs.

DISCUSSION

The HIV-infected population is growing older and with this increased age, a larger risk for age-associated disease (Niccoli and Partridge, 2012). Neurological issues are particularly troubling because even though HAART has decreased the prevalence of the more severe forms of HAND, the milder forms still remain. We hypothesized that one possible contributor to HAND is the premature induction in astrocytes of cellular senescence in response to HAART drugs. Our study provides

evidence for this hypothesis by demonstrating that combinations of HAART drugs are able to induce premature senescence, oxidative stress, mitochondrial dysfunction, and affect glycolysis in HAs. These results are novel since this is the first study to demonstrate HAART drug-induced senescence in a CNS cell type.

We evaluated the impact of widely used NRTIs, NNRTIs, and PIs on primary HAs. These drugs induced various aspects of the senescence program including decreased cellular proliferation, increased SA β-gal and expression of the cell cycle inhibitor p21. While astrocyte senescence has not been explicitly studied in HIV, there is evidence of astrocytes and other glial cells expressing cell cycle inhibitors common to senescence in HIV-infected patients (Jayadev et al., 2007). In concordance, we have previously demonstrated astrocyte senescence in association between AD (Bhat et al., 2012).

Senescent cells are irreversibly growth arrested and their resistance to apoptosis (Childs et al., 2014) allows them to persist in tissues, secreting inflammatory SASP components. The pro-inflammatory microenvironment produced by senescent cells can have major implications *in vivo* since inflammation may contribute to age-related decline in organ function (Freund et al., 2010). Significantly, CNS inflammation has been implicated in neurological disorders such as AD and Parkinson's disease (Jabbari Azad et al., 2014; Yan et al., 2014). Most importantly, CNS inflammation is found in HIV patients suffering from

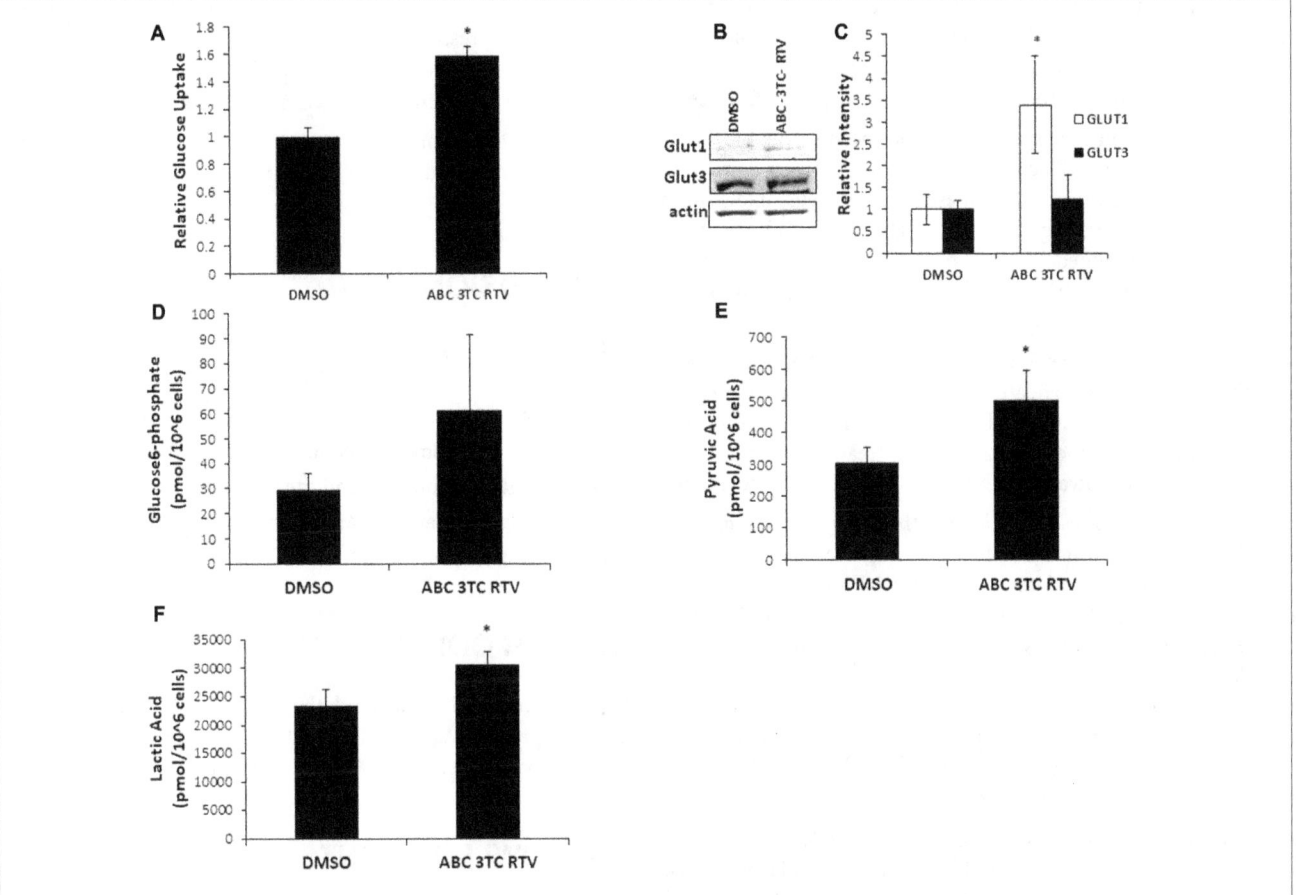

FIGURE 6 | Effect of HAART treatment on astrocyte metabolism. Human astrocytes were treated with ABC 10 μM, 3TC 5 μM, and RTV 1 μM for 1 week before assaying. (A) Glucose uptake was measured by flow cytometry after incubation with 2-NDBG for 30 min. (B) Western blot illustrating protein levels of Glut1 and Glut3. (C) Quantification of B. (D–F) Metabolite measurements, metabolites were quantified from HAART-treated cells and untreated by Human Metabolome Technologies Inc. using capillary electrophoresis mass spectrometry. *p-value < 0.05, $n = 3$, error bars are SD.

HAND, even without a productive brain infection (Tavazzi et al., 2014), suggesting that factors other than HIV may be at play. Indeed, our study demonstrates that HAART drug treatment induces the SASP in HAs characterized by the expression of several inflammatory cytokines. Until some of these cytokines are further validated the biological implications of this data are limited. However, IL-6 secretion was demonstrated by ELISA and is known to induce senescence in a paracrine manner (Acosta et al., 2013), potentially allowing for a chain reaction of senescence-inducing-senescence and a chronic inflammatory environment in HIV-infected patients extending beyond the initial insult. Co-culture experiments with HAART drug treated astrocytes and other CNS cell types are thus an important next step to examine this effect *in vitro*. In addition, attenuation of these secretions using anti-inflammatory, SASP modulating, or senescence-delaying drugs could be a potential therapy for HAND.

Dysfunctional mitochondria accumulate with age and can occur both in tissues that contain post-mitotic as well as mitotically active cells (Wallace, 2010). Mitochondrial dysfunction is known to induce cellular senescence both *in vitro*

and *in vivo* (Moiseeva et al., 2009), which made it worth looking at how HAART drugs affect the mitochondria in astrocytes. Our HAART drug treated astrocytes display changes in mitochondrial membrane potential, respiration, and mitochondrial ROS production. The production of mitochondrial ROS is particularly interesting as it is thought to be a causal factor in the induction of cellular senescence (Moiseeva et al., 2009). This warrants future studies using antioxidants as a treatment to potentially mitigate HAART drug induced mitochondrial dysfunction and senescence in HAs.

The CNS has extremely high-energy requirements. While only accounting for 2% of human body mass, the CNS is involved in 25% of glucose and 20% of oxygen consumption, indicating that metabolism in the CNS must be tightly controlled. Astrocytes are the key regulators of brain metabolic homeostasis providing nutrition to neurons (Stobart and Anderson, 2013) and changes in astrocyte metabolism can thus have a profound impact on the CNS. Highly active antiretroviral therapy drug treated astrocytes show increased glucose uptake and glycolysis, indicative of a high energy state. While we do not know if this is directly linked to the observed changes in mitochondria, increases in lactate have been observed in patients suffering from mitochondrial myopathies (Kaufmann et al., 2006). The increased utilization of glucose by astrocytes could also potentially impact neurons. While there has been little investigation to link HIV and glycolysis in CNS cells, the effect of HIV on T-cell metabolism has been studied. Glut1 is upregulated on $CD4^+$ T-cells of HIV infected patients compared to non-infected controls and these $Glut1^+$ cells likewise to our studies also have higher levels of glycolysis (Palmer et al., 2014). When examined *in vitro* not only does HIV infection increase glycolysis of $CD4^+$ T-cells, the increased glycolysis also associates with an improved virion production (Hegedus et al., 2014). These results are intriguing because enhanced aerobic glycolysis in the CNS correlates with impaired cognitive function in HIV infected patients (Dickens et al., 2015). Therefore, the increased glycolysis in our HAART drug treated astrocytes may have implications toward HAND. Further significance of our glycolysis results comes from the fact that alterations in glucose have been implicated in AD. Microglia treated with serum from AD patients were found to have increased levels of glycolytic enzymes (Jayasena et al., 2015). In addition, regions of the brain associated with high levels of glycolysis in healthy individuals correlate spatially with Aβ deposits in AD patients (Vlassenko et al., 2010). This suggests that increased levels of glycolysis in the CNS may lead later to Aβ deposits and neurodegeneration.

There are some caveats pertaining to our results. The blood brain barrier hinders penetration of HAART drugs into the CNS, meaning that the doses used to examine the effect of HAART drugs *in vitro* on CNS cells should be lower than plasma levels. However, the relevant parameter to determine physiological levels of HAART drugs in the CNS is a matter of debate. Our long-term (4 weeks) treatments were done using doses based on patient cerebral spinal fluid (CSF). However, HIV patients take these drugs for the rest of their lives and levels could accumulate in cells over a period of years to exceed that of CSF. Levels of ABC in brain homogenates from ABC-treated mice support this (Giri et al., 2008). Therefore, our higher dose 1-week treatments may still be physiological.

The use of HAART drug combinations instead of individual compounds means that the contributions of a specific component cannot be discerned. Since HIV patients take antiretroviral compounds as combinations, we decided to focus the scope of this manuscript accordingly. Interesting future directions include determining if individual antiretroviral drugs or classes are the sole contributors to our astrocyte senescence and dysfunction as well as determining if these adverse effects can be pharmacologically attenuated. In addition, since *in vitro* culturing of fetal astrocytes may not accurately reflect astrocytic function *in vivo*, we want to validate our results by using an *in vivo* model exposing mice to these HAART drugs.

While it is likely that other factors including HIV gene products and drugs of abuse may also contribute to the pathogenesis of HAND *in vivo*, our studies support HAART-induced cellular senescence as a mechanism implied in HAND development. In concordance with our results, clinical trial studies have indicated that drugs with greater CNS penetration

resulted in impaired neurocognitive performance, even though HIV was suppressed better (Marra et al., 2009), and that in HIV-infected patients who have preserved immune function, neurocognitive deficits improved after interruption of HAART treatment (Robertson et al., 2010). These results suggest that HAART drugs could still be a major factor in the development of HAND.

CONCLUSION

Our data demonstrate that HAs senesce in response to a combination of the HAART drugs ABC–3TC–RTV. This has implications for senescence in the CNS contributing to the neurological problems in patients with HAND. Changes in mitochondria, metabolism, and the secretory profile observed in this study suggest that these are potential targets for therapeutics, which could mitigate HAND.

AUTHOR CONTRIBUTIONS

JC conceived and performed the experiments and wrote the manuscript; LD, JW, and FT contributed to the experiments; and CT conceived the experiments and helped write the manuscript.

ACKNOWLEDGMENTS

We would like to thank Dr. Mark Zarella for help with the statistical analysis; Dr. Christian Sell and Dr. Timothy Nacarelli for assistance with mitochondrial experiments; and Dr. Jeffery Jacobson for advice about antiretroviral drug combinations.

SUPPLEMENTARY MATERIAL

FIGURE S1 | Expression of senescence markers in long-term HAART drug treated human astrocytes. Human astrocytes were treated with the following HAART combinations: abacavir (ABC) 3 μM, lamivudine (3TC) 1.9 μM, atazanavir (ATV) 50 nM, and ritonavir (RTV) 100 nM; or tenofovir (TDF) 100 nM, emtricitabine (FTC) 1.2 μM, ATV, and RTV; or TDF, FTC, and efavirenz (EFV) 125 nM for up to 4 weeks in complete astrocyte media. **(A)** Replicative life span curve showing cumulative population doublings as function of days of drug treatments. **(B)** Representative images of 3-week SA β-gal stained cells displayed at 20×. **(C)** SA β-gal activity quantitation. Human astrocytes were stained for SA β-gal 1–3-weeks after HAART treatment. **(D)** Western blot showing protein levels of p16 and p21 after 1 week of treatment. **(E)** IL-6 secretion, conditioned media was collected for 24 h after each treatment. IL-6 was measured by ELISA. *p-value < 0.05, $n = 3$, error bars are SD.

FIGURE S2 | Long-term HAART drug treatment induces total and mitochondrial ROS in human astrocytes. Human astrocytes were treated for up to 4 weeks with ABC 3 μM, 3TC 1.9 μM, atazanavir (ATV) 50 nM, and RTV 100 nM. **(A)** Mitochondrial ROS and **(B)** total ROS were measured by 30 min incubation with MitoSox and DCFDA, respectively, before quantification on a flow cytometer. *p-value < 0.05, $n = 3$, error bars are SD.

REFERENCES

Acosta, J. C., Banito, A., Wuestefeld, T., Georgilis, A., Janich, P., Morton, J. P., et al. (2013). A complex secretory program orchestrated by the inflammasome controls paracrine senescence. *Nat. Cell Biol.* 15, 978–990. doi: 10.1038/ncb2784

Afonso, P., Auclair, M., Boccara, F., Vantyghem, M. C., Katlama, C., Capeau, J., et al. (2015). LMNA mutations resulting in lipodystrophy and HIV protease inhibitors trigger vascular smooth muscle cell senescence and calcification: role of ZMPSTE24 downregulation. *Atherosclerosis* 245, 200–211. doi: 10.1016/j.atherosclerosis.2015.12.012

Arend, C., Brandmann, M., and Dringen, R. (2013). The antiretroviral protease inhibitor ritonavir accelerates glutathione export from cultured primary astrocytes. *Neurochem. Res.* 38, 732–741. doi: 10.1007/s11064-013-0971-x

Bhat, R., Crowe, E. P., Bitto, A., Moh, M., Katsetos, C. D., Garcia, F. U., et al. (2012). Astrocyte senescence as a component of Alzheimer's disease. *PLoS ONE* 7:e45069. doi: 10.1371/journal.pone.0045069

Bhatia, R., Ryscavage, P., and Taiwo, B. (2012). Accelerated aging and human immunodeficiency virus infection: emerging challenges of growing older in the era of successful antiretroviral therapy. *J. Neurovirol.* 18, 247–255. doi: 10.1007/s13365-011-0073-y

Bitto, A., Sell, C., Crowe, E., Lorenzini, A., Malaguti, M., Hrelia, S., et al. (2010). Stress-induced senescence in human and rodent astrocytes. *Exp. Cell Res.* 316, 2961–2968. doi: 10.1016/j.yexcr.2010.06.021

Bodnar, A. G., Ouellette, M., Frolkis, M., Holt, S. E., Chiu, C. P., Morin, G. B., et al. (1998). Extension of life-span by introduction of telomerase into normal human cells. *Science* 279, 349–352. doi: 10.1126/science.279.5349.349

Capeau, J. (2011). Premature aging and premature age-related comorbidities in HIV-infected patients: facts and hypotheses. *Clin. Infect. Dis.* 53, 1127–1129. doi: 10.1093/cid/cir628

Caron, M., Auclairt, M., Vissian, A., Vigouroux, C., and Capeau, J. (2008). Contribution of mitochondrial dysfunction and oxidative stress to cellular premature senescence induced by antiretroviral thymidine analogues. *Antivir. Ther.* 13, 27–38.

Chen, Q., Fischer, A., Reagan, J. D., Yan, L. J., and Ames, B. N. (1995). Oxidative DNA damage and senescence of human diploid fibroblast cells. *Proc. Natl. Acad. Sci. U.S.A.* 92, 4337–4341. doi: 10.1073/pnas.92.10.4337

Childs, B. G., Baker, D. J., Kirkland, J. L., Campisi, J., and van Deursen, J. M. (2014). Senescence and apoptosis: dueling or complementary cell fates? *EMBO Rep.* 15, 1139–1153. doi: 10.15252/embr.201439245

Chou, J. P., Ramirez, C. M., Wu, J. E., and Effros, R. B. (2013). Accelerated aging in HIV/AIDS: novel biomarkers of senescent human CD8+ T cells. *PLoS ONE* 8:e64702. doi: 10.1371/journal.pone.0064702

Crowe, E. P., Tuzer, F., Gregory, B. D., Donahue, G., Gosai, S. J., Cohen, J., et al. (2016). Changes in the transcriptome of human astrocytes accompanying oxidative stress-induced senescence. *Front. Aging Neurosci.* 8:208. doi: 10.3389/fnagi.2016.00208

de Almeida, S. M., Letendre, S., and Ellis, R. (2006). Human immunodeficiency virus and the central nervous system. *Braz. J. Infect. Dis.* 10, 41–50. doi: 10.1590/S1413-86702006000100009

Dickens, A. M., Anthony, D. C., Deutsch, R., Mielke, M. M., Claridge, T. D., Grant, I., et al. (2015). Cerebrospinal fluid metabolomics implicate bioenergetic adaptation as a neural mechanism regulating shifts in cognitive states of HIV-infected patients. *AIDS* 29, 559–569. doi: 10.1097/QAD.0000000000000580

Dimri, G. P., Lee, X., Basile, G., Acosta, M., Scott, G., Roskelley, C., et al. (1995). A biomarker that identifies senescent human cells in culture and in aging skin in vivo. *Proc. Natl. Acad. Sci. U.S.A.* 92, 9363–9367. doi: 10.1073/pnas.92.20.9363

Feeney, E. R., and Mallon, P. W. (2010). Impact of mitochondrial toxicity of HIV-1 antiretroviral drugs on lipodystrophy and metabolic dysregulation. *Curr. Pharm. Des.* 16, 3339–3351. doi: 10.2174/138161210793563482

Freund, A., Orjalo, A. V., Desprez, P. Y., and Campisi, J. (2010). Inflammatory networks during cellular senescence: causes and consequences. *Trends Mol. Med.* 16, 238–246. doi: 10.1016/j.molmed.2010.03.003

Freund, A., Patil, C. K., and Campisi, J. (2011). p38MAPK is a novel DNA damage response-independent regulator of the senescence-associated secretory phenotype. *EMBO J.* 30, 1536–1548. doi: 10.1038/emboj.2011.69

Giri, N., Shaik, N., Pan, G., Terasaki, T., Mukai, C., Kitagaki, S., et al. (2008). Investigation of the role of breast cancer resistance protein (Bcrp/Abcg2) on pharmacokinetics and central nervous system penetration of abacavir and zidovudine in the mouse. *Drug Metab. Dispos.* 36, 1476–1484. doi: 10.1124/dmd.108.020974

Heaton, R. K., Clifford, D. B., Franklin, D. R. Jr., Woods, S. P., Ake, C., Vaida, F., et al. (2010). HIV-associated neurocognitive disorders persist in the era of potent antiretroviral therapy: CHARTER Study. *Neurology* 75, 2087–2096. doi: 10.1212/WNL.0b013e318200d727

Hegedus, A., Kavanagh Williamson, M., and Huthoff, H. (2014). HIV-1 pathogenicity and virion production are dependent on the metabolic phenotype of activated CD4+ T cells. *Retrovirology* 11, 98. doi: 10.1186/s12977-014-0098-4

Hernandez-Vallejo, S. J., Beaupere, C., Larghero, J., Capeau, J., and Lagathu, C. (2013). HIV protease inhibitors induce senescence and alter osteoblastic potential of human bone marrow mesenchymal stem cells: beneficial effect of pravastatin. *Aging Cell* 12, 955–965. doi: 10.1111/acel.12119

Hill, B. G., Benavides, G. A., Lancaster, J. R. Jr., Ballinger, S., Dell'Italia, L., Jianhua, Z., et al. (2012). Integration of cellular bioenergetics with mitochondrial quality control and autophagy. *Biol. Chem.* 393, 1485–1512. doi: 10.1515/hsz-2012-0198

Jabbari Azad, F., Talaei, A., Rafatpanah, H., Yousefzadeh, H., Jafari, R., Talaei, A., et al. (2014). Association between cytokine production and disease severity in Alzheimer's disease. *Iran. J. Allergy Asthma Immunol.* 13, 433–439.

James, E. L., Michalek, R. D., Pitiyage, G. N., de Castro, A. M., Vignola, K. S., Jones, J., et al. (2015). Senescent human fibroblasts show increased glycolysis and redox homeostasis with extracellular metabolomes that overlap with those of irreparable DNA damage, aging, and disease. *J. Proteome Res.* 14, 1854–1871. doi: 10.1021/pr501221g

Jayadev, S., Yun, B., Nguyen, H., Yokoo, H., Morrison, R. S., and Garden, G. A. (2007). The glial response to CNS HIV infection includes p53 activation and increased expression of p53 target genes. *J. Neuroimmune Pharmacol.* 2, 359–370. doi: 10.1007/s11481-007-9095-x

Jayasena, T., Poljak, A., Braidy, N., Smythe, G., Raftery, M., Hill, M., et al. (2015). Upregulation of glycolytic enzymes, mitochondrial dysfunction and increased cytotoxicity in glial cells treated with Alzheimer's disease plasma. *PLoS ONE* 10:e0116092. doi: 10.1371/journal.pone.0116092

Kaufmann, P., Engelstad, K., Wei, Y., Jhung, S., Sano, M. C., Shungu, D. C., et al. (2006). Dichloroacetate causes toxic neuropathy in MELAS: a randomized, controlled clinical trial. *Neurology* 66, 324–330. doi: 10.1212/01.wnl.0000196641.05913.27

Lee, H. C., and Wei, Y. H. (2012). Mitochondria and aging. *Adv. Exp. Med. Biol.* 942, 311–327. doi: 10.1007/978-94-007-2869-1_14

Lefevre, C., Auclair, M., Boccara, F., Bastard, J. P., Capeau, J., Vigouroux, C., et al. (2010). Premature senescence of vascular cells is induced by HIV protease inhibitors: implication of prelamin A and reversion by statin. *Arterioscler. Thromb. Vasc. Biol.* 30, 2611–2620. doi: 10.1161/ATVBAHA.110.213603

Marra, C. M., Zhao, Y., Clifford, D. B., Letendre, S., Evans, S., Henry, K., et al. (2009). Impact of combination antiretroviral therapy on cerebrospinal fluid HIV RNA and neurocognitive performance. *AIDS* 23, 1359–1366. doi: 10.1097/QAD.0b013e32832c4152

Moiseeva, O., Bourdeau, V., Roux, A., Deschenes-Simard, X., and Ferbeyre, G. (2009). Mitochondrial dysfunction contributes to oncogene-induced senescence. *Mol. Cell. Biol.* 29, 4495–4507. doi: 10.1128/MCB.01868-08

Nacarelli, T., Azar, A., and Sell, C. (2016). Mitochondrial stress induces cellular senescence in an mTORC1-dependent manner. *Free Radic. Biol. Med.* 95, 133–154. doi: 10.1016/j.freeradbiomed.2016.03.008

Niccoli, T., and Partridge, L. (2012). Ageing as a risk factor for disease. *Curr. Biol.* 22, R741–R752. doi: 10.1016/j.cub.2012.07.024

Palmer, C. S., Ostrowski, M., Gouillou, M., Tsai, L., Yu, D., Zhou, J., et al. (2014). Increased glucose metabolic activity is associated with CD4+ T-cell activation and depletion during chronic HIV infection. *AIDS* 28, 297–309. doi: 10.1097/QAD.0000000000000128

Robertson, K. R., Su, Z., Margolis, D. M., Krambrink, A., Havlir, D. V., Evans, S., et al. (2010). Neurocognitive effects of treatment interruption in stable

HIV-positive patients in an observational cohort. *Neurology* 74, 1260–1266. doi: 10.1212/WNL.0b013e3181d9ed09

Rodier, F., and Campisi, J. (2011). Four faces of cellular senescence. *J. Cell Biol.* 192, 547–556. doi: 10.1083/jcb.201009094

Sato, A., Asano, T., Ito, K., and Asano, T. (2012). Ritonavir interacts with bortezomib to enhance protein ubiquitination and histone acetylation synergistically in renal cancer cells. *Urology* 79, e913–e921. doi: 10.1016/j.urology.2011.11.033

Serrano, M., Lin, A. W., McCurrach, M. E., Beach, D., and Lowe, S. W. (1997). Oncogenic ras provokes premature cell senescence associated with accumulation of p53 and p16INK4a. *Cell* 88, 593–602. doi: 10.1016/S0092-8674(00)81902-9

Stobart, J. L., and Anderson, C. M. (2013). Multifunctional role of astrocytes as gatekeepers of neuronal energy supply. *Front. Cell. Neurosci.* 7:38. doi: 10.3389/fncel.2013.00038

Suram, A., Kaplunov, J., Patel, P. L., Ruan, H., Cerutti, A., Boccardi, V., et al. (2012). Oncogene-induced telomere dysfunction enforces cellular senescence in human cancer precursor lesions. *EMBO J.* 31, 2839–2851. doi: 10.1038/emboj.2012.132

Tavazzi, E., Morrison, D., Sullivan, P., Morgello, S., and Fischer, T. (2014). Brain inflammation is a common feature of HIV-infected patients without HIV encephalitis or productive brain infection. *Curr. HIV Res.* 12, 97–110. doi: 10.2174/1570162X12666140526114956

Tilstra, J. S., Robinson, A. R., Wang, J., Gregg, S. Q., Clauson, C. L., Reay, D. P., et al. (2012). NF-kappaB inhibition delays DNA damage-induced senescence and aging in mice. *J. Clin. Invest.* 122, 2601–2612. doi: 10.1172/JCI45785

Vlassenko, A. G., Vaishnavi, S. N., Couture, L., Sacco, D., Shannon, B. J., Mach, R. H., et al. (2010). Spatial correlation between brain aerobic glycolysis and amyloid-beta (Abeta) deposition. *Proc. Natl. Acad. Sci. U.S.A.* 107, 17763–17767. doi: 10.1073/pnas.1010461107

Wallace, D. C. (2010). Mitochondrial DNA mutations in disease and aging. *Environ. Mol. Mutagen.* 51, 440–450. doi: 10.1002/em.20586

Wang, C. H., Wu, S. B., Wu, Y. T., and Wei, Y. H. (2013). Oxidative stress response elicited by mitochondrial dysfunction: implication in the pathophysiology of aging. *Exp. Biol. Med.* 238, 450–460. doi: 10.1177/1535370213493069

Yan, J., Fu, Q., Cheng, L., Zhai, M., Wu, W., Huang, L., et al. (2014). Inflammatory response in Parkinson's disease (Review). *Mol. Med. Rep.* 10, 2223–2233.

Zhou, H., Pandak, W. M. Jr., Lyall, V., Natarajan, R., and Hylemon, P. B. (2005). HIV protease inhibitors activate the unfolded protein response in macrophages: implication for atherosclerosis and cardiovascular disease. *Mol. Pharmacol.* 68, 690–700. doi: 10.1124/mol.105.012898

10

Function of B-Cell CLL/Lymphoma 11B in Glial Progenitor Proliferation and Oligodendrocyte Maturation

Chih-Yen Wang[1], Yuan-Ting Sun[2], Kuan-Min Fang[1], Chia-Hsin Ho[1], Chung-Shi Yang[3] and Shun-Fen Tzeng[1]*

[1] Institute of Life Sciences, College of Bioscience and Biotechnology, National Cheng Kung University, Tainan, Taiwan,
[2] Department of Neurology, National Cheng Kung University Hospital, College of Medicine, National Cheng Kung University, Tainan, Taiwan, [3] Institute of Biomedical Engineering and Nanomedicine, National Health Research Institutes, Zhunan, Taiwan

*Correspondence:
Shun-Fen Tzeng
stzeng@mail.ncku.edu.tw

B-cell CLL/lymphoma 11B (Bcl11b) – a C2H2 zinc finger transcriptional factor – is known to regulate neuronal differentiation and function in the development of the central nervous system (CNS). Although its expression is reduced during oligodendrocyte (OLG) differentiation, its biological role in OLGs remains unknown. In this study, we found that the downregulation of Bcl11b gene expression in glial progenitor cells (GPCs) by lentivirus-mediated gene knockdown (KD) causes a reduction in cell proliferation with inhibited expression of stemness-related genes, while increasing the expression of cell cyclin regulator p21. In contrast, OLG specific transcription factors (Olig1) and OLG cell markers, including myelin proteolipid protein (PLP) and myelin oligodendrocyte glycoprotein (MOG), were upregulated in Bcl11b-KD GPCs. Chromatin immunoprecipitation (ChIP) analysis indicated that Bcl11b bound to the promoters of Olig1 and PLP, suggesting that Bcl11b could act as a repressor for Olig1 and PLP, similar to its action on p21. An increase in the number of GC+- or PLP+- OLGs derived from Bcl11b-KD GPCs or OLG precursor cells was also observed. Moreover, myelin basic protein (MBP) expression in OLGs derived from Bcl11b-KD GPCs was enhanced in hippocampal neuron co-cultures and in cerebellar brain-slice cultures. The *in vivo* study using a lysolecithin-induced demyelinating animal model also indicated that larger amounts of MBP+-OLGs and PLP+-OLGs derived from implanted Bcl11b-KD GPCs were present at the lesioned site of the white matter than in the scramble group. Taken together, our results provide insight into the functional role of Bcl11b in the negative regulation of GPC differentiation through the repression of OLG differentiation-associated genes.

Main Points:

(1) Bcl11b regulates glial progenitor proliferation via inhibition of cell cycle regulator p21.
(2) Bcl11b downregulation in glial progenitors promotes their differentiation into mature oligodendrocytes *in vitro* and *in vivo*.
(3) Bcl11b could bind to the promotor regions of cell cycle regulator p21, Olig1, and PLP1 to control the proliferation and differentiation of glial progenitors.

Keywords: Bcl11b, glia, oligodendrocytes, glial progenitors, differentiation and proliferation

INTRODUCTION

Oligodendrocytes (OLGs), myelin-producing glia cells in the CNS, not only support the structure and energy metabolism of axons, but also facilitate the propagation of action potentials by extending their cellular processes to form multilayered myelin sheaths (Nave and Werner, 2014; Kremer et al., 2016). OLGs have been known to arise from OLG precursor cells (OPCs) during development and in adults (Herrera et al., 2001; Zuchero and Barres, 2013; Mayoral and Chan, 2016; Yang et al., 2017). In general, in the early phase of OLG differentiation, pre-myelinating OLGs with numerous complex elongating processes are generated from OPCs, and express O4, 2′,3′-cyclic-nucleotide 3′-phosphodiesterase (CNPase) and galactocerebroside (GC) (Zhang, 2001). The pre-myelinating OLGs further differentiate into mature myelinating OLGs, which are able to encircle axons with their extensions to form compact multilayered myelin consisting of lipids and myelin-associated proteins (Czepiel et al., 2015). The major myelin proteins in the CNS are MBP and PLP. Meanwhile, MOG and myelin-associated glycoprotein (MAG) only constitute 1% of all CNS myelin proteins (Nave and Werner, 2014). The differentiation of OPCs toward mature OLGs is controlled by extrinsic and transcriptional programs. It is well-documented that a pair of basic helix-loop-helix (bHLH) transcriptional factors, together with Olig1, Olig2, and Sox10, are required for OLG specification and maturation (Lu et al., 2002; Stolt et al., 2002; Zuchero and Barres, 2013). In addition, axonal/glial secreted factors and neuronal activity are important for OLG maturation and myelination (Emery, 2010; Zuchero and Barres, 2013). The progenitors isolated from embryonic and postnatal CNS tissues have been reported to give rise to astrocytes in serum-containing medium, but differentiate into OLGs in defined medium (Raff et al., 1983; Yang et al., 2011, 2017). For this reason, such bipotential progenitor cells have been termed as glial progenitor cells (GPCs). Accordingly, GPCs and OPCs are widely used as the culture models to study the molecular regulation of oligodendrocyte differentiation.

B-cell chronic lymphocytic leukemia/lymphoma 11B (Bcl11b), also named Ctip2, is a C2H2 zinc finger protein, originally discovered in T lymphoblastic leukemia (Bernard et al., 2001), and identified as a tumor suppressor in hematopoietic malignancies. The molecule can regulate cell-cycle progression by acting as a repressor for the expression of cyclin dependent kinase (CDK) inhibitors, such as p21 (Cherrier et al., 2009; Grabarczyk et al., 2010; Karanam et al., 2010; Chen et al., 2012). Expressed Bcl11b has been found in mouse brain regions, including the neocortex, hippocampus, and striatum (Arlotta et al., 2005, 2008; Chen et al., 2008; Simon et al., 2012), and is involved in the development of mouse cortical-projection neurons and striatal neurons (Arlotta et al., 2005, 2008; Chen et al., 2008). Findings indicating that proliferating progenitors and post-mitotic dentate granule cells in the dentate gyrus declined in Bcl11b mouse mutants during postnatal development point to the important role of Bcl11b in the regulation of progenitor proliferation (Simon et al.,

2012). Moreover, the lack of Bcl11b-impaired post-mitotic neuron differentiation in the hippocampus of a developing mouse (Simon et al., 2012) demonstrates that progenitor cell proliferation and differentiation in the developing CNS depend on Bcl11b expression. Despite Bcl11b expression being significantly higher in neurons than that detected in OPCs (Zhang et al., 2014), its levels in newly formed OLGs and myelinating OLGs were much lower than in OPCs. However, the involvement of Bcl11b in glial differentiation has not yet been identified.

In this study, we provide evidence in a rat model for the Bcl11b-mediated regulation of OLG maturation by using commercially available rat GPCs and OPCs prepared from rat embryonic cortex. To determine the functional role of Bcl11b in GPCs/OPCs and OLGs, lentivirus-mediated knockdown of Bcl11b (Bcl11b-KD) was first performed to effectively reduce Bcl11b expression in GPCs. The downregulation of Bcl11b not only suppressed GPC cell proliferation, but also reduced their stemness markers. Interestingly, Bcl11b-KD increased the expression of OLG cell markers (PLP and MOG), as well as the number of OLGs in the cultures. Moreover, MBP expression in OLGs derived from Bcl11b-KD GPCs was increased in the presence of neurons, which possibly could enhance neuronal interaction with OLGs from the Bcl11b-KD GPCs. These observations were also verified by implanting Bcl11b-KD GPCs into the lysolecithin-treated corpus callosum of adult rat brains. We also infer that the manipulation of Bcl11b expression in GPCs/OPCs could foster their differentiation toward mature OLGs in demyelinating CNS tissues.

MATERIALS AND METHODS

Materials

Media (DMEM/F12, MEM, Neurobasal medium), GlutaMAX™, StemPro® NSC SFM, B27 supplement, N2 supplement, poly-L-ornithine (PLO), and Lipofectamine 2000 were purchased from Invitrogen. Horse serum (HS) was obtained from HyClone Laboratories. Apotransferrin, biotin, bovine serum albumin (BSA), 5-bromo-2′-deoxyuridine (BrdU), cytosine arabinoside (Ara-C), diethylpyrocarbonate (DEPC), hydrocortisone, insulin, N-acetyl-cysteine (NAC), poly-D-lysine (PDL), sodium ampicillin, sodium pyruvate, sodium selenite, triiodothyronine (T3), and lysolecithin were obtained from Sigma. Ciliary neurotrophic factor (CNTF), epidermal growth factor (EGF), fibroblast growth factor-2 (FGF-2), and platelet derived growth factor-AA (PDGF-AA) were obtained from ProSpec. The antibodies used in this study are listed in **Table 1**.

Cell Culture

Rat glial progenitor cells (GPCs)

Glial progenitor cells prepared from cortical tissues of newborn Sprague-Dawley (SD) rats were purchased from Invitrogen (Cat no. N7746100). The cells, after passages, were seeded onto PLO-coated 100 mm petri dishes and maintained in the growth medium (GM) provided by the vendor (Wang et al., 2017). To induce the differentiation of GPCs into OLGs, the cultures were

TABLE 1 | The antibodies used in the study.

Antibodies	Manufacturer	Immunogen	Working dilution
Monoclonal mouse anti-APC (clone CC1)	Calbiochem (OP80)	Recombinant protein consisting of amino acids 1–226 of APC	1:50 (IF)
Monoclonal rat anti-Bcl11b	Abcam (ab18465)	Fusion protein of human Ctip2 amino acids 1–150 [25B6]	1:500 (WB)
Polyclonal rabbit anti-Bcl11b	Abcam (ab28448)	Synthetic peptide within residues 850–950 of human Bcl11b	1:200 (IF for rat brain)
Polyclonal rabbit anti-Bcl11b	Novus (NBP2-33549)	Recombinant protein: QGNPQHLSQRELITPEADHVEAAILEE DEGLEIEEPSGLGLMVGGPDPDLLTCG	1:200 (IF for mouse brain)
Monoclonal mouse anti-CNP	Covance (SMI-91R)	46 and 48 kDa subunit of 94 kDa myelin CNPase dimer (SMI-91)	1:2000 (WB) 1:200 (IF)
Monoclonal mouse anti-GAPDH	Millipore (MAB374)	Glyceraldehyde-3-phosphate dehydrogenase from rabbit muscle	1:2000 (WB)
Polyclonal rabbit anti-GC	Millipore (MAB342)	Synaptic plasma membranes from bovine hippocampus	1:200 (WB)
Monoclonal mouse anti-GFP	Millipore (MAB2510)	Bacterially expressed GFP fusion protein	1:200 (IF)
Polyclonal rabbit anti-GFP	Millipore (MAB3080)	Highly purified native GFP from Aequorea victoria	1:200 (IF)
Monoclonal mouse anti-MBP	Calbiochem (NE1018)	Purified human myelin basic protein with amino acids 70–89	1:1000 (WB) 1:200 (IF)
Polyclonal rabbit anti-MOG	Abcam (ab32760)	Synthetic peptide as within residues 200 to C-terminal of rat myelin oligodendrocyte glycoprotein	1:1000 (WB) 1:200 (IF)
Polyclonal rabbit anti-NG2	Millipore (AB5320)	Immunoaffinity purified NG2 Chondroitin Sulfate Proteoglycan from rat	1:200 (IF)
Polyclonal goat anti-NF200	Millipore (AB5539)	Purified bovine neurofilament-heavy	1:500 (IF)
Polyclonal rabbit anti-OLIG2	Millipore (AB9610)	Recombinant mouse Olig2	1:200 (IF)
Monoclonal mouse anti-p21	Calbiochem (OP79)	Recombinant mouse p21 protein (clone 22)	1:1000 (WB)
Rabbit antiserum anti-PDGFaR	Upstate (06-216)	GST-fusion protein corresponding to the 110 C-terminal amino acid residues of mouse PDGF type A receptor	1:200 (IF)
Polyclonal rabbit anti-PLP	Abcam (ab28486)	Synthetic peptide as amino acids 109–128 of mouse myelin PLP	1:1000 (WB) 1:200 (IF)

IF, immunofluorescence; WB, western blot.

maintained for 5 days in OLG differentiation medium (DM), which contains basal components in the GM without growth factors.

Rat oligodendrocyte progenitor cells (OPCs)

Oligodendrocyte progenitor cells were prepared by modifying the protocol in (Fancy et al., 2012). Animal use followed the National Institutes of Health (NIH) Guidelines for Animal Research (Guide for the Care and Use of Laboratory Animals) and was approved by the National Cheng Kung University Institutional Animal Care and Use Committee, Tainan, Taiwan (IACUC approval number: 103060). Briefly, cortical tissues from SD rat embryos at 14.5 days were dissected and passed through a 40-μm pore filter. The cells were replated onto petri dishes without PDL coating, and cultured for 5–7 days in oligosphere medium – consisting of DMEM/F12 medium, 2% B27, 1% N2 supplement, 10 ng/ml FGF2, 10 ng/ml EGF, and 10 ng/ml PDGF-AA. After the formation of oligospheres, the OPCs were dissociated and plated onto PDL-coated dishes in growth medium (GM), as previously described (Wang et al., 2017). The cells were maintained for 5 days in OLG differentiation medium (DM) containing DMEM medium, 4 mM L-glutamine, 1 mM sodium pyruvate, 0.1% BSA, 50 μg/ml apotransferrin, 5 μg/ml insulin, 30 nM sodium selenite, 10 nM biotin, 10 nM hydrocortisone, 15 nM T3, 10 ng/ml CNTF, and 5 μg/ml NAC.

Rat hippocampal neurons

Hippocampal tissues were isolated from newborn SD rat, and digested by 0.25% trypsin solution containing DNase at 37°C for 30 min. The hippocampal cells were seeded onto PDL-coated coverslips (1×10^4 cells/coverslip) in Neurobasal medium with 2% B27, 0.25% GlutaMAX™ and 10% HS. After a period of 1 h, the medium was replaced by Neurobasal medium with 2% B27 and 0.25% GlutaMAX™. Ara-C (1 mM) was added to the culture at day 4 to inhibit glia cell proliferation.

Lentivirus-Mediated shRNA Targeting Bcl11b

Previously, we found that the constructs of lentiviral vectors made in Biosettia (San Diego, CA, United States) were highly efficient in the inhibition of Bcl11b (NM_001277287) in rat glioma cells (Liao et al., 2016). Thus, we used the same lentiviral vector constructs for this study: pLV-mU6-EF1a-GFP-Puro-scramble (lenti-ctrl); and, pLV-mU6-EF1a-GFP-Puro-shBcl11b-916 (lenti-shBcl11b). The shRNA sequences are shown in **Table 2**. For gene transduction, GPCs/OPCs (1×10^6 cells/dish) were seeded onto 60-mm petri dishes in GM, and lentiviruses (300 μl/dish) carrying shRNA against Bcl11b and scramble lentiviruses were separately added into the medium for 24 h. The shRNA genes were allowed to express for 48 h, and the transfectants were selected in the presence of puromycin (3 μg/ml), also for 48 h. The efficiency of the lentiviral particles for Bcl11b downregulation in the GPCs was confirmed by quantitative polymerase chain reaction (QPCR) and western blotting. GPCs infected by lenti-ctrl were referred to as 'scramble,' while cells infected by lenti-Bcl11b-KD were called 'Bcl11b-KD'.

Quantitative Real-Time Polymerase Chain Reaction

The RNA (1 μg/sample) isolated from the GPCs and OPCs was reacted with M-MLV reverse transcriptase (Invitrogen) to generate cDNA, and then incubated with SYBR Green reagents (Roche) and specific primers (**Table 2**). The expression level of GAPDH was used as an internal control. StepOne Software v2.1 (Applied Biosystems) was used to determine the cycle-threshold (Ct) fluorescence values. The expression level of the target genes relative to the internal control was presented as $2^{-\Delta CT}$, where $\Delta CT = (Ct_{target} - Ct_{GAPDH})$.

Immunofluorescence

After harvesting, the cells were fixed in 4% paraformaldehyde for 10 min, and incubated in PBS containing 0.1% Triton-X100 for 30 min. The cultures then were incubated overnight at 4°C with primary antibodies (**Table 1**). To stain galactoceramide (GC), a major glycolipid of myelin, the cultures were directly incubated with anti-GC antibody after fixation with 4% paraformaldehyde, but without permeabilization by 0.1% Triton-X100. Alternatively, to carry out double immunofluorescence for NF200 (or GFP) and MBP (or PLP), the cultures were incubated with anti-NF200 (or anti-GFP) at 4°C overnight, and then with anti-MBP (or anti-PLP) at RT for 3 h. After reacting with the primary antibodies, appropriate secondary antibodies and FITC-avidin were added to the cultures at RT for 1 h and for 45 min, respectively. The immunostained cells were photographed under a Nikon E800 epifluorescence microscope equipped with a CCD camera and also under an Olympus FluoView laser scanning confocal microscope (FV1000, Japan).

Evaluation of Oligodendrocytic Differentiation

In addition to the morphological observations of OLG differentiation from GPCs or OPCs, OLG differentiation was evaluated by measurement of OLG cell-marker expression intensity, as described previously (Wang et al., 2017). MetaXpress software (Molecular Devices; Sunnyvale, CA, United States) and NIH ImageJ analysis software were used. Five randomly selected images (5–10 cells per image) were captured from each immunostained culture using the above-mentioned epifluorescence microscope with a 40X objective lens. The experiments were repeated in triplicate, and 70–100 cells in total per group were counted. The number of processes and branches per cell, the average length of the processes, and the total length of the outgrowing processes per cell in each field were quantified. The immunofluorescent intensity per cell and the number of immunostained cells in each field were also measured. The results are presented as the percentage of the data obtained from the Bcl11b-KD culture versus data from the scramble culture.

Western Blot Analysis

GPCs and OPCs were replated at a density of 1×10^6 cells/60-mm onto petri dishes for various experiments. After harvesting, the total protein content (100 μg) was extracted from the cultures and lysis buffered in 1% NP-40, 1% Triton-X100, and 0.1% SDS, which it was loaded onto 10% SDS polyacrylamide gel. After electrophoresis, the protein was transferred to a nitrocellulose membrane and immunoblotted overnight at 4°C with primary antibodies (**Table 1**). The immunoblotted

TABLE 2 | Primer sequences for QPCR analysis and Sequences for shRNA against rat Bcl11b.

Gene	Sequence
Rat Bcl11b (NM_001277287)	Forward (5′→3′): GCAGTCCAACCTAACCTGTGTC Reverse (5′→3′): GGGTGCCTTAATCAACCCTCAG
Rat CD133 (NM_021751)	Forward (5′→3′): CCAGCGGCAGAAGCAGAACGA Reverse (5′→3′): GTCAGGAGAGCCCGCAAGTCT
Rat Sox2 (NM_001109181)	Forward (5′→3′): CACAACTCGGAGATCAGCAA Reverse (5′→3′): CGGGGCCGGTATTTATAATC
Rat Bmi-1 (NM_001107368)	Forward (5′→3′): GCGTTACTTGGAGACCAGCA Reverse (5′→3′): CTTTCCGATCCGACCTGCTT
Rat Hes1 (NM_024360)	Forward (5′→3′): TACCCCAGCCAGTGTCAACA Reverse (5′→3′): TCCATGATAGGCTTTGATGACTTTC
Rat Hey1 (NM_001191845)	Forward (5′→3′): AGCGCAGACGAGAATGGAAA Reverse (5′→3′): CGCTTCTCGATGATGCCTCT
Rat Hey2 (NM_130417)	Forward (5′→3′): CTTGACAGAAGTGGCGAGGT Reverse (5′→3′): CATTGGGTTGGAGCAGGGAT
Rat p21 (NM_080782)	Forward (5′→3′): TGGACAGTGAGCAGTTGAGC Reverse (5′→3′): ACACGCTCCCAGACGTAGTT
Rat Olig1 (NM_021770)	Forward (5′→3′): GAGGGGCCTCTTTCCTTGTC Reverse (5′→3′): ACCGAGCTTCACAAGCCTAC
Rat Olig2 (NM_001100557)	Forward (5′→3′): GCTTAACAGAGACCCGAGCC Reverse (5′→3′): GTGGCGATCTTGGAGAGCTT
Rat MBP (NM_001025291)	Forward (5′→3′): GTGGGGGTAAGAGAAACGCA Reverse (5′→3′): CGAACACTCCTGTGGAACGA
Rat PLP (NM_030990)	Forward (5′→3′):GGCGACTACAAGACCACCAT Reverse (5′→3′):AATGACACACCCGCTCCAAA
Rat NFATc3 (NM_001108447)	Forward (5′→3′):TCTGACTTGGAACACCAGCC Reverse (5′→3′): AAGCAGTCAGAGCAGTTGGT
Rat MOG (NM_022668)	Forward (5′→3′): CCCAGCGCTTCAACATTACG Reverse (5′→3′): GCACCTAGCTTGTTTGTGTCTG
Rat GAPDH (NM_017008)	Forward (5′→3′): TCTACCCACGGCAAGTTC Reverse (5′→3′): GATGTTAGCGGGATCTCG
Scramble shRNA	GCAGTTATCTGGAAGATCAGGTTGGATCCAAC CTGATCTTCCAGATAACTGC
shBcl11b-916 (NM_001277287: nt637-656)	AAAAGAGCCTTCCAGCTACATTTGTTGGATCCA ACAAATGTAGCTGGAAGGCTC

membrane was incubated with secondary antibodies conjugated with peroxidase for 60 min at RT. The signal was detected by chemiluminescence using the ECL-Plus detection system (PerkinElmer Life Sciences).

Cell Growth Assays

MTT colorimetric assay, gliosphere formation assay, and colony formation assay were performed as described previously (Fang et al., 2014). GPCs and OPCs were maintained in GM either for different time periods (MTT assay), or for 7 days (gliosphere and colony formation assays). The cell proliferation of the GPCs in GM after 48 h was also examined using BrdU incorporation assay via the addition of BrdU (10 μM) into the culture 12 h before harvesting, following the previously described procedure (Wang et al., 2015). The number of gliospheres, colonies, and BrdU$^+$ cells in the culture were counted using ImageJ analysis software (NIH, United States).

TABLE 3 | Primer sequences for ChIP-QPCR analysis.

Promoter	Sequence
Rat $p21$ (+2862 ~ +2944) [Gene ID: 114851]	Forward (5'→3'): GCCCCTTTCTAGCTGTCTGG Reverse (5'→3'): GCTCCTTCACCCATCCCTG
Rat $Olig1$ (−1864 ~ −1763) [Gene ID: 60394]	Forward (5'→3'): CGTACCGCTTATGTGCAGGG Reverse (5'→3'): ACCCTACATTCCTAGCCATCG
Rat $Olig1$ (−1487 ~ −1264) [Gene ID: 60394]	Forward (5'→3'): CTGATAGCTGTGAGGGTGAAG Reverse (5'→3'): CCCAGATGCTGGGAATACAA
Rat $Olig1$ (−1121 ~ −1030) [Gene ID: 60394]	Forward (5'→3'): TGAGCCAGCCACTAAAAGACA Reverse (5'→3'): CTTCATCCTGGGGTGTCTGC
Rat $Olig1$ (−575 ~ −502) [Gene ID: 60394]	Forward (5'→3'): CAAAAGCTAACAAGTCCCGATCA Reverse (5'→3'): CGCAGTTCAGTCGTTAAAACACC
Rat $Olig1$ (−399 ~ −261) [Gene ID: 60394]	Forward (5'→3'): CAGCTACAGCAGTTCCCAGT Reverse (5'→3'): CTAGTTCAGCGGGTCATGCT
Rat $Olig1$ (−240 ~ −147) [Gene ID: 60394]	Forward (5'→3'): GCCCTATAAAGCTCCCTCCC Reverse (5'→3'): CAGCCAGAGTTGCCAGAGAT
Rat Plp (−1521 ~ −1423) [Gene ID: 24943]	Forward (5'→3'): GATCAGTGGGAGTGTGCAGG Reverse (5'→3'): CACTCTCCCCTGTCCCCTAA
Rat Plp (−1173 ~ −1065) [Gene ID: 24943]	Forward (5'→3'): AGTCCCAGAGATGCTCCTGA Reverse (5'→3'): GAGGGGAATCAAGCAGCCAA
Rat Plp (−982 ~ −862) [Gene ID: 24943]	Forward (5'→3'): GCTGCACTTTCGTAACAGGC Reverse (5'→3'): AGGTAGTAGCTTCCCAGGGT
Rat Plp (−625 ~ −433) [Gene ID: 24943]	Forward (5'→3'): TCTTGAGCCTGGTCACACAC Reverse (5'→3'): AGTTGGCCTTGACCATGGAA
Rat Plp (−368 ~ −266) [Gene ID: 24943]	Forward (5'→3'): TCCTCACCAGGGCTACCATT Reverse (5'→3'): AGGGGTCCTTAAATCCTCCCA
Rat Plp (−40 ~ −109) [Gene ID: 24943]	Forward (5'→3'): TTTAAGGGGGTTGGCTGTCA Reverse (5'→3'): AGTCTGTTTTGCGGCTGACT

Chromatin Immunoprecipitation

A chromatin immunoprecipitation (ChIP) assay kit (Millipore) was employed, the experimental procedure for which was based on the manual provided by the vendor. Briefly, GPCs were seeded at a density of 1×10^7 cells/100-mm petri dish for 2 days in GM. After 1% formaldehyde was added into the medium, the cells were removed and suspended in a cell-lysis buffer. After centrifugation, the resulting cell pellet was resuspended in a nuclear-lysis buffer for 10 min. The sample was then sonicated to produce DNA fragments at lengths of 200–600 bp, followed by incubation with anti-Bcl11b antibodies (or isotype IgG as negative control) and protein A/G magnetic beads at 4°C overnight. DNA-protein complexes were collected and treated by proteinase K at 62°C for 2 h. After DNA purification, the DNA fragments that potentially interacted with Bcl11b were analyzed by QPCR using specific primers (**Table 3**). Results from samples that had reacted with mouse isotype IgG (IgG) are referred to as negative control.

Assessments of *in Vitro* Co-culture of GPCs and Neurons

After hippocampal neurons were cultured for 7 days, GPCs at the density of 1×10^4 cells per coverslip were added into the hippocampal culture, and maintained in Neurobasal medium with 2% B27, 0.25% GlutaMAX^{TM} and T3 (30 ng/ml) for 7 days. The hippocampal neuron-OLG co-cultures were subjected to double immunofluorescence for NF200 and MBP. The assessments followed the methods described in our previous study (Wang et al., 2017). The intensity of MBP fluorescence, which overlapped with a neuronal fiber featuring immunoreactivity to NF200, was quantified using NIH ImageJ analysis software. Additionally, to further verify the overlap of the MBP$^+$-OLG process with the NF200-immunostained fiber, the cultures were subjected to confocal imaging analysis to acquire a z-stack reconstructed from 7 sequential images at 1-μm intervals. 3D images, including x-z and y-z views, were obtained from the same z-stack to identify the overlapping regions of MBP- and NF200-immunostaining.

Cerebellar Slice Culture

The *ex vivo* cerebellar slice culture was modified and performed according to a previous study (Lee et al., 2015). Briefly, the rat sagittal-cerebellar slices at P7 were dissected at a thickness of 350 μm using a Microslicer^{TM} DTK-1000 vibratory tissue slicer. The tissue slices were then plated on Millicell-CM culture inserts (Millipore, 0.4 μm) and maintained on the surface of the slice culture medium (50% MEM with Earle's salts, 35% Earle's balanced salt solution, 15% heat-inactivated horse serum, 1% GlutaMAX^{TM}) at 37°C for 9 days. The scramble and Bcl11b-KD GPCs were seeded onto a cerebellar slice at a number of 1×10^5 cells/slice. After 48 h, the scramble GPCs/cerebellar slice and Bcl11b-KD GPCs/cerebellar slice cultures were fixed with 4% paraformaldehyde and permeabilized by 0.3% Triton X-100 in PBS, followed by immunofluorescence for MBP and NF200.

GPC Transplantation Followed by Lysolecithin Injection

Adult male SD rats (250 ± 30 g) were anesthetized by intraperitoneal injection of chloral hydrate (50 mg/kg) and placed in a stereotaxic frame (Stoelting). A midline incision was made and the underlying tissue removed using a scalpel. A hole was drilled in the exposed skull by a dentist drill fitted with a 0.9 mm diameter carbide dental burr at 2 mm to the right of the sagittal suture. A Hamilton syringe with a 25-gauge needle was inserted 2.5 mm into the brain (corpus callosum). The fluid (5 μl) containing 1% lysolecithin was slowly injected into the brain. After injection, the needle was maintained in place for 2 min to prevent leakage. At 3 days post injection (dpi), the hole was re-exposed, and 1×10^5 GPCs in 5 μl PBS were injected into the brain at the same position. At 14 dpi, the rats were sacrificed and their brains removed. The brain tissues were fixed in 4% paraformaldehyde, and then cryoprotected in 30% (w/v) sucrose in PBS. The tissues were embedded in Tissue Tek OCT (Electron Microscopy Sciences), sectioned with a 15-μm thickness, and then subjected to immunofluorescence as described above.

Statistical Analysis

The statistical significance of the differences between the two groups of data was analyzed by a two-tailed unpaired Student's t-test, with all data expressed as means ± SEM. In all comparisons, differences were statistically significant at $p < 0.05$.

RESULTS

Reduced Stemness and Proliferation of Glial Progenitor Cells by Downregulation of Bcl11b Expression under the Growth Condition

The findings indicated that Bcl11b is highly expressed in mouse cortical neurons and hippocampal dentate gyrus granule neurons (Arlotta et al., 2005; Simon et al., 2012). Here, through immunofluorescence, we showed that Bcl11b expression was co-localized to $CC1^+$-OLGs in the corpus callosum, as well as $Olig2^+$-OLG lineage cells and $NG2^+$-glial progenitor cells (GPCs) in the subventricular zone (SVZ) of the rat brain (Supplementary Figure S1). *In vitro* examination showed that Bcl11b was produced in $A2B5^+$- and $NG2^+$-rat GPCs (Supplementary Figure S2A). Bcl11b expression was also detectable in $CNPase^+$-OLGs generated from rat GPCs (Supplementary Figure S2A). The mRNA levels of Bcl11b in neurons and distinct glial cells were also confirmed using primary cultures. Although Bcl11b showed considerable expression in neurons, OPCs were found to express only a moderate level of Bcl11b (Supplementary Figure S2B).

To determine the function of Bcl11b in GPCs and OLGs, we performed lentivirus-mediated shRNA delivery against Bcl11b expression (Bcl11b-KD) in GPCs. Lenti-Bcl11b shRNA efficiently downregulated Bcl11b mRNA expression (**Figure 1A**) and protein production (**Figure 1B**) in GPCs compared to what we

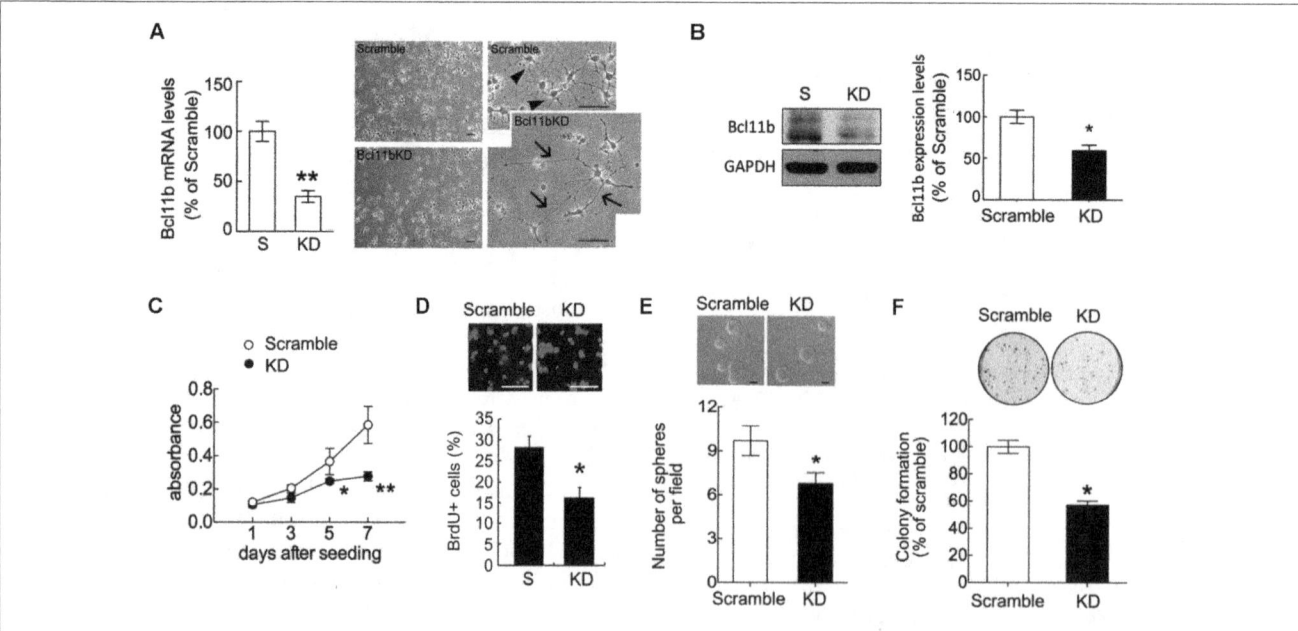

FIGURE 1 | Reduction in GPC cell proliferation after Bcl11b gene downregulation. GPCs were infected by lenti-ctrl (scramble) and lenti-shBcl11b (Bcl11b-KD) as described in section "Materials and Methods." The efficiency of Bcl11b downregulation in Bcl11b-KD GPCs was confirmed using QPCR for the measurement of Bcl11b mRNA expression **(A)** and western blot assay for Bcl11b protein production **(B)**. The phase-contrast images with low (left-panel) and high (right-panel) magnification were taken to examine the morphological change of GPCs after Bcl11b-KD **(A)**. Arrows in **(A)** indicate elongating processes extending from Bcl11b-KD GPCs compared to those of scramble GPCs (arrowheads). Scramble and Bcl11b-KD GPCs maintained in GM were subjected to MTT analysis **(C)**, BrdU incorporation assay **(D)**, sphere formation **(E)**, and colony formation assays **(F)**. Data are presented as means ± SEM of repeated independent experiments (n = 3–4). *p < 0.05, **p < 0.01 vs. scramble. Scale bar in **(A)** and **(D)** 50 μm; in **(E)** 100 μm.

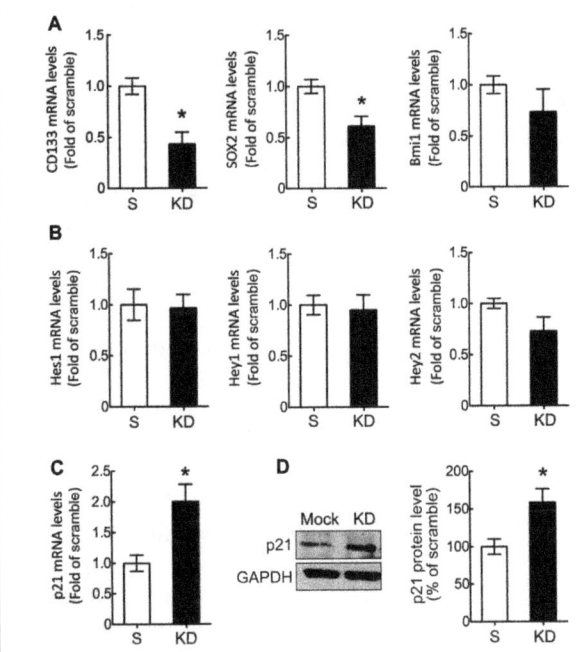

FIGURE 2 | Downregulation of stemness-related genes in GPCs by Bcl11b knockdown. Scramble and Bcl11b-KD GPCs were maintained in GM for 48 h, and then subjected to QPCR analysis for the measurement of stemness-associated genes including CD133, Sox2, and Bmi1 **(A)**, as well as the effector genes (Hes1, Hey1, and Hey 2) of Notch signaling **(B)**. The expression of p21 mRNA in scramble and Bcl11b-KD GPCs was examined using QPCR **(C)**. Moreover, total proteins were extracted from scramble and Bcl11b-KD GPCs in GM for 48 h, and then subjected to western blotting **(D)** to measure p21 protein production. The intensity of immunoreactive bands corresponding to p21 and GAPDH as a loading control was quantified by ImageJ software. Data shown in **(A–D)** are presented as means ± SEM of the three independent experiments. *$p < 0.05$ vs. the scramble.

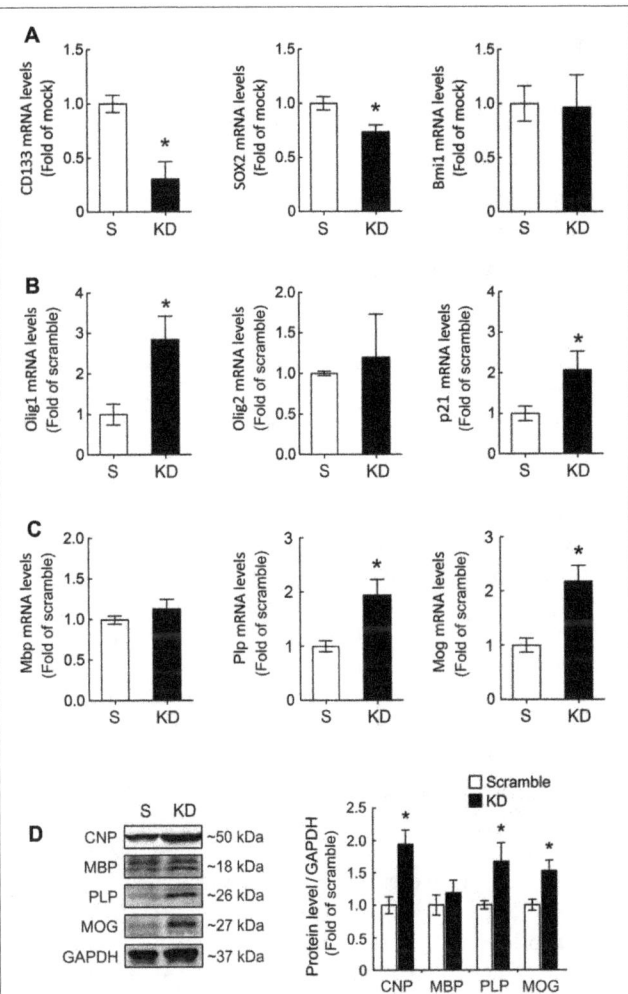

FIGURE 3 | Upregulation of Olig1 and p21 expression in GPCs after Bcl11b gene downregulation. **(A)** Scramble and Bcl11b-KD GPCs were cultured in DM for 48 h, and then subjected to QPCR analysis for the measurement of stemness-associated genes, CD133, Sox2, and Bmi1. **(B,C)** Moreover, transcription factor-associated OLG differentiation (i.e., Olig1 and Olig2), p21, and myelin-related genes were examined by QPCR analysis. **(D)** In addition, total proteins were extracted from scramble and Bcl11b-KD GPCs cultured in DM for 5 days, and then subjected to western blot analysis for the examination of OLG cell markers, CNPase, MBP, PLP, and MOG. The intensity of immunoreactive bands shown in the right panel was quantified by ImageJ software, and normalized by the level of GAPDH that is as a loading control. Data are presented as means ± SEM of the three independent experiments. *$p < 0.05$ vs. the scramble.

observed in GPCs infected by lenti-ctrl (scramble). In addition, it was observed that the morphology of the GPCs had altered into a shape with elongated processes after Bcl11b-shRNA transduction (**Figure 1A**, arrows). The examination of GPC cell growth at different times by MTT analysis or after 24 h by BrdU incorporation assay indicated that GPC cell growth was reduced in the GM after Bcl11b downregulation (**Figures 1C,D**). Moreover, the ability of GPCs to form glial spheres and colonies declined after Bcl11b-KD (**Figures 1E,F**).

Based on our previous findings that Bcl11b-KD reduced the expression of stemness-related genes (Sox2 and Bmi1) in glioma cells (Liao et al., 2016), we also examined the expression of stemness-related genes in the scramble and Bcl11b-KD cultures. As shown in **Figure 2**, the expression of Sox2 – but not of Bmi1 – was reduced in Bcl11b-KD GPCs compared to that detected in the scramble culture. Moreover, we found that CD133, a stem-cell marker, was downregulated in the GPCs after Bcl11b-KD (**Figure 2A**). These data reveal that Bcl11b actively participates in the regulation of GPC cell proliferation. Findings in a separate study indicated that Bcl11b can trigger early differentiation of epidermal keratinocytes by binding to the Notch1 promoter to promote the expression of Notch1 (Zhang et al., 2012). Thus,

we examined the expression of Notch downstream genes (Hes1, Hey1, and Hey2) in the GPCs cultured in GM for 48 h. The results indicated that the change in the expression of the three genes was insignificant in both the scramble and Bcl11b-KD GPCs (**Figure 2B**).

Given that Bcl11b can bind to the Sp1 promoter binding site of the cell cycle repressor p21 to cause p21 gene silencing (Cherrier et al., 2009), the p21 mRNA expression in Bcl11b-KD GPCs was compared to that detected in the scramble GPCs. The results showed that p21 mRNA expression and its protein production were significantly increased in the GPCs after Bcl11b-KD (**Figures 2C,D**). These results suggest that the reduced cell proliferation of GPCs in GM might be due to the increased effect of p21 after Bcl11b downregulation.

Upregulation of OLG-Specific Gene Expression in GPCs after Bcl11b Gene Knockdown

Our findings, as indicated above, raise the question of whether Bcl11b-KD induced the cell death of GPCs or instructed the GPCs toward the differentiation of OLGs. Since the cell death of GPCs with Bcl11b-KD in GM or in DM was not observed (data not shown), we next evaluated the differentiation ability of GPCs after Bcl11b-KD. When GPCs were cultured in DM for 48 h to induce OLG differentiation, the expression of stemness-related genes (i.e., CD133 and Sox2) was significantly reduced compared to the scramble culture (**Figure 3A**). No change was observed in Bmi1 expression after Bcl11b-KD. Examination of Olig1 and Olig2, two critical bHLH transcription factors specific for OLG differentiation, indicated that Bcl11b-KD caused the upregulation of Olig1, but not of Olig2 (**Figure 3B**). This also provides evidence that Bcl11b-KD can promote OLG lineage specification. Moreover, an increase in p21 gene expression was observed when Bcl11b-KD GPCs were cultured in DM (**Figure 3B**). Furthermore, our results showed that Bcl11b-KD induced upregulated gene and protein expression of myelin proteolipid protein (PLP) and myelin oligodendrocyte glycoprotein (MOG), which are myelin proteins in mature OLGs (**Figures 3C,D**). However, MBP mRNA and protein expression was not affected by Bcl11b-KD (**Figures 3C,D**). In addition, the protein level of CNPase, a myelin-associated enzyme and a marker for OLG differentiation, was upregulated in the Bcl11b-KD culture (**Figure 3D**). These results reveal that Bcl11b-KD in

the GPCs increased the expression of crucial proteins associated with the progression of the OLG lineage.

Interaction of Bcl11b with the Promoter Regions of Olig1 and PLP

It has been reported that Bcl11b is associated with the Sp1 binding sites containing the p21 promoter, and acts as a repressor for p21 expression (Cherrier et al., 2009). Through ChIP analysis, we verified that Bcl11b can bind to the Sp1 sequence located at the p21 promoter in the scramble GPCs (**Figure 4**). This finding reflects our observations of an increased expression of p21 in the GPCs after Bcl11b-KD (**Figure 2**). Through the ChIP method combined with high-throughput sequencing, the potential DNA binding sequences of Bcl11b have been identified (Tang et al., 2011). Accordingly, we used a nucleotide blast search for these potential Bcl11b binding sites on the Olig1 and Plp1 promoters, for which 6 and 7 potential segments were predicated for Bcl11b binding to within 2000 nt upstream sequences of the Olig1 and Plp1 promoters, respectively (**Figure 4**). To examine if Bcl11b can interact with the promoter of Olig1, ChIP experiments were performed using 6 designed primer pairs for the Bcl11b binding sites at the Olig1 promoter in both the scramble and Bcl11b-KD cultures (**Figure 4A**). The ChIP assay using anti-Bcl11b in combination with QPCR indicated that Bcl11b had a strong interaction with the segment ($-1487 \sim -1264$ bp) encompassing two predicted Bcl11b binding sites located at the Olig1 promoter (**Figure 4A**). This interaction, however, declined in Bcl11b-KD GPCs. In addition, Bcl11b was able to bind to the sequence at $-625 \sim -433$ bp of the Plp1 promoter in both cultures (**Figure 4B**), despite reduced binding being observed in Bcl11b-KD GPCs. Taken together, these findings, in conjunction with the results that Bcl11b-KD caused the upregulation of Olig1 and PLP in the GPCs, suggest

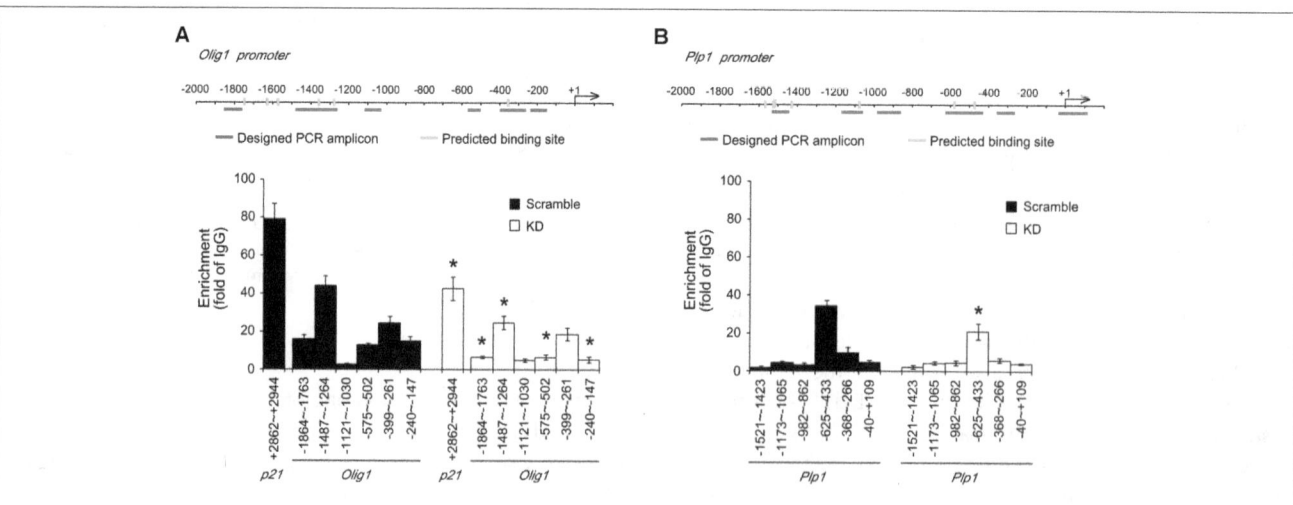

FIGURE 4 | Identification of potential Bcl11b binding regions located in Olig1 and Plp1 promoters. ChIP assay was conducted as described in Section "Materials and Methods" to verify the predicted Bcl11b binding sites located at the promoter regions of Olig1, Plp1 and p21. The predicted Bcl11b binding sites are labeled as green rods. The primers that match to the flanking regions indicated by red lines in Olig1 **(A)** and Plp1 **(B)** promoters were designed and synthesized. ChIP products prepared from scramble and Bcl11b-KD GPCs were analyzed by QPCR using the synthesized primers. The data shown in **(A,B)** are normalized over those obtained from IgG samples. Interaction of Bcl11b and p21 promoter is referred to be a positive control. Data are presented as means ± SEM of the four independent experiments. *$p < 0.05$ vs. the scramble.

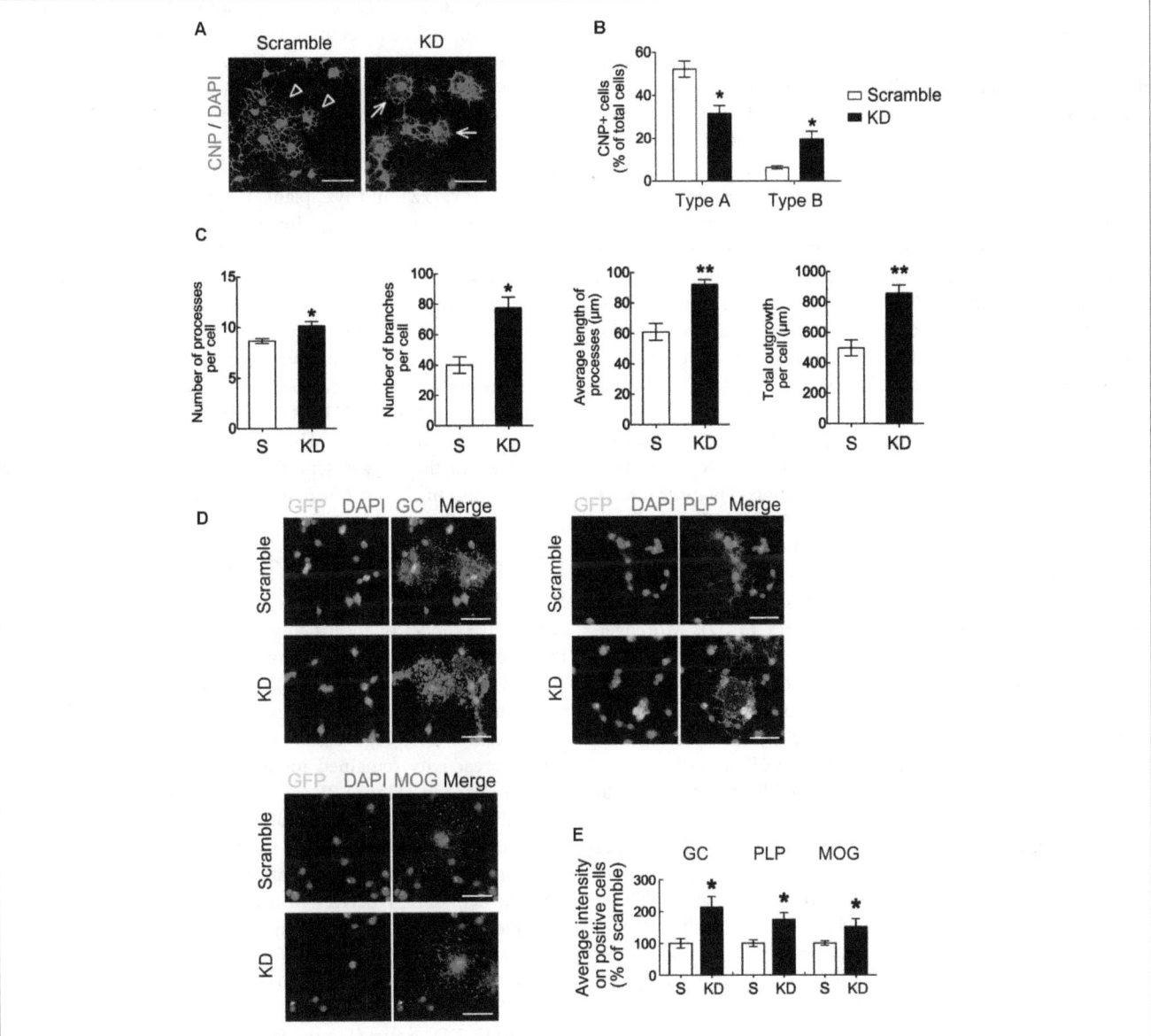

FIGURE 5 | Oligodendrocyte differentiation and maturation enhanced by Bcl11b gene knockdown. **(A,B)** Scramble and Bcl11b-KD GPCs were seeded and maintained in DM for 5 days. The cultures were subjected to immunofluorescence using anti-CNPase antibody **(A)**. Arrowheads indicate the differentiated cells with branching processes (type A). Arrows point to the cells extending the processes to form a ring-shaped network (type B). CNPase⁺-Type A and CNPase⁺-Type B cells in scramble and Bcl11b-KD cultures were quantified using ImageJ software **(B)**. **(C)** The images taken from scramble and Bcl11b-KD cultures were further analyzed by MetaXpress software to measure the number of cell processes per cell, the branching numbers per cell, the average length per process, and the total process length per cell. **(D,E)** Scramble and Bcl11b-KD cultures were subjected to double immunofluorescence for GC, PLP, and MOG, combined with GFP as well as DAPI nuclear counterstaining **(D)**. The immunofluorescent intensity of each immunoreactive cell in the cultures was measured using ImageJ software **(E)**. Data are presented as means ± SEM of at least three independent experiments (n = 70–100 cells per group). *$p < 0.05$ vs. the scramble. Scale bar in **(A,D)**, 50 μm.

that Bcl11b might act as a transcriptional repressor of these genes.

Enhanced Differentiation of GPCs toward OLGs by the Downregulation of Bcl11b Gene Expression

We used immunofluorescence to compare the morphological differences between the OLGs generated from the scramble GPCs and those from the Bcl11b-KD GPCs. The cultures were maintained in DM for 5 days to stimulate the differentiation of OLGs from GPCs. Through CNPase immunostaining, we observed that the OLGs generated from the scramble GPCs and those from the Bcl11b-KD GPCs had a shape with either multi-polar interconnected fine processes, termed as Type A (**Figure 5A**, arrowheads), or formed a ring-like structure of complex interwoven thick processes, named as Type B (**Figure 5A**, arrows). The process thickness was

1.60 ± 0.13 µm for Type A, and 3.10 ± 0.31 µm for Type B (Type A vs. Type B, $p < 0.001$). The number of CNPase$^+$-OLGs with the Type B shape was significantly higher in the Bcl11b-KD culture than in the scramble culture (**Figure 5B**). These results indicate that Bcl11b-KD changed the morphology of the OLGs derived from GPCs toward a more complex and mature shape. In addition, Bcl11b-KD increased the number of OLG processes and process branches per cell (**Figure 5C**). Increases in the average length of the cell processes and in the total process length per cell were observed (**Figure 5C**). In addition, a stronger GC immunoreactivity was observed in the OLGs derived from the Bcl11b-KD GPCs (**Figures 5D,E**). MOG and PLP immunofluorescence also showed enhanced differentiation of Bcl11b-KD GPCs toward mature OLGs when these cells were cultured in DM over 5 days.

We also used neural stem cells (NSCs) prepared from rat cortical tissues at E14 to generate OPCs. The efficiency of Bcl11b-KD in the OPCs by lentiviruses was examined (**Figures 6A,B**). Note that there was approximately 95% PDGFRα$^+$-cells in the scramble and Bcl11b-KD cultures maintained in GM (Supplementary Figure S3A). Bcl11b-KD causes decreased cell growth and sphere formation of the OPCs, which is comparable with the results from the GPCs (**Figures 6C,D**). Moreover, the findings were verified by Ki67 immunostaining, which showed lower amount of Ki67$^+$/GFP$^+$-cells in the Bcl11b-KD culture than that seen in the scramble culture (Supplementary Figure S3B). The morphology of CNPase$^+$-OLGs differentiated from the scramble NSCs is shown in **Figure 6E** (arrowheads). In comparison, the CNPase$^+$-OLGs generated from the NSCs that received Bcl11b-KD, showed elongated processes with a complex structure (**Figure 6E**, arrows). The total outgrowth length per CNPase$^+$-OLG from the Bcl11b-KD NSCs was greater than that of the CNPase$^+$-OLGs generated from the scramble culture (**Figure 6F**). In addition, the GC$^+$-OLGs had visible cell processes extending from their cell bodies (**Figure 6E**, arrows), and were more numerous in the Bcl11b-KD culture than in the scramble culture (**Figure 6F**). Increased GC immunoreactivity and GC$^+$-cell number in the Bcl11b-KD cultures were also observed (**Figures 6G,H**). Moreover, PLP$^+$-OLGs derived from Bcl11b-KD NSCs displayed a cell form with more complex processes (**Figure 6E**, arrows) compared to those observed in the scramble culture (**Figure 6E**, arrowheads). Furthermore, although an increase in PLP immunoreactivity and the number of PLP$^+$ cells was detected in the Bcl11b-KD culture (**Figures 6G,H**), Bcl11b-KD did not change the expression of MBP. These results demonstrate that the downregulation of Bcl11b gene expression in GPCs and NSCs caused an upregulation of GC and PLP, and promoted the differentiation of premyelinating OLGs.

A co-culture of GPCs with rat hippocampal neurons was established to further evaluate the effect of Bcl11b on OLG maturation. The co-culture was incubated for 7 days, and then subjected to double immunofluorescence for the identification of OLGs and neurons using anti-MBP and anti-NF200, respectively. The results displayed NF200$^+$-neuronal fibers covered by MBP$^+$-cell processes extending from the Bcl11b-KD OLGs (**Figures 7A,B**, arrows). By comparison, NF200$^+$-neuronal fibers were not covered to the same degree by the scramble OLGs (**Figures 7A,B**, arrows). Quantitative analysis also showed that the intensity of the MBP immunofluorescence overlapping onto the NF200$^+$-neuronal fibers in the Bcl11b-KD co-culture was higher than that in the scramble co-culture (**Figure 7B**). The 3D reconstruction imaging (xy, xz, and yz planes) from the serial confocal images further verified the intensive MBP immunoreactivity overlaid onto the NF200$^+$-neuronal fibers in the Bcl11b-KD co-culture (**Figure 7C**, arrows). In contrast, weak MBP immunofluorescence spotted around the NF200$^+$-neuronal fibers was seen in the scramble GPC/neuronal co-culture (**Figure 7C**, arrows). Moreover, the Bcl11b-KD GPCs were seeded onto *ex vivo* cerebellar-slice cultures for further examination. The Bcl11b-KD GPC/cerebellar tissue-slice co-culture displayed more MBP$^+$ processes over NF200$^+$ fibers compared to that in the scramble culture (**Figure 7D**). These observations further confirm that Bcl11b downregulation can progress OLG maturation.

Differentiation of Bcl11b-KD OLGs *in Vivo*

An *in vivo* model of focal demyelination that was induced by the injection of lysolecithin into the corpus callosum of adult rats was performed (**Figure 8A**). In this model, demyelination was detected at 3 dpi through examination of MBP$^+$-debris at the injection site (**Figure 8B**, arrows), while the dense MBP immunoreactivity remained in the corpus callosum receiving-vehicle injection (**Figure 8B**, arrowheads). To further examine the differentiation of GPCs at the demyelinated site, the scramble and Bcl11b-KD GPCs were separately implanted into the corpus callosum 3 days after the lysolecithin injection (**Figure 8C**). OLGs derived from the implanted GPCs were identified by double immunofluorescence for GFP and MBP. The GFP$^+$-scramble cells were found primarily at the injection site at 11 dpi (**Figure 8C**, arrowheads), whereas some of GFP$^+$/MBP$^+$-cells were found around the injection site in the Bcl11b-KD group (**Figure 8C**, arrows). The cell processes with intense MBP and PLP immunostaining were further detected in the GFP$^+$ Bcl11b-KD GPCs in the corpus callosum (**Figure 8D**, arrows). In contrast, the GFP$^+$-scramble cells expressed less MBP immunostaining (**Figure 8D**, arrowheads). The intensity of MBP and PLP immunostaining in the GFP$^+$-OLGs in the corpus callosum was higher when receiving Bcl11b-KD GPCs than with the scramble group (**Figure 8E**). In addition, the proportion of MBP and PLP expressing cells with respect to GFP$^+$-OLGs was also increased by Bcl11b-KD (**Figure 8E**). Notably, the amount of CC1$^+$/GFP$^+$-OLGs derived from Bcl11b-KD GPCs was higher than that observed in the scramble group (Supplementary Figures S4A,C), although the total amount of GFP$^+$-cells at the adjacent area to the injection site in the Bcl11b-KD group was lower than that analyzed in the scramble group (Supplementary Figure S4B), These results demonstrate that Bcl11b-KD GPCs effectively generate mature OLGs in the demyelinating region of the brain.

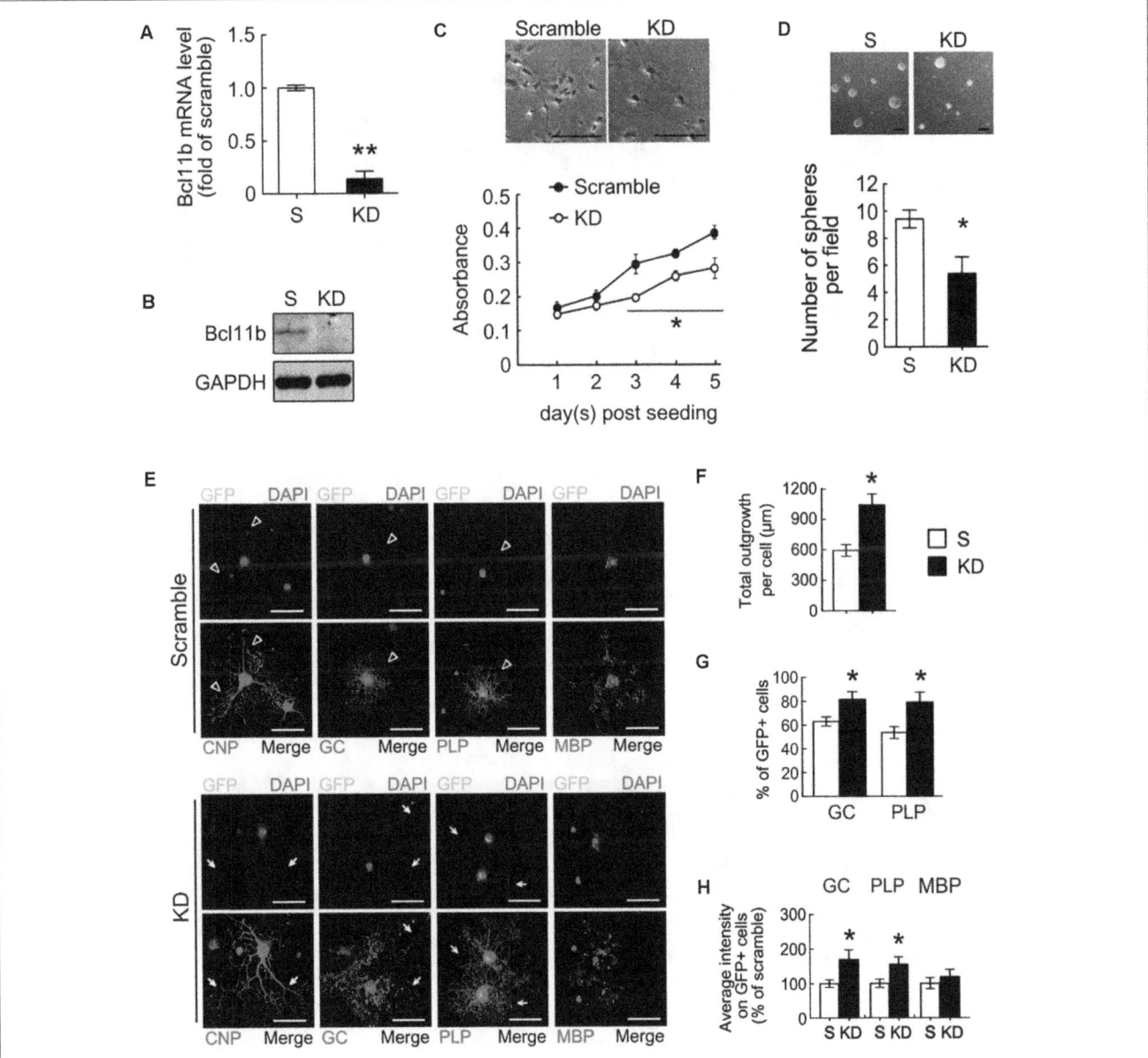

FIGURE 6 | Oligodendrocyte differentiation from OPCs promoted by Bcl11b gene knockdown. OPCs were derived from primary NSCs as described in Materials and Methods. OPCs were infected by lenti-ctrl (scramble) and lenti-sh-Bcl11b (Bcl11b-KD) and maintained in GM with puromycin for 2 days. The efficiency of Bcl11b-KD was measured by QPCR **(A)** and western blot **(B)**. The scramble- and Bcl11b-KD OPCs maintained in GM were subjected to MTT analysis **(C)**, and the sphere formation assay **(D)**. In addition, the scramble- and Bcl11b-KD OPCs were cultured in DM to generate OLGs for 5 days, followed by immunofluorescence for CNPase, GC, PLP, and MBP **(E)**. The total length of CNPase⁺-cell processes per OLG cultures was analyzed **(F)**. Moreover, the percentage of GC⁺-OLGs and PLP⁺-OLGs in the scramble and Bcl11b-KD was measured **(G)**. Moreover, the intensity of immunofluorescence for GC, PLP, and MBP in each cell was measured by ImageJ software. The results show the average fluorescence intensity of Bcl11b-KD cells over those of scramble cells **(H)**. Data are presented as means ± SEM of at least three independent experiments (n = 70–100 cells per group for **F,G**). *p < 0.05 vs. the scramble. Scale bar in **(C,D)**, 100 μm; in **(E)**, 50 μm.

DISCUSSION

Here, we show that Bcl11b expression can be detected in the GPCs within the SVZ of rat and mouse brains (Supplementary Figure S1). Moreover, Bcl11b expression declines in OLGs generated from OPCs. Further the *in vitro* study demonstrated that the downregulation of Bcl11b in GPCs caused the upregulation of the cell cycle inhibitor p21, and increased

premyelinating OLGs generated from the GPCs. In addition, we report that Bcl11b might function as a transcriptional repressor for Olig1 and PLP to downregulate their gene expression in GPCs. The observations from the co-culture system of GPCs with hippocampal neurons and with the *ex vivo* cerebellar slice cultures also indicated that MBP expression can be upregulated in OLGs derived from Bcl11b-KD GPCs. The *in vivo* study using a chemically induced demyelinating animal model indicated

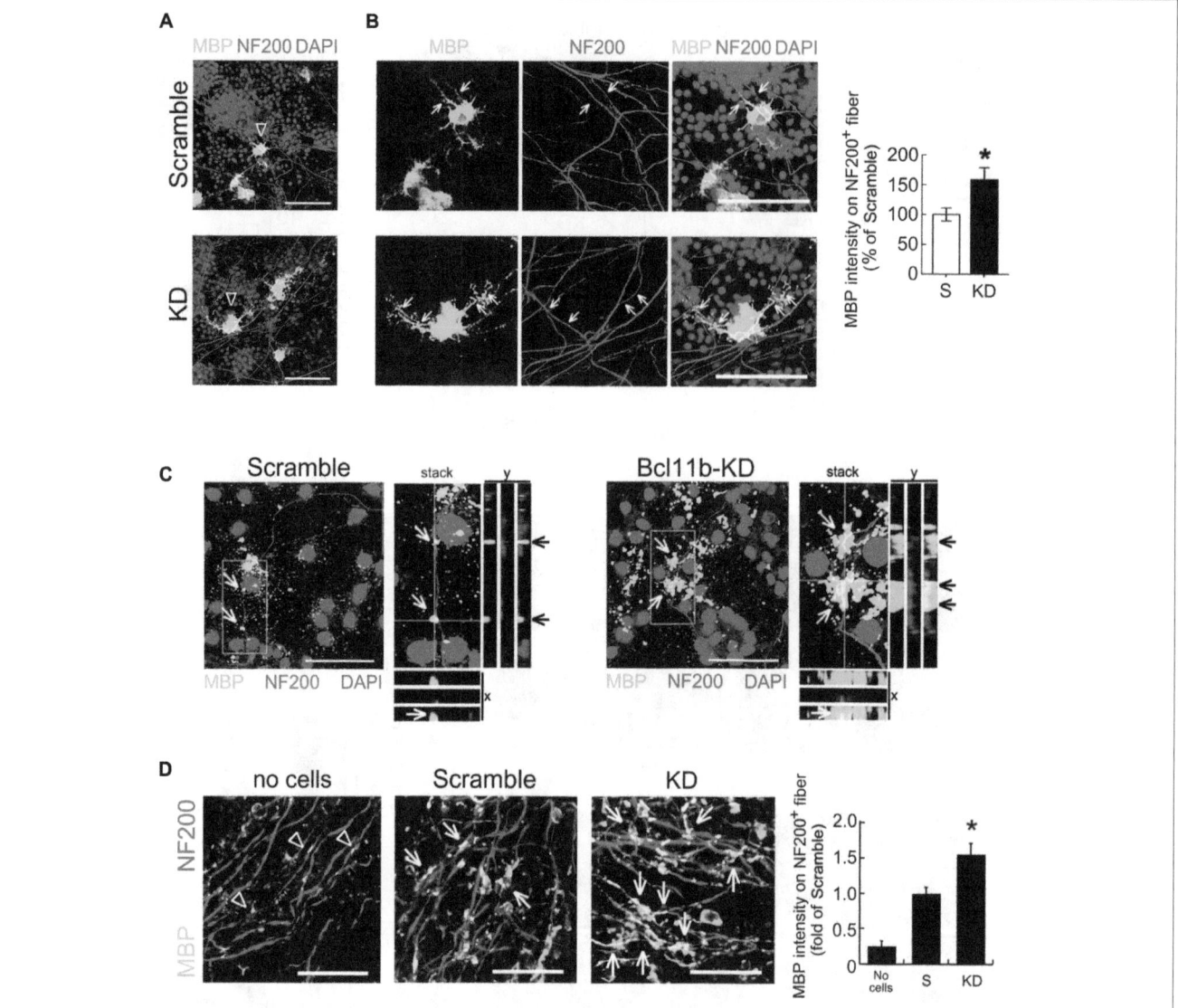

FIGURE 7 | Increased premyelinating processes of Bcl11b-KD GPCs along neuronal fibers. **(A)** Hippocampal neurons were co-cultured for 7 days with scramble and Bcl11b-KD GPCs. The cultures were then subjected to double immunofluorescence using anti-NF200 (red) and anti-MBP (green) antibodies. **(B)** The images with a higher magnification display the representative regions indicated by arrowheads in **(A)**. Arrows indicate the hippocampal fibers overlapping with MBP+-processes extending from scramble OLGs and Bcl11b-KD OLGs, respectively. The intensity of MBP immunostaining overlapping to NF200+- hippocampal fibers (arrowheads and arrows) in the scramble and Bcl11b-KD culture was quantified. **(C)** 3D-confocal imaging analysis was conducted to show the overlapping of MBP+-OLG processes (green) to NF200+-neuronal fibers (red). Arrows indicate the MBP immunostaining co-localized with NF200+-regions in the longitudinal y-z and transverse x–z views from a z-stack image. **(D)** Scramble and Bcl11b-KD GPCs were seeded onto rat cerebellar slice culture for 7 days. The cultures were then subjected to double immunofluorescence for MBP (green) and NF200 (red). The immunoreactive intensity of MBP overlapping to NF200+ fiber (arrows) was quantified by ImageJ software (right panel). Note that MBP+-cell debris was observed in the culture without the addition of GPCs (arrowheads). Data are presented as means ± SEM of three independent experiments (n = 20–30 cells per group). *p < 0.05 vs. scramble. Scale bar, 50 μm.

that Bcl11b downregulation can promote implanted GPCs to differentiate into OLGs in the demyelinating region. These findings demonstrate the positive role of Bcl11b in the cell growth of GPCs and OPCs, but point to its negative effect on the progression of the OLG lineage to maturation.

Bcl11b has been reported to act as a transcriptional repressor to silence the expression of p21 (Cherrier et al., 2009). Our previous study using human and rat glioma cells also showed that p21 expression was upregulated after the downregulation

of Bcl11b expression in glioma cell lines, suggesting that Bcl11b is an important regulator for glioma cell expansion through the repression of p21 action (Liao et al., 2016). The results from the present study showing that the proliferation of GPCs in GM was suppressed after Bcl11b-KD along with the upregulation of p21, also indicate that Bcl11b is involved in the maintenance of GPC stemness via the repression of p21 expression. Given that the ablation of p21 gene expression is known to inhibit OLG differentiation (Zezula et al., 2001),

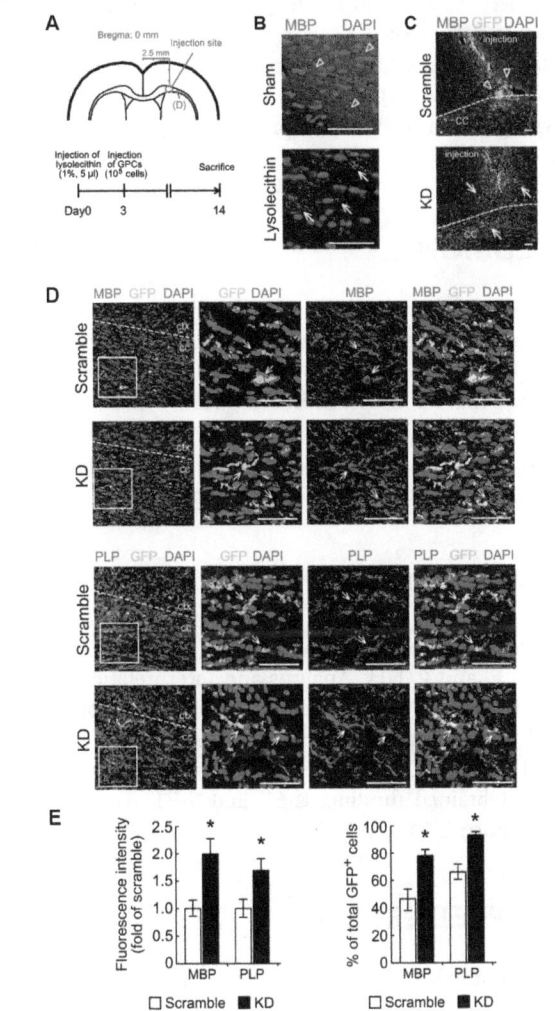

FIGURE 8 | Improved differentiation of OLGs derived from Bcl11b-KD GPCs under a demyelinating condition. **(A)** The diagrams show the injection site of lysolecithin and GPCs just above the corpus callosum (CC, upper panel), and indicate the time points (lower panel) for the injection of 1% lysolecithin at Day 0 and GPCs (scramble and Bcl11b-KD GPCs) at Day 3. The animals were sacrificed at Day 14 after lysolecithin injection to investigate the differentiation of implanted GPCs into OLGs at the demyelinating CC regions labeled by **(D)**. **(B)** The brain sections that were prepared from animals having received lysolecithin injection were subjected to MBP immunostaining (red) and DAPI nuclear counterstaining (blue). The sham-operated group was only receiving vehicle. MBP+ debris was clearly observed in lysolecithin-injected CC (arrows). Diffuse and dense MBP staining was seen in the sham-operated group. **(C)** The brain sections receiving lysolecithin injection were subjected to double immunofluorescence for GFP (green) and MBP (red) to identify OLGs derived from implanted GPCs at the injection site. **(D)** By double immunofluorescence, GFP+/MBP+-OLGs (upper panel) and GFP+/PLP+-OLGs (lower panel) derived from implanted scramble and Bcl11b-KD GPCs in the CC adjacent to the injection site **(A)** were indicated by arrows. **(E)** To examine OLGs derived from implanted GPCs in the CC, the immunoreactive intensity of MBP and PLP in GFP+-cells shown in **(D)** was measured by ImageJ software. The results are shown as the fold of the scramble group. The results are shown as the percentage of double immunostained OLGs (GFP+/MBP+, GFP+/PLP+) over the total number of GFP+-cells in the field. Data in **(E)** are presented as means ± SEM from at least 3 animals in each group. The dash lines shown in **(C)** designate the border between cortex and CC. *$p < 0.05$ vs. scramble. Scale bar, 50 μm.

the upregulation of p21 after Bcl11b-KD might be involved in the differentiation of GPCs into OLGs. In addition, although Sox2 as a transcription factor is required for the self-renewal of OPCs, it blocks OLG differentiation (Zhao et al., 2015). Our observations that Bcl11b-KD attenuated the expression of Sox2 in GPCs under the condition of GM or DM, raise the possibility that the downregulation of Bcl11b might facilitate OLG differentiation from GPCs. Indeed, the expression of OLG markers (CNPase, PLP, and MOG) can be increased in Bcl11b-KD GPCs or Bcl11b-KD OPCs, demonstrating that a decline in Bcl11b expression can promote GPC differentiation into mature OLGs.

It has been reported that neuronal progenitor proliferation and post-mitotic neuron differentiation in the hippocampus declined in Bcl11b mutant mice during development (Simon et al., 2012). In addition, the deficiency of Bcl11b can cause postmitotic vomeronasal sensory neurons to selectively undergo cell apoptosis (Enomoto et al., 2011). A lack of Bcl11b functionality has been reported to cause inefficient differentiation of spiny neurons in striatal medium, with reduced expression of mature striatal markers (Arlotta et al., 2008). Here, we show that the downregulation of Bcl11b expression in GPCs and OPCs can reduce their proliferation under the growth condition, but promote the OLG lineage progression. Our results, in conjunction with findings on neuronal development, demonstrate that functional Bcl11b is required for neural progenitor cell proliferation. Despite the expression of Bcl11b in rat and mouse OLGs located in the white matter being different, it remains to be further investigated if OLG differentiation could be impaired in the absence of functional Bcl11b.

Several Bcl11b binding sites have been assumed to occur at the promoters of striatal genes (Desplats et al., 2008). It has also been reported that the expression of Olig1, one of the key transcription factors for OLG differentiation, was induced in Bcl11b-deficient neurons (Enomoto et al., 2011). These findings, together with our observations of an increased expression of Olig1 in OLGs derived from Bcl11b-KD GPCs, raise the possibility that Bcl11b might be involved in the regulation of OLG-specific gene expression. By searching a transcription factor prediction database for potential Bcl11b binding sites in 2-kb long promoter regions of rat OLG-specific genes (Olig1, Olig2, MBP, and PLP), we validated six and seven (predicted) Bcl11b binding sites located at the Olig1 and PLP promoters, respectively. Our results from the ChIP assay further confirm that Bcl11b exhibits high binding affinity to the specific promoter regions of Olig1 ($-1487 \sim -1264$) and PLP ($-625 \sim -433$). Although the 2-kb long promoter regions of MBP contain 3 predicted binding sites with Bcl11b, the binding affinity of Bcl11b derived from the MBP promoters was extremely weak (data not shown). This is in accordance with our observation that MBP transcription was not affected by Bcl11b-KD. In other words, Bcl11b most likely acts as a transcriptional repressor of Olig1 and PLP, but not of MBP. Yet, the regulation of these genes by Bcl11b could also be due to the action of Bcl11b-induced chromatin remodeling via its interaction with nucleosome remodeling as well as histone deacetylase (NuRD) complex and SWI/SNF complex, which has been found in the control of T cells and neural development (Cismasiu et al., 2005; Topark-Ngarm et al., 2006; Son and Crabtree, 2014). Taken together, Bcl11b potentially deters the differentiation of GPCs into OLGs through the repression of Olig1 and PLP. Accordingly, the reduction of Bcl11b can decrease its inhibition in the expression of these two molecules, and subsequently promote OLG differentiation and maturation. Nevertheless, the upstream regulators that can mediate Bcl11b expression during OLG development remain to be determined.

Proteolipid protein is not only required for the stabilization of the myelin membrane and axonal structure (Uschkureit et al., 2000), but is also involved in the transportation of myelin membranes in the OLG secretory pathway (Nave and Werner, 2014). Thus, the upregulation of PLP expression in OLGs derived from Bcl11b-KD GPCs could promote the maintenance of OLG interaction with neuronal fibers in hippocampal neuron co-cultures and cerebellar-slice systems, as well as in chemically induced demyelination of the corpus callosum. Sox10 can directly induce MBP expression (Liu et al., 2007), or cooperate with Olig1 to increase MBP expression (Li et al., 2007). However, based on our *in vitro* study, we found no change in the Sox10 gene expression after Bcl11b-KD. This may explain why MBP gene expression and its protein production did not increase in our *in vitro* study. Yet, MBP expression has been known to be upregulated in the presence of neurons and neuronal factors (Bologa et al., 1986; Jensen et al., 2000; Simons and Trajkovic, 2006). Here, we provide important evidence that MBP expression in OLGs derived from Bcl11b-KD GPCs was increased under the condition of hippocampal neuron co-culture and cerebellar-slice culture. In addition, our *in vivo* results demonstrate that an increase in MBP immunoreactivity was detected in OLGs derived from implanted Bcl11b-KD GPCs toward the demyelination of white matter

regions. These *ex vivo* and *in vivo* results further revealed that the downregulation of Bcl11b expression can facilitate glial progenitors to induce OLG maturation. Moreover, in response to neuronal signals, OLGs derived from GPCs, in tandem with the downregulation of Bcl11b expression, could possibly foster myelination.

CONCLUSION

Bcl11b is essential for maintaining the proliferation of glial progenitors and their stemness properties. Reduced expression of Bcl11b in glial progenitors can contribute to the differentiation of GPCs toward mature OLGs. Moreover, our *in vitro* and *in vivo* findings provide important information for the development of an effective cell therapeutic strategy for demyelinating disorders via the downregulation of Bcl11b expression.

AUTHOR CONTRIBUTIONS

Study concept and design: C-YW and S-FT. Acquisition of data: C-YW, K-MF, and C-HH. Analysis and interpretation of data: C-YW, Y-TS, and S-FT. Drafting of the manuscript: C-YW, Y-TS, and S-FT. Critical revision of the article for important intellectual content: C-YW and S-FT. Statistical analysis: C-YW and K-MF. Obtained funding: C-SY and S-FT. Technical and material support: C-SY and S-FT.

ACKNOWLEDGMENTS

The authors thank Ms. Yun Ting Hsieh for technical assistance.

REFERENCES

Arlotta, P., Molyneaux, B. J., Chen, J., Inoue, J., Kominami, R., and Macklis, J. D. (2005). Neuronal subtype-specific genes that control corticospinal motor neuron development in vivo. *Neuron* 45, 207–221. doi: 10.1016/j.neuron.2004. 12.036

Arlotta, P., Molyneaux, B. J., Jabaudon, D., Yoshida, Y., and Macklis, J. D. (2008). Ctip2 controls the differentiation of medium spiny neurons and the establishment of the cellular architecture of the striatum. *J. Neurosci.* 28, 622–632. doi: 10.1523/JNEUROSCI.2986-07.2008

Bernard, O. A., Busson-LeConiat, M., Ballerini, P., Mauchauffe, M., Della Valle, V., Monni, R., et al. (2001). A new recurrent and specific cryptic translocation, t(5;14)(q35;q32), is associated with expression of the Hox11L2 gene in T acute lymphoblastic leukemia. *Leukemia* 15, 1495–1504. doi: 10.1038/sj.leu. 2402249

Bologa, L., Aizenman, Y., Chiappelli, F., and de Vellis, J. (1986). Regulation of myelin basic protein in oligodendrocytes by a soluble neuronal factor. *J. Neurosci. Res.* 15, 521–528. doi: 10.1002/jnr.49015 0409

Chen, B., Wang, S. S., Hattox, A. M., Rayburn, H., Nelson, S. B., and McConnell, S. K. (2008). The Fezf2-Ctip2 genetic pathway regulates the fate choice of subcortical projection neurons in the developing cerebral cortex. *Proc. Natl. Acad. Sci. U.S.A.* 105, 11382–11387. doi: 10.1073/pnas.080491 8105

Chen, S., Huang, X., Chen, S., Yang, L., Shen, Q., Zheng, H., et al. (2012). The role of BCL11B in regulating the proliferation of human naive T cells. *Hum. Immunol.* 73, 456–464. doi: 10.1016/j.humimm.2012.02.018

Cherrier, T., Suzanne, S., Redel, L., Calao, M., Marban, C., Samah, B., et al. (2009). p21(WAF1) gene promoter is epigenetically silenced by CTIP2 and SUV39H1. *Oncogene* 28, 3380–3389. doi: 10.1038/onc.2009.193

Cismasiu, V. B., Adamo, K., Gecewicz, J., Duque, J., Lin, Q., and Avram, D. (2005). BCL11B functionally associates with the NuRD complex in T lymphocytes to repress targeted promoter. *Oncogene* 24, 6753–6764. doi: 10.1038/sj.onc. 1208904

Czepiel, M., Boddeke, E., and Copray, S. (2015). Human oligodendrocytes in remyelination research. *Glia* 63, 513–530. doi: 10.1002/glia.22769

Desplats, P. A., Lambert, J. R., and Thomas, E. A. (2008). Functional roles for the striatal-enriched transcription factor, Bcl11b, in the control of striatal gene expression and transcriptional dysregulation in Huntington's disease. *Neurobiol. Dis.* 31, 298–308. doi: 10.1016/j.nbd.2008.05.005

Emery, B. (2010). Regulation of oligodendrocyte differentiation and myelination. *Science* 330, 779–782. doi: 10.1126/science.1190927

Enomoto, T., Ohmoto, M., Iwata, T., Uno, A., Saitou, M., Yamaguchi, T., et al. (2011). Bcl11b/Ctip2 controls the differentiation of vomeronasal sensory neurons in mice. *J. Neurosci.* 31, 10159–10173. doi: 10.1523/JNEUROSCI.1245-11.2011

Fancy, S. P., Glasgow, S. M., Finley, M., Rowitch, D. H., and Deneen, B. (2012). Evidence that nuclear factor IA inhibits repair after white matter injury. *Ann. Neurol.* 72, 224–233. doi: 10.1002/ana.23590

Fang, K. M., Yang, C. S., Lin, T. C., Chan, T. C., and Tzeng, S. F. (2014). Induced interleukin-33 expression enhances the tumorigenic activity of rat glioma cells. *Neuro Oncol.* 16, 552–566. doi: 10.1093/neuonc/not234

Grabarczyk, P., Nahse, V., Delin, M., Przybylski, G., Depke, M., Hildebrandt, P., et al. (2010). Increased expression of bcl11b leads to chemoresistance accompanied by G1 accumulation. *PLOS ONE* 5:e12532. doi: 10.1371/journal. pone.0012532

Herrera, J., Yang, H., Zhang, S. C., Proschel, C., Tresco, P., Duncan, I. D., et al. (2001). Embryonic-derived glial-restricted precursor cells (GRP cells) can differentiate into astrocytes and oligodendrocytes *in vivo*. *Exp. Neurol.* 171, 11–21. doi: 10.1006/exnr.2001.7729

Jensen, M. B., Poulsen, F. R., and Finsen, B. (2000). Axonal sprouting regulates myelin basic protein gene expression in denervated mouse hippocampus. *Int. J. Dev. Neurosci.* 18, 221–235. doi: 10.1016/S0736-5748(99)00091-X

Karanam, N. K., Grabarczyk, P., Hammer, E., Scharf, C., Venz, S., Gesell-Salazar, M., et al. (2010). Proteome analysis reveals new mechanisms of Bcl11b-loss driven apoptosis. *J. Proteome Res.* 9, 3799–3811. doi: 10.1021/pr901096u

Kremer, D., Gottle, P., Hartung, H. P., and Kury, P. (2016). Pushing forward: remyelination as the new frontier in CNS diseases. *Trends Neurosci.* 39, 246–263. doi: 10.1016/j.tins.2016.02.004

Lee, H. K., Laug, D., Zhu, W., Patel, J. M., Ung, K., Arenkiel, B. R., et al. (2015). Apcdd1 stimulates oligodendrocyte differentiation after white matter injury. *Glia* 63, 1840–1849. doi: 10.1002/glia.22848

Li, H., Lu, Y., Smith, H. K., and Richardson, W. D. (2007). Olig1 and Sox10 interact synergistically to drive myelin basic protein transcription in oligodendrocytes. *J. Neurosci.* 27, 14375–14382. doi: 10.1523/JNEUROSCI.4456-07.2007

Liao, C. K., Fang, K. M., Chai, K., Wu, C. H., Ho, C. H., Yang, C. S., et al. (2016). Depletion of B cell CLL/lymphoma 11B gene expression represses glioma cell growth. *Mol. Neurobiol.* 53, 3528–3539. doi: 10.1007/s12035-015-9231-1

Liu, Z., Hu, X., Cai, J., Liu, B., Peng, X., Wegner, M., et al. (2007). Induction of oligodendrocyte differentiation by Olig2 and Sox10: evidence for reciprocal interactions and dosage-dependent mechanisms. *Dev. Biol.* 302, 683–693. doi: 10.1016/j.ydbio.2006.10.007

Lu, Q. R., Sun, T., Zhu, Z., Ma, N., Garcia, M., Stiles, C. D., et al. (2002). Common developmental requirement for Olig function indicates a motor neuron/oligodendrocyte connection. *Cell* 109, 75–86. doi: 10.1016/S0092-8674(02)00678-5

Mayoral, S. R., and Chan, J. R. (2016). The environment rules: spatiotemporal regulation of oligodendrocyte differentiation. *Curr. Opin. Neurobiol.* 39, 47–52. doi: 10.1016/j.conb.2016.04.002

Nave, K. A., and Werner, H. B. (2014). Myelination of the nervous system: mechanisms and functions. *Annu. Rev. Cell Dev. Biol.* 30, 503–533. doi: 10.1146/annurev-cellbio-100913-013101

Raff, M. C., Miller, R. H., and Noble, M. (1983). A glial progenitor cell that develops *in vitro* into an astrocyte or an oligodendrocyte depending on culture medium. *Nature* 303, 390–396. doi: 10.1038/303390a0

Simon, R., Brylka, H., Schwegler, H., Venkataramanappa, S., Andratschke, J., Wiegreffe, C., et al. (2012). A dual function of Bcl11b/Ctip2 in hippocampal neurogenesis. *EMBO J.* 31, 2922–2936. doi: 10.1038/emboj.2012.142

Simons, M., and Trajkovic, K. (2006). Neuron-glia communication in the control of oligodendrocyte function and myelin biogenesis. *J. Cell Sci.* 119, 4381–4389. doi: 10.1242/jcs.03242

Son, E. Y., and Crabtree, G. R. (2014). The role of BAF (mSWI/SNF) complexes in mammalian neural development. *Am. J. Med. Genet. C Semin. Med. Genet.* 166C, 333–349. doi: 10.1002/ajmg.c.31416

Stolt, C. C., Rehberg, S., Ader, M., Lommes, P., Riethmacher, D., Schachner, M., et al. (2002). Terminal differentiation of myelin-forming oligodendrocytes depends on the transcription factor Sox10. *Genes Dev.* 16, 165–170. doi: 10.1101/gad.215802

Tang, B., Di Lena, P., Schaffer, L., Head, S. R., Baldi, P., and Thomas, E. A. (2011). Genome-wide identification of Bcl11b gene targets reveals role in brain-derived neurotrophic factor signaling. *PLOS ONE* 6:e23691. doi: 10.1371/journal.pone.0023691

Topark-Ngarm, A., Golonzhka, O., Peterson, V. J., Barrett, B. Jr., Martinez, B., Crofoot, K., et al. (2006). CTIP2 associates with the NuRD complex on the promoter of p57KIP2, a newly identified CTIP2 target gene. *J. Biol. Chem.* 281, 32272–32283. doi: 10.1074/jbc.M602776200

Uschkureit, T., Sporkel, O., Stracke, J., Bussow, H., and Stoffel, W. (2000). Early onset of axonal degeneration in double (plp−/−mag−/−) and hypomyelinosis in triple (plp−/−mbp−/−mag−/−) mutant mice. *J. Neurosci.* 20, 5225–5233.

Wang, C. Y., Deneen, B., and Tzeng, S. F. (2017). MicroRNA-212 inhibits oligodendrocytes during maturation by down-regulation of differentiation-associated gene expression. *J. Neurochem.* 143, 112–125. doi: 10.1111/jnc.14138

Wang, C. Y., Yang, S. H., and Tzeng, S. F. (2015). MicroRNA-145 as one negative regulator of astrogliosis. *Glia* 63, 194–205. doi: 10.1002/glia.22743

Yang, J., Cheng, X., Qi, J., Xie, B., Zhao, X., Zheng, K., et al. (2017). EGF enhances oligodendrogenesis from glial progenitor cells. *Front. Mol. Neurosci.* 10:106. doi: 10.3389/fnmol.2017.00106

Yang, Y., Lewis, R., and Miller, R. H. (2011). Interactions between oligodendrocyte precursors control the onset of CNS myelination. *Dev. Biol.* 350, 127–138. doi: 10.1016/j.ydbio.2010.11.028

Zezula, J., Casaccia-Bonnefil, P., Ezhevsky, S. A., Osterhout, D. J., Levine, J. M., Dowdy, S. F., et al. (2001). p21cip1 is required for the differentiation of oligodendrocytes independently of cell cycle withdrawal. *EMBO Rep.* 2, 27–34. doi: 10.1093/embo-reports/kve008

Zhang, L. J., Bhattacharya, S., Leid, M., Ganguli-Indra, G., and Indra, A. K. (2012). Ctip2 is a dynamic regulator of epidermal proliferation and differentiation by integrating EGFR and Notch signaling. *J. Cell Sci.* 125, 5733–5744. doi: 10.1242/jcs.108969

Zhang, S. C. (2001). Defining glial cells during CNS development. *Nat. Rev. Neurosci.* 2, 840–843. doi: 10.1038/35097593

Zhang, Y., Chen, K., Sloan, S. A., Bennett, M. L., Scholze, A. R., O'Keeffe, S., et al. (2014). An RNA-sequencing transcriptome and splicing database of glia, neurons, and vascular cells of the cerebral cortex. *J. Neurosci.* 34, 11929–11947. doi: 10.1523/JNEUROSCI.1860-14.2014

Zhao, C., Ma, D., Zawadzka, M., Fancy, S. P., Elis-Williams, L., Bouvier, G., et al. (2015). Sox2 sustains recruitment of oligodendrocyte progenitor cells following CNS demyelination and primes them for differentiation during remyelination. *J. Neurosci.* 35, 11482–11499. doi: 10.1523/JNEUROSCI.3655-14.2015

Zuchero, J. B., and Barres, B. A. (2013). Intrinsic and extrinsic control of oligodendrocyte development. *Curr. Opin. Neurobiol.* 23, 914–920. doi: 10.1016/j.conb.2013.06.005

11

Diversity of Astroglial Effects on Aging- and Experience-Related Cortical Metaplasticity

Ulyana Lalo[1], Alexander Bogdanov[2] and Yuriy Pankratov[1]*

[1]School of Life Sciences, University of Warwick, Coventry, United Kingdom, [2]Institute for Chemistry and Biology, Immanuel Kant Baltic Federal University, Kaliningrad, Russia

*Correspondence:
Yuriy Pankratov
y.pankratov@warwick.ac.uk

Activity-dependent regulation of synaptic plasticity, or metaplasticity, plays a key role in the adaptation of neuronal networks to physiological and biochemical changes in aging brain. There is a growing evidence that experience-related alterations in the mechanisms of synaptic plasticity can underlie beneficial effects of physical exercise and caloric restriction (CR) on brain health and cognition. Astrocytes, which form neuro-vascular interface and can modulate synaptic plasticity by release of gliotransmitters, attract an increasing attention as important element of brain metaplasticity. We investigated the age- and experience-related alterations in astroglial calcium signaling and stimulus-dependence of long-term synaptic plasticity in the neocortex of mice exposed to the mild CR and environmental enrichment (EE) which included ad libitum physical exercise. We found out that astrocytic Ca^{2+}-signaling underwent considerable age-related decline but EE and CR enhanced astroglial signaling, in particular mediated by noradrenaline (NA) and endocannabinoid receptors. The release of ATP and D-Serine from astrocytes followed the same trends of age-related declined and EE-induced increase. Our data also showed that astrocyte-derived ATP and D-Serine can have diverse effects on the threshold and magnitude of long-term changes in the strength of neocortical synapses; these effects were age-dependent. The CR- and EE-induced enhancement of astroglial Ca^{2+}-signaling had more stronger effect on synaptic plasticity in the old (14–18 months) than in the young (2–5 months) wild-type (WT) mice. The effects of CR and EE on synaptic plasticity were significantly altered in both young and aged dnSNARE mice. Combined, our data suggest astrocyte-neuron interactions are important for dynamic regulation of cortical synaptic plasticity. This interaction can significantly decline with aging and thus contributes to the age-related cognitive impairment. On another hand, experience-related increase in the astroglial Ca^{2+}-signaling can ameliorate the age-related decline.

Keywords: ATP release, D-serine, caloric restriction, exocytosis, metaplasticy, CB1 receptors, BCM model

INTRODUCTION

Although an age-related cognitive decline is widely recognized a major societal and scientific problem, fundamental mechanisms of brain longevity are not fully understood. Synaptic plasticity enables the mammalian brain to adapt to environmental challenges during development, adulthood and aging. Nowadays the age-related change in the cognitive functions is viewed not as complete loss of synaptic plasticity but as alteration of its mechanisms (Hillman et al., 2008; Nithianantharajah and Hannan, 2009; van Praag, 2009; Mercken et al., 2012; Merzenich et al., 2014).

Importantly, neural networks and synapses are remarkably responsive to environmental stimuli, physiological modifications, and experience (Hillman et al., 2008; Nithianantharajah and Hannan, 2009; van Praag, 2009; Mercken et al., 2012; Rodríguez et al., 2013; Merzenich et al., 2014). Physical exercise and environmental enrichment (EE) can have beneficial effects on aging brain, both in animal models and human patients (Hillman et al., 2008; Nithianantharajah and Hannan, 2009; van Praag, 2009; Mercken et al., 2012; Merzenich et al., 2014). Also, caloric restriction (CR), usually defined as a reduced intake of calories not causing malnutrition, can have life-extending effects, linked to improvement of brain health and plasticity (Mercken et al., 2012; Park et al., 2012; Madeo et al., 2014; López-Otín et al., 2016).

Aging- and experience-related alterations in synaptic plasticity are closely linked to brain metaplasticity, which is usually defined as "the plasticity of synaptic plasticity." Metaplasticity can occur when priming synaptic or cellular activity or inactivity leads to persistent change in the direction or degree of synaptic plasticity (Abraham and Bear, 1996; Hulme et al., 2013a). Astrocytes are gaining an increasing attention as a very important element of brain cellular networks regulating metaplasticity (Hulme et al., 2013b; Monai et al., 2016; Boué-Grabot and Pankratov, 2017; Singh and Abraham, 2017). Astrocytes form interface between the synapses and brain vasculature (Gourine et al., 2010; Halassa and Haydon, 2010; Araque et al., 2014) and therefore are strategically positioned to couple the enriched mental and physical activity to the brain longevity. Astrocytes can also respond to both high-fat and calorie-restricted diet (Seidel et al., 2006; Lin et al., 2014; Metna-Laurent and Marsicano, 2015).

Importantly, astrocytes can exert bi-directional effects on synaptic plasticity by releasing different gliotransmitters (Pascual et al., 2005; Henneberger et al., 2010; Araque et al., 2014; Pougnet et al., 2014; Rasooli-Nejad et al., 2014; Pankratov and Lalo, 2015; Lalo et al., 2016; Boué-Grabot and Pankratov, 2017; Papouin et al., 2017a). It is conceivable that overall effect of astroglia on plasticity of particular type of synapses would depend on physiological (or pathological) context, i.e., pattern of local neural activity, repertoire of transmitters released from neurons and repertoire of post- and pre-synaptic receptors expressed in the synapse. Such dependence of astroglial modulation of synaptic plasticity on prior activity of network can render an important role for astrocytes in metaplasticity (Hulme et al., 2013b; Monai et al., 2016; Boué-Grabot and Pankratov, 2017; Singh and Abraham, 2017).

The responsiveness of synaptic plasticity to EE, physical activity and CR provides an opportunity to ameliorate the negative consequences of aging on cognitive function. Still, fundamental cellular and molecular mechanisms underlying the impact of CR and exercise on brain metaplasticity remain largely unexplored. There are also many uncertainties in the mechanisms of neuro-glial interaction (Bazargani and Attwell, 2016; Papouin et al., 2017b; Singh and Abraham, 2017; Fiacco and McCarthy, 2018; Savtchouk and Volterra, 2018), particularly in the aging brain (Rodríguez et al., 2014; Verkhratsky et al., 2014, 2017). Until recently, most studies of brain aging have been focused on functional changes in neural networks and alterations in neuronal morphology and gene expression whereas glia-neuron interactions remained largely overlooked. Recent reports of aging-related changes of Ca^{2+}-signaling, morphology and gene expression in astrocytes (Lalo et al., 2011, 2014b; Rodríguez et al., 2014; Verkhratsky et al., 2014, 2017; Soreq et al., 2017) highlighted a crucial importance of study of brain aging and neurodegeneration in the context of complex cellular interactions which maintain synaptic dynamics and homeostasis (De Strooper and Karran, 2016). Still, changes in the glia-driven modulation of synaptic metaplasticity in aging brain remain almost unexplored.

In the present article, we explored role for astroglial Ca^{2+}-signaling and release of gliotransmitters in aging- and environment-related cortical metaplasticity. As a model of impaired astroglial exocytosis, we used dnSNARE mice whose validity, in particular the lack of neuronal expression of dnSNARE, has been recently verified (Pankratov and Lalo, 2015; Sultan et al., 2015; Lalo et al., 2016).

MATERIALS AND METHODS

All animal work has been carried out in accordance with UK legislation (ASPA) and "3R" strategy; all experimental protocols were approved by University of Warwick Ethical Review Committee and Animal Welfare Committee.

Slice and Cell Preparation

Mice of two aged groups, 2–5 (average 3.3) months and 14–18 (average 16.8) months were anesthetized by halothane and then decapitated, in accordance with UK legislation. Brains were removed rapidly after decapitation and placed into ice-cold physiological saline containing (mM): NaCl 130, KCl 3, $CaCl_2$ 0.5, $MgCl_2$ 2.5, NaH_2PO_4 1, $NaHCO_3$ 25, glucose 15, pH of 7.4 gassed with 95% O_2–5% CO_2. Transverse slices (280 μm) were cut at 4° C and then placed in physiological saline containing (mM): NaCl 130, KCl 3, $CaCl_2$ 2.5, $MgCl_2$ 1, NaH_2PO_4 1, $NaHCO_3$ 22, glucose 15, pH of 7.4 gassed with 95% O_2–5% CO_2 and kept for 1–4 h prior to cell isolation and recording.

Astrocytes were identified by their morphology under DIC observation, EGFP fluorescence (astrocytes from dn-SNARE and GFAP-EGFP mice) or staining with sulforhodamine 101 (astrocytes from WT mice). After recording, the identification of astrocyte was confirmed via functional properties (high potassium conductance, low input resistance, strong activity of glutamate transporters) as described previously (Lalo et al., 2011, 2014a; Rasooli-Nejad et al., 2014; Pankratov and Lalo, 2015).

Electrophysiological Recordings

Whole-cell voltage clamp recordings from cortical neurones and astrocytes cells were made with patch pipettes (4–5 MΩ) filled with intracellular solution (in mM): 110 CsCl, 10 NaCl, 10 HEPES, 5 MgATP, 1 D-Serine, 0.1 EGTA, pH 7.35; Currents were monitored using an MultiClamp 700B patch-clamp amplifier (Axon Instruments, USA) filtered at 2 kHz and digitized at 4 kHz. Experiments were controlled by Digidata1440A data acquisition board (Axon Instruments, USA) and WinWCP software (Strathclyde University, UK); data were analyzed by self-designed software. Liquid junction potentials were compensated with the patch-clamp amplifier. The series and input resistances were respectively 5–7 MΩ and 600–1100 MΩ; both series and input resistance varied by less than 20% in the cells accepted for analysis.

Field excitatory postsynaptic potentials (fEPSPs) were measured via a glass micropipette filled with extracellular solution (0.5–1 MΩ resistance) placed in neocortical layer II/III. The fEPSPs were evoked by the stimulation of neuronal afferents descending from layers IV–V. For activation of synaptic inputs, axons originating from layer IV–VI neurons were stimulated with a bipolar coaxial electrode (WPI, USA) placed in layer V close to the layer IV border, approximately opposite the site of recording; stimulus duration was 300 μs. The stimulus magnitude was set 3–4 times higher than the minimal stimulus necessary to elicit a response in layer II pyramidal neurons (Rasooli-Nejad et al., 2014; Pankratov and Lalo, 2015; Lalo et al., 2016).

The long-term potentiation/depression (LTP/LTD) was induced by different number of trains of high-frequency theta-burst stimulation (HFS-trains); each train (100 ms-long) consisted of 10 pulses stimulated at 100 Hz, trains were delivered with 200 ms intervals, every 10 trains were separated by 10 s-long intervals.

Multi-photon Fluorescent Ca^{2+}-Imaging in Astrocytes

To monitor the cytoplasmic free Ca^{2+}concentraton ([Ca^{2+}]$_i$) *in situ*, astrocytes of neocortical slices were loaded via 30 min incubation with 1 μM of Rhod-2AM or Oregon Green Bapta-2AM and sulphorhodamine 101 (wild-type (WT) mice) at 33°C. Two-photon images of neurons and astrocytes were acquired at 5 Hz frame-rate using a Zeiss LSM-7MP multi-photon microscope coupled to a SpectraPhysics MaiTai pulsing laser; experiments were controlled by ZEN LSM software (Carl Zeiss, Germany). Images were further analyzed offline using ZEN LSM (Carl Zeiss) and ImageJ (NIH) software. The [Ca^{2+}]$_i$ levels were expressed as ΔF/F ratio averaged over a region of interest (ROI). For analysis of spontaneous Ca^{2+}–transients in astrocytes, three ROIs located over dendrites and one ROI located over the soma were chosen. Overall Ca^{2+}-response to receptors agonists or synaptic stimulation was quantified using an ROI covering the whole cell image.

Measurement of Extracellular Concentration of ATP and D-Serine in the Brain Tissue

The concentration of ATP within cortical slices was measured using microelectrode biosensors obtained from Sarissa Biomedical Ltd (Coventry, UK). A detailed description of the properties of biosensors and recording procedure has been published previously (Frenguelli et al., 2007). Briefly, biosensors consisted of ATP or D-Serine metabolizing enzymes

immobilized within a matrix on thin (25 μM) Pt/Ir wire. This allowed insertion of the sensors into the cortical slice and minimized the influence of a layer of dead surface tissue. ATP and D-serine biosensors were used simultaneously. A third, null, biosensor was also used. This sensor is identical to the ATP and D-serine sensors and has a matrix, but lacks enzymes. The signal from the null sensor was subtracted from the signal obtained on the ATP and D-serine sensor. This allows the contribution of any non-specific electroactive substances that bypass the sensor screening layer to be eliminated. Biosensors show a linear response to increasing concentration of ATP and D-Serine and have a rise time less than 10 s (Frenguelli et al., 2007). Biosensors were calibrated with known concentrations (10 μM) of ATP and D-Serine before the slice was present in the perfusion chamber and after the slice had been removed. This allowed compensation of any reduction in sensitivity during the experiment. The integrity of the screening layer was assessed with 10 μM 5-HT. Biosensor signals were acquired at 1 kHz with a 1400 CED interface and analyzed using Spike 6.1 software (Cambridge Electronics Design, Cambridge, UK).

Data Analysis

All data are presented as mean $\pm$ SD and the statistical significance of differences between data groups was tested by two-tailed unpaired t-test, unless indicated otherwise. For all cases of statistical significance reported, the statistical power of the test was 0.8–0.9. Each neocortical slice was used only for one experiment (e.g., fluorescent recordings in single astrocyte or single LTP induction experiment). The number of experiments/cells reported is therefore equal to the number of slices used. The experimental protocols were allocated randomly so the data in any group were drawn from at least from four animals, typically from 5 to 12 mice. The average ratio of experimental unit per animal was 1.3 for the LTP experiments and 1.5 for biosensor and fluorescent Ca^{2+}-measurements.

The spontaneous transmembrane currents recorded in neurons were analyzed offline using methods described previously (Lalo et al., 2011, 2016). The amplitude distributions of spontaneous and evoked currents were analyzed with the aid of probability density functions and likelihood maximization techniques; all histograms shown were calculated as probability density functions. The amplitude distributions were fitted with either multi-quantal binomial model or bi-modal function consisting of two Gaussians with variable peak location, width and amplitude. Parameters of models were fit using likelihood maximization routine.

RESULTS

Age- and Environment-Related Alterations in Astroglial Ca^{2+}-Signaling

Previously, we have reported the aging-related decline in the density of purinergic and glutamatergic receptors and their contribution to Ca^{2+}-signaling in neocortical astrocytes (Lalo et al., 2011, 2014b). In accordance with decrease in the Ca^{2+}-signaling, the exocytosis of gliotransmitters, such ATP, D-Serine

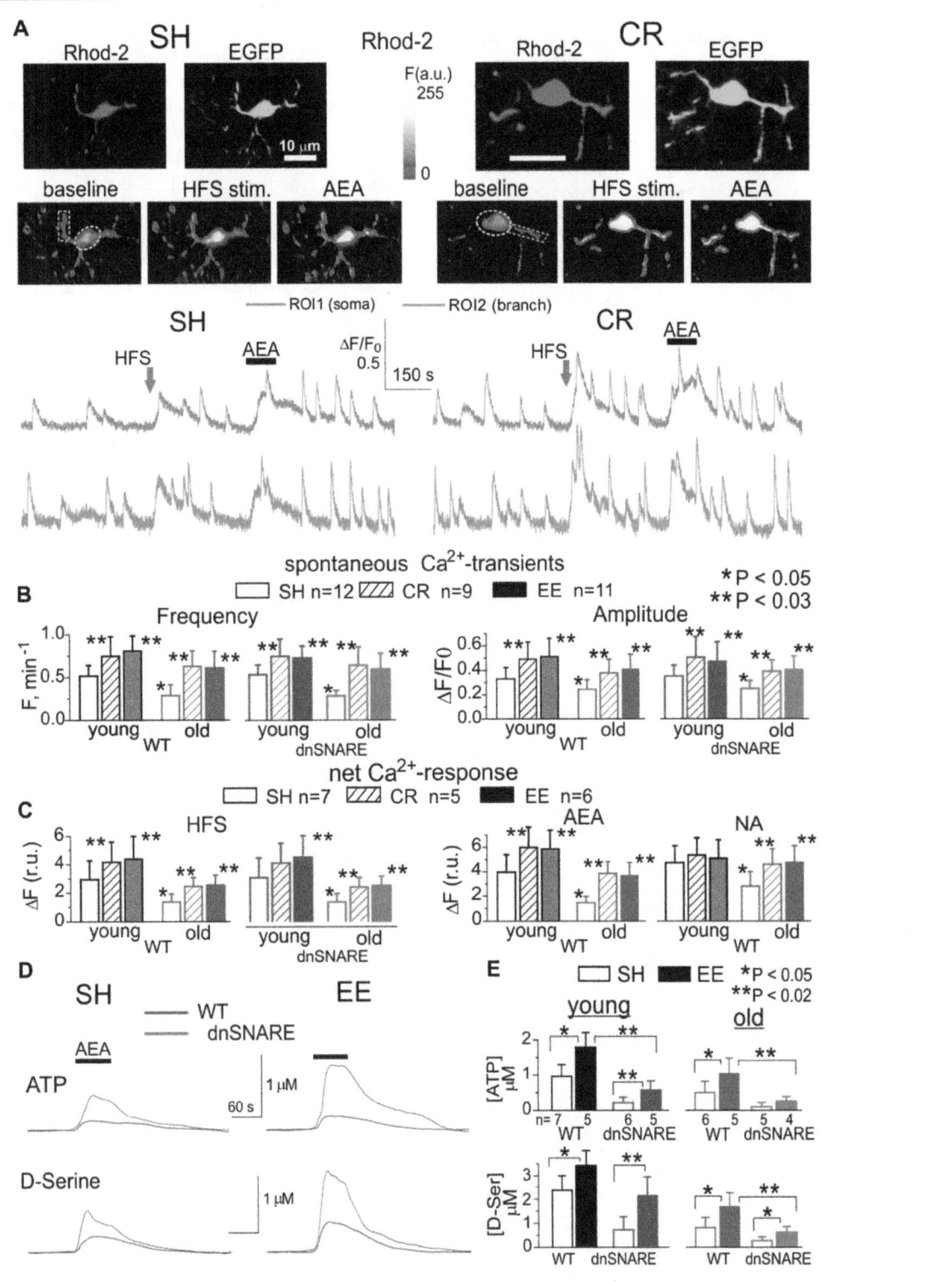

FIGURE 1 | Age- and experience-related changes in the adrenergic Ca^{2+} signaling and release of gliotransmitters in the neocortex. Astroglial Ca^{2+} -signaling **(A–C)** and release of ATP and D-Serine were evaluated in the neocortex of 2–5 month-old (young) and 14–18 month-old mice (old) as described previously in Lalo et al. (2014a), Rasooli-Nejad et al. (2014) and Pankratov and Lalo (2015). The dnSNARE mice and their wild-type (WT) littermates were kept either in standard housing (SH) or exposed to environmental enrichment (EE) or caloric restriction (CR) as described in the text. **(A)** Representative multi-photon images of EGFP fluorescence and presudo-color images of Rhod-2 fluorescence recorded in the astrocytes of old dn-SNARE mouse before and after the 100 ms-long episode of high-frequency stimulation (HFS) of cortical afferents and application of CB1 receptor agonist anandamide (AEA). Graphs below show the time course of Rhod-2 fluorescence

(Continued)

FIGURE 1 | Continued
averaged over regions indicated in the fluorescent images. Note the increase in the spontaneous in the Ca^{2+} -elevations and responses to HFS and application of AEA. **(B)** The pooled data on peak amplitude and frequency of the baseline spontaneous Ca^{2+} - transients recorded in astrocytes of WT and dn-SNARE mice of different age and treatment groups. Number and size of spontaneous events were pooled for the whole cell image. **(C)** The pooled data on the net responses to the HFS and application of AEA (500 nM) and noradrenaline (NA, 1 µM). Net response was evaluated as an integral Ca^{2+}-signal measured within 3 min after stimulation, averaged over the whole cell image and normalized to the baseline integral Ca^{2+} signal. Data in the panels **(B,C)** shown as mean ± SD for the number of cells indicated. Note the lack of the difference in the Ca^{2+}-signaling in the WT and dnSNARE mice. Asterisks (*,**) indicate statistical significance of the effect of EE- or CR-treatment (as compared to SH) and difference between the old and young mice of the same treatment group. **(D,E)** AEA-activated release of ATP and D-Serine in the neocortical slices of SH and EE mice was detected using microelectrode sensors as described previously (Lalo et al., 2014a,b; Rasooli-Nejad et al., 2014). **(E)** The representative responses of cortical slices of the old WT and dn-SNARE mice to the application of 500 nM AEA were recorded using microelectrode sensors to ATP and D-Serine placed in the layer II/III. The data are shown as an elevation relative to the resting concentration. **(E)** The pooled data on the peak magnitude of ATP- and D-Serine transients evoked by application of AEA; data shown as mean ± SD for number of experiments indicated. Asterisks (*,**) indicate statistical significance of difference in the magnitude of ATP- and D-Serine responses between WT and dn-SNARE mice (unpaired *t*-test) and SH- and EE-mice of similar genotype. The significant reduction in the AEA-evoked responses in the cortical slices from dn-SNARE mice strongly supports the vesicular mechanism of ATP and D-Serine release from astrocytes. Note the decrease in the ATP- and D-Serine transients in the old mice and EE-induced increase.

and Glutamate, from neocortical astrocytes also declined with aging (Lalo et al., 2011, 2014b; Lalo and Pankratov, 2017). In the present work, we tried to elucidate whether EE or CR can mitigate the negative effects of aging. We have focused primarily on receptors for NA and endocannabinoids since recent results highlighted the importance of these receptors for glia-neuron communications (Min and Nevian, 2012; Ding et al., 2013; Paukert et al., 2014; Metna-Laurent and Marsicano, 2015; Pankratov and Lalo, 2015; Oliveira da Cruz et al., 2016). In particular, our data (Rasooli-Nejad et al., 2014; Pankratov and Lalo, 2015) have shown that both α1-adrenoreceptors and CB1 receptors can trigger release of gliotransmitters from neocortical astrocytes.

We explored the difference in the spontaneous and synaptically-evoked cytosolic Ca^{2+}-transients in the neocortical layer 2/3 astrocytes of 2–4 months old (young adults) and 14–18 months old (old) WT and dnSNAREs mice. Astroglial Ca^{2+} signaling was monitored using multi-photon fluorescent microscopy as described previously (Lalo et al., 2014a; Rasooli-Nejad et al., 2014; Pankratov and Lalo, 2015). We compared animals kept under standard housing conditions (SH) vs. animals exposed to the EE from birth (Correa et al., 2012), including *ad libitum* access to running wheel, or kept on mild CR(CR) diet (food intake individually regulated to maintain the body weight loss of 10%–15%) for 4–6 weeks. We also assessed the impact of exogenous activation of adrenergic and eCB receptors on astroglial Ca^{2+} signaling under these conditions (**Figures 1A,B**).

There was no significant difference in the astroglial Ca^{2+}-signaling between dnSNARE mice and their WT littermates. In both WT and dnSNARE mice of SH, the amplitude and frequency of spontaneous Ca^{2+}-transients in the neocortical astrocytes underwent significant decrease with aging. The EE and CR had significant positive effect on the astroglial Ca^{2+}-signaling both in the young and old mice (**Figures 1B,C**). Interestingly, effects of EE and CR on the amplitude and frequency of Ca^{2+}-transients in the old age (60%–95% and 70%–110% correspondingly) were more profound than in the young mice (35%–50% and 30%–45%).

To probe the responses of astrocyte to the stimulation of neighboring synapses, we evaluated astrocytic Ca^{2+}-transients evoked by the short episode of high-frequency stimulation (HFS) of thalamo-cortical afferents, as described previously (Lalo et al., 2011, 2016; Rasooli-Nejad et al., 2014). The HFS-evoked Ca^{2+}-responses followed the similar trends as spontaneous astroglial activity: significant reduction in the old age was opposed by EE- and CR-induced increase (**Figure 1C**).

Similarly to our previous reports (Rasooli-Nejad et al., 2014; Pankratov and Lalo, 2015), activation of CB1 receptors by 500 nM anandamide (AEA) or α1-adrenoreceptors by 1 µM NA evoked profound elevation in the cytosolic Ca^{2+} concentration in the astrocytes of young but not old mice (**Figures 1A,D**). The EE and CR had a moderate effect on the amplitudes of NA- and AEA-evoked responses in the young mice but caused the significant enhancement of astroglial responses in the old mice (**Figure 1D**).

Apart from direct activation of Ca^{2+}-responses, astroglial CB1 receptors and α1-adrenoreceptors are capable to enhance the spontaneous signaling, as reported previously (Min and Nevian, 2012; Rasooli-Nejad et al., 2014; Pankratov and Lalo, 2015). Indeed, application of NA or AEA caused significant increase in the amplitude and frequency of spontaneous Ca^{2+}-transients in the astrocytes of mice of SH (**Figure 1C**). Rather surprisingly, the relative effects of both NA and AEA were much smaller in the EE-mice or CR-mice. This was most likely related to the higher baseline spontaneous activity in astrocytes of EE- and CR-mice as compared to SH.

To evaluate the age-related alterations in gliotransmission, we measured extracellular concentrations of ATP and D-Serine in the neocortical tissue of WT and dnSNARE mice with microelectrode biosensors and activated astrocytes with AEA. In the old age and environment groups, the CB1 receptor-activated elevations in the ATP and D-Serine concentrations were significantly reduced in the dnSNARE mice, supporting the previous data (Lalo et al., 2014a,b; Rasooli-Nejad et al., 2014) on important contribution of astroglial exocytosis in the release of these transmitters. Consistent with changes in the Ca^{2+}-signaling, release of ATP and D-Serine significantly decreased in the old age but was strongly up-regulated by the EE (**Figures 1D,E**). These results closely agree with our previous data on adrenoceptor-activated release of gliotransmitters (Lalo and Pankratov, 2017).

Our results suggest a major role of astrocytes in the release of D-Serine, which is in line with numerous previous

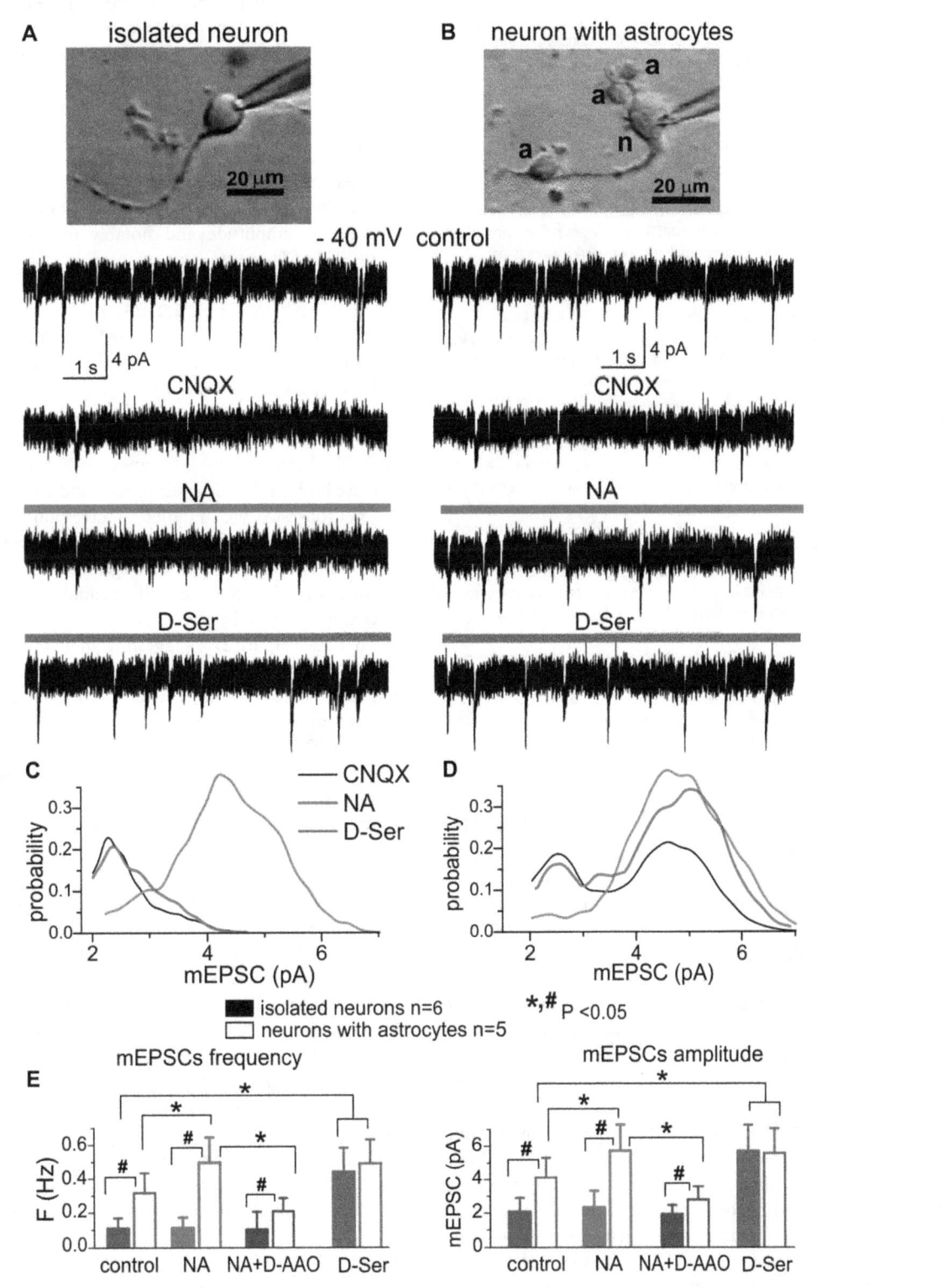

FIGURE 2 | The lack of efficient release of D-Serine from neurons. **(A,C)** Modulation of NMDA receptors by the D-Serine was evaluated in acutely isolated neocortical neurons retaining the functional glutamatergic synapses (Lalo et al., 2006; Lalo and Pankratov, 2017). **(B,D)** Similar recordings were made in the neuron attached to few astrocytes (marked as **"a"**) after dissociation of this neuron-astrocyte "bundle" from neocortex of WT mice; identification of astrocytes was confirmed by their electrophysiological properties after recordings (Lalo et al., 2014a; Rasooli-Nejad et al., 2014; Lalo and Pankratov, 2017). **(A,B)** Representative whole-cell currents recorded at −40 mV in the presence of picrotoxin, TTX, PPADS and 5-BDBD (control) before and after consecutive application of CNQX, (NA, 1 μM) and

(*Continued*)

exogenous D-Serine (10 μM) after washout of NA. The transient events which can be seen in the neurons **(A,B)** in the control are miniature excitatory synaptic currents, as demonstrated in Lalo et al. (2006) and Lalo and Pankratov (2017). Note that amplitude and frequency of the events recorded in the control in fully isolated neurone are similar to the dissociated neurone with attached astrocytes **(B)**. Inhibition of AMPA receptors with CNQX markedly suppressed the spontaneous activity in the isolated neuron as compared to its counterpart with attached astrocytes **(B)**. The CNQX–insensitive currents were completely eliminated by application D-APV in both cases confirming they were mediated by NMDARs (data are not shown). **(C,D)** The corresponding amplitude distributions (probability density functions) for the spontaneous NMDAR currents recorded in neurons shown in **(A,B)**. **(E)** Pooled data (mean ± SD for number of cells indicated) on the amplitude and frequency of NMDAR-mediated mEPSCs recorded in fully isolated neurons (open bars) and dissociated neurons with attached astrocytes (closed bars) at −40 mV in presence of CNQX (control) and after application of NA alone, NA in presence of D-amino acid oxidase (DAAO, 0.15 U/ml) or exogenous D-Serine. The statistical significance (un-paired t-test) of the difference between fully isolated neurons and neurons with astrocytes is indicated by the hash symbol (#); asterisks (*) indicate statistically significance of the effect of NA and D-Serine in comparison to control conditions (−40 mV, CNQX). Note that the amplitude and frequency of NMDAR-mediated mEPSCs in fully isolated neurons were much lower in comparison to the dissociated neuron with astrocytes **(A,E)**. However, mEPSCs in the isolated neurons were restored in the presence of exogenous D-Serine, as evidenced by appearance of events with larger amplitude **(C)**. The most straightforward explanation is that most NMDA receptors on the membrane of fully isolated neurons were not exposed to sufficient concentrations of co-agonist until exogenous D-Serine was applied. Application of NA led to increase in the number of mEPSCs with larger amplitudes **(D)** in the neurons with attached astrocytes but not in fully isolated neurons, suggesting that effect of NA on synaptic transmission is mediated by release of gliotransmitters. The effect of NA on the amplitude and frequency of mEPSCs was strongly attenuated by D-AAO, suggesting the involvement of D-Serine.

reports (Panatier et al., 2006; Henneberger et al., 2010; Sultan et al., 2015; Papouin et al., 2017b). Since the importance and specificity of glial release of D-Serine is hotly debated currently (Wolosker et al., 2016; Papouin et al., 2017b; Savtchouk and Volterra, 2018), we tried to directly assess contribution of neurons and astrocytes in the release of D-serine. For this purpose, we used acutely-isolated neocortical neurons which were devoid of the influence of glial cells (**Figure 2**). We used non-enzymatic vibro-dissociation which retains functional synapses on the dendrites of isolated neurons, which can be verified by the presence of miniature spontaneous synaptic currents (Duguid et al., 2007; Rasooli-Nejad et al., 2014; Lalo et al., 2016; Lalo and Pankratov, 2017). Apart from the pure isolated neurons, the vibro-dissociation technique allows, upon some adjustment, to dissociate neurons with few astrocytes attached. Such neuron-astrocyte "bundle," retaining a certain proportion of intimate contacts between astrocytic and neuronal membranes, could serve as a "minimalistic" model of glia-neuron interaction unit (Rasooli-Nejad et al., 2014; Lalo and Pankratov, 2017).

We recorded the NMDA receptor-mediated spontaneous currents (NMDAR mEPSCs) in the acutely-dissociated neocortical pyramidal neurons at membrane potential of −40 mV in the presence of 100 μM picrotoxin, 30 μM CNQX and 20 μM PPADS (**Figures 2A,B**). To compensate

for the putative depletion of neuronal D-Serine content due to intracellular perfusion, intracellular solution was supplemented with 1 mM D-Serine; intracellular concentration of EGTA was set to 0.1 mM. In the absence of external D-Serine or glycine, fully isolated neurons exhibited a low baseline frequency (0.11 ± 0.04 Hz, $n = 6$) of NMDAR-mediated mEPSCs and the application of NA did not cause notable changes in their amplitude or frequency (**Figures 2A,E**). Application of exogenous D-Serine dramatically increased the average amplitude and notably increased the frequency of mEPSCs. This was accompanied by the shift of mEPSCs amplitudes (**Figure 2C**) towards higher quantal size (from 2.3 ± 0.7 pA to 4.4 ± 1.2 pA, $n = 6$). In contrast to the fully isolated neurons, the NMDAR mEPSCs recorded in the "neuron-astrocyte bundles," could be observed at relative high frequency (0.32 ± 0.11 Hz) even in the absence of exogenous D-Serine (**Figure 2B**). They had bimodal amplitude distributions with peaks at 2.4 ± 0.8 pA and 4.7 ± 1.4 pA ($n = 5$). Activation of astrocytes by NA cause a significant increase both in the mean amplitude and frequency of mEPSCs in "neuron-astrocyte bundles" (correspondingly 49 ± 24% and 77 ± 38%, $n = 5$); which was accompanied by increase in the number of mEPSCs of larger quantal size (**Figure 2D**). Importantly, the NA-induced facilitation of mEPSCs was efficiently blocked by the application of exogenous D-amino acid oxidase (**Figure 2E**), suggesting an involvement of D-Serine in the action of NA. The effect of exogenous D-Serine on NMDAR mEPSCs in neurons with attached astrocytes did not differ considerably from the fully isolated neurons (**Figures 2D,E**).

The most parsimonious explanation of the above results would be a lack of D-Serine (or other co-agonists) in the vicinity of glutamatergic synapses devoid astroglial influence, which drives the amplitude of NMDAR mEPSCs below the threshold of detection. Activation of astrocytes by NA triggers release of D-serine, which diffuses to nearby synapses where it enhances the NMDAR mEPSCs. Application of exogenous D-Serine restores the amplitudes of NMDAR-mediated currents in majority of synapses. Our results argue against the predominant role of neurons in release of D-Serine (Wolosker et al., 2016) and agree with recent transcriptomic data on the expression of serine racemase (SRR) gene in cortical astrocytes (Chai et al., 2017) and in vivo data on the physiological role for glial-derived D-Serine (Papouin et al., 2017a,b).

Combined, the above results demonstrate that spontaneous and evoked Ca^{2+}-signaling in astrocytes and release of gliotransmitters can undergo significant decrease with aging but can be enhanced by EE or CR. It is also worth to note that astroglial CB1 and adrenergic receptors were capable to activate Ca^{2+}-elevation and release of gliotransmitters even in old mice of SH supporting the notion that glia-neuron communications do not stop with aging. Our data also show a common trend of EE and CR to enhance astroglial signaling mainly in case when it is "weakened," e.g., in the old mice and suggest an existence of "optimal" level of astroglial Ca^{2+}-signaling which may be reached in the younger age.

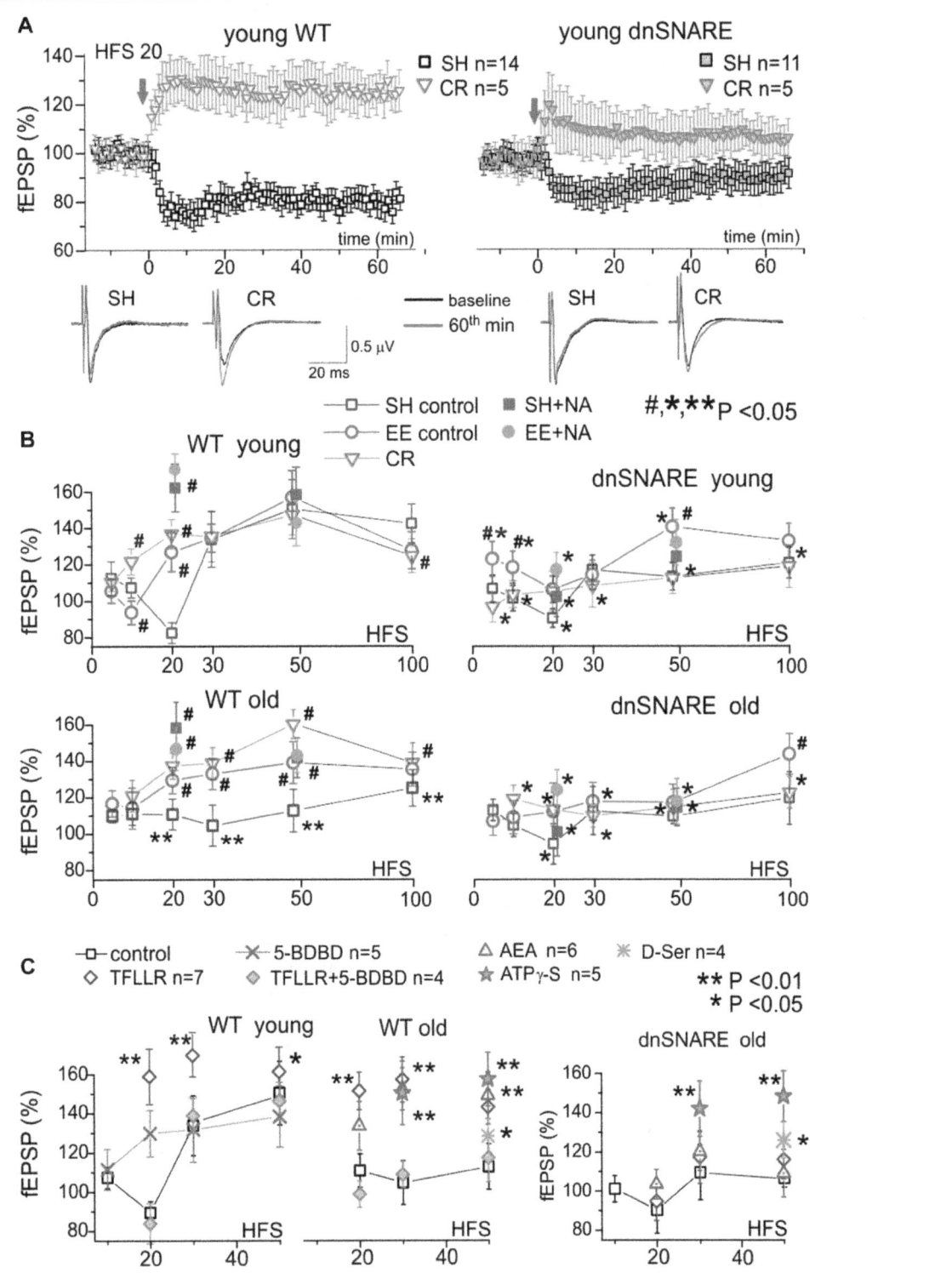

FIGURE 3 | Impact of gliotransmission on synaptic plasticity in the neocortex. Long-term potentiation (LTP) of field excitatory postsynaptic potentials (fEPSPs) was induced in the layer 2/3 of somatosensory cortex of the WT and dnSNARE mice by different number of pulses of high-frequency theta-burst stimulation (100 ms-long trains of 100 Hz pulses delivered with 200 ms intervals, every 10 trains were separated by 10 s-long intervals), as previously described (Lalo et al., 2014a, 2016; Rasooli-Nejad et al., 2014; Pankratov and Lalo, 2015). **(A)** The representative time course of changes in the fEPSP induced by 20 HFS trains delivered at zero time, measured in the young mice of SH and mice exposed to CR. Dots in the graphs represent the average of six consecutive fEPSPs; data are shown as mean ± SD for number of experiments indicated. Data were normalized to the fEPSP slope averaged over 10 min period prior to the HFS. The insets show

(Continued)

Impact of Astrocytes on Synaptic Plasticity in the Neocortex

Role for heterosynaptic cortical metaplasticity in brain computation, learning and memory is often studied in the framework of BCM-model (Bienenstock et al., 1982) whose two essential prerequisites are: bidirectional synaptic modification and sliding modification threshold. Accordingly, the general paradigm of BCM model, sub-threshold stimulation should induce long-term depression (LTD) whereas more stronger stimuli should induce potentiation (LTP) and prior stimulation (experience) can shift the LTD/LTP threshold one way or another (Bienenstock et al., 1982; Hulme et al., 2013b).

There is accumulating evidence of both positive and negative effects of astrocytes on the strength of excitatory synapses mediated by glia-derived ATP and adenosine (Hulme et al., 2013b; Boué-Grabot and Pankratov, 2017). Furthermore, our recent data demonstrated that glia-derived ATP can cause down-regulation of NMDA receptors and affect the threshold of LTP induction (Lalo et al., 2016). This effect contrasts with widely accepted positive effect of glia-derived NMDA receptor co-agonist D-Serine on LTP (Henneberger et al., 2010; Papouin et al., 2017a). Hence, one could expect the gliotransmission

to be important, if not quintessential, element of cortical metaplasticity. To test this hypothesis, we explored the stimulus-dependence of the LTP of fEPSPs in the neocortical layer 2/3 of WT and dnSNARE mice. The fEPSPs were evoked by the stimulation of neuronal afferents descending from layers IV-V (Rasooli-Nejad et al., 2014; Pankratov and Lalo, 2015; Lalo et al., 2016); LTP was induced by different number of pulses of high-frequency theta-burst stimulation (100 ms-long trains of 100 Hz pulses delivered with 200 ms intervals, every 10 trains were separated by 10 s-long intervals). Similarly to Ca^{2+}-signaling experiments, we compared effects of EE and CR on LTP in the WT and dnSNARE mice (**Figure 3**).

Under control conditions, the long-term plasticity of fEPSPs in the young WT mice exhibited a characteristic stimulus-strength dependence: no potentiation at very weak stimulus, depression at moderate strengths and then potentiation which magnitude grew sharply upon reaching certain threshold but decline slightly in a bell-shaped manner at very strong stimuli. Such dependence goes very well in line with predictions of the BCM theory.

Importantly, impairment of glial exocytosis in the dnSNARE mice dramatically "flattened" the stimulus-dependence inhibiting both the depression at moderate stimuli and potentiation at strong stimuli (**Figure 3B**). The similar "flattened" pattern of LTP/LTD dependence on stimulus strength was observed in the old mice (**Figure 3B**). Taking into account the significant age-related decrease in astroglial Ca^{2+}-signaling, one can suggest that decrease in the Ca^{2+}-dependent release of gliotransmitters (either aging-related or due to dnSNARE expression) strongly attenuates the extent of bi-directional modulation of plasticity of neocortical synapses.

This hypothesis was corroborated by the effects of activation of astrocytes by exogenously applied NA (1 μM) and AEA (200 nM). Previously, we have verified that effects of these agonists on synaptic transmission and plasticity are predominantly related to their ability to activate the astroglial Ca^{2+}-signaling and release of gliotransmitters (Lalo et al., 2014a, 2016; Rasooli-Nejad et al., 2014; Pankratov and Lalo, 2015). The 5 min-long application of NA (started 2 min prior to the induction protocol) significantly increased the magnitude of LTP induced in the neurons of younger WT mice by weak stimulation (20 HFS pulses) and the magnitude of LTP in older mice both at weak and strong (50 HFS pulses) stimuli. The NA-induced LTP enhancement was strongly attenuated in the dnSNARE mice supporting the importance of astroglial exocytosis for this effect. The similar action was shown by application of AEA or agonist of glia-specific PAR-1 receptors (TFLLR, 10 μM; **Figure 3C**).

These data closely agree with our previous results that by astroglial α1-ARs and CB1 receptors can facilitate LTP induction by triggering the release of ATP and D-Serine (Lalo et al., 2014a, 2016; Rasooli-Nejad et al., 2014; Pankratov and Lalo, 2015).

To verify that same molecular mechanisms can rescue the LTP in old mice, we investigated the effect of inhibition of ATP receptors and exogenous application of gliotransmitters (**Figure 3C**). First, selective inhibition of neuronal P2×4 receptors, which are not strongly expressed by neocortical astrocytes (Lalo et al., 2008), by 5-BDBD (5 μM) decreased both the magnitude and threshold of LTP; similar effects were recently observed in the P2×4 knock-out mice (Lalo et al., 2016). Second, inhibition of P2×4 receptors occluded the positive effects of AEA and TFLLR both in the young and old mice (**Figure 3C**). Finally, application of non-hydrolyzable ATP analog ATP-γS (10 µM) or D-Serine (10 µM) enhanced the LTP in the neocortex of both WT and dnSNARE mice.

Based on these results one could suggest that Ca^{2+} dependent release of ATP and D-Serine from astrocytes may be implicated in cortical metaplasticity by modifying both threshold and magnitude of long-term alterations of synaptic strength. Hence, aging-related decline in gliotransmission can lead to deficit in the LTP which could, in turn, be rescued by enhancement of astroglial signaling, for example by EE or CR (**Figure 1**). This notion was strongly supported by our observations of LTP modifications in the EE- and CR-exposed WT and dnSNARE mice (**Figure 3B**).

In the young WT mice, the main effect of EE and CR was a marked leftward shift in the threshold of LTP induction without increase in the maximal LTP amplitude (**Figure 3B**). So, both EE and CR caused statistically significant alterations in the LTP only at weaker stimuli (10 and 20 HSF). However, in the old WT mice, the EE- and CR-induced enhancement in astroglial signaling was accompanied by the leftward shift in the LTP induction threshold and significant increase in the magnitude of LTP at moderate and strong stimuli. Generally, the exposure of old mice to the EE and CR lead to a "younger" LTP phenotype. In the dnSNARE mice, both young and old, the positive action of CR on the LTP magnitude was effectively occluded strongly suggesting the crucial importance of release of gliotransmitters for CR-related metaplasticity. At the same time, dnSNARE expression had a less straightforward effect on the EE-induced modifications of LTP.

Although there was statistically significant difference in the LTP magnitude between WT and dnSNARE EE-exposed young mice, there also was significant EE-induce increase in the LTP in dnSNARE mice (as compared to SH mice) at most induction protocols. This rather surprising observation of "rescue" of LTP in the young dnSNARE mice by EE have two important implications. First, lack of LTP deficit in the EE dn-SNARE mice strongly argues against the non-selective "leaky" expression of dn-SNARE transgene in neurons. Second, the LTP deficit in the dn-SNARE mice kept in SH clearly demonstrated the physiological relevance of exocytotic release of gliotransmitters from astrocytes. The marked responsiveness of dn-SNARE mice to EE suggested the existence of some compensatory mechanisms which could overcome the deficit in vesicular gliotransmission. The most straightforward explanation might be a mosaic expression of dnSNARE transgene across the astrocyte population (Pascual et al., 2005; Sultan et al., 2015) and EE-induced enhancement of Ca^{2+}-signaling and the release of gliotransmitters from non-dnSNARE astrocytes. In the neocortex and hippocampus, just 50%–60% of astrocytes express dnSNARE (Halassa et al., 2009). Due to the presence of significant proportion of un-affected astrocytes in the dnSNARE mice, the EE-induced enhancement of Ca^{2+}-signaling can, very likely, lead to overall

increase of gliotransmitter release even in dnSNARE animals (albeit in a lesser extent than in their WT littermates). Combined with threshold-like behavior of LTP, EE-induced increase in gliotransmitter release can be capable to rescue the LTP in dnSNARE mice. Also, impact of EE on synaptic transmission may involve other mechanisms of neuronal origin, like BDNF/MSK1-dependent homeostatic synaptic scaling (Correa et al., 2012).

We would like to emphasize that the pattern of EE-induced alterations in stimulus-dependence of LTP was different in the young dnSNARE and WT mice. The increase in the LTP both at weak and strong stimuli in the dnSNARE mice contrasted to the alterations of the threshold, but not in the maximal magnitude, in their WT littermates. Furthermore, the EE was not very efficient in the old dnSNARE mice (**Figure 3B**), except at very strong stimulus (100 HFS) suggesting that role of glial exocytosis can increase with age. Hence, one way or another, our data strongly suggest the involvement of glial exocytosis in the EE-induced metaplasticity.

Interestingly, the EE modified the effect of additional action of astrocytes on the LTP. Application of NA caused significant enhancement of LTP in the neocortex of WT mice only at weaker stimulus (20 HFS pulses) but did not have notable effect on the LTP induced by stronger stimulus (50 HFS) in both age groups. One could suggest that exposure to EE, causing increase in astrocytic Ca^{2+}-signaling and gliotransmission, "pre-empted" the effect of activation of adrenoceptors.

Taken together, the above results strongly support the importance of astroglial Ca^{2+}-signaling and gliotransmission for the bi-directional modulation of synaptic plasticity in the neocortex. Our data also suggest an importance of astroglial eCB and NA receptors for metaplasticity-related effects of aging, EE and CR.

DISCUSSION

During last two decades a plethora of experimental results has been obtained showing that various gliotransmitters can exert different, and sometimes opposing, effects on neuronal signaling (Gordon et al., 2009; Araque et al., 2014; Pougnet et al., 2014; Bazargani and Attwell, 2016; Boué-Grabot and Pankratov, 2017). In particular, release of glutamate and D-serine from astrocytes can enhance the activity of NMDA component of excitatory synapses and thereby the long-term synaptic plasticity (Panatier et al., 2006; Henneberger et al., 2010; Rasooli-Nejad et al., 2014). There is also an evidence of the positive impact of glia-derived ATP on activity and trafficking of AMPA receptors (Boué-Grabot and Pankratov, 2017). Our previous data showed that glia-derived ATP can down-regulate the GABAergic tonic and phasic inhibitory transmission (Lalo et al., 2014a). Hence, one might expect the glia-derived ATP to shift the balance between excitatory and inhibitory synaptic inputs towards excitation and thereby facilitate the induction of LTP. At the same time, our recent data have shown that ATP, acting via postsynaptic P2X receptors, can down-regulate NMDA receptors and decrease the magnitude of LTP (Lalo et al., 2016). Thus, release of ATP and D-Serine from astrocytes can lead to bi-directional modifications of synaptic strength and thereby play an important role in the mechanisms of brain metaplasticity. Our results present several lines of evidence to support this notion.

We observed that positive effects of EE and CR can counterbalance the aging-related decrease in the Ca^{2+}-signaling and release of gliotransmitters from neocortical astrocytes. The EE- and CR-induced enhancement in astroglial function was accompanied by marked alterations in the stimulus-dependence of the neocortical LTP. These alterations different considerably in the WT and dnSNARE mice supporting the importance of release of gliotransmitters for the beneficial effects of EE and CR on synaptic plasticity in the old age. The EE- and CR-induced enhancement of the LTP in old mice could also be mimicked, irrespective of dnSNARE expression, by short-term application of ATP analog (ATPγS) and D-Serine during LTP induction period, indicating the importance of these gliotransmitters.

Our results also show that additional activation of astrocytes via eCB, adrenergic or PAR-1 receptors during LTP induction protocol can increase the magnitude of the resulting LTP. Importantly, such activation of astrocytes caused significant effect only in cases when LTP magnitude was small, i.e., at weaker stimuli in the younger mice of SH or at stronger stimuli in the old mice. The α1-AR mediated facilitation of LTP was also occluded by the treatments (EE and CR) which already increased astrocytic Ca^{2+}-signaling and LTP.

Combined, our results suggest that activation of astrocytes performs a "balancing act," facilitating the LTP in case of deficit in signaling and gliotransmission but attenuating the LTP magnitude in case of stronger stimuli and higher levels of astroglial activity. A strong dependence of glia-driven modulation, of both the threshold and magnitude of LTP, on the activity of neuronal P2×4 receptors (**Figure 3**, see also Lalo et al., 2016) may imply an important role for ATP receptor-mediated down-regulation of GABA and NMDA receptors in the mechanisms of metaplasticity.

Overall, our observations of LTP modifications in the EE- and CR-exposed WT and dnSNARE mice go in line with our hypothesis that Ca^{2+} dependent release of ATP and D-Serine from astrocytes can affect both threshold and magnitude of long-term alterations of synaptic strength and thereby can be very important for cortical metaplasticity. Our data also indicate that modulation of neocortical plasticity, in particular via glia-neuron communications, can be analyzed in the framework of BCM theory.

It becomes increasingly evident that net effect of activation of astroglial Ca^{2+}-signaling on firing rate or plasticity of neural networks cannot be a priori considered as solely positive (or negative). This might explain why some attempts to genetically modify Ca^{2+} signaling in astrocytes did not yield the expected results (Bazargani and Attwell, 2016; Fiacco and McCarthy, 2018; Savtchouk and Volterra, 2018). Our data also suggest that release of gliotransmitters, in particular ATP and D-Serine, can be affected by environment, physical exercise and diet, in addition to the recently reported effects of sleep cycle (Papouin et al., 2017a). A high degree of experience- and environment-related plasticity

of astroglial function may further add to the variability of the results obtained in different laboratory settings.

Our results, in particular shown in **Figure 2**, can also be important for the current debate on mechanisms D-Serine release in the brain. Although a large body of evidence, obtained in the last decade using different experimental approaches supports an importance of astroglia in the release of D-Serine and in the D-Serine-mediated modulation of NMDA receptors, an alternative (and rather extreme) view of predominantly neuronal release of D-Serine at physiological conditions has been recently suggested (Wolosker et al., 2016). One should note that notion of predomimant role of neurons in release of D-Serine originates mainly from the data suggesting a high level of expression of SRR in neurons rather than astroglia. However, this contradicts to recent transcriptomic data showing relatively high level of Srr gene expression in cortical and hippocampal astrocytes (Chai et al., 2017). There are several other flaws in the "neuronal D-Serine" theory which have been thoroughly addressed by Papouin et al. (2017b).

One should emphasize that even the studies, questioning astroglia-specific expression of serine racemase, do not deny an ability of astrocytes to accumulate D-serine (Wolosker et al., 2016) and a reconciliating theory of cooperation between neuronal and glial release has been recently suggested by Ivanov and Mothet (2018). Whatever is concentration of D-Serine in neurons and astrocytes, their contribution into extracellular D-serine level will depend mainly on efficiency of neuronal vs. glial mechanisms of release. So far, there is a lack direct evidence that neurons possess an efficient mechanism of release of D-serine (Papouin et al., 2017b), with Asc-1 being suggest to be a main pathway (Papouin et al., 2017b). Release of D-serine from strocytes can occur via Ca^{2+}-dependent exocytosis, as an alternative pathway one might also suggest Ca^{2+}-dependent large conductance chloride channels (Woo et al., 2012). These channels have been shown to be permeable to glutamate (Woo et al., 2012) which has even larger molecular size than D-Serine.

Compared to vesicular or channel-mediated release allowing the movement of multiple molecules, transporter, that releases a single molecule per single act of conformational change, is intrinsically slow. So, one could hardly expect the putative Asc-1 transporter-mediated release of D-Serine from neurons to be more than Ca^{2+}-dependent release of D-Serine from astrocytes.

Indeed, our experiments in isolated neurons and neuron-astrocyte "bundles," which allowed to dis-entangle glial and neuronal sources of D-Serine, showed that neuronal release on its own could not provide enough D-Serine to maintain the activity of synaptic NMDAR (**Figure 2**). We would like to stress that our preparation of isolated neurons provide a very good accessibility of synapses to D-amino acid oxidase, however we did not observe any marked effect of DAAO on synaptic NMDAR-mediated currents in conditions when neurons were perfused with 1 mM D-Serine via whole-cell recording pipette. This observation on its own argues against efficient release of D-Serine from neurons. Combined with results of biosensor measurements of D-serine release in brain slices of WT and dnSNARE mice (**Figure 1D**, see also Rasooli-Nejad et al., 2014), our data strongly support an importance of astrocytes as a source of D-serine at physiological conditions.

Our data on age-related changes in astroglial Ca^{2+}-signaling and release of ATP and D-Serine (**Figure 1**) go in line with accumulating evidence that alterations in extracellular levels of ATP, adenosine and D-Serine can play an important role in physiological and pathological brain aging (Mothet et al., 2006; Orellana et al., 2012; Gundersen et al., 2015; Rodrigues et al., 2015; Piacentini et al., 2017). Our present (**Figures 2**, **3**) and previous data (Lalo et al., 2014b, 2016; Pankratov and Lalo, 2015) on astroglial-driven modulation of synaptic transmission and plasticity supports the recently emerged view that molecular and functional alteration in astrocytes, such as release of gliotransmitters, can precede or even cause, changes in synaptic dynamics and homeostasis during aging and progression of neurodegenerative diseases (De Strooper and Karran, 2016; Soreq et al., 2017). Our results also suggest that widely reported beneficial effects (Hillman et al., 2008; Nithianantharajah and Hannan, 2009; van Praag, 2009; Mercken et al., 2012; Merzenich et al., 2014) of enriched environment on synaptic plasticity and memory in mice and, possibly, effects of active life style in humans, can be mediated by enhancement of release of gliotransmitters, very likely due to increased endocannabinoid and adrenergic Ca^{2+}-signaling in astrocytes. Since both EE and CR upregulated Ca^{2+}signaling in astrocytes and affected synaptic plasticity in the dnSNARE-dependent manner, one can hypothesize that beneficial effects of CR on synaptic plasticity also originate from enhanced gliotransmission and glia-neuron communications. Still, alternative mechanisms of action of CR might be suggested, which can be underlined by various metabolic effects of CR (Mercken et al., 2012; Park et al., 2012; Madeo et al., 2014; López-Otín et al., 2016), for example enhancement mitochondrial function in astrocytes. Specific roles for gliotransmission- and metabolism-related mechanisms in beneficial effects of CR on aging brain are yet to be studied.

To conclude, our results strongly support physiological importance of astroglial cannabinoid and adrenergic signaling and glia-derived ATP for communication between astrocytes and neurons and experience-related modulation of synaptic plasticity across a lifetime.

AUTHOR CONTRIBUTIONS

UL, AB and YP contributed to the design and interpretation of experiments and commented on the manuscript. UL and AB performed experiments and data analysis. YP and UL conceived the study and wrote the manuscript.

ACKNOWLEDGMENTS

We thank Dr. Mark Wall for help with biosensor experiments.

REFERENCES

Abraham, W. C., and Bear, M. F. (1996). Metaplasticity: the plasticity of synaptic plasticity. *Trends Neurosci.* 19, 126–130. doi: 10.1016/s0166-2236(96)80018-x

Araque, A., Carmignoto, G., Haydon, P. G., Oliet, S. H., Robitaille, R., and Volterra, A. (2014). Gliotransmitters travel in time and space. *Neuron* 81, 728–739. doi: 10.1016/j.neuron.2014.02.007

Bazargani, N., and Attwell, D. (2016). Astrocyte calcium signaling: the third wave. *Nat. Neurosci.* 19, 182–189. doi: 10.1038/nn.4201

Bienenstock, E. L., Cooper, L. N., and Munro, P. W. (1982). Theory for the development of neuron selectivity: orientation specificity and binocular interaction in visual cortex. *J. Neurosci.* 2, 32–48. doi: 10.1523/JNEUROSCI.02-01-00032.1982

Boué-Grabot, E., and Pankratov, Y. (2017). Modulation of central synapses by astrocyte-released ATP and postsynaptic P2X receptors. *Neural Plast.* 2017:9454275. doi: 10.1155/2017/9454275

Chai, H., Diaz-Castro, B., Shigetomi, E., Monte, E., Octeau, J. C., Yu, X., et al. (2017). Neural circuit-specialized astrocytes: transcriptomic, proteomic, morphological, and functional evidence. *Neuron* 95, 531.e9–549.e9. doi: 10.1016/j.neuron.2017.06.029

Correa, S. A., Hunter, C. J., Palygin, O., Wauters, S. C., Martin, K. J., McKenzie, C., et al. (2012). MSK1 regulates homeostatic and experience-dependent synaptic plasticity. *J. Neurosci.* 32, 13039–13051. doi: 10.1523/JNEUROSCI.0930-12.2012

De Strooper, B., and Karran, E. (2016). The cellular phase of Alzheimer's disease. *Cell* 164, 603–615. doi: 10.1016/j.cell.2015.12.056

Ding, F., O'Donnell, J., Thrane, A. S., Zeppenfeld, D., Kang, H., Xie, L., et al. (2013). α1-Adrenergic receptors mediate coordinated Ca^{2+} signaling of cortical astrocytes in awake, behaving mice. *Cell Calcium* 54, 387–394. doi: 10.1016/j.ceca.2013.09.001

Duguid, I. C., Pankratov, Y., Moss, G. W., and Smart, T. G. (2007). Somatodendritic release of glutamate regulates synaptic inhibition in cerebellar Purkinje cells via autocrine mGluR1 activation. *J. Neurosci.* 27, 12464–12474. doi: 10.1523/JNEUROSCI.0178-07.2007

Fiacco, T. A., and McCarthy, K. D. (2018). Multiple lines of evidence indicate that gliotransmission does not occur under physiological conditions. *J. Neurosci.* 38, 3–13. doi: 10.1523/JNEUROSCI.0016-17.2017

Frenguelli, B. G., Wigmore, G., Llaudet, E., and Dale, N. (2007). Temporal and mechanistic dissociation of ATP and adenosine release during ischaemia in the mammalian hippocampus. *J. Neurochem.* 101, 1400–1413. doi: 10.1111/j.1471-4159.2007.04425.x

Gordon, G. R., Iremonger, K. J., Kantevari, S., Ellis-Davies, G. C., MacVicar, B. A., and Bains, J. S. (2009). Astrocyte-mediated distributed plasticity at hypothalamic glutamate synapses. *Neuron* 64, 391–403. doi: 10.1016/j.neuron.2009.10.021

Gourine, A. V., Kasymov, V., Marina, N., Tang, F., Figueiredo, M. F., Lane, S., et al. (2010). Astrocytes control breathing through pH-dependent release of ATP. *Science* 329, 571–575. doi: 10.1126/science.1190721

Gundersen, V., Storm-Mathisen, J., and Bergersen, L. H. (2015). Neuroglial transmission. *Physiol. Rev.* 95, 695–726. doi: 10.1152/physrev.00024.2014

Halassa, M. M., Florian, C., Fellin, T., Munoz, J. R., Lee, S. Y., Abel, T., et al. (2009). Astrocytic modulation of sleep homeostasis and cognitive consequences of sleep loss. *Neuron* 61, 213–219. doi: 10.1016/j.neuron.2008.11.024

Halassa, M. M., and Haydon, P. G. (2010). Integrated brain circuits: astrocytic networks modulate neuronal activity and behavior. *Annu. Rev. Physiol.* 72, 335–355. doi: 10.1146/annurev-physiol-021909-135843

Henneberger, C., Papouin, T., Oliet, S. H., and Rusakov, D. A. (2010). Long-term potentiation depends on release of D-serine from astrocytes. *Nature* 463, 232–236. doi: 10.1038/nature08673

Hillman, C. H., Erickson, K. I., and Kramer, A. F. (2008). Be smart, exercise your heart: exercise effects on brain and cognition. *Nat. Rev. Neurosci.* 9, 58–65. doi: 10.1038/nrn2298

Hulme, S. R., Jones, O. D., and Abraham, W. C. (2013a). Emerging roles of metaplasticity in behaviour and disease. *Trends Neurosci.* 36, 353–362. doi: 10.1016/j.tins.2013.03.007

Hulme, S. R., Jones, O. D., Raymond, C. R., Sah, P., and Abraham, W. C. (2013b). Mechanisms of heterosynaptic metaplasticity. *Philos. Trans. R. Soc. Lond. B Biol. Sci.* 369:20130148. doi: 10.1098/rstb.2013.0148

Ivanov, A. D., and Mothet, J. P. (2018). The plastic D-serine signaling pathway: sliding from neurons to glia and vice-versa. *Neurosci. Lett.* doi: 10.1016/j.neulet. 2018.05.039 [Epub ahead of print].

Lalo, U., Palygin, O., North, R. A., Verkhratsky, A., and Pankratov, Y. (2011). Age-dependent remodelling of ionotropic signalling in cortical astroglia. *Aging Cell* 10, 392–402. doi: 10.1111/j.1474-9726.2011.00682.x

Lalo, U., Palygin, O., Rasooli-Nejad, S., Andrew, J., Haydon, P. G., and Pankratov, Y. (2014a). Exocytosis of ATP from astrocytes modulates phasic and tonic inhibition in the neocortex. *PLoS Biol.* 12:e1001747. doi: 10.1371/journal. pbio.1001857

Lalo, U., Rasooli-Nejad, S., and Pankratov, Y. (2014b). Exocytosis of gliotransmitters from cortical astrocytes: implications for synaptic plasticity and aging. *Biochem. Soc. Trans.* 42, 1275–1281. doi: 10.1042/BST20140163

Lalo, U., Palygin, O., Verkhratsky, A., Grant, S. G., and Pankratov, Y. (2016). ATP from synaptic terminals and astrocytes regulates NMDA receptors and synaptic plasticity through PSD-95 multi-protein complex. *Sci. Rep.* 6:33609. doi: 10.1038/srep33609

Lalo, U., and Pankratov, Y. (2017). Exploring the Ca^{2+}-dependent synaptic dynamics *in vibro*-dissociated cells. *Cell Calcium* 64, 91–101. doi: 10.1016/j. ceca.2017.01.008

Lalo, U., Pankratov, Y., Kirchhoff, F., North, R. A., and Verkhratsky, A. (2006). NMDA receptors mediate neuron-to-glia signaling in mouse cortical astrocytes. *J. Neurosci.* 26, 2673–2683. doi: 10.1523/JNEUROSCI.4689-05.2006

Lalo, U., Pankratov, Y., Wichert, S. P., Rossner, M. J., North, R. A., Kirchhoff, F., et al. (2008). P2X1 and P2X5 subunits form the functional P2X receptor in mouse cortical astrocytes. *J. Neurosci.* 28, 5473–5480. doi: 10.1523/JNEUROSCI.1149-08.2008

Lin, A. L., Coman, D., Jiang, L., Rothman, D. L., and Hyder, F. (2014). Caloric restriction impedes age-related decline of mitochondrial function and neuronal activity. *J. Cereb. Blood Flow Metab.* 34, 1440–1443. doi: 10.1038/jcbfm. 2014.114

López-Otín, C., Galluzzi, L., Freije, J. M., Madeo, F., and Kroemer, G. (2016). Metabolic control of longevity. *Cell* 166, 802–821. doi: 10.1016/j.cell. 2016.07.031

Madeo, F., Pietrocola, F., Eisenberg, T., and Kroemer, G. (2014). Caloric restriction mimetics: towards a molecular definition. *Nat. Rev. Drug Discov.* 13, 727–740. doi: 10.1038/nrd4391

Mercken, E. M., Carboneau, B. A., Krzysik-Walker, S. M., and de Cabo, R. (2012). Of mice and men: the benefits of caloric restriction, exercise, and mimetics. *Ageing Res. Rev.* 11, 390–398. doi: 10.1016/j.arr.2011. 11.005

Merzenich, M. M., Van Vleet, T. M., and Nahum, M. (2014). Brain plasticity based therapeutics. *Front. Hum. Neurosci.* 8:385. doi: 10.3389/fnhum.2014.00385

Metna-Laurent, M., and Marsicano, G. (2015). Rising stars: modulation of brain functions by astroglial type-1 cannabinoid receptors. *Glia* 63, 353–364. doi: 10.1002/glia.22773

Min, R., and Nevian, T. (2012). Astrocyte signaling controls spike timing-dependent depression at neocortical synapses. *Nat. Neurosci.* 15, 746–753. doi: 10.1038/nn.3075

Monai, H., Ohkura, M., Tanaka, M., Oe, Y., Konno, A., Hirai, H., et al. (2016). Calcium imaging reveals glial involvement in transcranial direct current stimulation-induced plasticity in mouse brain. *Nat. Commun.* 7:11100. doi: 10.1038/ncomms11100

Mothet, J. P., Rouaud, E., Sinet, P. M., Potier, B., Jouvenceau, A., Dutar, P., et al. (2006). A critical role for the glial-derived neuromodulator D-serine in the age-related deficits of cellular mechanisms of learning and memory. *Aging Cell* 5, 267–274. doi: 10.1111/j.1474-9726.2006.00216.x

Nithianantharajah, J., and Hannan, A. J. (2009). The neurobiology of brain and cognitive reserve: mental and physical activity as modulators of brain disorders. *Prog. Neurobiol.* 89, 369–382. doi: 10.1016/j.pneurobio.2009.10.001

Oliveira da Cruz, J. F., Robin, L. M., Drago, F., Marsicano, G., and Metna-Laurent, M. (2016). Astroglial type-1 cannabinoid receptor (CB1): a new player in the tripartite synapse. *Neuroscience* 323, 35–42. doi: 10.1016/j.neuroscience. 2015.05.002

Orellana, J. A., von Bernhardi, R., Giaume, C., and Saez, J. C. (2012). Glial hemichannels and their involvement in aging and neurodegenerative diseases. *Rev. Neurosci.* 23, 163–177. doi: 10.1515/revneuro-2011-0065

Panatier, A., Theodosis, D. T., Mothet, J. P., Touquet, B., Pollegioni, L., Poulain, D. A., et al. (2006). Glia-derived D-serine controls NMDA receptor activity and synaptic memory. *Cell* 125, 775–784. doi: 10.1016/j.cell.2006. 02.051

Pankratov, Y., and Lalo, U. (2015). Role for astroglial α1-adrenoreceptors in gliotransmission and control of synaptic plasticity in the neocortex. *Front. Cell. Neurosci.* 9:230. doi: 10.3389/fncel.2015.00230

Papouin, T., Dunphy, J. M., Tolman, M., Dineley, K. T., and Haydon, P. G. (2017a). Septal cholinergic neuromodulation tunes the astrocyte-dependent gating of hippocampal NMDA receptors to wakefulness. *Neuron* 94, 840.e7–854.e7. doi: 10.1016/j.neuron.2017.04.021

Papouin, T., Henneberger, C., Rusakov, D. A., and Oliet, S. H. R. (2017b). Astroglial versus neuronal D-serine: fact checking. *Trends Neurosci.* 40, 517–520. doi: 10.1016/j.tins.2017.05.007

Park, S. J., Ahmad, F., Philp, A., Baar, K., Williams, T., Luo, H., et al. (2012). Resveratrol ameliorates aging-related metabolic phenotypes by inhibiting cAMP phosphodiesterases. *Cell* 148, 421–433. doi: 10.1016/j.cell.2012.01.017

Pascual, O., Casper, K. B., Kubera, C., Zhang, J., Revilla-Sanchez, R., Sul, J. Y., et al. (2005). Astrocytic purinergic signaling coordinates synaptic networks. *Science* 310, 113–116. doi: 10.1126/science.1116916

Paukert, M., Agarwal, A., Cha, J., Doze, V. A., Kang, J. U., and Bergles, D. E. (2014). Norepinephrine controls astroglial responsiveness to local circuit activity. *Neuron* 82, 1263–1270. doi: 10.1016/j.neuron.2014.04.038

Piacentini, R., Li Puma, D. D., Mainardi, M., Lazzarino, G., Tavazzi, B., Arancio, O., et al. (2017). Reduced gliotransmitter release from astrocytes mediates tau-induced synaptic dysfunction in cultured hippocampal neurons. *Glia* 65, 1302–1316. doi: 10.1002/glia.23163

Pougnet, J. T., Toulme, E., Martinez, A., Choquet, D., Hosy, E., and Boué-Grabot, E. (2014). ATP P2X receptors downregulate AMPA receptor trafficking and postsynaptic efficacy in hippocampal neurons. *Neuron* 83, 417–430. doi: 10.1016/j.neuron.2014.06.005

Rasooli-Nejad, S., Palygin, O., Lalo, U., and Pankratov, Y. (2014). Cannabinoid receptors contribute to astroglial Ca^{2+}-signalling and control of synaptic plasticity in the neocortex. *Philos. Trans. R. Soc. Lond. B Biol. Sci.* 369:20140077. doi: 10.1098/rstb.2014.0077

Rodrigues, R. J., Tome, A. R., and Cunha, R. A. (2015). ATP as a multi-target danger signal in the brain. *Front. Neurosci.* 9:148. doi: 10.3389/fnins.2015. 00148

Rodríguez, J. J., Terzieva, S., Olabarria, M., Lanza, R. G., and Verkhratsky, A. (2013). Enriched environment and physical activity reverse astrogliodegeneration in the hippocampus of AD transgenic mice. *Cell Death Dis.* 4:e678. doi: 10.1038/cddis.2013.194

Rodríguez, J. J., Yeh, C. Y., Terzieva, S., Olabarria, M., Kulijewicz-Nawrot, M., and Verkhratsky, A. (2014). Complex and region-specific changes in astroglial markers in the aging brain. *Neurobiol. Aging* 35, 15–23. doi: 10.1016/j. neurobiolaging.2013.07.002

Savtchouk, I., and Volterra, A. (2018). Gliotransmission: beyond black-and-white. *J. Neurosci.* 38, 14–25. doi: 10.1523/JNEUROSCI.0017-17.2017

Seidel, B., Bigl, M., Franke, H., Kittner, H., Kiess, W., Illes, P., et al. (2006). Expression of purinergic receptors in the hypothalamus of the rat is modified by reduced food availability. *Brain Res.* 1089, 143–152. doi: 10.1016/j.brainres. 2006.03.038

Singh, A., and Abraham, W. C. (2017). Astrocytes and synaptic plasticity in health and disease. *Exp. Brain Res.* 235, 1645–1655. doi: 10.1007/s00221-017-4928-1

Soreq, L., UK Brain Expression Consortium, North American Brain Expression Consortium, Rose, J., Soreq, E., Hardy, J., et al. (2017). Major shifts in glial regional identity are a transcriptional hallmark of human brain aging. *Cell Rep.* 18, 557–570. doi: 10.1016/j.celrep.2016.12.011

Sultan, S., Li, L., Moss, J., Petrelli, F., Cassé, F., Gebara, E., et al. (2015). Synaptic integration of adult-born hippocampal neurons is locally controlled by astrocytes. *Neuron* 88, 957–972. doi: 10.1016/j.neuron.2015. 10.037

van Praag, H. (2009). Exercise and the brain: something to chew on. *Trends Neurosci.* 32, 283–290. doi: 10.1016/j.tins.2008.12.007

Verkhratsky, A., Rodríguez, J. J., and Parpura, V. (2014). Neuroglia in ageing and disease. *Cell Tissue Res.* 357, 493–503. doi: 10.1007/s00441-014-1814-z

Verkhratsky, A., Rodríguez-Arellano, J. J., Parpura, V., and Zorec, R. (2017). Astroglial calcium signalling in Alzheimer's disease. *Biochem. Biophys. Res. Commun.* 483, 1005–1012. doi: 10.1016/j.bbrc.2016.08.088

Wolosker, H., Balu, D. T., and Coyle, J. T. (2016). The rise and fall of the d-serine-mediated gliotransmission hypothesis. *Trends Neurosci.* 39, 712–721. doi: 10.1016/j.tins.2016.09.007

Woo, D. H., Han, K. S., Shim, J. W., Yoon, B. E., Kim, E., Bae, J. Y., et al. (2012). TREK-1 and Best1 channels mediate fast and slow glutamate release in astrocytes upon GPCR activation. *Cell* 151, 25–40. doi: 10.1016/j.cell.2012.09.005

12

In Utero Administration of Drugs Targeting Microglia Improves the Neurodevelopmental Outcome Following Cytomegalovirus Infection of the Rat Fetal Brain

Robin Cloarec [1,2†], Sylvian Bauer [1†], Natacha Teissier [3,4], Fabienne Schaller [1,5],
Hervé Luche [6], Sandra Courtens [1], Manal Salmi [1], Vanessa Pauly [7], Emilie Bois [3,4],
Emilie Pallesi-Pocachard [1,8], Emmanuelle Buhler [1,5], François J. Michel [1,9],
Pierre Gressens [3,4], Marie Malissen [6], Thomas Stamminger [10], Daniel N. Streblow [11],
Nadine Bruneau [1] and Pierre Szepetowski [1*]

[1] INMED, French National Institute of Health and Medical Research INSERM U1249, Aix-Marseille University, Marseille, France, [2] Neurochlore, Marseille, France, [3] French National Institute of Health and Medical Research INSERM U1141, Paris Diderot University, Sorbonne Paris Cité, Paris, France, [4] PremUP, Paris, France, [5] PPGI Platform, INMED, Marseille, France, [6] Centre National de la Recherche Scientifique CNRS UMS3367, CIPHE (Centre D'Immunophénomique), French National Institute of Health and Medical Research INSERM US012, PHENOMIN, Aix-Marseille University, Marseille, France, [7] Laboratoire de Santé Publique EA 3279, Faculté de Médecine Centre d'Evaluation de la Pharmacodépendance-Addictovigilance de Marseille (PACA-Corse) Associé, Aix-Marseille University, Marseille, France, [8] PBMC platform, INMED, Marseille, France, [9] InMAGIC platform, INMED, Marseille, France, [10] Institute for Clinical and Molecular Virology, University of Erlangen-Nuremberg, Erlangen, Germany, [11] Vaccine and Gene Therapy Institute, Oregon Health and Science University, Portland, OR, United States

*Correspondence:
Pierre Szepetowski
pierre.szepetowski@inserm.fr

[†] These authors have contributed equally to this work.

Congenital cytomegalovirus (CMV) infections represent one leading cause of neurodevelopmental disorders. Recently, we reported on a rat model of CMV infection of the developing brain in utero, characterized by early and prominent infection and alteration of microglia—the brain-resident mononuclear phagocytes. Besides their canonical function against pathogens, microglia are also pivotal to brain development. Here we show that CMV infection of the rat fetal brain recapitulated key postnatal phenotypes of human congenital CMV including increased mortality, sensorimotor impairment reminiscent of cerebral palsy, hearing defects, and epileptic seizures. The possible influence of early microglia alteration on those phenotypes was then questioned by pharmacological targeting of microglia during pregnancy. One single administration of clodronate liposomes in the embryonic brains at the time of CMV injection to deplete microglia, and maternal feeding with doxycyxline throughout pregnancy to modify microglia in the litters' brains, were both associated with dramatic improvements of survival, body weight gain, sensorimotor development and with decreased risk of epileptic seizures. Improvement of microglia activation status did not persist postnatally after doxycycline discontinuation; also, active brain infection remained unchanged by doxycycline. Altogether our data indicate that early microglia alteration, rather than brain CMV load per se, is instrumental in influencing survival and the neurological outcomes of CMV-infected rats, and suggest that microglia might participate in the neurological outcome of congenital CMV in humans. Furthermore this study represents a first proof-of-principle for the design of microglia-targeted preventive strategies in the context of congenital CMV infection of the brain.

Keywords: cytomegalovirus, microglia, rat model, fetal brain, neurological outcome

INTRODUCTION

Perinatal and congenital infections cause morbidity and mortality throughout the world. Some pathogens are of considerable public health impact, such as *Toxoplasma gondii*, rubella, human immunodeficiency virus, Zika virus, and human cytomegalovirus (CMV). CMVs belong to the *Herpesviridae* family. In humans, congenital CMV infection can cause severe neurological diseases and defects (Adler and Nigro, 2013). These include microcephaly, polymicrogyria, hearing loss, cerebral palsy, epileptic seizures and intellectual disability, as well as the as-yet elusive influence on the emergence of schizophrenia, autism or epilepsy. Despite the incidence and the medical and socioeconomical burden of congenital CMV, which represents about 1% of all live births, the pathophysiological mechanisms underlying the emergence of neurodevelopmental disorders remain elusive (Cheeran et al., 2009). In the absence of effective preventive or curative therapies, understanding the pathogenesis is mandatory before strategies for early interventions can be designed and tested. The pathophysiology of congenital CMV disease is inherently complicated and involves different stages, from maternal CMV primary infection or reactivation and the associated maternal immune responses, to infection and dissemination within the developing brain—not to mention the crossing of the placental and blood-brain barriers.

Insights into the early events following CMV infection of the developing brain are particularly needed. CMVs are generally species-specific; thus, the development of relevant animal models has been, and will continue to be, critical to our understanding of the mechanisms involved in CMV congenital brain disease (Britt et al., 2013; Cekinovic et al., 2014). Whereas multiple routes (intracranial, intraperitoneal or intraplacental) and developmental timepoints (antenatal or neonatal) of CMV inoculation were used, and despite the lack of materno-fetal transmission of CMV infection in rodents, convergent insights into the alteration of innate and adaptive immune responses have emerged from such models (Kosmac et al., 2013; Sakao-Suzuki et al., 2014; Bradford et al., 2015; Slavuljica et al., 2015; Cloarec et al., 2016; Seleme et al., 2017). The production of cytokines by glial cells, the recruitment of peripheral immune cells, and the altered status of microglia, are all likely to influence neuropathogenesis. Microglia are targeted by CMV during human congenital disease (Teissier et al., 2014) and in murine models of intraplacental or neonatal infections (Kosugi et al., 2002; Sakao-Suzuki et al., 2014). Recently, we reported on a rat model of CMV infection of the developing brain displaying prominent infection of brain myelomonocytic cells and early alteration of microglia (Cloarec et al., 2016). Microglial cells originate from erythromyeloid progenitors located in the yolk sac

during embryogenesis (Ginhoux et al., 2010) and represent the resident mononuclear phagocytes of the brain (Ginhoux et al., 2013; Ginhoux and Jung, 2014). These cells play crucial roles not only in immune defense, maintenance of the neural environment, injury, and repair, but also in neurogenesis, synaptogenesis, synaptic pruning, connectivity, and modulation of synaptic and neuronal activity (Frost and Schafer, 2016). Importantly, early microglial responses might well combat against CMV infection; but these responses might likely have detrimental effects by interacting with important neurodevelopmental processes.

To which extent and to which direction—favorable or detrimental—early microglia alteration would influence the emergence and severity of neurodevelopmental phenotypes in the developing brain *in utero* in the context of CMV infection represent an important pathophysiological question. Herein, we have tested whether early pharmacological targeting of microglia during pregnancy impacts postnatal neurological manifestations in our previously reported rat model of CMV infection of the embryonic brain (Cloarec et al., 2016) and have identified a critical role for microglia.

MATERIALS AND METHODS

Experimental Design

In this study, we explored whether neuroimmune events associated with brain CMV infection *in utero* could be involved in the emergence of postnatal neurological consequences. We first explored whether infected rats would display phenotypes related to the human pathology. We then tested two independent methods in order to target microglia: (1) doxycycline treatment, which is known to attenuate microglia activation in the developing brain (Cunningham et al., 2013) and (2) liposomes containing clodronate to deplete microglia by uptake and release into the cytosol of a non-hydrolysable ATP analog leading to cell death. Finally, we determined whether animals would still display postnatal neurological consequences following each treatment.

Ethical Statement

Animal experimentations were performed in accordance with the French legislation and in compliance with the European Communities Council Directives (2010/63/UE). Depending on the age of the animals, euthanasia were performed after anesthesia with 4% isoflurane with overdose of pentobarbital (120 mg/kg) or with decapitation. This study was approved under the French department of agriculture and the local veterinary authorities by the Animal Experimentation Ethics Committee (*Comité d'Ethique en Expérimentation Animale*) n°14 under licenses n°01010.02 and n°2016100715494790.

CMV Infection and Pharmacological Treatments

Wistar rats (Janvier Labs, France) were raised and mated at INMED Post Genomic Platform (PPGI) animal facility. Rat CMV recombinant Maastricht strain (RCMV-Δ145-147-gfp) with a green fluorescent protein (GFP) expression cassette, and its production, purification and titration, were reported previously (Baca Jones et al., 2009). *In utero* intracerebroventricular (icv) injections were performed at embryonic day 15 (E15) as previously described (Salmi et al., 2013; Cloarec et al., 2016) in embryos from timed pregnant rats that were anaesthetized with ketamine (100 mg/kg)/xylazine (10 mg/kg). Microglia were depleted *in vivo* with clodronate liposomes icv (0.5 μL/injection, Encapsula Nanosciences) co-injected with 1.75 10^3 pfu of rat CMV; alternatively, phosphate-buffered saline (PBS)-containing liposomes (0.5 μL/injection) were co-injected as a control (untreated condition). Microglia status was modified in the embryos *in vivo* with doxycycline *per os* given to the mother throughout pregnancy (200 mg/kg in food pellet chow, Safe).

Immunohistochemistry Experiments

Immunohistochemistry experiments on coronal brain sections (50–100 μm, vibratome, Microm; 14 μm, cryostat, Leica) were carried out as described previously (Cloarec et al., 2016) with the following primary (anti-Iba1: 1/500, Wako; anti-Cd68, Ed1 clone: 1/200, Millipore) and secondary (Alexa Fluor 568 or 647-conjugated goat anti-rabbit or anti-mouse IgGs; Life Technologies) antibodies. Hoescht 33258 (1:2000, Sigma) was used for nuclei staining.

For tissue clearing experiments (see next subsection), whole infected brains were first immunostained as follows. Tissue samples were dehydrated in methanol/1X PBS series: 20, 40, 60, 80, 100 $\times$ 2 for 1h each at room temperature (RT) and then incubated overnight at RT on a platform shaker in a solution of PBSG-T [PBS 1X containing 0.2% gelatin (Sigma-Aldrich), 0.5% Triton X-100 (Sigma-Aldrich) and 0.02% Sodium-Azide (Sigma-Aldrich)]. Next, samples were transferred to PBSG-T containing anti-GFP antibodies (AVES, 1:2,000) and placed at 37°C, with rotation at 100 rpm, for 10 days. This was followed by six washes of 1 h in PBSG-T at RT. Samples were then incubated in secondary antibodies (Donkey anti-chicken Alexa-Fluor 647, Jackson ImmunoResearch, 1:500) diluted in PBSG-T for 2 days at 37°C. After six washes of 1 h in PBSG-T at RT, samples were stored at 4°C in PBS until clearing.

Tissue Clearing

Tissue clearing was performed according to previously reported 3DISCO and iDISCO+ clearing procedures (Erturk et al., 2012; Belle et al., 2014, 2017; Renier et al., 2014) with slight modifications. Briefly, all incubation steps were performed at RT in a fume hood using a 15 ml centrifuge tube (TPP, Dutscher). Samples were first dehydrated in a graded series (20, 40, 60, 80, and 100%) of methanol (Sigma-Aldrich) for 1 h. This was followed by a delipidation step of 20 min in dichloromethane (DCM; Sigma-Aldrich). Samples were transferred to 100% DCM and finally cleared overnight at RT in dibenzylether (DBE; Sigma-Aldrich).

Quantitative Reverse Transcription Polymerase Chain Reaction (qRT-PCR)

Total RNA samples were extracted from whole CMV-infected brains at P1 using TRIZOL reagent (Life Technology). cDNA was synthesized from 1 μg of total RNA using Quantitect Reverse Transcription Kit according to manufacturer protocol (Qiagen). RT-PCRs were carried out using SYBR-Green chemistry (Roche Diagnostics) and Roche amplification technology (Light Cycler 480). PCR primers were designed for GFP transcripts (forward: 5′-gggcacaagctggagtaca; reverse: 5′-cttgatgccgttcttctgc) and for control gene *Rpl19* (ribosomal protein L19) (Cloarec et al., 2016). Primer pairs were optimized to ensure specific amplification of the PCR product and the absence of any primer dimer. Quantitative PCR standard curves were set up for all analyses.

Microscopy, Cell Counting, and Image Analyses

Images of brain sections were acquired with a Stereo Microscope Olympus SZX16 equipped with digital camera DP73, or a Zeiss Axio Imager Z2 microscope with structured illumination (ApoTome) equipped with Zeiss AxioCam MRm camera and processed using Axiovision software, or with a confocal laser scanning microscope Leica TCS SP5X equipped with a white light laser, a 405 nm diode for ultra-violet excitation, and 2 HyD detectors. For cell counting analyses on immunostained brain sections, at least three adjacent brain sections were analyzed throughout the entire z-dimension for each sample using confocal microscopy, according to previously reported procedures (Cloarec et al., 2016) and as further detailed in Supplementary Materials. A phagocytic activation index (PAI) was defined as the ratio of Iba1$^+$ Ed1$^+$ cells to the total number of Iba1$^+$ cells. ImageJ software was used to quantify fluorescence areas on coronal brain sections selected according to their coordinates, as indicated in the rat brain atlas (Khazipov et al., 2015), excluding the meninges. Percentage of fluorescence area of a given brain section was obtained by normalizing to the total area of this brain section.

Whole-brain imaging after tissue clearing was performed using ImspectorPro software (LaVision BioTec) with a binocular stereomicroscope (MXV10, Olympus) equipped with a 2X objective (MVPLAPO, Olympus) used at magnifications 0.8X. A laser (NKT Photonics SuperK extrem) with a 640/30 nm emission filtre and two cylindrical lenses was used to generate a light sheet. Samples were placed in an imaging chamber made of 100% quartz (LaVision BioTec) filled with DBE and illuminated from the side by the laser. Images were acquired with an Andor NEO sCMOS camera (2,560 $\times$ 2,160 pixel size, LaVision BioTec). The Z-step size between each image was fixed at 3 μm. 3D (three-dimensional)-images, quantifications and movies were generated using Imaris x64 software (version 8.4.1, Bitplane). Stack images were first converted to Imaris file (.ims) using ImarisFileConverter. 3D reconstruction of the sample was performed using "volume rendering." Brain segmentation was performed using "surface" tool by creating a mask around each volume to remove meninges and peripheral tissue containing antibody aggregates, allowing GFP intensity quantification only in brain parenchyma. GFP mean intensity obtained was then normalized to the total volume of segmented brain. 3D pictures and movies were generated using the "snapshot" and "animation" tools.

Flow Cytometry

Leukocytes from brains obtained from anesthetized P1 or P7 pups were isolated as previously described (Cloarec et al., 2016). Approximately, 1–3 $\times$ 10^6 leukocytes were incubated with Zombie NIR Fixable Viability kit (1:200; Biolegend) for 20 min at RT. Fc receptors were blocked using mouse anti-rat CD32 antibody (FcγII receptor, clone D34-485) for 10 min. at 4°C to reduce nonspecific binding. Blocked cell samples were stained with antibodies against combinations of cell surface markers as previously described (Cloarec et al., 2016). An average of 1.3 $\times$ 10^5 living singlet cells were analyzed per brain equivalent on a BD LSRFortessa cell cytometer and raw data were analyzed using FACSDiva V8.0 software (BD Biosciences).

Phenotyping

Acquisition of classical developmental righting and cliff aversion reflexes was monitored daily between postnatal day 1 (P1) and P20 as further detailed in Supplementary Materials. The presence of hindlimb paralysis was determined visually in animals that had a postural misplacement and immobility of their hindlimbs. Generalized tonic-clonic epileptic seizures (GTCS) were detected visually, usually after animal handling, especially during cage changing and behavioral testing. They consisted in a classical behavioral sequence including (i) movement arrest and loss of postural equilibrium (ii) hypertonic posture of the trunk, limbs and tail, symmetrically, and (iii) repeated, large clonic movements of all limbs, often with respiratory arrest, incontinence, motor automatisms such as chewing and grooming, terminated by a catatonic phase. Auditory experiments were performed and monitored as detailed in Supplementary Figure 1.

Data Analysis and Statistics

Data were expressed as means $\pm$ s.e.m. unless otherwise stated. Non-parametric Mann Whitney test (two-tailed) followed by Bonferroni correction, if needed, and non-parametric Kruskall-Wallis test were used to detect heterogeneous distribution between groups followed by Dunn's post-hoc test for multiple comparisons. Univariate Cox analysis and Fisher's exact test (two-tailed) were used to compare between survival distributions and rates. Parametric Student's t-test was used to compare between body weight gains, whereas Chi-square test with Bonferroni correction and mixed model for repeated data were used for all other phenotypic comparisons between groups of animals. Generalized mixed models allow incorporating correlations while observations are collected over time. The group of animal was entered as a covariate and the time point (time when measures were repeated) was introduced as a random effect with autogressive covariance structure to account for the within-subjects correlations, assuming that two measurements timely close to each other are closely correlated, and less correlated when they get farther apart. As we are modeling

binary data, we specified the binomial distribution and the logit function as the link one using the PROC GLIMMIX with sas 9.4. Significance threshold was set at 0.05 unless otherwise stated in the figure legends.

RESULTS

CMV Infection of the Rat Fetal Brain Leads to Postnatal Mortality and to Neurological Manifestations

To determine whether CMV infection and the accompanying immune responses in the developing rat brain could be associated with the emergence of postnatal consequences, recalling those seen in the corresponding human congenital disorder, recombinant rat CMV expressing GFP was injected icv in embryos from timed-pregnant rats at E15 as previously described (Cloarec et al., 2016).

During the period of evaluation, i.e., until P20, CMV infection significantly decreased postnatal survival as compared with control animals, which were injected icv with the vehicle (MEM) ($p = 0.0017$) (Figures 1A, 2; Supplementary Table 1). Indeed, 70.4% (38 out of 54 pups) of CMV-infected newborns died in the first three postnatal weeks, compared to 2.9% (one out of 34) controls ($p < 10^{-4}$, Chi2 test). In contrast, antenatal mortality was not affected as no significant difference was observed in the ratio of live animals at birth (72.8%, $n = 103$) as compared to controls (68.5%, $n = 54$) (Fisher's exact test, two-tailed). Body weight was similar at P1 between CMV-infected (7.20 g $\pm$ 0.10) and control (7.40 g $\pm$ 0.11) newborns (Student's t-test), but CMV infection *in utero* significantly impacted the postnatal evolution of body weight gain ($p < 10^{-4}$) (Figure 1B; Supplementary Table 1).

Congenital CMV infection is the leading cause of non-hereditary congenital sensorineural hearing loss (Smith et al., 2005). In order to assess whether CMV infection *in utero* causes hearing loss in rats, auditory tests were performed at P40. Hearing thresholds were significantly higher for clicks ($p = 0.0146$) and for 24 kHz bursts ($p = 0.0017$) in the CMV infected group compared to the control group (Supplementary Figure 1; Supplementary Table 1). As hearing loss caused by congenital CMV infection may be progressive in children, we aimed at evaluating hearing thresholds in CMV-infected animals at later ages. However, a significant deterioration of hearing thresholds was also detected in control (MEM-injected) rats between P40 and P60 (data not shown); this restrained us from performing such a longitudinal analysis in CMV-infected rats as the deterioration of hearing thresholds seen in control rats would likely preclude reliable interpretation of the overall data.

Human congenital CMV infection is also well known to be a contributing cause of cerebral palsy, a group of disorders involving variable degree of sensorimotor disabilities (Colver et al., 2014). Sensorimotor development was evaluated by daily monitoring pups between P1 and P20 for the acquisition of the classic righting and cliff aversion reflexes (Rousset et al., 2013). The righting reflex consisted in assessing the ability of rodent pups to coordinate the necessary movement to roll over from

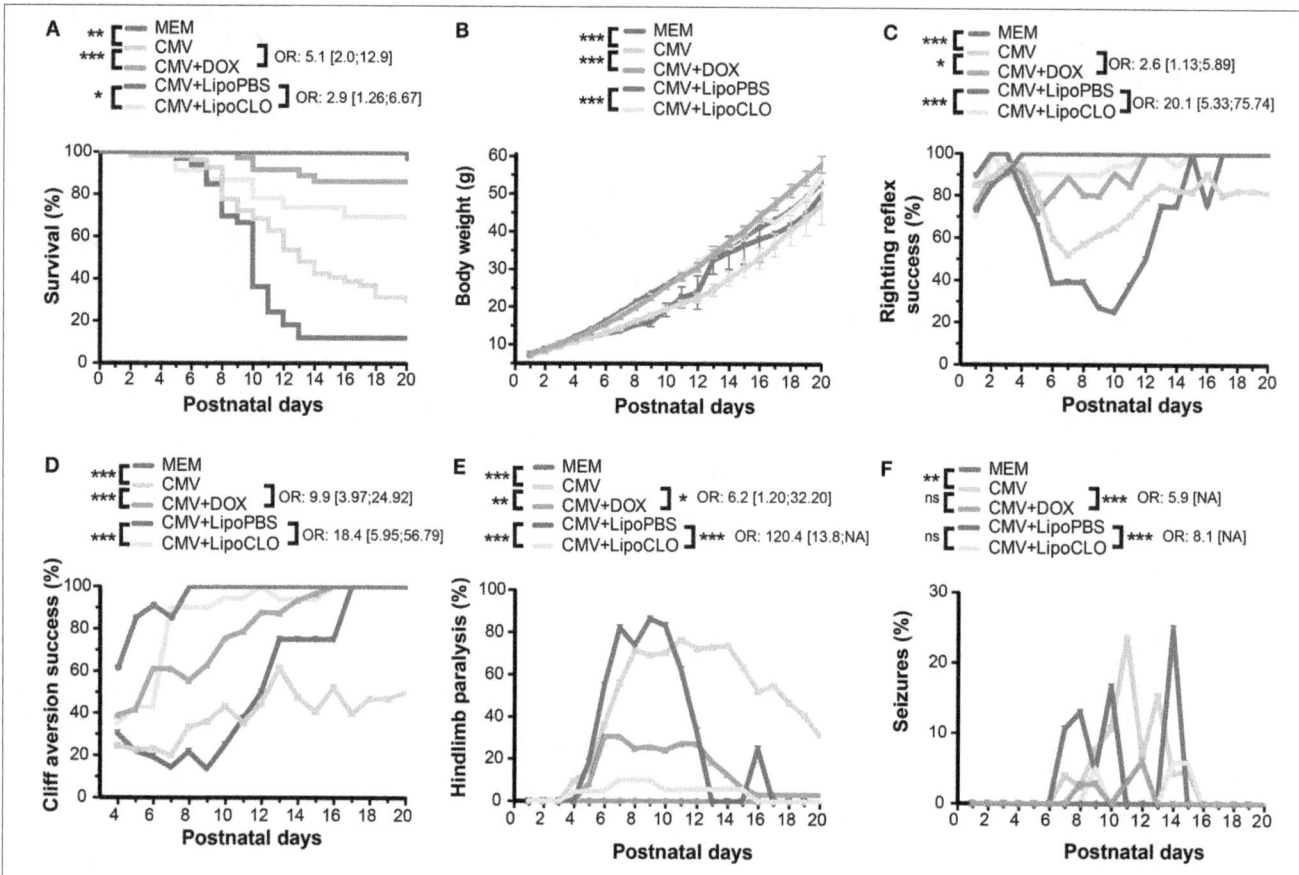

FIGURE 1 | CMV infection of the embryonic rat brain causes decreased survival and neurodevelopmental defects that can be prevented *in utero* by microglia-targeted drug-based strategies. Phenotypic investigations (see **Figure 2**) in the three first postnatal weeks in pups previously subjected to the following procedures during pregnancy: vehicle intraventricularly (icv) (MEM; $n = 34$, four litters); CMV icv, pregnant rat fed with doxycycline (DOX) during pregnancy (CMV+DOX; $n = 36$, four litters); CMV icv, untreated pregnant rat (CMV; $n = 54$, six litters); CMV with clodronate liposomes icv (CMV+LipoCLO; $n = 23$, two litters); CMV with PBS liposomes icv (CMV+LipoPBS; $n = 33$, four litters). Sex ratio did not differ between the five groups at birth ($p = 0.15$, Chi-square test). **(A)** Kaplan-Meier survival curves indicated significant decreased survival in the CMV vs. MEM group, and significant rescues in the CMV+DOX and CMV+LipoCLO groups vs. their untreated counterparts (CMV and CMV+LipoPBS, respectively). **(B)** Cumulative body weight decreased significantly in the CMV vs. MEM group. Significant improvements were seen in the CMV+DOX and CMV+LipoCLO groups vs. their untreated counterparts (CMV and CMV+LipoPBS, respectively). **(C,D)** The proportion of animals succeeding to the righting reflex **(C)** or the cliff aversion reflex **(D)** sensorimotor tests decreased significantly in the CMV vs. MEM group, and significantly improved in the CMV+DOX vs. CMV group, and in the CMV+LipoCLO vs. CMV+LipoPBS group. **(E,F)** Hindlimb paralysis **(E)** and generalized tonic-clonic epileptic seizures (GTCS) **(F)** occurred significantly more frequently in the CMV vs. MEM group. This improved significantly in the CMV+DOX vs. untreated CMV group and in the CMV+LipoCLO vs. untreated CMV+LipoPBS group. Note that because the most severely affected pups were at higher risk to death, a misleading effect of apparent improvement with time could be perceived from curve shapes even for untreated conditions. Left sides of the cohorts legend: univariate Cox analysis **(A)**, mixed model for repeated data **(B–D)**, or Chi2 test with Bonferroni correction **(E,F)**. Right sides of the cohorts legend: odds ratio (OR) ± confidence intervals correspond to the risks in the untreated vs. their corresponding treated groups: postnatal mortality **(A)**; failure at test **(C,D)**; hindlimb paralysis **(E)**; GTCS **(F)**. ***$p < 0.001$; **$p < 0.01$; *$p < 0.05$; ns, not significant, for all figure panels, except **(E,F)** where ***$p < 0.0002$; **$p < 0.002$. NA, not available.

their backs onto their paws. A significant proportion of CMV-infected pups showed inability to right, as compared with MEM-injected pups, which all performed the test successfully ($p < 10^{-4}$; **Figures 1C, 2**; Supplementary Figure 2A; Supplementary Table 1). In the cliff aversion test, pups were placed with their forepaws overhanging the edge of a board and the time required to turn away from the edge was recorded. Control MEM-injected rats showed a clear performance improvement from P4 as all succeeded the test after P8 (**Figures 1D, 2**; Supplementary Table 1). In contrast, a significant proportion of CMV-infected pups was unable to complete the cliff aversion test all along the first three postnatal weeks ($p < 10^{-4}$; **Figure 1D, 2**; Supplementary

Figure 2B; Supplementary Table 1). Hindlimb paralysis was detected in the first three postnatal weeks in 83.3% of CMV-infected pups, likely preventing animals from turning away from the cliff. Hindlimb paralysis was not seen in any non-infected pup ($p < 10^{-4}$; **Figures 1E, 2**; Supplementary Table 1).

Patients with human congenital CMV are also at high risk of postnatal epileptic seizures (Suzuki et al., 2008). Consistently, 24.1% of CMV-infected rat pups exhibited GTCS at least once in the first three postnatal weeks whereas none of MEM-injected pups ever showed any GTCS ($p = 0.002$; **Figure 1F, 2**; Supplementary Table 1). Interestingly, a clear relationship between epileptic seizures and death was observed wherein 85%

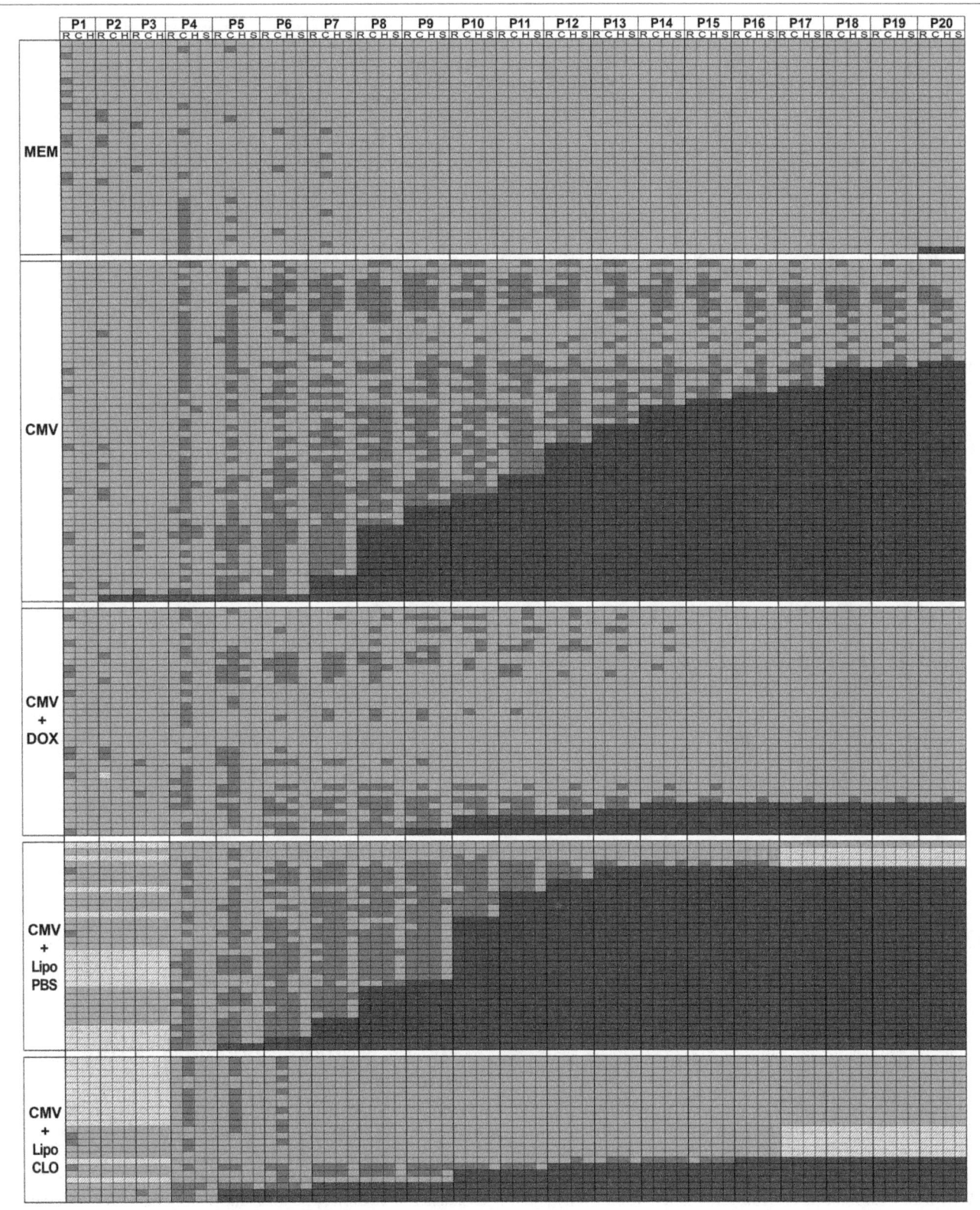

FIGURE 2 | Overview of day-per-day phenotype assessment in the first three postnatal weeks. Color-encoded overview of postnatal test evaluations and observations performed as from postnatal day P1 (see also **Figure 1**; Supplementary Table 1). Five experimental cohorts of rats were analyzed postnatally: i/control embryos were injected intraventricularly (icv) with vehicle at E15 (MEM; n = 34 from four litters); ii/embryos were injected icv with rat CMV at E15 and were born from dams fed with

(Continued)

FIGURE 2 | control chow all over pregnancy (CMV; $n = 54$ from six litters); iii/embryos were injected icv with rat CMV at E15 and were born from dams fed with doxycycline-supplemented chow all over pregnancy (CMV+DOX; $n = 36$ from four litters); iv/embryos were injected icv with rat CMV and with control (PBS) liposomes at E15 (CMV+LipoPBS; $n = 33$ from 4 litters); v/embryos were injected icv with rat CMV and with clodronate liposomes at E15 (CMV+LipoCLO; $n = 23$ from 2 litters). The table shows color-encoded results for each individual pup evaluated on a daily basis (P1 to P20 columns) for the phenotypic parameters as detailed below. Performances at the righting (R sub-column) and cliff aversion (C sub-column) reflexes were measured from P1 or from P4, respectively; performances to right (R) or to turn away from the cliff (C) were color-encoded following a binary rule (blue: success; red: failure). For righting reflex evaluation, rat pups were placed in a supine position, and the time required to flip to the prone position was measured. For the cliff aversion reflex, animals were placed with their forepaws overhanging the edge of a board; the time required to turn >90° away from the edge was recorded. For both tests, a maximum observation time of 30 s. was used. Each pup was also monitored daily for the appearance of hindlimb paralysis (H sub-column) and of generalized tonic-clonic seizures (GTCS) (S sub-column); the data were color-encoded following a binary rule where blue reflected the absence of either event, whereas red indicated the occurrence of paralysis or seizure. When pups died, this was also annotated on the table and color-encoded in dark gray. Hatched cells: data unavailable.

of seizing rats deceased, as compared to 37% of non-seizing rats ($p < 10^{-4}$, Chi2 test). Similarly, 72% of rats with hindlimb paralysis deceased during the period of evaluation, as compared to 13% of non-paralyzed rats ($p < 10^{-4}$).

Acute icv Injection of Clodronate Liposomes *in Utero* Depletes Microglia and Improves the Postnatal Outcome

The consequences of rat CMV infection as described above, recapitulated several cardinal clinical features of the human pathology. The early and prominent infection and alteration of microglia upon CMV infection of the developing rat brain *in utero* (Cloarec et al., 2016) suggested that microglia might contribute to the pathophysiology of congenital CMV. In order to evaluate the possible impact of microglia on the emergence of the neurological and other developmental defects, experiments were designed to target microglia with drugs during pregnancy.

In a first series of experiments, microglia were acutely depleted with clodronate liposomes co-injected icv together with rat CMV at E15. Immunohistochemistry experiments confirmed that clodronate liposomes triggered a significant decrease in the total number of Iba1$^+$ microglial cells in the dorsolateral part of the striatal wall of the lateral ventricles taken as the region of interest (ROI) at P1 (112.3 cells/ROI $\pm$ 9.37), as compared to the condition where PBS liposomes were used (558.5 cells/ROI $\pm$ 128.3; $p = 0.0022$; **Figure 3A**; Supplementary Table 2). A significant decrease in the absolute number of phagocytically active, Iba1$^+$ Ed1(CD68)$^+$ microglia was also oberved (clodronate liposomes: 27.17 cells/ROI $\pm$ 2.80; PBS liposomes: 141 cells/ROI $\pm$ 23.82) ($p = 0.0022$). This was associated with a dramatic reduction of CMV spreading, since GFP$^+$ infected cells were barely visible within the brains of treated embryos. The reduction in CMV infection was confirmed by quantifying the percentage of fluorescent areas in coronal brain sections taken from treated and untreated pups at P1 ($p = 0.0022$; **Figure 3B**; Supplementary Table 3).

Microglia depletion and reduction of brain CMV infection were associated with a significant improvement of survival and neurodevelopmental outcomes. Postnatal mortality was reduced by 2.9 fold ($p = 0.012$) in clodronate-treated, CMV-infected pups, relative to untreated, infected pups (**Figures 1A, 2**; Supplementary Table 1). Indeed, 69.6% (16 out of 23 newborns) of clodronate-treated, CMV-infected newborns survived at P16, as compared to 12.1% (4 out of 33) of untreated, CMV-infected

newborns ($p < 10^{-4}$, Chi2 test). A similar significant increase in body weight gain was observed in the clodronate-treated pups ($p < 10^{-4}$; **Figure 1B**; Supplementary Table 1).

CMV-infected pups treated with clodronate liposomes *in utero* also performed significantly better at the righting and the cliff aversion reflexes when compared to infected pups, which had received control (PBS) liposomes *in utero* (**Figures 1C,D, 2**; Supplementary Figure 2; Supplementary Table 1). With clodronate the odds in favor of success to righting and to cliff aversion tests were at 20.1:1 and 18.4:1, respectively ($p < 10^{-4}$ for both). Hindlimb paralysis also improved significantly ($p < 10^{-4}$) with clodronate liposomes (17% of rat pups) as compared with control liposomes (88%) (**Figures 1E, 2**; Supplementary Table 1). There was a dramatic, 120.4-fold decrease in the risk to hindlimb paralysis in clodronate-treated CMV-infected rats ($p < 10^{-4}$). Epileptic seizures also occurred less frequently in the first three weeks of life upon treatment with clodronate liposomes (8.7% of rats experiencing at least one seizure), as compared to PBS liposomes (24.2% of epileptic rats) but this difference did not reach statistical significance ($p = 0.14$) (**Figures 1F, 2**; Supplementary Table 1). However, when the risk of seizures was considered, it decreased significantly by 8.1-fold in clodronate-treated, CMV-infected rats when compared to their PBS-liposomes counterparts ($p < 10^{-4}$).

Hence a single injection of clodronate liposomes at the time of CMV icv infection not only led to depletion of microglia and to a dramatic reduction of active CMV infection in the rat developing brain, but the treatment also led to a stunning improvement of survival, body weight gain, sensorimotor development and epileptic seizures in early postnatal life.

Chronic Administration of Doxycycline to Pregnant Mothers Improves Microglia Phenotype

Tetracyclines can efficiently modify microglia status in the brains of rat pups after chronic maternal administration during pregnancy (Cunningham et al., 2013). Tetracyclines impact on microglia phenotype independently of their canonical antibacterial action (Tikka et al., 2001). In order to confirm the possible role of microglia in the emergence of phenotypes associated with brain CMV infection, doxycycline, a second-generation, lipophilic tetracycline that crosses blood-brain and placental barriers, was administered *per os* to pregnant dams from E0 to birth. The effects of doxycycline treatment on

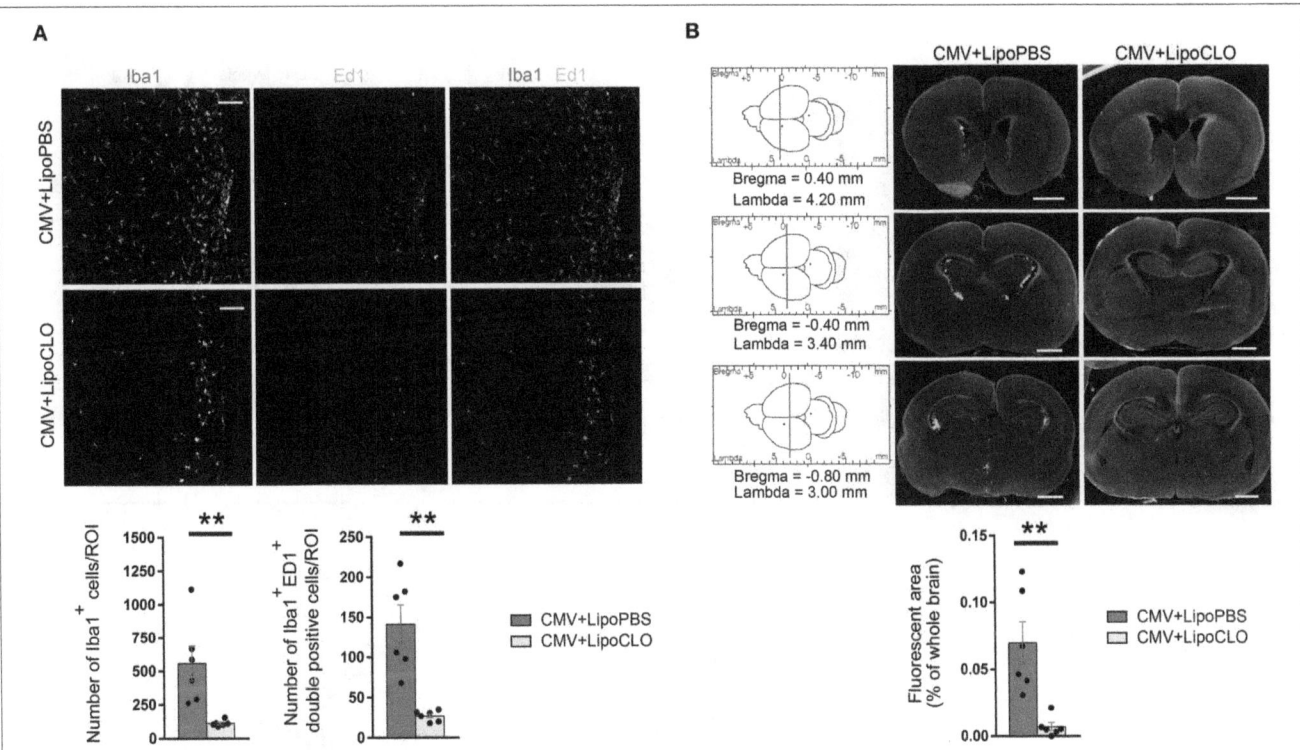

FIGURE 3 | Clodronate liposomes deplete microglia and reduce CMV spreading in the developing brain. Recombinant rat CMV allowing expression of GFP (green) in the infected cells was injected icv in rat embryos at E15 together with either clodronate liposomes (LipoCLO) to deplete microglia, or PBS liposomes (LipoPBS) as control. **(A)** LipoCLO decreased the number of Iba1$^+$ microglia (red) and of Iba1$^+$, Ed1$^+$ (cyan) phagocytically active microglia. Three to four adjacent coronal brain sections were analyzed by confocal microscopy throughout the entire z-dimension (n = 6 brains in each condition). ROI: region of interest (775 × 775 μm^2). Bar: 100 μm. Mann-Whitney test, two-tailed, with Bonferroni correction. **p < 0.005. **(B)** LipoCLO reduced rat CMV infection of the brain. Brains were analyzed at P1 using fluorescent binocular microscopy. Three sections were selected according to their coordinates in the rostrocaudal axis (left). CMV infection decreased dramatically after treatment with LipoCLO (n = 6 brains in each condition), as quantified by measuring the proportion of fluorescent (GFP) areas. Bar: 1 mm. Mann-Whitney test, two-tailed. **p < 0.01.

microglia phenotype was tested by immunohistochemistry in the dorsolateral part of the striatal wall of the lateral ventricles, a region where active CMV infection was frequently detected (Cloarec et al., 2016), and by multicolor flow cytometry analysis of the whole brain.

In line with previously reported experiments (Cloarec et al., 2016), rat CMV infection at E15 led to a significant increase in the proportion of phagocytically active, Iba1$^+$ Ed1(CD68)$^+$ microglia/macrophages cells at P1. Indeed, the phagocytic activation index (PAI), defined as the ratio of Iba1$^+$ Ed1$^+$ cells to the total number of Iba1$^+$ cells, was significantly increased (42.74% ± 6.16) as compared with control (MEM-injected) rats (PAI = 14.28% ± 4.29) (p = 0.0174), and was significantly improved by doxycycline (17.32% ± 6.44; p = 0.0321) (**Figure 4A**; Supplementary Table 2). Consistent data were obtained when whole brains were analyzed at P1 by flow cytometry (**Figure 4B**; Supplementary Table 4). As previously reported (Cloarec et al., 2016), a significant increase in the proportion of major histocompatibility (MHC) class II-positive microglia (fraction IV: CD45$^{low/int}$, CD11b/c$^+$, RT1B$^+$) was detected in CMV-infected brains (13.98% ± 2.12) as compared with controls (0.73% ± 0.05; p < 0.0001). This increase in activated microglia was significantly

counteracted in doxycycline-treated, CMV-infected pups (5.66% ± 1.13) (p = 0.0483). The proportions of other types of CD45$^+$ hematopoietic cells were not changed by doxycycline (**Figure 4B**; Supplementary Figure 3; Supplementary Table 4). Importantly, the favorable impact of the doxycycline *in utero* treatment on reactive microglia seen at P1 did not last after treatment discontinuation. The proportion of reactive, CD45$^{low/int}$, CD11b/c$^+$, RT1B$^+$ microglia (fraction IV) at P7 were similar in treated (55.76% ± 4.60) and in untreated (58.93% ± 5.54) CMV-infected rats (**Figure 4C**; Supplementary Figure 3; Supplementary Table 4).

The consequences of *in utero* treatment with doxycycline on brain CMV infection *per se* were also evaluated. Independent series of brains were submitted to tissue clearing using the iDisco method (Renier et al., 2014). Fluorescence-based 3D quantification of the infected areas of whole brains showed no difference between treated and untreated rats at P1 (**Figure 5A**; Supplementary Table 3; Supplementary Videos 1, 2). Consistently, no significant difference in GFP gene expression was found by qRT-PCR between treated and untreated CMV-infected brains at P1 (**Figure 5B**; Supplementary Table 5). Moreover, the proportions of CMV-infected, GFP$^+$ cells as detected by flow cytometry analysis performed on CD45$^+$

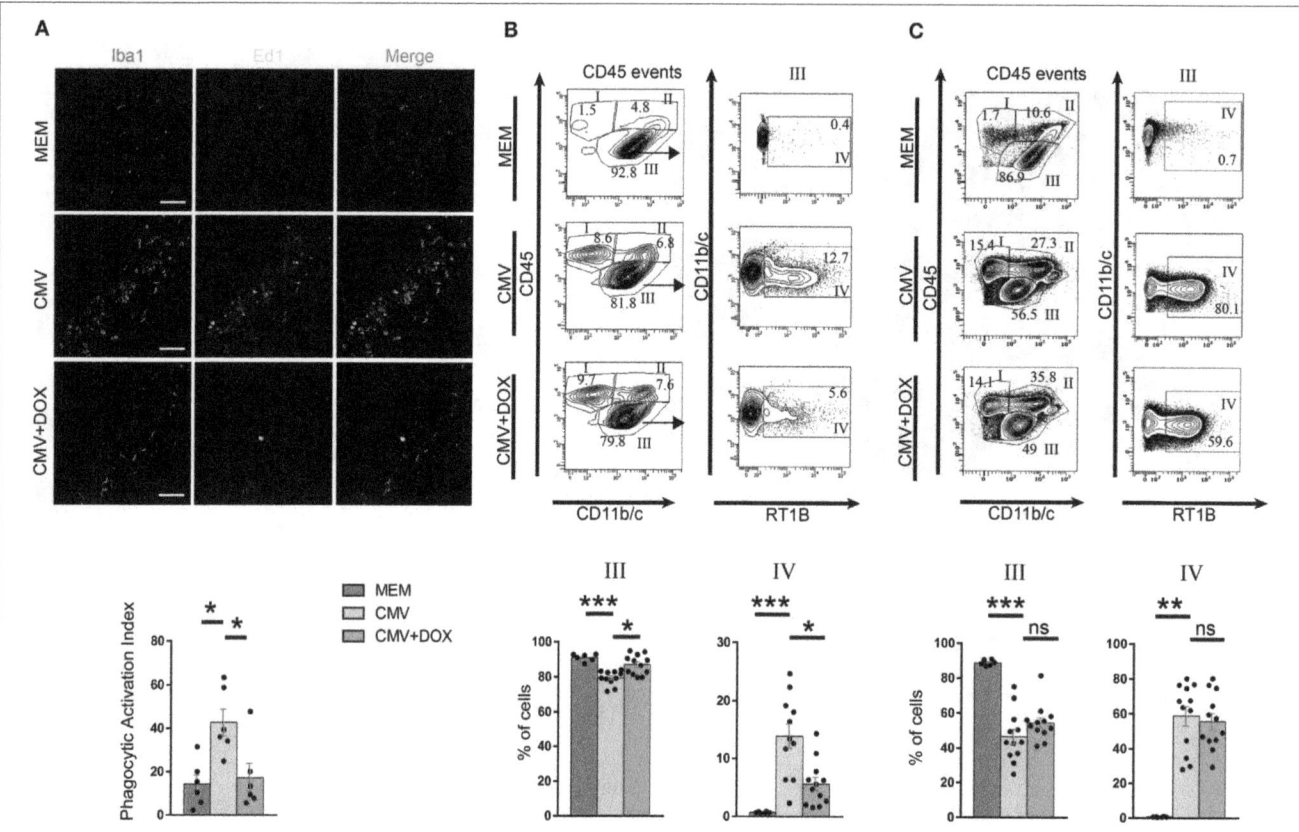

FIGURE 4 | CMV infection of the embryonic rat brain leads to altered microglia phenotype, which is rescued by maternal feeding with doxycycline throughout pregnancy. **(A)** Microglia status in the lateral ventricles was assessed on coronal sections observed by confocal microscopy at P1 by quantifying microglial cells (red, Iba1$^+$) and phagocytically active microglial cells (Iba1$^+$, red; Cd68/Ed1$^+$, cyan). Those values were then used to calculate the phagocytic activation index (PAI) defined as the ratio of the number of Iba1$^+$, Ed1$^+$ activated cells to the total number of Iba1$^+$ cells. PAI increased in the CMV vs. the MEM group. Doxycycline (CMV+DOX) significantly counteracted PAI increase ($n = 6$ brains in each condition). ROI: region of interest ($387 \times 200\ \mu m^2$). Bar: $50\,\mu m$. Kruskall-Wallis test followed by Dunn's *post-hoc* test. *$p < 0.05$. **(B,C)** Flow cytometry analysis of leukocytes collected at P1 **(B)** and P7 **(C)**. Total leukocytes (CD45 events) were gated for CD45 and Cd11b/c, thus defining fractions I (lymphocytic cells), II (myelo-monocytic cells) and III (resident microglial cells) (see also Supplementary Figure 3). Fraction III corresponding to CD45$^{low/int}$, CD11b/c$^+$ microglial cells was further characterized for RT1B expression to identify reactive microglia (RT1B$^+$; fraction IV). Representative flow cytometry plots and the corresponding quantifications are shown for each group. **(B)** At P1, the increased proportion of reactive microglia triggered by CMV infection was significantly counteracted by doxycycline ($n = 6$ MEM; $n = 11$ CMV; $n = 12$ CMV+DOX). **(C)** At P7, i.e., 7 days after discontinuation of doxycycline, the decreased proportion of total microglial cells (fraction III) and, within that fraction, the increased proportion of reactive microglia (fraction IV) triggered by CMV infection, were not counteracted by doxycycline treatment given *in utero* ($n = 6$ MEM; $n = 12$ CMV; $n = 12$ CMV+DOX). Kruskall Wallis test followed by Dunn's post test. ***$p < 0.001$; **$p < 0.01$; *$p < 0.05$; ns, not significant.

hematopoietic cells isolated from CMV-infected brains at P1, were not significantly different between doxycycline-treated and untreated pups (**Figure 5C**; Supplementary Table 4). Also, no significant difference in GFP$^+$ infected cells was found by flow cytometry at P7 between doxycycline-treated and untreated pups. Hence the early and transient impact of doxycycline on microglia phenotype was not accompanied by a parallel decrease in CMV infection of the brain.

Chronic Doxycycline Administration to the Mother Throughout Pregnancy Improves the Postnatal Outcome

Owing to the transient improvement of microglia status upon maternal administration of doxycycline during pregnancy, we then tested whether this would in turn impact the postnatal

outcomes. In doxycycline-treated, CMV-infected pups, survival rate improved significantly as compared to infected pups from untreated dams (**Figure 1A, 2**; Supplementary Table 1). CMV-infected pups treated with doxycycline *in utero* had a risk 5.1 lower to postnatal death in the first three postnatal weeks, than their untreated, CMV-infected counterparts ($p = 0.0006$). At P20, 86.1% ($n = 31$ out of 36 newborns) of infected pups from doxycycline-treated dams had survived as compared to 29.6% ($n = 16$ out of 54 newborns) of infected pups from untreated dams ($p < 0.001$, Chi2 test). Significant improvement in body weight gain was also observed in CMV-infected pups after doxycycline treatment *in utero* ($p < 10^{-4}$; **Figure 1B**; Supplementary Table 1).

Whereas no rescue could be obtained at P40 on hearing threshold after doxycycline treatment (Supplementary Figure 1; Supplementary Table 1), responses to sensorimotor tests (righting and cliff aversion reflexes) improved significantly

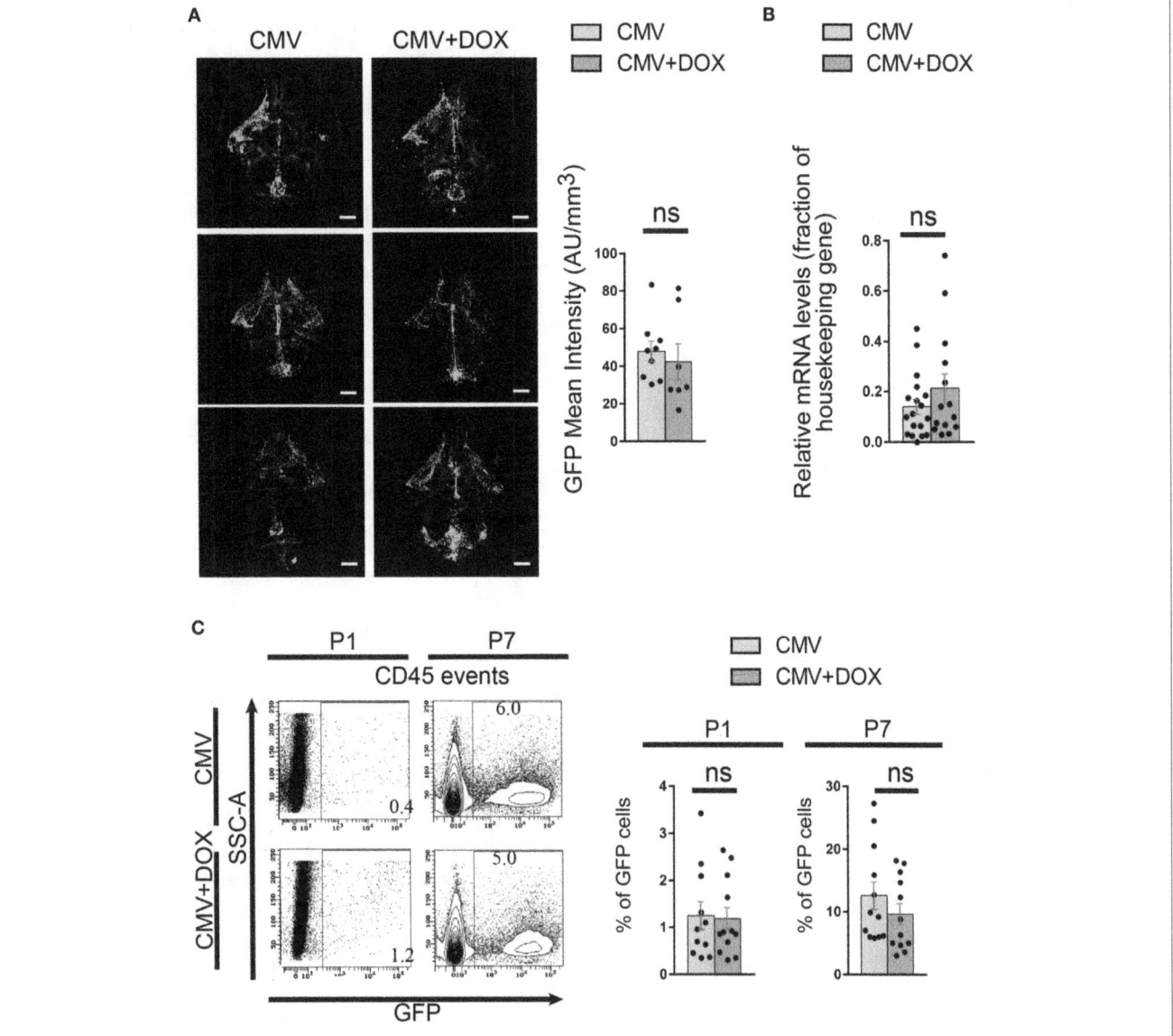

FIGURE 5 | Maternal feeding with doxycycline throughout pregnancy does not impact rat CMV infection of the brain. Green fluorescent protein (GFP) expression was used to monitor infection of the pups' brains by rat CMV with three independent methods. **(A)** Tissue clearing. Whole brains from P1 rat pups were submitted to tissue clearing and GFP expression was quantified by measuring mean fluoresence intensity normalized to each corresponding total brain volume. Representative examples of clarified brains are shown (light sheet microscopy) (see also Supplementary Videos 1, 2). No significant difference was found between the CMV+DOX group ($n = 7$) and its untreated counterpart (CMV; $n = 9$). Bar: 2 mm. Mann Whitney test, two-tailed. ns, not significant. **(B)** qRT-PCR. Relative mRNA expression of GFP was assessed by quantitative RT-PCR in CMV-infected brains at P1, using *RPL-19* as reference gene. No difference was found between the CMV+DOX group ($n = 14$) and its untreated counterpart (CMV; $n = 18$). Values of fold change represent averages from triplicate measurements for each sample. Mann Whitney test, two-tailed. ns: not significant. **(C)** Flow cytometry. Relative proportion of GFP$^+$, CMV-infected cells was estimated by flow cytometry analysis of CD45$^+$ cells isolated from CMV-infected brains at P1 and at P7. No difference was found between the CMV+DOX group ($n = 12$) and its untreated counterpart (CMV; $n = 11$). SSC, side-scattered light. Mann Whitney test, two-tailed. Ns, not significant.

($p = 0.025$ and $p < 10^{-4}$, respectively) in infected pups treated with doxycycline *in utero* (success to righting: odds ratio 2.6:1; success to cliff aversion: odds ratio 9.9:1) (**Figures 1C,D, 2**; Supplementary Figure 2; Supplementary Table 1).

Hindlimb paralysis was also observed less frequently after doxycycline administration (**Figures 1E, 2**; Supplementary Table 1). Hence, 50% ($n = 18$ out of 36) of doxycycline-treated,

CMV-infected pups displayed paralysis during the first three postnatal weeks, as compared to 83.3% ($n = 45$ out of 54) in the untreated counterparts ($p = 0.001$). Consistently, the risk of hindlimb paralysis reduced by 6.2 fold ($p = 0.03$). Whereas the proportion of epileptic, CMV-infected rats decreased, albeit not significantly ($p = 0.12$), upon treatment with doxycycline (11%) as compared with untreated, infected rats (24%), the risk of epileptic seizures during the period of evaluation decreased significantly by 5.9-fold in CMV-infected rats receiving doxycycline *in utero* ($p < 10^{-4}$; **Figures 1F, 2**; Supplementary Table 1).

Hence, while doxycycline administration during pregnancy did not significantly modify the amount of active CMV infection in the developing brain after birth, it led to a transient modification of microglia phenotype that was associated with long-term favorable effects on survival, body weight gain and neurodevelopmental outcomes.

DISCUSSION

Phenotypes in the Rat and in Humans

Studies on different rodent models that showed similarities with the human disease at the neuroanatomical, cellular and molecular levels, have suggested that neuroimmune alterations might play an important pathophysiological role. In a rat model of CMV infection of the developing brain *in utero*, we reported recently the detection of early neuroimmune anomalies including microglia alteration (Cloarec et al., 2016). Utilizing this model, we herein demonstrate altered postnatal oucomes in CMV infected neonates including increased postnatal lethality, decreased body weight gain, early-onset epileptic seizures, sensorimotor impairment reminiscent of cerebral palsy, and hearing defects. Despite the fact that maternal, placental and peripheral embryonic events were purposedly bypassed by directly infecting the embryonic brain *in utero*, this model recapitulated several fundamental phenotypic features of the human pathology following CMV infection. Moreover, the postnatal outcomes varied dramatically among the rat pups, as in the human disease. This indicates that in addition to the aforementioned upstream factors, the somehow unpredictable and highly diverse outcome of congenital CMV infection might rely, at least partly, on fetal brain-related events. The consequences of CMV infections of the brain on the future neurological phenotypes had been hardly addressed in rodent models until recently. Auditory features associated with inflammation of the inner ear recalling the hearing losses seen in human congenital infections were recently reported in a neonatal mouse model (Bradford et al., 2015). Also, early and long-term neurological dysfunctions including acquisition of the righting and cliff aversion reflexes as well as motor performances and social behavior have been reported very recently in a murine model of neonatal infection (Ornaghi et al., 2017). Interestingly, in the present rat model of rat CMV infection *in utero*, more dramatic neurological consequences were observed, probably because CMV was inoculated directly into the ventricles and at an earlier developmental stage.

Targeting Microglia With Doxycycline and Clodronate

Generally, microglia-targeted rescue strategies had already been successfully used in a range of models for various postnatal neurological insults and disorders, such as neonatal excitotoxic brain damage (Dommergues et al., 2003), experimental autoimmune encephalitis (Ponomarev et al., 2011), cerebral

palsy (Kannan et al., 2012), or Alzheimer disease (Hong et al., 2016). In the present model of prenatal cerebral infection, two previously reported microglia-targeted drug-based strategies both led to the successful rescue of the neurological phenotypes. While the respective modes of action of clodronate liposomes on the one hand, and doxycycline on the other hand, are unrelated at the molecular and subcellular levels, we cannot firmly exclude the possibility that they would both have influenced the phenotypes by microglia-independent, off-target mechanisms. Doxycycline might have other biological consequences (Yrjanheikki et al., 1999) such as a broader anti-inflammatory action on other immune and non-immune cells (Yrjanheikki et al., 1998). Hence, it was shown in a murine model of multiple sclerosis that minocycline, another second-generation tetracycline, can decrease leukocytes infiltration into the spinal cord (Brundula et al., 2002). However, in our model doxycycline did not modify the different fractions corresponding to the non-microglial immune cells detected in the CMV-infected brains (e.g., monocytes/macrophages, dendritic cells, T cells and B cells), as analyzed by flow cytometry (see Supplementary Figure 3). Moreover tetracyclines have already been used to decrease microglia activation in NMDA-induced excitotoxicity (Tikka et al., 2001), experimental autoimmune encephalitis (Popovic et al., 2002), or epileptogenesis (Abraham et al., 2012).

Influence of Early Microglia Alteration on the Postnatal Phenotypes

Whereas clodronate liposomes led to microglia depletion associated with a dramatic decrease in brain CMV infection, doxycycline promoted microglia modification while leaving CMV infection of the brain unchanged. Moreover, the early improvement of microglia was associated with long-term beneficial effects. Indeed, the effect of doxycycline on microglia was transient and was no longer detected at P7. Despite this observation, the favorable impact on the clinical phenotypes lasted long after treatment was stopped. This suggests that the early alteration of microglia following CMV infection had long-term detrimental effects on the neurological and other outcome. Although we cannot exclude the existence of more subtle changes such as a modification in the distribution of CMV in different brain areas, our data indicate that early microglia alteration, rather than CMV infection and viral cytopathic effects *per se*, promoted the emergence of most neurological and other postnatal manifestations. The relationship between CMV load and the occurrence and severity of CMV-related phenotypes has indeed been challenged in a murine model of neonatal CMV infection (Kosmac et al., 2013; Seleme et al., 2017) and in human congenital CMV (Revello et al., 1999; Boppana et al., 2013). Our data are congruent with recent reports on inflammatory processes in CMV infection of the murine neonatal brain (Kosmac et al., 2013; Bradford et al., 2015; Kawasaki et al., 2015; Slavuljica et al., 2015; Seleme et al., 2017).

The microglia-targeted strategies used in the present study led to significant improvements of sensorimotor impairments and of epilepsy. The alteration and possible role of microglia in epileptic disorders and in cerebral palsy have already been discussed

(Fleiss and Gressens, 2012; Vezzani et al., 2012; Devinsky et al., 2013). The presence of hearing impairment in the present model might be difficult to interpret. First, a progressive deterioration of auditory performances was seen in the control rats; this is consistent with previous reports on the existence of age-related hearing loss in control Wistar rats (Alvarado et al., 2014), which was likely related with the lack of strial melanin in albino rodents (Ohlemiller, 2009). Second, a detrimental effect of doxycycline on the development of the auditory system cannot be excluded, notably given its known potential auditory toxicity. Third, whereas hearing loss caused by murine neonatal CMV infection was recently associated with persistent inflammation of the inner ear (Bradford et al., 2015), the lack of significant improvement after doxycycline treatment *in utero* might indicate that microglia does not play an important role in hearing impairment in our model.

Microglia in Brain Development and in Neurodevelopmental Disorders

A crucial pathogenic impact of early microglia alteration in the context of congenital CMV is consistent with the role of microglia in the development, the functioning and the pathology of the central nervous system (Prinz and Priller, 2014). Microglia undergo several programmed changes during brain development in rodents (Matcovitch-Natan et al., 2016), impact synaptic transmission, as well as synapse formation and elimination. Microglia also shape embryonic and postnatal brain circuits (Paolicelli et al., 2011; Schafer et al., 2012). The alteration of microglia by CMV infection early during brain development might well have disrupted the timing of such developmental programs, leading to altered development of neuronal networks and the subsequent detrimental neurological outcome. As an example, susceptibility to epileptic seizures could be caused by several microglia-dependent processes, such as the alteration of migration and neocortical laminar distribution of interneurons, the altered control of synaptic development and regulation, or an increased glutamate release (Paolicelli and Ferretti, 2017). As a matter of fact, the transient improvement of microglia status as obtained here with doxycycline was sufficient to obtain dramatic improvements of the neurological phenotypes. Prenatal alteration of microglia and the accompanying microglia priming can have long-term consequences on synaptic functioning (Roumier et al., 2008) and could participate in the susceptibility to neurodevelopmental disorders occurring as consequences of viral infections of the developing brain.

Besides Microglia

Whereas early microglia improvement was sufficient to impact significantly the prenatal phenotypes, this of course does not exclude the important roles likely played in parallel by other components of the neuroimmune system, either at the molecular level such as cytokines and chemokines, or at the cellular level such as infiltrating monocytes or lymphocytes. Hence, the role of T lymphocytes infiltration in CMV infection has been proposed (Bantug et al., 2008); T cells control viral spread within the brain and are pivotal in resolving acute CMV infection of the neonatal murine brain (Slavuljica et al., 2015).

However, the amount of T cells was not impacted by doxycycline in the present model, which is consistent with the lack of any significant change in CMV infection observed here after doxycycline treatment. Similarly, doxycycline did not modify the proportions of dendritic cells or monocytes within the infected brains. Other types of glial cells such as astrocytes are also likely to play a pathophysiological role, as reported in other models of CMV infection (Lokensgard et al., 2016; Ornaghi et al., 2017), and notably given the known reciprocal interactions between microglia and astrocytes.

Besides CMV

Apart from congenital CMV, microglia infection and alteration can occur in a wide range of viral encephalitis. Microglia might participate in neuronal damage seen in encephalitis caused by Japanese encephalitis virus (Thongtan et al., 2012). Microglia status might inform on neuronal dysfunctioning following murine leukemia virus-induced encephalopathy (Xue et al., 2010). Microglia might also participate in HIV-associated cognitive and behavioral impairments (Hong and Banks, 2015). Il34$^{-/-}$ mice with fewer microglia were protected from synaptic defects triggered by West Nile virus infection (Vasek et al., 2016). Interestingly also, acute administration of minocycline prevented against long-term behavioral outcome in a model of temporal lobe epilepsy caused by infection with Theiler's virus (Barker-Haliski et al., 2016). Whether microglia-targeted strategies could be successfully applied to these and other congenital infections of the brain remains to be demonstrated.

From the Rat Model to the Human Disease

Although extrapolation to the corresponding human pathology should be approached with caution, it is tempting to speculate that microglia might also play an important role in human neurological disease following congenital CMV infection. As a matter of fact, microglial nodules and infected microglial cells were detected in human CMV-infected fetal brains (Teissier et al., 2014). While different innovative strategies to combat against CMV infection and its consequences have been proposed recently in complementary rodent models (see Ornaghi et al., 2017), microglia represent another promising target for various neuropathological conditions (Cartier et al., 2014) Strategies directed toward the immune system have already been proposed with the use of corticosteroids in a mouse model of neonatal CMV infection (Kosmac et al., 2013) and with the administration of immunoglobulins to prevent against human CMV congenital infection (Revello et al., 2014). Pharmacological interventions *in utero* have already been successful in rodent models of motor impairment (Yamada et al., 2009) and of epilepsy (Salmi et al., 2013). The use of doxycycline in pregnancy could also be revisited in line with its potential benefits (Cross et al., 2016). Generally risk reduction of fetal infection and subsequent disease still remains a crucial issue (Hamilton et al., 2014). Together with progress in fetal medicine (Bianchi, 2012) and in the identification of reliable biomarkers to predict severity (Desveaux et al., 2016), the present study, which now warrants confirmation and expansion using complementary rodent models of CMV infection and other microglia modifying tools, represents a first proof-of-principle for the future design of microglia-targeted strategies to prevent against the severe neurological outcome of congenital CMV infection of the brain.

CONCLUDING REMARKS

In conclusion, we have shown here that the neurological and other phenotypes caused by CMV infection of the rat brain *in utero*, are caused by the early activation of microglia, rather than by the virus load by itself, at least in this model - suggesting that microglia might also influence the corresponding human disease.

AUTHOR CONTRIBUTIONS

RC and SB performed most experiments and data analyzes, with equal contribution. NT supervised, performed and validated auditory tests and participated in the overall strategy. FS performed most *in utero* injections. HL supervised and participated in flow cytometry. SC performed a subset of immunohistochemistry, confocal microscopy and image analyzes. MS participated in immunohistochemistry and pilot rescue experiments at the initial stages of the project. VP designed and performed most statistical analyzes. EBo performed and analyzed auditory tests. EP-P performed and analyzed qRT-PCR. EBu supervised *in utero* injections and performed a subset of them. FM headed the InMAGIC imaging platform and participated in images acquisition and analyzes. PG participated in the follow-up of the project and in the overall strategy. MM provided strong scientific support on flow

cytometry analyzes. TS and DS provided and purified the RCMV strain and provided strong expertise in virology. NB provided scientific support and advices all along the duration of the project and participated in the overall strategy. PS was the project leader. He decided on the overall strategy, directed the follow-up of experiments, supervised data analysis, and wrote the manuscript with help of RC and SB. All authors contributed the final version of the manuscript.

FUNDING

This work was supported by INSERM (Institut National de la Santé et de la Recherche Médicale), by ANR (Agence Nationale de la Recherche) grant EPILAND to PS (ANR-2010-BLAN-1405 01), and by the PACA (Provence-Alpes-Côte d'Azur) Regional Council (CERVIR to PS, and IM3D3C grants). TS was supported by the Deutsche Forschungsgemeinschaft (SFB796). This work was also carried out within the FHU EPINEXT thanks to the support of the AMIDEX project (ANR-11-IDEX-0001-02) funded by the Investissements d'Avenir French Governement program managed by the French National Research Agency (ANR).

ACKNOWLEDGMENTS

We thank P. Grenot at CIPHE, F. Bader, and S. Corby at INMED animal core facilities, B. Riffault at Neurochlore for his help in iDisco experiments, and H. Child at Oxford University for his help as a trainee *via* the BIOTRAIL international student exchange program.

REFERENCES

Abraham, J., Fox, P. D., Condello, C., Bartolini, A., and Koh, S. (2012). Minocycline attenuates microglia activation and blocks the long-term epileptogenic effects of early-life seizures. *Neurobiol. Dis.* 46, 425–430. doi: 10.1016/j.nbd.2012.02.006

Adler, S. P., and Nigro, G. (2013). "Clinical cytomegalovirus research: congenital infection," in *Cytomegaloviruses: From Molecular Pathogenesis to Intervention*, Vol. 2, ed M. J. Reddehase (Mainz: Caister Academic Press), 55–73.

Alvarado, J. C., Fuentes-Santamaria, V., Gabaldon-Ull, M. C., Blanco, J. L., and Juiz, J. M. (2014). Wistar rats: a forgotten model of age-related hearing loss. *Front. Aging Neurosci.* 6:29. doi: 10.3389/fnagi.2014.00029

Baca Jones, C. C., Kreklywich, C. N., Messaoudi, I., Vomaske, J., Mccartney, E., Orloff, S. L., et al. (2009). Rat cytomegalovirus infection depletes MHC II in bone marrow derived dendritic cells. *Virology* 388, 78–90. doi: 10.1016/j.virol.2009.02.050

Bantug, G. R., Cekinovic, D., Bradford, R., Koontz, T., Jonjic, S., and Britt, W. J. (2008). CD8+ T lymphocytes control murine cytomegalovirus replication in the central nervous system of newborn animals. *J. Immunol.* 181, 2111–2123. doi: 10.4049/jimmunol.181.3.2111

Barker-Haliski, M. L., Vanegas, F., Mau, M. J., Underwood, T. K., and White, H. S. (2016). Acute cognitive impact of antiseizure drugs in naive rodents and corneal-kindled mice. *Epilepsia* 57, 1386–1397. doi: 10.1111/epi.13476

Belle, M., Godefroy, D., Couly, G., Malone, S. A., Collier, F., Giacobini, P., et al. (2017). Tridimensional visualization and analysis of early human development. *Cell* 169, 161–173 e112. doi: 10.1016/j.cell.2017.03.008

Belle, M., Godefroy, D., Dominici, C., Heitz-Marchaland, C., Zelina, P., Hellal, F., et al. (2014). A simple method for 3D analysis of immunolabeled

axonal tracts in a transparent nervous system. *Cell Rep.* 9, 1191–1201. doi: 10.1016/j.celrep.2014.10.037

Bianchi, D. W. (2012). From prenatal genomic diagnosis to fetal personalized medicine: progress and challenges. *Nat. Med.* 18, 1041–1051. doi: 10.1038/nm.2829

Boppana, S. B., Ross, S. A., and Fowler, K. B. (2013). Congenital cytomegalovirus infection: clinical outcome. *Clin Infect Dis* 57(Suppl. 4), S178–S181. doi: 10.1093/cid/cit629

Bradford, R. D., Yoo, Y. G., Golemac, M., Pugel, E. P., Jonjic, S., and Britt, W. J. (2015). Murine CMV-induced hearing loss is associated with inner ear inflammation and loss of spiral ganglia neurons. *PLoS Pathog.* 11:e1004774. doi: 10.1371/journal.ppat.1004774

Britt, W.J., Cekinovic, D., and Jonjic, S. (2013). "Murine model of neonatal cytomegalovirus infection," in *Cytomegaloviruses: From Molecular Pathogenesis to Intervention*, Vol. 2, ed. M. J. Reddehase (Mainz: Caister Academic Press), 119–141.

Brundula, V., Rewcastle, N. B., Metz, L. M., Bernard, C. C., and Yong, V. W. (2002). Targeting leukocyte MMPs and transmigration: minocycline as a potential therapy for multiple sclerosis. *Brain* 125, 1297–1308. doi: 10.1093/brain/awf133

Cartier, N., Lewis, C. A., Zhang, R., and Rossi, F. M. (2014). The role of microglia in human disease: therapeutic tool or target? *Acta Neuropathol.* 128, 363–380. doi: 10.1007/s00401-014-1330-y

Cekinovic, D., Lisnic, V. J., and Jonjic, S. (2014). Rodent models of congenital cytomegalovirus infection. *Methods Mol. Biol.* 1119, 289–310. doi: 10.1007/978-1-62703-788-4_16

Cheeran, M. C., Lokensgard, J. R., and Schleiss, M. R. (2009). Neuropathogenesis of congenital cytomegalovirus infection: disease mechanisms and prospects for intervention. *Clin. Microbiol. Rev.* 22, 99–126, Table of Contents. doi: 10.1128/CMR.00023-08

Cloarec, R., Bauer, S., Luche, H., Buhler, E., Pallesi-Pocachard, E., Salmi, M., et al. (2016). Cytomegalovirus infection of the rat developing brain in utero prominently targets immune cells and promotes early microglial activation. *PLoS ONE* 11:e0160176. doi: 10.1371/journal.pone.0160176

Colver, A., Fairhurst, C., and Pharoah, P. O. (2014). Cerebral palsy. *Lancet* 383, 1240–1249. doi: 10.1016/S0140-6736(13)61835-8

Cross, R., Ling, C., Day, N. P., Mcgready, R., and Paris, D. H. (2016). Revisiting doxycycline in pregnancy and early childhood–time to rebuild its reputation? *Expert Opin. Drug Saf.* 15, 367–382. doi: 10.1517/14740338.2016.1133584

Cunningham, C. L., Martinez-Cerdeno, V., and Noctor, S. C. (2013). Microglia regulate the number of neural precursor cells in the developing cerebral cortex. *J. Neurosci.* 33, 4216–4233. doi: 10.1523/JNEUROSCI.3441-12.2013

Desveaux, C., Klein, J., Leruez-Ville, M., Ramirez-Torres, A., Lacroix, C., Breuil, B., et al. (2016). Identification of Symptomatic Fetuses Infected with Cytomegalovirus Using Amniotic Fluid Peptide Biomarkers. *PLoS Pathog.* 12:e1005395. doi: 10.1371/journal.ppat.1005395

Devinsky, O., Vezzani, A., Najjar, S., De Lanerolle, N. C., and Rogawski, M. A. (2013). Glia and epilepsy: excitability and inflammation. *Trends Neurosci.* 36, 174–184. doi: 10.1016/j.tins.2012.11.008

Dommergues, M. A., Plaisant, F., Verney, C., and Gressens, P. (2003). Early microglial activation following neonatal excitotoxic brain damage in mice: a potential target for neuroprotection. *Neuroscience* 121, 619–628. doi: 10.1016/S0306-4522(03)00558-X

Erturk, A., Becker, K., Jahrling, N., Mauch, C. P., Hojer, C. D., Egen, J. G., et al. (2012). Three-dimensional imaging of solvent-cleared organs using 3DISCO. *Nat. Protoc.* 7, 1983–1995. doi: 10.1038/nprot.2012.119

Fleiss, B., and Gressens, P. (2012). Tertiary mechanisms of brain damage: a new hope for treatment of cerebral palsy? *Lancet Neurol.* 11, 556–566. doi: 10.1016/S1474-4422(12)70058-3

Frost, J. L., and Schafer, D. P. (2016). Microglia: architects of the developing nervous system. *Trends Cell Biol.* 26, 587–597. doi: 10.1016/j.tcb.2016.02.006

Ginhoux, F., Greter, M., Leboeuf, M., Nandi, S., See, P., Gokhan, S., et al. (2010). Fate mapping analysis reveals that adult microglia derive from primitive macrophages. *Science* 330, 841–845. doi: 10.1126/science.1194637

Ginhoux, F., and Jung, S. (2014). Monocytes and macrophages: developmental pathways and tissue homeostasis. *Nat. Rev. Immunol.* 14, 392–404. doi: 10.1038/nri3671

Ginhoux, F., Lim, S., Hoeffel, G., Low, D., and Huber, T. (2013). Origin and differentiation of microglia. *Front. Cell. Neurosci.* 7:45. doi: 10.3389/fncel.2013.00045

Hamilton, S. T., Van Zuylen, W., Shand, A., Scott, G. M., Naing, Z., Hall, B., et al. (2014). Prevention of congenital cytomegalovirus complications by maternal and neonatal treatments: a systematic review. *Rev. Med. Virol.* 24, 420–433. doi: 10.1002/rmv.1814

Hong, S., and Banks, W. A. (2015). Role of the immune system in HIV-associated neuroinflammation and neurocognitive implications. *Brain Behav. Immun.* 45, 1–12. doi: 10.1016/j.bbi.2014.10.008

Hong, S., Beja-Glasser, V. F., Nfonoyim, B. M., Frouin, A., Li, S., Ramakrishnan, S., et al. (2016). Complement and microglia mediate early synapse loss in Alzheimer mouse models. *Science* 352, 712–716. doi: 10.1126/science.aad8373

Kannan, S., Dai, H., Navath, R. S., Balakrishnan, B., Jyoti, A., Janisse, J., et al. (2012). Dendrimer-based postnatal therapy for neuroinflammation and cerebral palsy in a rabbit model. *Sci. Transl. Med.* 4:130ra146. doi: 10.1126/scitranslmed.3003162

Kawasaki, H., Kosugi, I., Sakao-Suzuki, M., Meguro, S., Arai, Y., Tsutsui, Y., et al. (2015). Cytomegalovirus initiates infection selectively from high-level beta1 integrin-expressing cells in the brain. *Am. J. Pathol.* 185, 1304–1323. doi: 10.1016/j.ajpath.2015.01.032

Khazipov, R., Zaynutdinova, D., Ogievetsky, E., Valeeva, G., Mitrukhina, O., Manent, J. B., et al. (2015). Atlas of the postnatal rat brain in stereotaxic coordinates. *Front. Neuroanat.* 9:161. doi: 10.3389/fnana.2015.00161

Kosmac, K., Bantug, G. R., Pugel, E. P., Cekinovic, D., Jonjic, S., and Britt, W. J. (2013). Glucocorticoid treatment of MCMV infected newborn mice attenuates CNS inflammation and limits deficits in cerebellar development. *PLoS Pathog.* 9:e1003200. doi: 10.1371/journal.ppat.1003200

Kosugi, I., Kawasaki, H., Arai, Y., and Tsutsui, Y. (2002). Innate immune responses to cytomegalovirus infection in the developing mouse brain and their evasion by virus-infected neurons. *Am. J. Pathol.* 161, 919–928. doi: 10.1016/S0002-9440(10)64252-6

Lokensgard, J. R., Mutnal, M. B., Prasad, S., Sheng, W., and Hu, S. (2016). Glial cell activation, recruitment, and survival of B-lineage cells following MCMV brain infection. *J. Neuroinflamm.* 13:114. doi: 10.1186/s12974-016-0582-y

Matcovitch-Natan, O., Winter, D. R., Giladi, A., Vargas Aguilar, S., Spinrad, A., Sarrazin, S., et al. (2016). Microglia development follows a stepwise program to regulate brain homeostasis. *Science* 353:aad8670. doi: 10.1126/science.aad8670

Ohlemiller, K. K. (2009). Mechanisms and genes in human strial presbycusis from animal models. *Brain Res.* 1277, 70–83. doi: 10.1016/j.brainres.2009.02.079

Ornaghi, S., Hsieh, L. S., Bordey, A., Vergani, P., Paidas, M. J., and Van Den Pol, A. N. (2017). Valnoctamide inhibits cytomegalovirus infection in developing brain and attenuates neurobehavioral dysfunctions and brain abnormalities. *J. Neurosci.* 37, 6877–6893. doi: 10.1523/JNEUROSCI.0970-17.2017

Paolicelli, R. C., Bolasco, G., Pagani, F., Maggi, L., Scianni, M., Panzanelli, P., et al. (2011). Synaptic pruning by microglia is necessary for normal brain development. *Science* 333, 1456–1458. doi: 10.1126/science.1202529

Paolicelli, R. C., and Ferretti, M. T. (2017). Function and dysfunction of microglia during brain development: consequences for synapses and neural circuits. *Front. Synaptic Neurosci.* 9:9. doi: 10.3389/fnsyn.2017.00009

Ponomarev, E. D., Veremeyko, T., Barteneva, N., Krichevsky, A. M., and Weiner, H. L. (2011). MicroRNA-124 promotes microglia quiescence and suppresses EAE by deactivating macrophages via the C/EBP-alpha-PU.1 pathway. *Nat. Med.* 17, 64–70. doi: 10.1038/nm.2266

Popovic, N., Schubart, A., Goetz, B. D., Zhang, S. C., Linington, C., and Duncan, I. D. (2002). Inhibition of autoimmune encephalomyelitis by a tetracycline. *Ann. Neurol.* 51, 215–223. doi: 10.1002/ana.10092

Prinz, M., and Priller, J. (2014). Microglia and brain macrophages in the molecular age: from origin to neuropsychiatric disease. *Nat. Rev. Neurosci.* 15, 300–312. doi: 10.1038/nrn3722

Renier, N., Wu, Z., Simon, D. J., Yang, J., Ariel, P., and Tessier-Lavigne, M. (2014). iDISCO: a simple, rapid method to immunolabel large tissue samples for volume imaging. *Cell* 159, 896–910. doi: 10.1016/j.cell.2014.10.010

Revello, M. G., Lazzarotto, T., Guerra, B., Spinillo, A., Ferrazzi, E., Kustermann, A., et al. (2014). A randomized trial of hyperimmune globulin to prevent congenital cytomegalovirus. *N. Engl. J. Med.* 370, 1316–1326. doi: 10.1056/NEJMoa1310214

Revello, M. G., Zavattoni, M., Baldanti, F., Sarasini, A., Paolucci, S., and Gerna, G. (1999). Diagnostic and prognostic value of human cytomegalovirus load and IgM antibody in blood of congenitally infected newborns. *J. Clin. Virol.* 14, 57–66. doi: 10.1016/S1386-6532(99)00016-5

Roumier, A., Pascual, O., Bechade, C., Wakselman, S., Poncer, J. C., Real, E., et al. (2008). Prenatal activation of microglia induces delayed impairment of glutamatergic synaptic function. *PLoS ONE* 3:e2595. doi: 10.1371/journal.pone.0002595

Rousset, C. I., Kassem, J., Aubert, A., Planchenault, D., Gressens, P., Chalon, S., et al. (2013). Maternal exposure to lipopolysaccharide leads to transient motor dysfunction in neonatal rats. *Dev. Neurosci.* 35, 172–181. doi: 10.1159/000346579

Sakao-Suzuki, M., Kawasaki, H., Akamatsu, T., Meguro, S., Miyajima, H., Iwashita, T., et al. (2014). Aberrant fetal macrophage/microglial reactions to cytomegalovirus infection. *Ann. Clin. Trans. Neurol.* 1, 570–588. doi: 10.1002/acn3.88

Salmi, M., Bruneau, N., Cillario, J., Lozovaya, N., Massacrier, A., Buhler, E., et al. (2013). Tubacin prevents neuronal migration defects and epileptic activity caused by rat Srpx2 silencing in utero. *Brain* 136, 2457–2473. doi: 10.1093/brain/awt161

Schafer, D. P., Lehrman, E. K., Kautzman, A. G., Koyama, R., Mardinly, A. R., Yamasaki, R., et al. (2012). Microglia sculpt postnatal neural circuits in an activity and complement-dependent manner. *Neuron* 74, 691–705. doi: 10.1016/j.neuron.2012.03.026

Seleme, M. C., Kosmac, K., Jonjic, S., and Britt, W. J. (2017). Tumor necrosis factor alpha-induced recruitment of inflammatory mononuclear cells leads to inflammation and altered brain development in murine cytomegalovirus-infected newborn mice. *J. Virol.* 91:pii: e01983-16. doi: 10.1128/JVI.01983-16

Slavuljica, I., Kvestak, D., Huszthy, P. C., Kosmac, K., Britt, W. J., and Jonjic, S. (2015). Immunobiology of congenital cytomegalovirus infection of the central nervous system-the murine cytomegalovirus model. *Cell. Mol. Immunol.* 12, 180–191. doi: 10.1038/cmi.2014.51

Smith, R. J., Bale, J. F. Jr., and White, K. R. (2005). Sensorineural hearing loss in children. *Lancet* 365, 879–890. doi: 10.1016/S0140-6736(05)71047-3

Suzuki, Y., Toribe, Y., Mogami, Y., Yanagihara, K., and Nishikawa, M. (2008). Epilepsy in patients with congenital cytomegalovirus infection. *Brain Dev.* 30, 420–424. doi: 10.1016/j.braindev.2007.12.004

Teissier, N., Fallet-Bianco, C., Delezoide, A. L., Laquerriere, A., Marcorelles, P., Khung-Savatovsky, S., et al. (2014). Cytomegalovirus-induced brain malformations in fetuses. *J. Neuropathol. Exp. Neurol.* 73, 143–158. doi: 10.1097/NEN.0000000000000038

Thongtan, T., Thepparit, C., and Smith, D. R. (2012). The involvement of microglial cells in Japanese encephalitis infections. *Clin. Dev. Immunol.* 2012:890586. doi: 10.1155/2012/890586

Tikka, T., Fiebich, B. L., Goldsteins, G., Keinanen, R., and Koistinaho, J. (2001). Minocycline, a tetracycline derivative, is neuroprotective against excitotoxicity by inhibiting activation and proliferation of microglia. *J. Neurosci.* 21, 2580–2588. Available online at: http://www.jneurosci.org/content/21/8/2580.long

Vasek, M. J., Garber, C., Dorsey, D., Durrant, D. M., Bollman, B., Soung, A., et al. (2016). A complement-microglial axis drives synapse loss during virus-induced memory impairment. *Nature* 534, 538–543. doi: 10.1038/nature18283

Vezzani, A., Auvin, S., Ravizza, T., and Aronica, E. (2012). "Glia-neuronal interactions in ictogenesis and epileptogenesis: role of inflammatory mediators," in *Jasper's Basic Mechanisms of the Epilepsies. 4th Edn.*, eds J. L. Noebels, M. Avoli, M. A. Rogawski (Bethesda, MD: National Center for Biotechnology Information).

Xue, Q. S., Yang, C., Hoffman, P. M., and Streit, W. J. (2010). Microglial response to murine leukemia virus-induced encephalopathy is a good indicator of neuronal perturbations. *Brain Res.* 1319, 131–141. doi: 10.1016/j.brainres.2009.12.089

Yamada, M., Yoshida, Y., Mori, D., Takitoh, T., Kengaku, M., Umeshima, H., et al. (2009). Inhibition of calpain increases LIS1 expression and partially rescues *in vivo* phenotypes in a mouse model of lissencephaly. *Nat. Med.* 15, 1202–1207. doi: 10.1038/nm.2023

Yrjanheikki, J., Keinanen, R., Pellikka, M., Hokfelt, T., and Koistinaho, J. (1998). Tetracyclines inhibit microglial activation and are neuroprotective in global brain ischemia. *Proc. Natl. Acad. Sci. U.S.A.* 95, 15769–15774. doi: 10.1073/pnas.95.26.15769

Yrjanheikki, J., Tikka, T., Keinanen, R., Goldsteins, G., Chan, P. H., and Koistinaho, J. (1999). A tetracycline derivative, minocycline, reduces inflammation and protects against focal cerebral ischemia with a wide therapeutic window. *Proc. Natl. Acad. Sci. U.S.A.* 96, 13496–13500. doi: 10.1073/pnas.96.23.13496

13

Early Microglia Activation Precedes Photoreceptor Degeneration in a Mouse Model of CNGB1-Linked Retinitis Pigmentosa

Thomas Blank [1*†], Tobias Goldmann [1,2*†], Mirja Koch [3†], Lukas Amann [1,4], Christian Schön [3], Michael Bonin [5,6], Shengru Pang [1,4], Marco Prinz [1,7], Michael Burnet [2], Johanna E. Wagner [3], Martin Biel [3] and Stylianos Michalakis [3*]

[1] Institute of Neuropathology, Faculty of Medicine, University of Freiburg, Freiburg, Germany, [2] In Vivo Pharmacology, Synovo GmbH, Tübingen, Germany, [3] Center for Integrated Protein Science Munich CiPSM and Department of Pharmacy, Center for Drug Research, Ludwig-Maximilians-Universität München, Munich, Germany, [4] Faculty of Biology, University of Freiburg, Freiburg, Germany, [5] Institute for Medical Genetics and Applied Genomics Transcriptomics, University of Tübingen, Tübingen, Germany, [6] IMGM Laboratories GmbH, Planegg, Germany, [7] BIOSS Centre for Biological Signalling Studies, University of Freiburg, Freiburg, Germany

*Correspondence:
Thomas Blank
thomas.blank@uniklinik-freiburg.de;
Tobias Goldmann
tobias.goldmann@gmail.com;
Stylianos Michalakis
stylianos.michalakis@cup.
uni-muenchen.de

†These authors have contributed
equally to this work.

Retinitis pigmentosa (RP) denotes a family of inherited blinding eye diseases characterized by progressive degeneration of rod and cone photoreceptors in the retina. In most cases, a rod-specific genetic defect results in early functional loss and degeneration of rods, which is followed by degeneration of cones and loss of daylight vision at later stages. Microglial cells, the immune cells of the central nervous system, are activated in retinas of RP patients and in several RP mouse models. However, it is still a matter of debate whether activated microglial cells may be responsible for the amplification of the typical degenerative processes. Here, we used $Cngb1^{-/-}$ mice, which represent a slow degenerative mouse model of RP, to investigate the extent of microglia activation in retinal degeneration. With a combination of FACS analysis, immunohistochemistry and gene expression analysis we established that microglia in the $Cngb1^{-/-}$ retina were already activated in an early, predegenerative stage of the disease. The evidence available so far suggests that early retinal microglia activation represents a first step in RP, which might initiate or accelerate photoreceptor degeneration.

Keywords: retinitis pigmentosa, retinal degeneration, cyclic nucleotide-gated channel, microglia, innate immune response

INTRODUCTION

It is generally accepted that immune responses follow injury and damage to tissues and organs. Microglia are the resident immune cells within the brain and retina, commonly known as the macrophages of the central nervous system (CNS). In response to injury or inflammatory stimuli, the resting microglia can be rapidly activated to participate in pathological responses, including migration to the affected site, release of various inflammatory molecules, and clearing of cellular debris (1–3). Although microglia are essential for maintaining a healthy CNS, paradoxically they may undergo phenotypic changes to influence several neurodegenerative diseases and psychiatric disorders including Alzheimer's disease (AD), Parkinson's disease, and Rett syndrome (4). Moreover, activation of microglia has also been detected in several retinal degenerative mouse models (5, 6) and in patients suffering from retinitis pigmentosa (RP) (7). RP describes a heterogeneous group of hereditary retinal degenerations with a world-wide prevalence of 1:4,000 (8). To date, more than 50 different genetic mutations have been detected, which cause non-syndromic RP (9). RP is characterized by an initial progressive degeneration of rods and followed by the loss of cones leading to severe visual impairment (8, 10). It should be noted that the disease severity, rate of disease progression, age of onset and clinical findings may differ significantly among patients based on the fact that RP represents a heterogeneous group of inherited retinal disorders (11). Typically, the earliest clinical symptom of RP is an initial night blindness caused by the dysfunctional rod system. Subsequent degeneration of cones leads to a gradual loss of the visual field, which initially impairs the periphery and spreads to the macula. The consequences include so-called "tunnel vision" and eventually complete blindness (10).

Here, we used the *Cngb1* knockout (*Cngb1⁻/⁻*) mouse to study the activation of immune cells in a model of RP with slowly progressing photoreceptor degeneration. *Cngb1* encodes the B subunit of the cyclic nucleotide-gated channel in rod photoreceptors. *Cngb1⁻/⁻* mice show initial signs of rod degeneration including gliosis already between 14 and 21 days of age (12) while the peak of neuronal cell death occurs around 4 weeks of age (13). Even though the degenerative process begins already at this early age, the degeneration advances very slowly and shows a slower progression of the disease when compared with other RP mouse models like the rd1 mice (14). For RP a general feature is that cone photoreceptors deteriorate secondary to rods with a considerable slower progression rate (12). In the present study, we found that microglia in the *Cngb1⁻/⁻* retina showed already increased cell numbers and pronounced activation in 4-week-old mice. At this time point only a minor photoreceptor cell loss was detected. Our data suggest that *Cngb1⁻/⁻* microglia are potentially an early driving force, which substantially contributes to the retinal degeneration and long-term visual impairments found in RP.

MATERIALS AND METHODS

Animals

Cngb1⁻/⁻ were generated by us (12). All mice used in the study were bred on a mixed genetic background of the 129Sv and C57BL/6N strain. Animals were housed under standard white light (200 lux, 12 h dark–light periods) with free access to food and water. Both male and female mice were used in equal shares. Age-matched wild-type mice were used as controls. Day of birth was considered as postnatal day 1 (P1). All procedures concerning animals were performed with permission of the local authority (Regierung von Oberbayern and RP Freiburg).

Optical Coherence Tomography (OCT) Analysis

For OCT examinations, mice received intraperitoneal injections of ketamin (0.1 mg/g) and xylazin (0.02 mg/g). Before the scanning procedure, Tropicamid eye drops were instilled into the eye for pupil dilation (Mydriadicum Stulln, Pharma Stulln GmbH, Stulln, Germany). Subsequently, hydroxylpropyl methylcellulose (Methocel 2%; OmniVision, Puchheim, Germany) was applied to keep the eyes moist. The examination was performed with an adapted Spectralis HRA + OCT system by Heidelberg Engineering (Dossenheim, Germany) in combination with optic lenses described previously (15). OCT scans were conducted using a 12 circular scan mode centered at the optic nerve head. This procedure allowed for measurements of the photoreceptor layer thickness at a comparable distance from the optic nerve head. In detail, outer nuclear layer (ONL) thickness was measured between the clearly visible outer limiting membrane and the outer plexiform layer (OPL). For statistical analysis, the mean ONL thickness was calculated from single values measured in the dorsal, temporal, nasal, and ventral region around the optic nerve.

Microarray Analysis

For microarray experiments, retinal tissue was obtained from mice of two different age groups (P12 and P28). For differential gene expression analysis of *Cngb1⁻/⁻* and wt animals, an Affymetrix platform was used according to the manufacturer's instructions as described before (16). In short, retinas were dissected, shock-frozen in liquid nitrogen and stored at −80°C until further use. RNA was extracted using RNeasy Minikit (Qiagen, Hilden, Germany) according to the manufacturer's instructions. RNA concentration and purity were determined using NanoDrop2000 (Thermo Scientific®). Fragmented and labeled cRNA of three wild-type and three *Cngb1⁻/⁻* retinas was hybridized on Affymetrix Mouse Genome 430 2.0 Arrays, respectively. A probe-level summary was determined with the help of Affymetrix GeneChip Operating Software using the MAS5 algorithm. Raw data were normalized using the Array Assist Software 4.0 (Stratagene, La Jolla, CA, USA) in combination with the GC-robust multichip average algorithm. Significance was determined by a *t*-test without multiple testing correction (Array Assist software), selecting all transcripts with a minimum change in expression level of 1.5-fold together with a *p*–value <0.05.

Quantitative PCR

cDNA synthesis was performed with the RevertAid First Strand cDNA Synthesis Kit (Thermo Scientific) according to the manufacturer's manual. PCR was performed on a StepOnePlus Real-Time PCR System (Applied Biosystems) using SYBR Select Master Mix (Applied Biosystems). For quantitative PCR, two technical replicates per gene were generated and normalized to the housekeeping gene aminolevulinic acid synthase. The following primer sets were used:

Gene	Forward primer (5'→3')	Reverse primer (5'→3')
Irf8	GCTGATCAAGGAACCTTGTG	CAGGCCTGCACTGGGCTG
Aif1/Iba-1	ATCAACAAGCAATTCCTCGATGA	CAGCATTCGCTTCAAGGACATA
C1qc	CCCAGTTGCCAGCCTCAAT	GGAGTCCATCATGCCCGTC
Cx3cr1	GAGTATGACGATTCTGCTGAGG	CAGACCGAACGTGAAGACGAG

Flow Cytometry

$Cngb1^{-/-}$ and wt mice were euthanized at P28 and perfused with phosphate-buffered saline. Retinas were removed and mechanically dissociated into single cell suspensions by pipetting. Dissociated cells were stained with live/dead dye (1:1,000, eBioscience) in PBS for 30 min at 4°C. In order to prevent unspecific binding to Fc receptors, their binding domains were blocked by unstained CD16/32 (1:250, 2.4G2, Becton Dickinson) in FACS-Buffer (2% FCS, 5 mM EDTA in PBS) for 20 min at 4°C. Cells were stained with CD11b (BV421, 1:300, M1/70, eBioscience), CD45 (APC-eF780, 1:200, 30-F11, eBioscience), F4/80 (PE; 1:200, BM8, eBioscience), CD44 (PE, 1:200 IM7, Becton Dickinson), and MHC class II (PE, 1:200, M5/114.15.2, eBioscience) in FACS-Buffer at 4°C for 20 min and analyzed using a FACSCanto II (Becton Dickinson). Viable cells were gated by forward and side scatter pattern. Data were acquired with FACSdiva software (Becton Dickinson). Postacquisition analysis was performed using FlowJo software (Tree Star, Inc.).

Retina Preparation and Immunohistochemistry

Retinas were dissected at P28 and further processed as described for immunohistology and whole mount preparation (17–19). Primary antibodies were added overnight at a dilution of 1:500 for Iba-1 (019-19741, WACO, Japan), 1:250 for Lamp 2 (ab13524, Abcam, Cambridge, UK), 1:200 for cleaved caspase-3 (9661, Cell Signaling Technology, Danvers, MA, USA) and 1:100 for Mhc class II (ab23990, Abcam), at 4°C. Secondary antibodies were added at the following dilution: Alexa Flour 488 1:500, Alexa Flour 555 1:500 and Alexa Fluor 568 1:500 for 2 h at room temperature. Nuclei were counterstained with DAPI. The examined area was determined microscopically by a TCS SP8 confocal scan microscope (Leica) or a conventional fluorescence microscope (Olympus BX-61).

Visual Cliff

The visual cliff behavior was analyzed in an open-top Plexiglas chamber. Half of the box protruded from the counter to provide a 3-foot depth. The box on the counter displayed a base with a checkerboard pattern and the box off the counter showed the base with the same checkerboard pattern, except for the 3 feet of depth. The mouse was placed on the dividing line between both halves of the chamber at the edge of the counter and was allowed to choose between the two sides. If the mouse stepped to the shallow side, time was scored as time spent on the "safe side." Each mouse performed this task twice for 10 min with a time window of 1 h between trials. The visual cliff behavior was averaged to generate mean percentage of time in which the mouse chose to stay at the shallow side ($n = 3$–5 mice per group).

Statistical Analysis

All graphical data represent mean ± SEM. Sample sizes are provided in the figure legends. In order to test for significant differences, an unpaired t-test was applied. Differences were considered as significant when p-value <0.05.

RESULTS

At 28 days after birth (P28), minor (~15%) but significant retina degeneration was observed in $Cngb1^{-/-}$ mice (**Figures 1A,B**) (12) and more than 1,000 genes were dysregulated (>1.5-fold dysregulated, $p < 0.05$, STab.1) as seen from Affymetrix gene chip arrays. Already at this early time point $Cngb1^{-/-}$ mice displayed substantial visual impairment (**Figure 1C**). We were analyzing gene expression data with the help of Ingenuity Pathway Analysis software, to identify potential shifts in biological functions or in canonical pathways at early and predegenerative stages. Interestingly, activation of the immune system was already apparent 12 days after birth (P12) as indicated by upregulated genes that were involved in processes like antigen presentation, immune cell trafficking, immunological diseases, humoral immune response, and inflammatory disease (**Figures 1D,E**). At P28 many of the dysregulated genes were attributed to cell death, survival, or neurological diseases and also to pathways and signaling cascades that are assigned to immunological processes (**Figures 1F,G**; Table S1 in Supplementary Material). Particularly genes linked to inflammatory responses, inflammatory diseases and immune cell trafficking were significantly altered in $Cngb1^{-/-}$ retinas at P28 when compared to wt retinas. A detailed analysis of upstream regulators (URs), which are not directly altered in their expression level but are responsible for expression changes of their target genes, revealed the activation of diverse proinflammatory mediators like TNF, IL6, and NF-κB in $Cngb1^{-/-}$ retinas at P28 (**Figure 1H**).

The generated microarray data clearly suggested the presence of an activated immune system in $Cngb1^{-/-}$ retinas. That is why we focused on microglial cells, which represent the immune competent cells of the CNS and retina (5). Microglia tend to proliferate upon tissue destruction during neurodegeneration in order to clear the cellular debris and to restore tissue homeostasis (4). Thus, we first determined microglia cell numbers using flow cytometry (FACS), histological and qPCR approaches. For FACS analysis, we gated microglia as live $CD45^{lo}CD11b^+$ cells (**Figures 2A,B**). Quantification of single cell suspensions prepared from $Cngb1^{-/-}$ and wt retinas 4 weeks after birth revealed significantly increased microglia cell numbers in $Cngb1$-deficient mice compared to age-matched wt (**Figure 2B**). $CD45^{hi}CD11b^+$ cell numbers were not increased (**Figure 2B**). Immunofluorescence of Iba-1, a specific marker for microglia and macrophages, confirmed a strong elevation of this immune cell population in retinas of $Cngb1^{-/-}$ mice (**Figures 2C,D**). In addition to increased Iba-1$^+$ cell numbers in degenerating retinas of $Cngb1^{-/-}$ mice, microglial cells also changed their localization. In wt retinas, microglia cells were mainly found in the inner plexiform layer (IPL) or OPL, while Iba-1-positive cells of $Cngb1^{-/-}$ retinas were additionally found in the ONL and in the photoreceptor layer close to the retinal epithelium (**Figure 2C**, asterisk; **Figures 3A,C**). Further analysis of wt and $Cngb1^{-/-}$ retina microarray data indicated that several microglia-specific genes like $Cx3cr1$, $Aif1$, $Irf8$, $C1qc$ (20–22) were upregulated in the $Cngb1^{-/-}$ group (Table S1 in Supplementary Material). Subsequent RT-qPCR analysis of these microglia cell-specific genes confirmed their increased expression levels (**Figure 2E**). In summary, microglial cell numbers were strongly

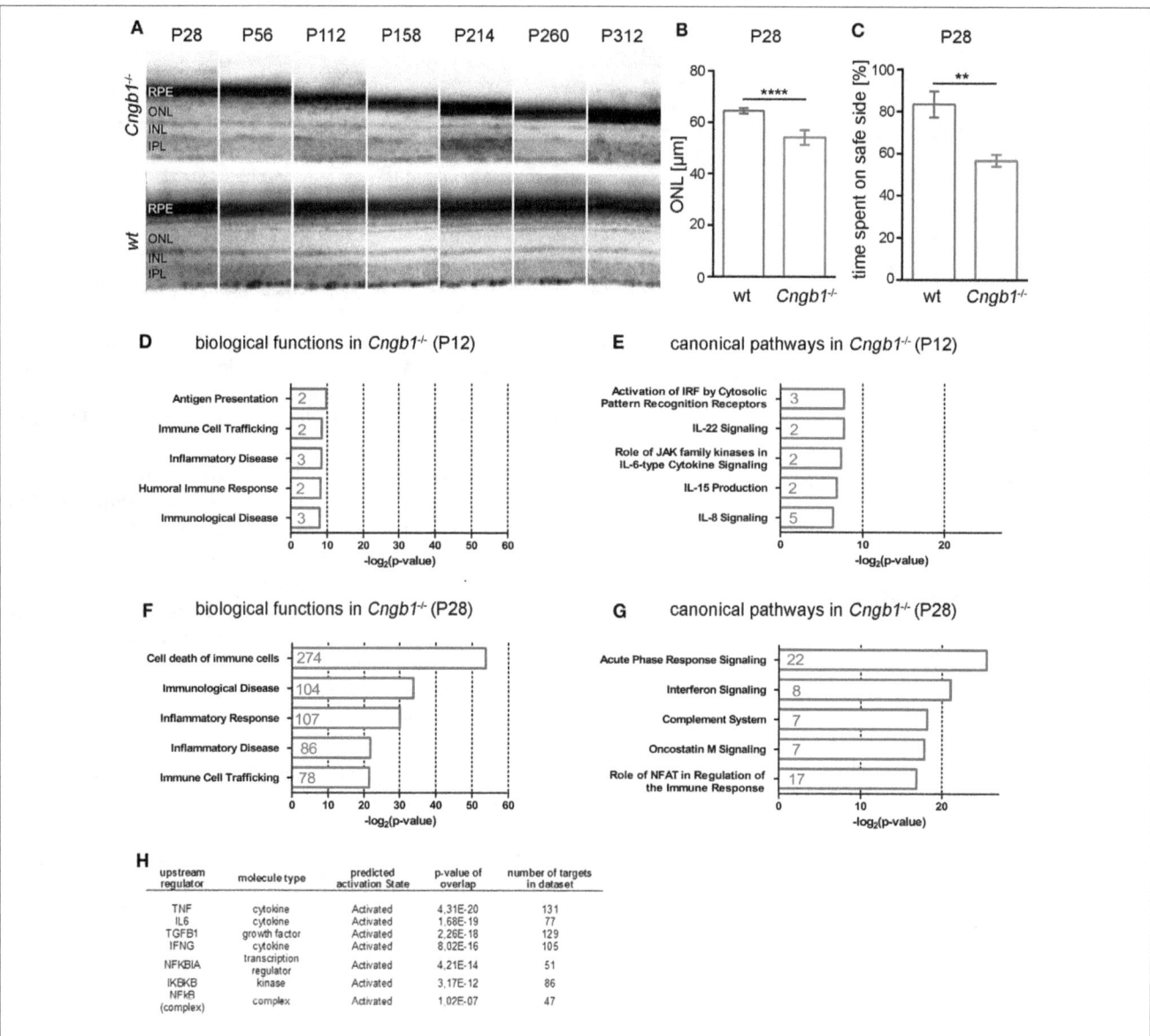

FIGURE 1 | Activation of immunological pathways in *Cngb1⁻ᐟ⁻* retinas at 12 and 28 days after birth. **(A)** Representative optical coherence tomography (OCT) images from *Cngb1⁻ᐟ⁻* (upper panel) and wild-type retinas (lower panel), which display the slow progression (P28-P312) of outer nuclear layer (ONL) thinning over time. **(B)** Quantification of ONL thickness from OCT data at P28 (****$p < 0.0001$, $n = 6$ for each genotype). **(C)** Performance in the visual cliff test for the study of visual depth perception (**$p < 0.01$, $n = 3$ for each genotype). **(D–G)** Biological functions and canonical pathways were significantly altered in *Cngb1⁻ᐟ⁻* mice compared to age-matched controls. A high number of genes were related to the immune system or to immune responses. **(H)** The indicated upstream regulators for proinflammatory cytokines and the NF-κB pathway were predicted to have a significantly higher activation state in *Cngb1⁻ᐟ⁻* retinas when compared to age-matched wild-type controls ($p < 0.05$).

increased in *Cngb1⁻ᐟ⁻* mice at 4 weeks of age, which corresponds to an early, degenerative stage of the disease.

In response to disrupted tissue homeostasis, microglial cells get activated and change their morphology together with the expression of surface markers (4). In *Cngb1⁻ᐟ⁻* retinas, microglia showed a transition from a resting to an activated state (**Figure 3A**). The cells underwent morphological changes to take on an amoeboid shape with fewer branches compared to the resting state phenotype in wt retinas (23, 24). As specialized phagocytes, one of the functions microglia have is to remove

debris of dying or dead cells (25). In mice, CD44 is a competent receptor for phagocytosis in macrophages (26) and an increase of CD44 expression was detected in the initial microarray data analysis (STab.1). Subsequent FACS expression analysis of CD45^lo CD11b⁺ cells could link CD44 to microglia, as the number of CD45^lo CD11b⁺CD44⁺ cells as well as the expression levels of CD44 in microglia in *Cngb1⁻ᐟ⁻* retinas were elevated (**Figure 3B**). Active phagocytosis of microglial cells can also be monitored *in situ* by immunohistological staining of lysosome-associated membrane protein (lamp)-2 (19). In P28 *Cngb1⁻ᐟ⁻* retinas, we

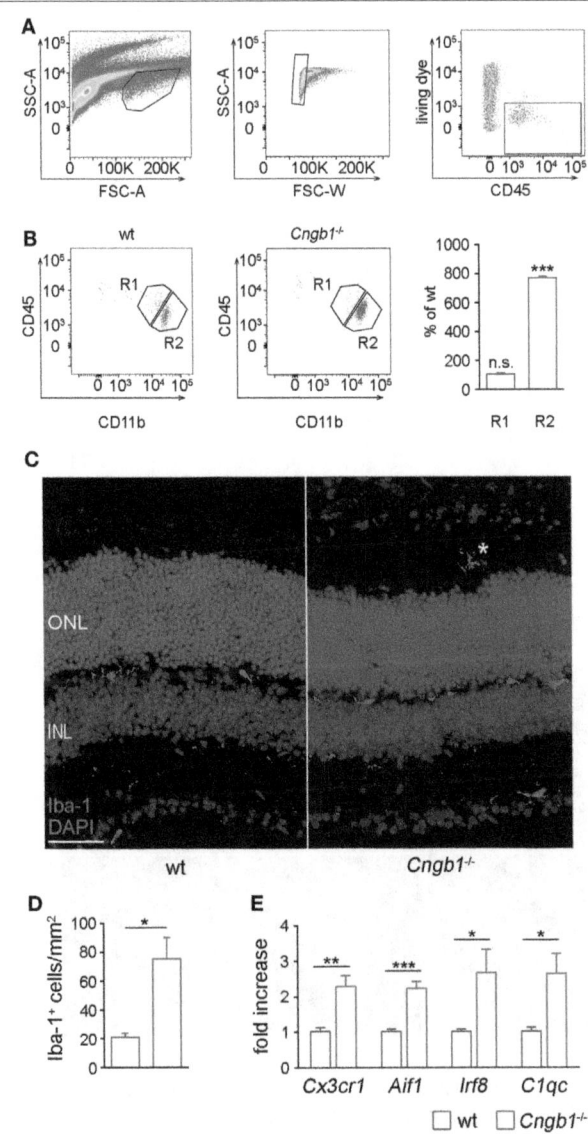

FIGURE 2 | Strong increase of retinal microglia cells in *Cngb1*$^{-/-}$ retinas 28 days after birth. **(A)** Gating strategy for identifying microglia cells in retina lysates by size (left panel), single cells (middle panel), and living CD45$^+$ cells (right panel) expression. **(B)** Representative gating and quantification of CD11b$^+$CD45^{low} retinal microglia (R2) and CD11b$^+$CD45^{hi} cells (R1) from wild-type or *Cngb1*$^{-/-}$ retinas. (Statistical significance was determined vs. percentage of wild-type cell counts, ns = non significant; ***$p < 0.001$, $n = 3$ for each genotype.) **(C)** Immunofluorescence of Iba-1 (red) in the retina of *Cngb1*$^{-/-}$ and age-matched wt demonstrating migration of microglia into the photoreceptor layer of *Cngb1*$^{-/-}$ mice (asterisk). Nuclei were counterstained with DAPI. ONL, outer nuclear layer, INL, inner nuclear layer, scale bar 50 µm. **(D)** Quantification of Iba-1-positive cells in *Cngb1*$^{-/-}$ mice compared to wt animals (*$p < 0.05$, $n = 4$ for each genotype). **(E)** Gene expression of microglia-specific genes in *Cngb1*$^{-/-}$ and wt retinas. Significant increase in the expression of *Cx3cr1*, *Aif1*, *Irf8*, and *C1qc* in retinas of *Cngb1*-deficient mice (*$p < 0.05$, **$p < 0.01$, ***$p < 0.001$, $n = 4$ for each genotype).

detected Lamp-2-positive microglia particularly in the ONL and photoreceptor layer (**Figure 3C**). Increased MHC class II expression, which indicates microglial activation, was further observed by FACS and immunohistochemistry in *Cngb1*$^{-/-}$ retinas when

compared to wt retinas (**Figure 3D**), while the expression levels of the macrophage marker F4/80 remained unchanged in both genotypes (**Figure 3B**). Although the neurodegenerative process became evident at P28, microglia were activated already at P12 (**Figures 4A,C–E**). At this time point no apoptotic cells as indicated by the absence of positive signals for cleaved caspase 3 were present (**Figures 4B,F**).

The Ingenuity UR analysis indicated increased activity of diverse proinflammatory signaling cascades (**Figure 1H**). One of these cascades was the NF-κB pathway, which can be induced by a variety of signals to finally induce a specific pattern of transcription. In this classical pathway, activated IKK-β, which is part of an IKK-α–IKK-β–IKK-γ complex, phosphorylates the inhibitory subunits IkB-α, IkB-β, or IkB-ε, leading to their proteasomal degradation. As a result, NF-κB homodimers and heterodimers, mainly composed of RelA, RelC, and p50, accumulate in the nucleus (27). Here, we confirmed the presence of an activated NF-κB-signaling pathway in microglia by the immunofluorescent detection of phosphorylated IκB colocalized with Iba-1-positive microglia, which were misplaced in the outer segment of *Cngb1*$^{-/-}$ retinas (**Figure 5**).

DISCUSSION

Our present results link retinal degeneration to immune system activation, and here, more precisely, to the activation of microglia. In *Cngb1*$^{-/-}$ mice, neuronal cell death does not start before postnatal day 15 (P15) (12). Already before this early stage of degeneration, retinal gene expression analysis at P12 indicated an immune response in biological and canonical pathways. These data clearly indicated that activation of the immune system starts prior to the actual retinal degenerative process in *Cngb1*$^{-/-}$ mice. Between P21 and P28 retinal degeneration reaches its maximum (12). We determined activation of the immune system by gene expression pathway analyses and immunohistochemical detection. At this early disease stage, microglia had already migrated entirely through the various layers of the *Cngb1*$^{-/-}$ retina toward the photoreceptors. Microglial cells are the local immune cells of the CNS and normally reside at the IPL/OPL of the retina (5, 28). Upon activation, microglia migrate toward the injury site, change their morphology from ramified cells to amoeboid phagocytes and start expressing several surface markers including F4/80, MHCII, and complement receptor 3 (CD11b/18, Ox42) (1, 5, 29). Our findings suggest that microglial activation occurs before the onset of neurodegeneration. This early microglia activation might be responsible for the observed high CD44 representation in *Cngb1*$^{-/-}$ retinas. CD44 is implicated in the pathogenesis of inflammation and contributes to the recruitment of inflammatory cells as well as to increased phagocytosis (26, 30, 31). Increased relative expression of the cell surface adhesion receptor CD44 seems to be a very general feature of retinal degeneration considering that it was also present in rd10 mice (32). Previous work using the rd10 mouse model of RP had already suggested a contribution of microglia in retinal degeneration (33). In this mouse model activated microglia infiltrate into the photoreceptor layer and contribute actively to photoreceptor demise *via* the

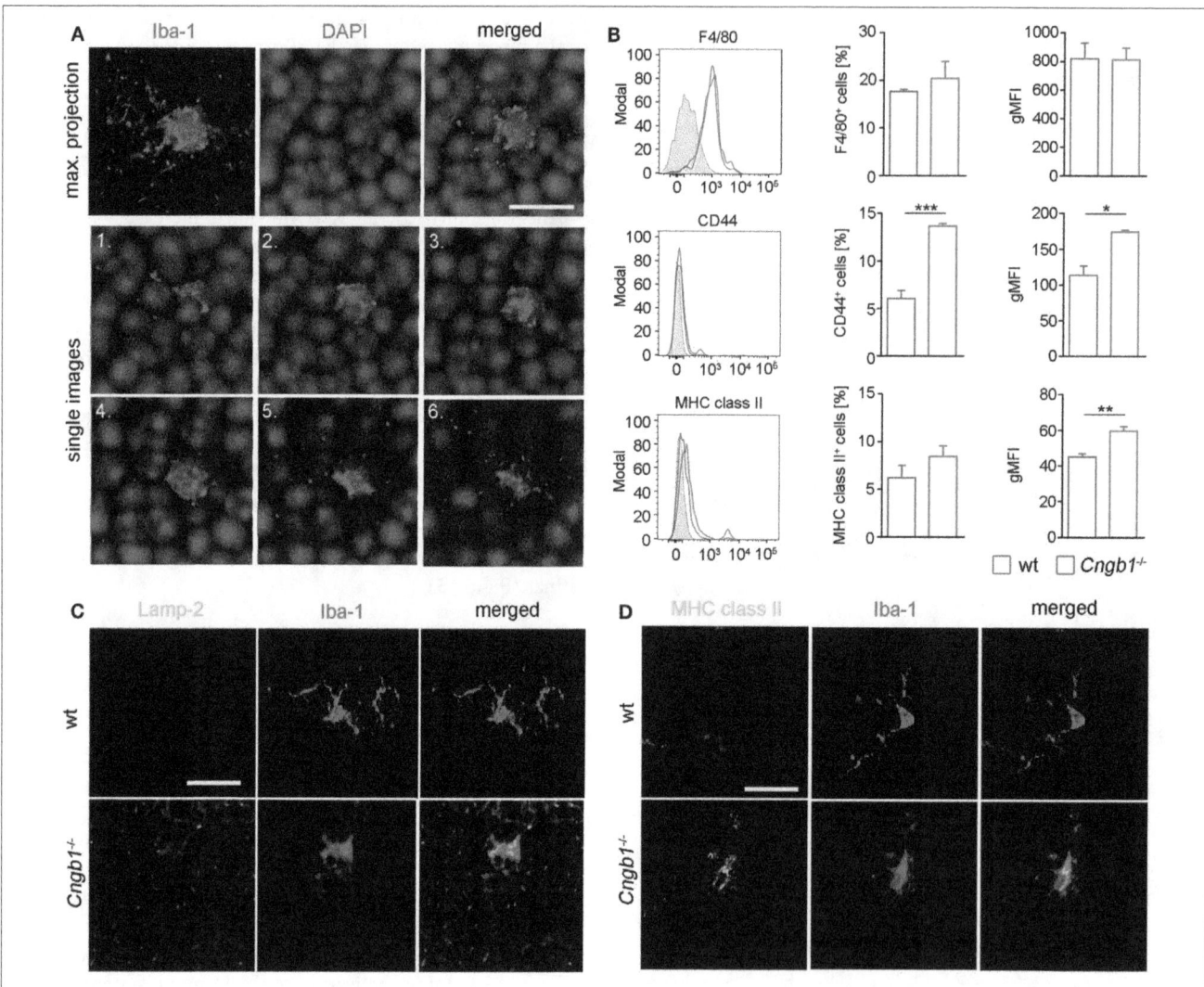

FIGURE 3 | Amoeboid microglia morphology in *Cngb1*-/- is accompanied by the expression of activation markers. **(A)** Whole mount images of Iba-1-stained microglia in the outer nuclear layer (ONL) of *Cngb1*-/- mice induced morphological changes from a resting to an amoeboid phenotype at P28. Nuclei were counterstained with DAPI. Upper panel: maximum projection. Lower two panels; single images with an increment of 1.3 μm. Scale bar. 10 μm. **(B)** Flow cytometric analysis of retinal microglia for the expression of F4/80, CD44, and MHC class II (left panel). Quantification of the numbers (middle panel) and geometric mean fluorescent intensities (gMFI, right panel) of F4/80, CD44, and MHC class II are depicted. Results were obtained from two independent experiments with at least three replicates (*$p < 0.05$, **$p < 0.01$, ***$p < 0.001$, $n = 3$ for each genotype, blue line = wt, red line = *Cngb1*-/-, grey = isotype control). **(C)** Misplaced microglia in the ONL in *Cngb1*-/- coexpressed activation marker Lamp-2. Scale bar 25 μm. **(D)** Confirmation of MHC II expression by costaining of Iba-1 (red) and MHC class II (green) in wt and *Cngb1*-/- retinas. Scale bar 25 μm.

phagocytotic clearance of viable photoreceptors and the secretion of proinflammatory cytokines that potentiate photoreceptor apoptosis (33, 34). It still remained unclear whether microglial activation was responsible for further photoreceptor cell death. Even though, genetic depletion of microglia slowed down the degenerative process in rd10 mice (33). In follow-up experiments, it would be interesting to investigate whether microglial cells are actually the main detrimental force in *Cngb1*-deficient retinas. This could either be achieved by allowing CX3CR1+ retinal microglia to express diphtheria toxin and be specifically ablated upon tamoxifen administration (33, 35) or by pharmacological ablation using the CSF1R inhibitor (36).

Our results indicate that microglia activation is an important step in the degenerative process of rods in RP. The intriguing question however is whether microglia get activated during a predegenerative state or whether signals from a small number of degenerating cells is sufficient to initiate the activation of microglia before the actual "degeneration peak." In the rd10 retina activated microglia infiltrate the ONL at P16. Since photoreceptor apoptosis started only at P19 microglia activation preceded the initiation of photoreceptor apoptosis (37). Comparable findings of microglia proliferation and activation at early time points were also described in rd1 and rd10 mice, which represent further mouse models of RP (37–39). Both RD models are induced by a

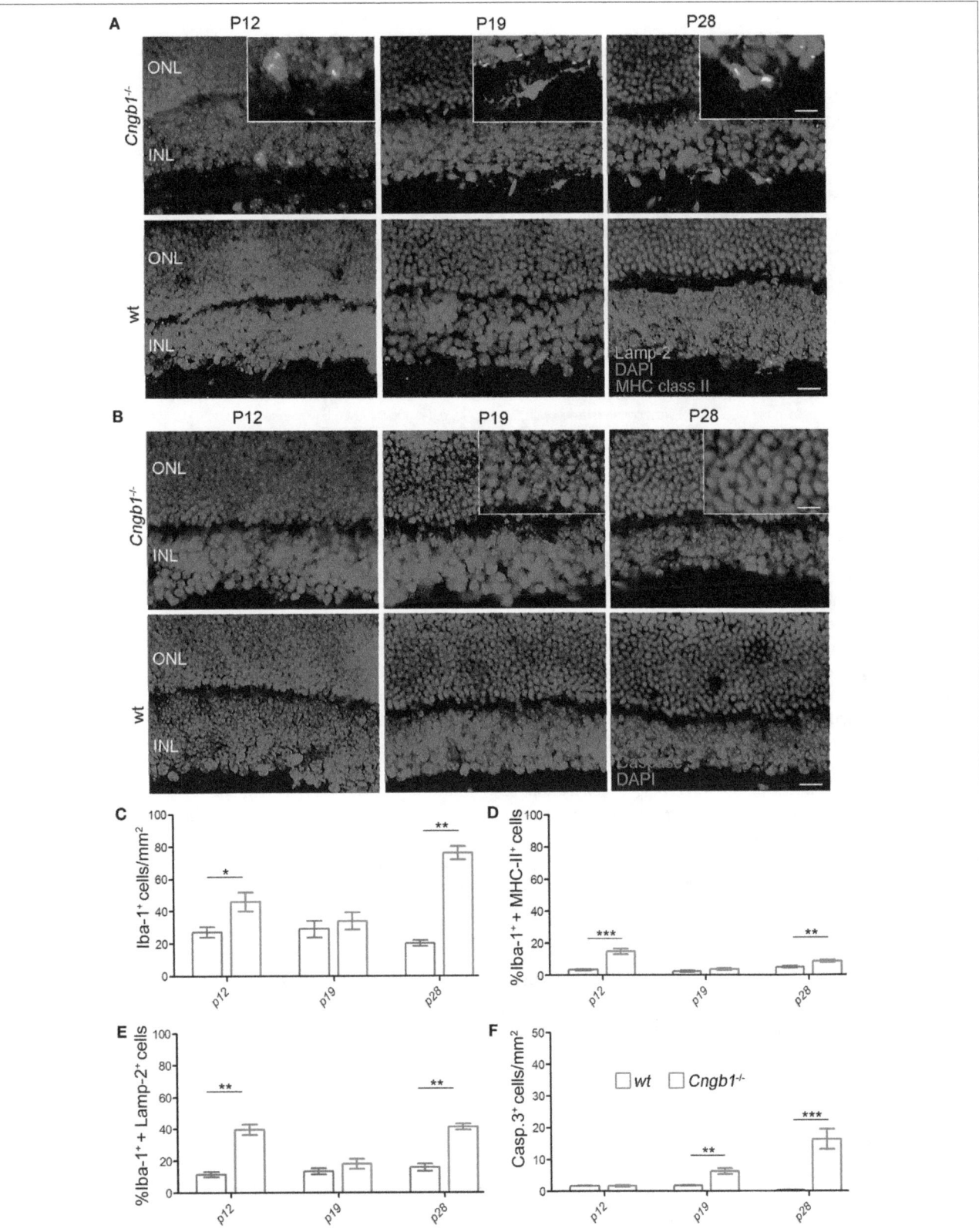

FIGURE 4 | Time course of early photoreceptor apoptosis and microglia activation. **(A)** Costaining of Iba-1-positive microglia (pink) in the ONL and INL of *Cngb1*−/− and wt mice at P12, P19 and P28 with the activation marker Lamp-2 (green) and MHC class II (red). **(B)** Immunofluorescence of cleaved caspase 3 (red) in *Cngb1*−/− and wt mice. Nuclei were counterstained with DAPI (blue). **(C)** Quantification of Iba-1-positive cells and the percentage of Iba-1+ MHC class II− **(D)** or Iba-1+ Lamp-2-positive cells **(E)** in *Cngb1*−/− mice compared to wt animals. **(F)** Cleaved caspase 3-positive cells in *Cngb1*−/− and wt mice at indicated time points. ONL, outer nuclear layer, INL, inner nuclear layer, scale bar 20 µm; insert 10 µm (*$p < 0.05$, **$p < 0.01$, ***$p < 0.001$, $n = 5$ for each genotype).

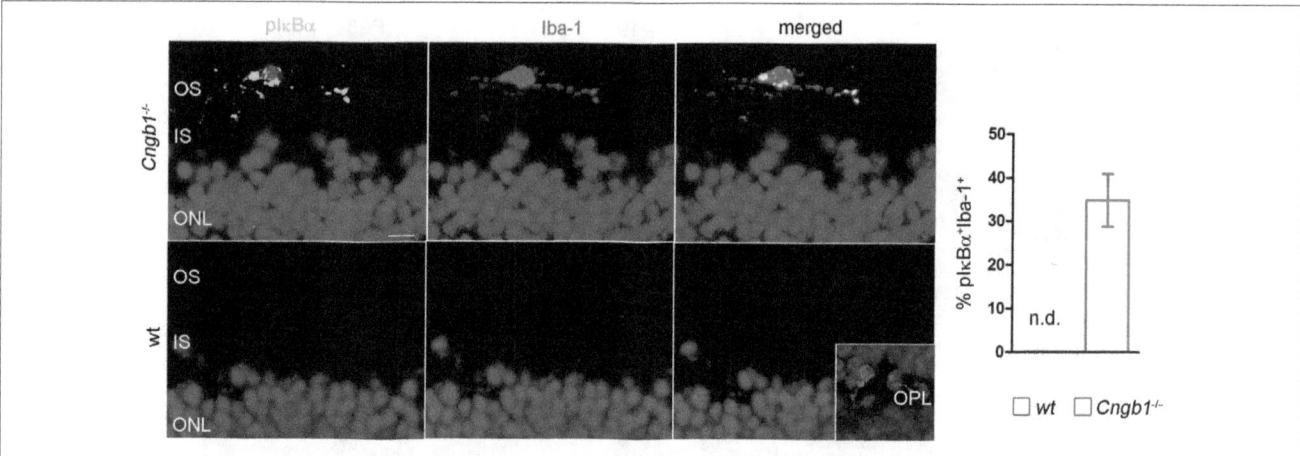

FIGURE 5 | Activation of proinflammatory signaling in *Cngb1*−/− mice. Immunofluorescence of Iba-1 (red) and pIκBα (green) in *Cngb1*−/− and wt mice at P28. Quantification of double-positive cells in wt and KO revealed exclusive presence of active NF-κB-signaling in *Cngb1*−/− mice (n = 3 for each genotype). Insert: Iba-1-positive resting microglia in the OPL. Nuclei were counterstained with DAPI. ONL, outer nuclear layer, IS, inner segment, OS, outer segment, OPL outer plexiform layer, scale bar 20 µm, nd = not detectable.

mutation in the rod photoreceptor-specific Pde6b gene (40, 41). In fact, retinal architecture in rd10 mice displayed alterations from as early as P5, which is at least 13 days before photoreceptor loss (39). These alterations included increased proliferation of microglia within the retina, which ultimately led to increased numbers of activated microglia. At the same time there was a significant decrease of glutamine synthetase in Müller glia followed by an increase in glial fibrillary acidic protein immunofluorescence, which is expressed in Müller glia and astrocytes (39). A similar activation of astrocytes might be present in Cngb1-deficient retinas when one considers the abundant LAMP-2 labeling outside microglial cells. To what extent this reactive gliosis contributes to photoreceptor degeneration in both mouse models is not clear. The observed microglia activation can play a critical role in neuroinflammation and impose subsequent damage with progressive photoreceptor loss (42). In contrast, microglia might also be beneficial during retinal degeneration. This assumption is based on studies showing that microglia-derived trophic support protects photoreceptors *in vivo* under stressful conditions (43). Our data suggest that resident microglia and not monocyte-derived macrophages are mainly involved in the neurodegenerative process. Both cell populations are phenotypically distinguishable with a unique microglial CD45^{lo} CD11c^{lo} F4/80^{lo} I-A/I-E^{lo} profile and a monocyte-derived macrophage CD45^{hi}CD11b^{hi} signature (44). However, it has also been shown that activated retinal microglia upregulate CD45 (45) and that differentiation of monocytes into macrophages may be associated with downregulation of CD45 sometimes to levels that make the two cell populations indistinguishable (46). The small CD45^{hi} CD11b$^-$ cell population found in Cngb1-wt and Cngb1-ko retinas presumably represents circulating retina-specific T cells (47), which have been reported to protect against spontaneous organ-specific autoimmunity (48). At the molecular level, inflammation is often regulated by numerous molecules and factors, including the transcription factor NF-κB (49). The activation of NF-κB in microglia, as seen in

our present RP mouse model, is often associated with the release of reactive oxygen species and proinflammatory cytokines (such as IL-1β, interferon-γ, and TNF-α) that can cause secondary neurotoxicity and neuronal cell death including the degeneration of photoreceptors (50). Dying photoreceptor cells, in turn, induce NF-κB in microglial cells and thereby further their activation (51). Detrimental NF-κB-signaling in microglia has a key role in several degenerative processes of the CNS as documented for aging including AD (52), amyotrophic lateral sclerosis (53), and multiple sclerosis (54). When mice and rats express mutant rhodopsin, they experience photoreceptor cell death and, much as humans, develop the clinical signs of autosomal dominant retinitis pigmentosa (ADRP). During the progression of ADRP, microglia get activated and display heightened NF-κB-signaling (55). Increased expression of NF-κB protein and NF-κB DNA-binding activity in microglia of the retina has also been reported during photoreceptor degeneration of *rd* mice. In this model, the neurotoxic role of microglial NF-κB activation in photoreceptor apoptosis was mediated by increased TNF-α production in microglial cells (56). Several studies have also indicated that NF-κB activation leads to enhanced IL-1β secretion by microglia, which makes them contribute to rod degeneration in RP by potentiating apoptosis (33).

In terms of therapy, targeting microglia may reduce the production of several proinflammatory mediators and may therefore result in broader therapeutic effects than inhibition of single cytokines. However, chemical or genetic depletion of microglia would provide an approach with only short-term beneficial effects since microglia has been shown to repopulate once the treatment ends (35, 36). Particular attention should be paid to unwanted depletion or damage to other cells like optic nerve oligodendrocyte precursor cells. As an example, secondary to microglia depletion by the CSF-1R inhibitor BLZ945, oligodendrocyte precursor cells are reduced in early, postnatal mouse brains (57). As recently described, tamoxifen, a selective estrogen receptor modulator approved for the treatment of breast cancer and previously linked to a low incidence of retinal toxicity, was unexpectedly found to exert marked protective effects against photoreceptor degeneration. Tamoxifen treatment decreased retinal microglia activation in a genetic (*Pde6b*^rd10) model of RP and limited the production of inflammatory cytokines and as a consequence reduced microglial-mediated toxicity to photoreceptors (58). Minocycline, a semi-synthetic tetracycline derivative, prevents NF-κB activation by blockade of Toll-like receptor signaling and counteracts microglial release of TNF-α and IL-1β. This is probably why there are also good indications that minocycline is effective in dampening microglial neurotoxicity and to prevent photoreceptor apoptosis (37, 59). Like minocycline, sulforaphane, a naturally occurring isothiocyanate, also inhibits the proteolytic cleavage of NF-κB and inhibits light-induced photoreceptor apoptosis (60). In a similar manner, polysaccharides were effective in preserving photoreceptors against degeneration in rd10 mice partly through inhibition of NF-κB (61).

In conclusion, both strategies, inhibiting microglial activation and/or inhibition of NF-κB-signaling, can provide useful approaches to prevent retinal degeneration in RP.

AUTHOR CONTRIBUTIONS

TB, TG, and SM designed research. TB, TG, MK, LA, CS, MBo SP, MP, MBu, MBi, SM, and JW performed experiments, analyzed, and interpreted the data. TG and MK designed the figures. TG, TB, and SM wrote the manuscript. All authors edited the manuscript.

ACKNOWLEDGMENTS

We would like to thank Maria Oberle and Kerstin Skokann for their excellent technical assistance during all experiments.

REFERENCES

Kreutzberg GW. Microglia: a sensor for pathological events in the CNS. *Trends Neurosci* (1996) 19:312–8. doi:10.1016/0166-2236(96)10049-7

Koizumi S, Shigemoto-Mogami Y, Nasu-Tada K, Shinozaki Y, Ohsawa K, Tsuda M, et al. UDP acting at P2Y6 receptors is a mediator of microglial phagocytosis. *Nature* (2007) 446:1091–5. doi:10.1038/nature05704

Tay TL, Mai D, Dautzenberg J, Fernandez-Klett F, Lin G, Sagar, et al. A new fate mapping system reveals context-dependent random or clonal expansion of microglia. *Nat Neurosci* (2017) 20:793–803. doi:10.1038/nn.4547

Prinz M, Priller J. Microglia and brain macrophages in the molecular age: from origin to neuropsychiatric disease. *Nat Rev Neurosci* (2014) 15:300–12. doi:10.1038/nrn3722

Karlstetter M, Scholz R, Rutar M, Wong WT, Provis JM, Langmann T. Retinal microglia: just bystander or target for therapy? *Prog Retin Eye Res* (2015) 45:30–57. doi:10.1016/j.preteyeres.2014.11.004

Reyes NJ, O'koren EG, Saban DR. New insights into mononuclear phagocyte biology from the visual system. *Nat Rev Immunol* (2017) 17:322–32. doi:10.1038/nri.2017.13

Gupta N, Brown KE, Milam AH. Activated microglia in human retinitis pigmentosa, late-onset retinal degeneration, and age-related macular degeneration. *Exp Eye Res* (2003) 76:463–71. doi:10.1016/S0014-4835(02)00332-9

den Hollander AI, Black A, Bennett J, Cremers FP. Lighting a candle in the dark: advances in genetics and gene therapy of recessive retinal dystrophies. *J Clin Invest* (2010) 120(9):3042–53. doi:10.1172/jci42258

Daiger SP, Sullivan LS, Bowne SJ. Genes and mutations causing retinitis pigmentosa. *Clin Genet* (2013) 84:132–41. doi:10.1111/cge.12203

Sahel JA, Marazova K, Audo I. Clinical characteristics and current therapies for inherited retinal degenerations. *Cold Spring Harb Perspect Med* (2014) 5:a017111. doi:10.1101/cshperspect.a017111

Sorrentino FS, Gallenga CE, Bonifazzi C, Perri P. A challenge to the striking genotypic heterogeneity of retinitis pigmentosa: a better understanding of the pathophysiology using the newest genetic strategies. *Eye (Lond)* (2016) 30:1542–8. doi:10.1038/eye.2016.197

Huttl S, Michalakis S, Seeliger M, Luo DG, Acar N, Geiger H, et al. Impaired channel targeting and retinal degeneration in mice lacking the cyclic nucleotide-gated channel subunit CNGB1. *J Neurosci* (2005) 25:130–8. doi:10.1523/JNEUROSCI.3764-04.2005

Arango-Gonzalez B, Trifunovic D, Sahaboglu A, Kranz K, Michalakis S, Farinelli P, et al. Identification of a common non-apoptotic cell death mechanism in hereditary retinal degeneration. *PLoS One* (2014) 9:e112142. doi:10.1371/journal.pone.0112142

Rivas MA, Vecino E. Animal models and different therapies for treatment of retinitis pigmentosa. *Histol Histopathol* (2009) 24:1295–322. doi:10.14670/ HH-24.1295

Schon C, Asteriti S, Koch S, Sothilingam V, Garcia Garrido M, Tanimoto N, et al. Loss of HCN1 enhances disease progression in mouse models of CNG channel-

linked retinitis pigmentosa and achromatopsia. *Hum Mol Genet* (2016) 25:1165–75. doi:10.1093/hmg/ddv639

Michalakis S, Schaferhoff K, Spiwoks-Becker I, Zabouri N, Koch S, Koch F, et al. Characterization of neurite outgrowth and ectopic synaptogenesis in response to photoreceptor dysfunction. *Cell Mol Life Sci* (2013) 70:1831–47. doi:10.1007/s00018-012-1230-z

Claes E, Seeliger M, Michalakis S, Biel M, Humphries P, Haverkamp S. Morphological characterization of the retina of the CNGA3(-/-)Rho(-/-) mutant mouse lacking functional cones and rods. *Invest Ophthalmol Vis Sci* (2004) 45:2039–48. doi:10.1167/iovs.03-0741

Michalakis S, Geiger H, Haverkamp S, Hofmann F, Gerstner A, Biel M. Impaired opsin targeting and cone photoreceptor migration in the retina of mice lacking the cyclic nucleotide-gated channel CNGA3. *Invest Ophthalmol Vis Sci* (2005) 46:1516–24. doi:10.1167/iovs.04-1503

Goldmann T, Zeller N, Raasch J, Kierdorf K, Frenzel K, Ketscher L, et al. USP18 lack in microglia causes destructive interferonopathy of the mouse brain. *EMBO J* (2015) 34:1612–29. doi:10.15252/embj.201490791

Jung S, Aliberti J, Graemmel P, Sunshine MJ, Kreutzberg GW, Sher A, et al. Analysis of fractalkine receptor CX(3)CR1 function by targeted deletion and green fluorescent protein reporter gene insertion. *Mol Cell Biol* (2000) 20:4106–14. doi:10.1128/MCB.20.11.4106-4114.2000

Kierdorf K, Erny D, Goldmann T, Sander V, Schulz C, Perdiguero EG, et al. Microglia emerge from erythromyeloid precursors via Pu.1- and Irf8-dependent pathways. *Nat Neurosci* (2013) 16:273–80. doi:10.1038/nn.3318

Crotti A, Ransohoff RM. Microglial physiology and pathophysiology: insights from genome-wide transcriptional profiling. *Immunity* (2016) 44:505–15. doi:10.1016/j.immuni.2016.02.013

Karlstetter M, Sorusch N, Caramoy A, Dannhausen K, Aslanidis A, Fauser S, et al. Disruption of the retinitis pigmentosa 28 gene Fam161a in mice affects photoreceptor ciliary structure and leads to progressive retinal degeneration. *Hum Mol Genet* (2014) 23:5197–210. doi:10.1093/hmg/ddu242

Zhou T, Huang Z, Sun X, Zhu X, Zhou L, Li M, et al. Microglia polarization with M1/M2 phenotype changes in rd1 mouse model of retinal degeneration. *Front Neuroanat* (2017) 11:77. doi:10.3389/fnana.2017.00077

Diaz-Aparicio I, Beccari S, Abiega O, Sierra A. Clearing the corpses: regulatory mechanisms, novel tools, and therapeutic potential of harnessing microglial phagocytosis in the diseased brain. *Neural Regen Res* (2016) 11:1533–9. doi:10.4103/1673-5374.193220

Vachon E, Martin R, Plumb J, Kwok V, Vandivier RW, Glogauer M, et al. CD44 is a phagocytic receptor. *Blood* (2006) 107:4149–58. doi:10.1182/blood-2005-09-3808

Oeckinghaus A, Ghosh S. The NF-kappaB family of transcription factors and its regulation. *Cold Spring Harb Perspect Biol* (2009) 1:a000034. doi:10.1101/cshperspect.a000034

Lawson LJ, Perry VH, Dri P, Gordon S. Heterogeneity in the distribution and morphology of microglia in the normal adult mouse brain. *Neuroscience* (1990) 39:151–70. doi:10.1016/0306-4522(90)90229-W

Streit WJ, Walter SA, Pennell NA. Reactive microgliosis. *Prog Neurobiol* (1999) 57:563–81. doi:10.1016/S0301-0082(98)00069-0

Brennan FR, O'neill JK, Allen SJ, Butter C, Nuki G, Baker D. CD44 is involved in selective leucocyte extravasation during inflammatory central nervous system disease. *Immunology* (1999) 98:427–35. doi:10.1046/j.1365-2567. 1999.00894.x

Brocke S, Piercy C, Steinman L, Weissman IL, Veromaa T. Antibodies to CD44 and integrin alpha4, but not L-selectin, prevent central nervous system inflammation and experimental encephalomyelitis by blocking secondary leukocyte recruitment. *Proc Natl Acad Sci U S A* (1999) 96:6896–901. doi:10.1073/pnas.96.12.6896

Uren PJ, Lee JT, Doroudchi MM, Smith AD, Horsager A. A profile of transcriptomic changes in the rd10 mouse model of retinitis pigmentosa. *Mol Vis* (2014) 20:1612–28.

Zhao L, Zabel MK, Wang X, Ma W, Shah P, Fariss RN, et al. Microglial phagocytosis of living photoreceptors contributes to inherited retinal degeneration. *EMBO Mol Med* (2015) 7:1179–97. doi:10.15252/emmm.201505298

Zabel MK, Zhao L, Zhang Y, Gonzalez SR, Ma W, Wang X, et al. Microglial phagocytosis and activation underlying photoreceptor degeneration is regulated by CX3CL1-CX3CR1 signaling in a mouse model of retinitis pigmentosa. *Glia* (2016) 64:1479–91. doi:10.1002/glia.23016

Bruttger J, Karram K, Wortge S, Regen T, Marini F, Hoppmann N, et al. Genetic cell ablation reveals clusters of local self-renewing microglia in the mammalian central nervous system. *Immunity* (2015) 43:92–106. doi:10.1016/j. immuni.2015.06.012

Elmore MR, Najafi AR, Koike MA, Dagher NN, Spangenberg EE, Rice RA, et al. Colony-stimulating factor 1 receptor signaling is necessary for microglia viability, unmasking a microglia progenitor cell in the adult brain. *Neuron* (2014) 82:380–97. doi:10.1016/j.neuron.2014.02.040

Peng B, Xiao J, Wang K, So KF, Tipoe GL, Lin B. Suppression of microglial activation is neuroprotective in a mouse model of human retinitis pigmentosa. *J Neurosci* (2014) 34:8139–50. doi:10.1523/JNEUROSCI.5200-13.2014

Zeiss CJ, Johnson EA. Proliferation of microglia, but not photoreceptors, in the outer nuclear layer of the rd-1 mouse. *Invest Ophthalmol Vis Sci* (2004) 45:971–6. doi:10.1167/iovs.03-0301

Roche SL, Wyse-Jackson AC, Byrne AM, Ruiz-Lopez AM, Cotter TG. Alterations to retinal architecture prior to photoreceptor loss in a mouse model of retinitis pigmentosa. *Int J Dev Biol* (2016) 60:127–39. doi:10.1387/ ijdb.150400tc

Chang B, Hawes NL, Hurd RE, Davisson MT, Nusinowitz S, Heckenlively JR. Retinal degeneration mutants in the mouse. *Vision Res* (2002) 42:517–25. doi:10.1016/S0042-6989(01)00146-8

Chang B, Hawes NL, Pardue MT, German AM, Hurd RE, Davisson MT, et al. Two mouse retinal degenerations caused by missense mutations in the beta-subunit of rod cGMP phosphodiesterase gene. *Vision Res* (2007) 47:624–33. doi:10.1016/j.visres.2006.11.020

Amor S, Peferoen LA, Vogel DY, Breur M, Van Der Valk P, Baker D, et al. Inflammation in neurodegenerative diseases – an update. *Immunology* (2014) 142:151–66. doi:10.1111/imm.12233

Harada T, Harada C, Kohsaka S, Wada E, Yoshida K, Ohno S, et al. Microglia- Muller glia cell interactions control neurotrophic factor production during light-induced retinal degeneration. *J Neurosci* (2002) 22:9228–36.

O'Koren EG, Mathew R, Saban DR. Fate mapping reveals that microglia and recruited monocyte-derived macrophages are definitively distinguishable by phenotype in the retina. *Sci Rep* (2016) 6:20636. doi:10.1038/srep20636

Maneu V, Noailles A, Megias J, Gomez-Vicente V, Carpena N, Gil ML, et al. Retinal microglia are activated by systemic fungal infection. *Invest Ophthalmol Vis Sci* (2014) 55:3578–85. doi:10.1167/iovs.14-14051

Muller A, Brandenburg S, Turkowski K, Muller S, Vajkoczy P. Resident microglia, and not peripheral macrophages, are the main source of brain tumor mononuclear cells. *Int J Cancer* (2015) 137:278–88. doi:10.1002/ijc. 29379

Horai R, Silver PB, Chen J, Agarwal RK, Chong WP, Jittayasothorn Y, et al. Breakdown of immune privilege and spontaneous autoimmunity in mice expressing a transgenic T cell receptor specific for a retinal autoantigen. *J Autoimmun* (2013) 44:21–33. doi:10.1016/j.jaut.2013.06.003

McPherson SW, Heuss ND, Pierson MJ, Gregerson DS. Retinal antigen-specifi regulatory T cells protect against spontaneous and induced autoimmunity and require local dendritic cells. *J Neuroinflammation* (2014) 11:205. doi:10.1186/s12974-014-0205-4

Aggarwal BB. Nuclear factor-kappaB: the enemy within. *Cancer Cell* (2004) 6:203–8. doi:10.1016/j.ccr.2004.09.003

Block ML, Zecca L, Hong JS. Microglia-mediated neurotoxicity: uncovering the molecular mechanisms. *Nat Rev Neurosci* (2007) 8:57–69. doi:10.1038/ nrn2038

Yang LP, Zhu XA, Tso MO. A possible mechanism of microglia-photoreceptor crosstalk. *Mol Vis* (2007) 13:2048–57.

von Bernhardi R, Eugenin-Von Bernhardi L, Eugenin J. Microglial cell dysregulation in brain aging and neurodegeneration. *Front Aging Neurosci* (2015) 7:124. doi:10.3389/fnagi.2015.00124

Frakes AE, Ferraiuolo L, Haidet-Phillips AM, Schmelzer L, Braun L, Miranda CJ, et al. Microglia induce motor neuron death via the classical NF-kappaB pathway in amyotrophic lateral sclerosis. *Neuron* (2014) 81: 1009–23. doi:10.1016/j.neuron.2014.01.013

Goldmann T, Wieghofer P, Muller PF, Wolf Y, Varol D, Yona S, et al. A new type of microglia gene targeting shows TAK1 to be pivotal in CNS autoimmune inflammation. *Nat Neurosci* (2013) 16:1618–26. doi:10.1038/nn.3531

Rana T, Shinde VM, Starr CR, Kruglov AA, Boitet ER, Kotla P, et al. An acti- vated unfolded protein response promotes retinal degeneration and triggers an inflammatory response in the mouse retina. *Cell Death Dis* (2014) 5:e1578. doi:10.1038/cddis.2014.539

Zeng HY, Tso MO, Lai S, Lai H. Activation of nuclear factor-kappaB during retinal degeneration in rd mice. *Mol Vis* (2008) 14:1075–80.

Hagemeyer N, Hanft KM, Akriditou MA, Unger N, Park ES, Stanley ER, et al. Microglia contribute to normal myelinogenesis and to oligodendrocyte progenitor maintenance during adulthood. *Acta Neuropathol* (2017) 134:441–58. doi:10.1007/s00401-017-1747-1

Wang X, Zhao L, Zhang Y, Ma W, Gonzalez SR, Fan J, et al. Tamoxifen provides structural and functional rescue in murine models of photoreceptor degeneration. *J Neurosci* (2017) 37:3294–310. doi:10.1523/JNEUROSCI.2717-16.2017

Scholz R, Sobotka M, Caramoy A, Stempfl T, Moehle C, Langmann T. Minocycline counter-regulates pro-inflammatory microglia responses in the retina and protects from degeneration. *J Neuroinflammation* (2015) 12:209. doi:10.1186/ s12974-015-0431-4

Yang LP, Zhu XA, Tso MO. Role of NF-kappaB and MAPKs in light-induced photoreceptor apoptosis. *Invest Ophthalmol Vis Sci* (2007) 48:4766–76. doi:10.1167/iovs.06-0871

Wang K, Xiao J, Peng B, Xing F, So KF, Tipoe GL, et al. Retinal structure and function preservation by polysaccharides of wolfberry in a mouse model of retinal degeneration. *Sci Rep* (2014) 4:7601. doi:10.1038/srep07601

miR-145-5p/Nurr1/TNF-α Signaling-Induced Microglia Activation Regulates Neuron Injury of Acute Cerebral Ischemic/Reperfusion in Rats

Xuemei Xie [1,2], Li Peng [1,2], Jin Zhu [1,2], Yang Zhou [1,2], Lingyu Li [1,2], Yanlin Chen [1,2], Shanshan Yu [1,2] and Yong Zhao [1,2]*

[1]Department of Pathology, Chongqing Medical University, Chongqing, China, [2]Key Laboratory of Neurobiology, Chongqing Medical University, Chongqing, China

*Correspondence:
Yong Zhao
zhaoyong668@cqmu.edu.cn

Nurr1 is a member of the nuclear receptor 4 family of orphan nuclear receptors that is decreased in inflammatory responses and leads to neurons death in Parkinson's disease. Abnormal expression of Nurr1 have been attributed to various signaling pathways, but little is known about microRNAs (miRNAs) regulation of Nurr1 in ischemia/ reperfusion injury. To investigate the post transcriptional regulatory networks of Nurr1, we used a miRNA screening approach and identified miR-145-5p as a putative regulator of Nurr1. By using computer predictions, we identified and confirmed a miRNA recognition element in the 3′UTR of Nurr1 that was responsible for miR-145-5p-mediated suppression. We next demonstrated that overexpression of Nurr1 inhibited TNF-α expression in microglia by trans-repression and finally attenuated ischemia/reperfusion-induced inflammatory and cytotoxic response of neurons. Results of further *in vivo* study revealed that anti-miR-145-5p administration brought about increasing expression of Nurr1 and reduction of infarct volume in acute cerebral ischemia. Administration of anti-miR-145-5p promotes neurological outcome of rats post MCAO/R. It might be an effective therapeutic strategy to relieve neurons injury upon ischemia/reperfusion of rats through interrupting the axis signaling of miR-145-5p- Nurr1-TNF-α in acute phase.

Keywords: Mir-145-5p, Nurr1, TNF-α, cerebral, ischemic/reperfusion

INTRODUCTION

Cerebral ischemia/reperfusion (I/R) injury-induced neuronal cell death is the most difficult problem to solve in clinical stroke (Al-Mufti et al., 2017). Activated microglia have been reported to act as sensors that detect abnormal metabolic changes following I/R, including reactive oxygen species and inflammatory cytokines (Yuan et al., 2014; Fumagalli et al., 2015). Microglia can release excessive proinflammatory cytokines and/or cytotoxic factors, such as tumor necrosis factor-α (TNF-α), interleukin-1β (IL-1β) and nitric oxide (NO), which have been shown to contribute to neuronal damage (Yuan et al., 2014; Roqué et al., 2016; Gullo et al., 2017). Therefore, suppressing overreaction of the microglial inflammatory response may be an efficacious therapeutic strategy to alleviate progression of stroke.

Nurr1 (NR4A2) is a member of the nuclear receptor 4 family of orphan nuclear receptors (Kim et al., 2015; Zou et al., 2017), and has been studied extensively in Parkinson's disease recently. Nurr1 has been shown to inhibit expression of proinflammatory neurotoxic mediators by docking to NF-κB-p65 on target inflammatory gene promoters in microglia (Saijo et al., 2009; Kim et al., 2015). Contra-directional coupling of Nur77 and Nurr1 are involved in neurodegeneration or injury by regulating endoplasmic reticulum stress and mitochondrial impairment (Gao et al., 2016; Wei et al., 2016). Cystatin C induces VEGF expression and attenuates PC12 cell degeneration by regulating p-PKC-α/p-ERK1/2-NURR1 signaling and inducing autophagy in Parkinson's disease (Zou et al., 2017). Taken together, these findings indicate that Nurr1 could be a promising therapeutic target in neuronal diseases. However, the specific functions and/or mechanisms of Nurr1 in I/R injury are still unknown.

Increasing recent evidence suggests important roles for microRNAs (miRNAs) in molecular processes of cerebral ischemia pathogenesis, which involve fast post-transcriptional effects and simultaneous regulation of various target genes (Ouyang et al., 2015; Minhas et al., 2017). Although several miRNAs have been reported to target Nurr1 in cancer cells (Yang et al., 2012; Wu et al., 2015; Beard et al., 2016), interaction between miRNAs and Nurr1 in cerebral I/R injury is still poorly understood.

In the present study, we therefore sought to determine: (1) how Nurr1 level and Nurr1-related microRNAs change in the brain of MCAO/R rats, an *in vivo* model; (2) in *in vitro* studies, what are the post-transcriptional regulatory network of Nurr1 and the specific mechanisms of Nurr1 on microglia activation; (3) in an *in vivo* study, whether miR-145-5p/Nurr1/TNF-α signal exerts neuron injury function upon cerebral ischemia-reperfusion in rats.

MATERIALS AND METHODS

Experimental Animals and Ethics Statement

About 170 adult male Sprague-Dawley (SD) rats (weight between 250–300 g, 60–80 d) were purchased from the Laboratory Animal Centre of Chongqing Medical University used for the *in vivo* study. Brain tissues from newborn SD rats (0–24 h) were used to culture primary neurons of cerebral cortex and glia cells. This study has been carried out within an appropriate ethical framework. All experimental procedures were performed in accordance with the National Institutes of Health (NIH) Guide for the Care and Use of Laboratory Animals and approved by Biomedical Ethics Committee of Chongqing Medical University. Maximum efforts have been made to minimize the number of animals used and their suffering. Adult male Sprague-Dawley rats were used that is exempted from ethics approval.

Cell Culture of Primary Microglia and Neurons and OGD/R Treatment

Primary glial cells were isolated from the cerebrum and cerebellum of rats (1-day-old) and placed in 6-well plates at a density of 1.2×10^6 cells/ml of DMEM supplemented with 10% fetal bovine serum (FBS), non-essential amino acids, and insulin. Plates were then placed in a humidified incubator with 5% CO_2/balanced with air (result: 20% O_2) at 37°C with medium change per 48 h. Microglia were isolated from the mixed glial population after confluent (about 2 weeks) by a method previously described (Jose et al., 2014). Microglia cultures with more than 96% purity were used for the study.

Primary neurons were obtained from the cerebral cortex of 1 day-old rats and cultured as described in our previous studies (Zhou et al., 2015; Chen et al., 2016). Approximately 2.0×10^6 cells per 2 mL of Neurobasal Medium containing 1% penicillin (Pen, 100 U/mL) and 1% streptomycin (Strep, 100 U/mL), and 2% B27 supplement were seeded per well. Neurons were cultured in a humidified incubator with 5% CO_2 at 37°C. After 6–7 days of culture *in vitro*, cells were examined to ensure more than 90% purity of neurons which could be used for further study.

The thorough method of OGD/R was conducted as previously described (Tauskela et al., 2003). Briefly, microglia/neurons were washed and cultured with glucose-free DMEM, which had been previously equilibrated with 1% O_2 + 5% CO_2 + 94% N_2 at 37°C for 2.0 h in an incubator. Cells were exposed to hypoxia by placing them in an incubator filled with a gas mixture of 1% O_2 + 5% CO_2 + 94% N_2 for 1.5 h at 37°C. The glucose-free DMEM were then changed back to their special mediums and cultures were returned to the normal incubator for recovery times of 0.5 h, 1.0 h, 2 h, 6 h or 12.0 h. An appropriate time of reoxygenation was selected for subsequent studies. Cells exposed to normoxia were used as negative control.

Coculture of Microglia and Neurons

Primary microglia were seeded onto Transwell permeable support membrane inserts (Corning, NY, USA) at a density 1.0×10^6/well in DMEM medium supplemented with 10% FBS and allowed to settle and grow for 24 h, which constituted the upper chamber in a 2-chambered microglia-neuronal cell coculture system. The cultured neurons were seeded into the bottom of 6-well plates at a density of 2.0×10^6 cells/well in specific Neurobasal medium as described above until confluence at 24 h. These neurons were then cocultured with microglia-containing inserts for 48 h in the following conFigureureration: (1) neurons incubated with empty inserts lacking microglia (neurons only); (2) neurons incubated with inserts containing microglia that had not been exposed to OGD/R (microglia normoxia (nor.) + neurons); and (3) neurons incubated with inserts containing activated microglia that had been previously exposed to OGD/R for 2 h (microglia OGD/R 2 h + neurons). Following coculture, the two coculture chambers were disassembled, and exposed neurons were washed with PBS and harvested for micro-RNA, mRNA, and/or cell viability analysis (Chen et al., 2012, 2016).

Construction of Middle Cerebral Artery Occlusion and Reperfusion (MCAO/R) Model for *in Vivo* Experiments

SD rats were fed and housed under standard conditions before operation, and the room temperature was monitored at 24–28°C throughout the surgical procedure. MCAO model of rats were construct by method of intraluminal vascular occlusion as previously described in our laboratory (Chen et al., 2012, 2016). All the surgical procedures were performed successfully under anesthesia with 3.5% chloral hydrate (350 mg/Kg, intraperitoneal injection). The nylon filament with its tip rounded (diameter 0.24–0.28 mm), determined by the animal weight, was inserted into the middle cerebral artery for 1.5 h. Reperfusion of the ischemic artery was established by withdrawal of the filament until the tip cleared the lumen of the external carotid artery. Regional cerebral blood flow was monitored by an ultrasonic blood flow meter during the operation.

A successful occlusion and reperfusion of MCAO model was evaluated by methods of Neurological Outcome Assays and infarction volume assays as described in references (Lourbopoulos et al., 2008; Wang et al., 2014). Rats of sham-operated were subjected to the same surgical procedures as MCAO/R group except for occlusion of the external carotid artery. Animals which had blood reperfusion below 70% or died during reperfusion were excluded from analysis.

After MCAO, the experimental rats in each group ($n = 5$) were euthanized randomly at the end of reperfusion for 2, 6, 12, 24 and 48 h to detect the alterations of various index. SD rats were divided randomly into 15 groups. Before suffered MCAO/R, the groups were: null group, scramble group, anti-miR-145-5p group, miR-145-5p mimic group, Nurr1-siRNA group, Nurr1 activation plasmid group (Nurr1; intra-cerebroventricular injection). After MCAO/R, the groups were: sham-operated (sham) group, MCAO/R group, null + MCAO/R group, scramble + MCAO/R group, anti-miR-145-5p + MCAO/R group, miR-145-5p mimic + MCAO/R group, Nurr1-siRNA + MCAO/R group, Nurr1 + MCAO/R group. miR-145-5p mimic and anti-miR-145-5p were administered to the animals before they underwent any surgery via intracerebroventricular infusion.

Neurological Outcome Assays

Two different assays were used to assess the neurological deficit of the rat on days 1, 7 and 14 after MCAO/R after the induction of stroke. A modified Neurological Severity Score (mNSS) was applied, three motor tests and two sensory tests were included, which are evaluated by total score 18 (Lourbopoulos et al., 2008; Wang et al., 2014). The scores are added up to a score between 12 (severe impairment) and 6 (no neurological impairment). Additionally, the foot fault test was performed to assay fine motor skills and proprioception. The amount of foot faults was computed via ((number of left forelimb faults) + (number of left hindlimb faults))/((total number of left forelimb steps) + (total number of left hindlimb steps)) (Lourbopoulos et al., 2008; Wang et al., 2014).

Infarct Volume Analysis

SD rats were sacrificed upon different treatment, and the brain tissue was prepared into 3 mm sections for 2,3,5-triphenyltetrazolium chloride (TTC) staining (Sigma-Aldrich, St. Louis, MO, USA). Cerebral tissue sections were incubated with 2% TCC solution in the dark room at 37°C for 5 min; the tissue sections were then fixed in 4% paraformaldehyde. The living brain tissues was bright red, however the ischemic or necrotic tissues were pale. The digital images were further analyzed by Sigma Scan Pro 5.0 Software. The real infarct volume (cortex, striatum and hemisphere), after excluding edema, was calculated by subtraction of the ipsilateral non-infarcted regional volume from the contralateral regional volume. The real infarct volume was then divided by the contralateral regional volume as a percent of the contralateral region (McCarter et al., 2017).

RNA Extraction, Reverse Transcription and RT-qPCR

Total RNA and miRNA in the brain tissues or primary cultured cells were extracted using TRIzol® with miRcute miRNA isolation kits (Tiangen Biotech, Beijing, China). The concentration and integrity of sample RNA were determined using Nanodrop ND-1000 spectrophotometry (Nanodrop Tech, Rockland, DE, USA) and denaturing agarose gel electrophoresis. The cDNA was obtained by reverse transcription with an miRcute miRNA First-Strand cDNA Synthesis kit (Tiangen Biotech, Beijing, China). The RT-qPCR was performed using iQ5 RT-qPCR detection system (Bio-Rad Laboratories, Inc., Hercules, CA, USA).

Quantitation of Nurr1 and TNF-α mRNAs was performed using SYBR green assay (Zhou et al., 2015). Specific primer sequences were designed and generated by Sangon Biotech (Shanghai, China). For miRNA detection, the cDNA was obtained by reverse transcription with an miRcute miRNA First-Strand cDNA Synthesis kit (Tiangen Biotech, Beijing, China). Stem-loop qRT-PCR reactions were performed according to manufacturer's protocols using miRNA specific stem-loop primers (Sangon Biotech, Shanghai, Co., Ltd.). Real-Time PCR reactions were conducted on a PCR amplifier (CFX-96 Content Real-time System). The endogenous control of both mRNAs and miRNAs was Ribosomal RNA (18S rRNA) for the quantitative PCR (qPCR) assays because it is known to expressed stably in cerebral ischemic conditions (Sepramaniam et al., 2010; Liu F. J. et al., 2015).

miRNA Profiling

A miRNA microarray was performed according to the MicroRNA Expression Profiling Assay Guide (Illumina Inc., San Diego, CA, USA). The extracted RNA from the cerebral cortex of rats was labeled with Hy3 dye at 3′-end using the miRCURY LNA Power Labeling Kit (Illumina Inc.) in the 500 ng intact total RNA sample (Liu C. et al., 2015). The labeled miRNAs were hybridized to the BeadChip for 16–18 h, according to manufacturer's instructions (Illumina Inc.). The microarray chips were then washed and scanned by InnoScan700, microarray scanner and analyzed by GenomeStudio™ Gene Expression Module v1.0 software (Illumina Inc.). The p value < 0.01 was considered to be accurately detected and selected for further analysis.

miRNA Target Prediction and Mutagenesis

Six miRNA target-prediction algorithms were used to identify putative miRNA regulators of Nurr1: http://www.mirbase.org/; http://mirtar.mbc.nctu.edu.tw/human/index.php; http://www.targetscan.org; http://www.microrna.org; http://mirdb.org/cgi-bin/search.cgi; and http://pictar.mdc-berlin.de/. By using these algorithms, a putative seed region was determined and mutated using site-directed mutagenesis (Mutagenex, Inc., Piscataway, NJ, USA; Jeyaseelan et al., 2008). Reporter constructs containing either the wild-type or mutated 3′UTR were used to demonstrate miR-145-5p specificity in the Nurr1 3′UTR.

Cloning of Nurr1 3′UTR and Dual Luciferase Reporter Assay

The 3′UTR of *Nurr1* was amplified by PCR using gene specific primers and cloned into firefly-luciferase-expressing vector pMIR-REPORT (Ambion, Austin, TX, USA). HEK293T cell was used in this study for its high transfection efficiency (>90%; Karra and Dahm, 2010). Briefly, HEK293T cells were cultured in 24-well plates and transfected with 40 nM anti- or mimic microRNAs for 3 h followed by 200 ng/well of pMIR-REPORT constructs for 3–5 h. Before lysed for measurement of luciferase activity, cells were then left to recover in CO_2 incubator at 37°C for 48 h. The effect of miRNA binding with 3′UTR of Nurr1 quantified using Dual luciferase assay according to manufacturer's protocol (Cat # E1910; Promega, USA). Transfection efficiencies were normalized to those of cells by co-transfecting with the Renilla-luc-expressing vector pRL-CMV (Cat # E2261; Promega, Madison, WI, USA) at 5 ng/well.

Gene Transfection and siRNA Interference Experiments *in Vivo* and *in Vitro*

The pcDNA-Nurr1 overexpression lentivirus was constructed and packaged by Neuron Biotech (Shanghai, China). The Nurr1-siRNA, scramble siRNA and siRNA reagent system were purchased from Santa Cruz Biotechnology Inc. (Cat # sc-36111). For *in vitro* assays, the confluent microglia and neuron were transfected with pcDNA- Nurr1 using Lipofectamine 2000 (Invitrogen, Carlsbad, CA, USA) according to the manufacturer's instructions. The siRNA interference of Nurr1 was performed according to the manufacturer's instructions. Cells transfected with empty vectors (EV) and un-transfected cells (Null) served as negative control groups. Sustained Nurr1 overexpression or down-regulation were confirmed by qRT-PCR and western blot analysis 72 h after transfection. Scramble siRNA and Nurr1-siRNA were dissolved in RNase-free water to a final concentration of 2 μg/μL, and injected ipsilaterally into the left lateral cerebral ventricle 48 h before MCAO.

Cell Viability Analysis

The cultured neurons with different treatments were grown in 96-well plates separately at a density of 2.0×10^5 cells/mL and then harvested to analyze cell viability using The Cell Titer 96 Aqueous One Solution cell proliferation assay with 3-(4,5-dimethylthiazol-2-yl)-2,5-diphenyltetrazolium bromide (MTS; Cat # G3582, Promega, USA). Absorbance was measured at 490 nm using a microplate reader.

Chromatin Immunoprecipitation (ChIP) Assay

ChIP assay were performed using EZChIP kit which contains all necessary reagents to perform 22 individual (ChIP) reactions using inexpensive protein G agarose beads (Cat # 17-371, EZ-ChIP™, Merck Millipore, USA). Briefly, for each experimental condition, 2×10^7 rat primary microglia were needed. Cells were cross-linked for 10 min with 1% formaldehyde. Samples were incubated with Anti-Nurr1 antibody [N1404]—ChIP Grade (ab41917; Cat # ab41917, 1:100, Abcam, Cambridge, MA, USA; 1:100 in ChIP dilution buffer - 0.01% SDS, 1.1% Triton X-100, 1.2 mM EDTA, 16.7 mM Tris-HCl (pH 8.1), 167 mM NaCl, proteanase inhibitor cocktail) for 16 h at 4°C. A 150-bp region of the rat proximal TNF-α promoter was amplified spanning the NF-κB site. Protein binding was detected using RT-qPCR.

Immunofluorescence Staining

Fresh freeze brain sections (10 μm) were incubated with 10% normal goat serum/0.3% Triton-X 100 diluted in PBS blocking solution at 37°C for 1 h. Slides were then incubated with the following corresponding primary antibodies: Polyclonal rabbit anti-rat Nurr1 (1:100 dilution; sc-991, Santa Cruz, CA, USA) and monoclonal mouse anti-goat CD68 (1:50 dilution; Cat # ab1211, Abcam, Cambridge, MA, USA) overnight at 4°C. Following incubation, slides were then washed in 0.1 M PBS and incubated for 1 h with the following secondary antibodies: goat anti-rabbit Immunoglobulin G (1:100 dilution; zhongshanjinqiao, China) and goat anti-mouse Immunoglobulin G (1:200 dilution; zhongshanjinqiao, China). DAPI was used to stain the nuclei (Sigma-Aldrich). All the sections were visualized under a Nikon ECLIPSE Ti fluorescence microscope, loaded with a CoolSNAP photometrics camera at 400× magnification.

Western Immunoblot Analysis

Cell lysis buffer of cultured neurons and the rat cortex supplemented with proteinase inhibitors were used to extract total protein. The quantified proteins were separated by 8% SDS-PAGE and transferred onto polyvinylidene fluoride (PVDF) membranes. The membranes were subsequently blocked with 5% skim-milk for 2.0 h at room temperature and incubated in primary antibody overnight at 4°C. Dilutions for primary antibodies were as follows: Polyclonal rabbit anti-Nurr1 and anti-TNF-α (1:500 dilution; 6945S, Cell Signaling Technology, USA), Polyclonal mouse anti-β-actin (1:5000, ABclonal, Wuhan, China, AC004) and anti-IL1β (1:1000, Bioworld technology, USA). The membranes were then incubated with appropriate secondary antibodies at 37°C for 2 h (dilution 1:5000, Sangon Biotech, Shanghai, Co., Ltd.). The density of bands was detected

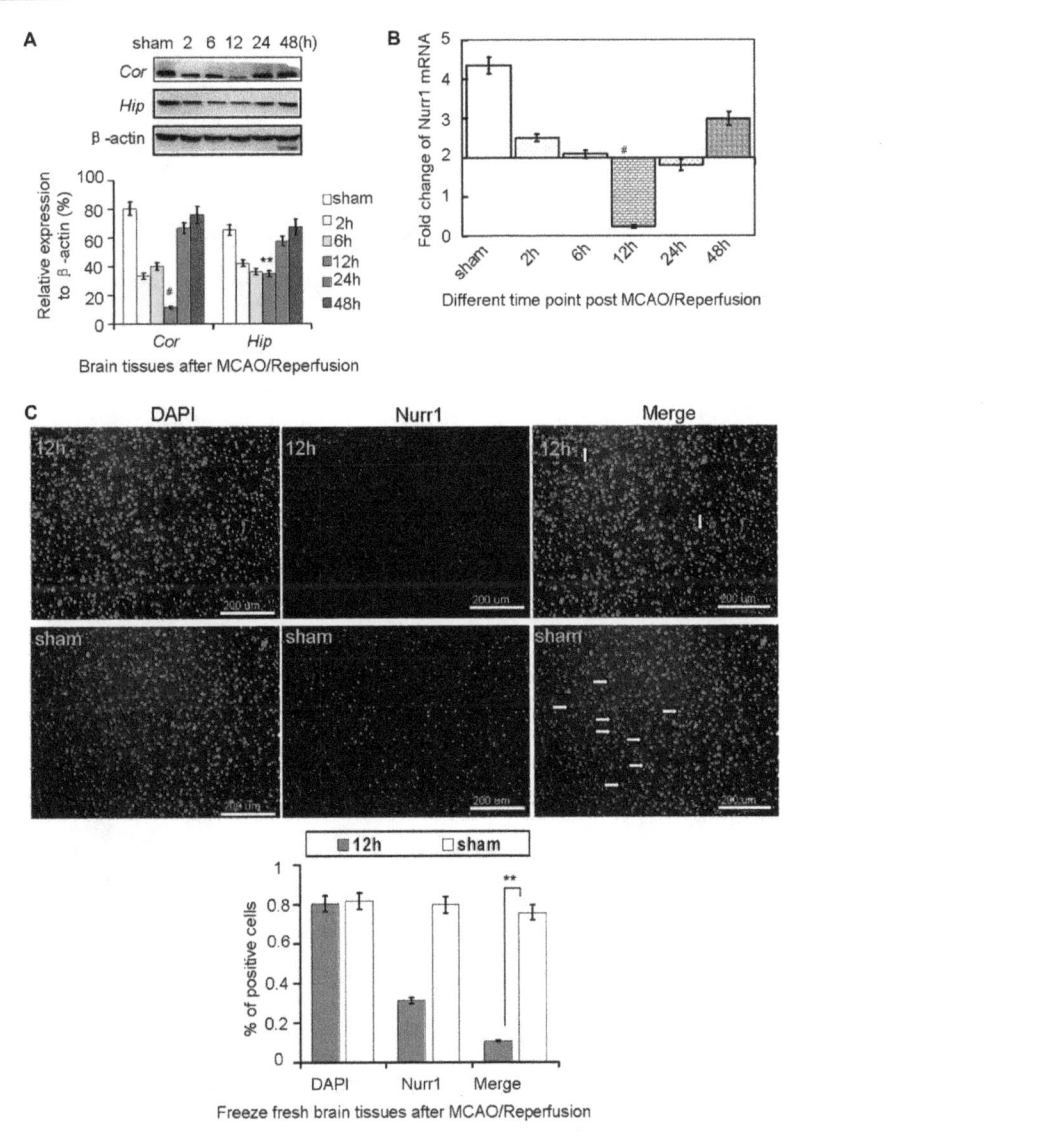

FIGURE 1 | Nurr1 expression at 2, 6, 12, 24 and 48 h after MCAO/R. **(A)** Nurr1 protein was measured by Western blot at each time point upon cerebral ischemia/reperfusion. The gray intensities of the bands were quantified using ImageJ software and presented as percentage of β-actin (internal control, %). Nurr1 expression decreased sharply and reached a minimum at 12 h in cerebral cortex (*Cor*) and hippocampus (*Hip*) of rats. **(B)** Nurr1 mRNA was measured by RT-qPCR and presented as relative expression (mean ± SD, $n = 3$). **(C)** Nurr1 protein expression was significantly declined in the infarct side of cortex by immunofluorescence staining from fresh freeze brain sections 12 h after MCAO/R. **$p < 0.01$, #$p < 0.001$.

using an enhanced chemiluminescence system (Cat # 32132, Pierce™ ECL Plus Western Blotting Substrate ECL Plus, USA), and the gray value of bands was quantified using ImageJ analysis software. The relative expression quantity of protein was scored as the ratio of target protein intensity to β-actin staining intensity.

Statistical Analysis

All data are expressed as mean ± SD. One-way analysis of variance (ANOVA) followed by Bonferroni test was used to compare results among all groups. The Spearman's correlation test was used to examine the correlations of relative expression levels between protein and miRNA. The SPSS 17.0 software

package was used to execute all statistics. All experiments were independently repeated at least three times. p-values < 0.05 were considered statistically significant.

RESULTS

Nurr1 Expression Decreased in Both the Cerebral Cortex and Hippocampus of Rats after MCAO/R

Every three rats were euthanized randomly in each group at different time point and different treatment after MCAO/R. Rats

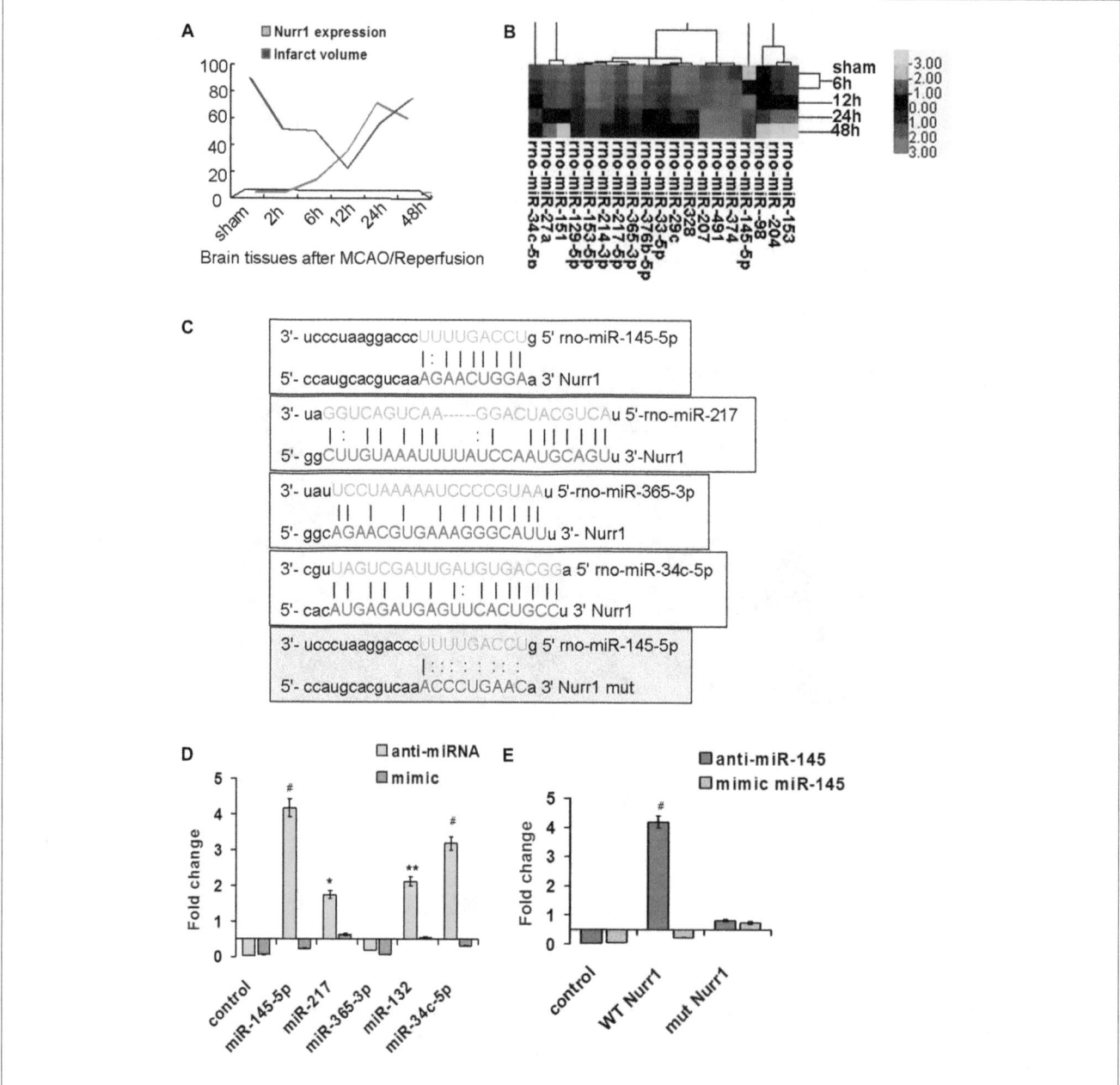

FIGURE 2 | Screening for miRNAs that directly target the 3′UTR of Nurr1. **(A)** The infarct volume was negatively correlated with Nurr1 expression some extent by Pearson correlation test, but was not statistically significant (correlation coefficient −0.457, P = 0.362). **(B)** Heat map of predicted miRNAs expression in cerebral cortex tissue from sham and MCAO/R animals (one-way analysis of variance (ANOVA), $p < 0.05$). **(C)** The predicted binding sites of selected miRNAs (in green color) to the 3′UTR of Nurr1 (in red color) is mapped in this figure. Nucleotides which were altered for mutational studies are marked in gray color of background. **(D)** Quantitation of the effects of anti miRNAs and miRNA mimics interaction with the 3′UTR of Nurr1. **(E)** Quantitation of the effects of anti-miR-145-5p and miR-145-5p mimic interactions with the normal binding sites (WT) and mutated binding sites (mut) in 3′UTR of Nurr1. $^*p < 0.05$, $^{**}p < 0.01$, $^\#p < 0.001$.

in the sham group were used as negative control. Approximately 138 rats were euthanized in the whole study, with the 82.4% (138/170) survival rate of the animals after MCAO surgery. Both mRNA and protein levels of Nurr1 decreased from the start of ischemic stroke, reached a minimum at 12 h, and then increased until 48 h (**Figures 1A,B**). Similarly, Nurr1 expression significantly decreased on the infarct side of the cortex and hippocampus of fresh frozen brain tissue sections as shown

by immunofluroescence assays (**Figure 1C**). In addition, brain tissues were stained with 2,3,5-triphenyl-tetrazolium chloride (TTC) to measure infarct size on the ipsilateral side. Infarct size peaked at ∼24 h and then decreased progressively until 48 h after occlusion (**Figure 2A**). Infarct volume was negatively correlated with Nurr1 expression in the cortex, but this was not statistically significant using the Pearson correlation test (correlation coefficient −0.457, P = 0.362).

TABLE 1 | microRNAs that potentially regulate Nurr1 through its 3'UTR by bioinformatic analysis prediction in three foremost databases in this field.

miRNA	Mature sequence	Target Score of Nurr1 prediction			Other target genes
		MIRDB	TARGETSCAN (Pct)	MICRORNA (mirSVR)	
>rno-miR-145-5p	GUCCAGUUUUCCCAGGAAUCCCU	81	0.30	−1.03	SRGAP2, Abhd17c, Ebf3, Add3, Ythdf2, Kcna4, Elmo1, Snx8, Abhd17b, Cbfb, Ino80, Ap1g1, Rev3l, Scamp3, Mdfic, Spop, Zfyve9, Mpzl2, Prkx, Pikfyve, Angpt2, RGD1562865, Zbtb6, Cachd1, Zfhx4, Zfyve26, Dusp6, Actg1, Fam135a, H1f0, H2afx, Spsb4, Lox, Rapgef2, Gosr2, Coro2b, Snx27, Cdo1, Spats2, Nol9, Csmd3
>rno-miR-365-3p	UAAUGCCCCUAAAAAUCCUUAU	—	0.28	−0.55	Ing3, Fads1, Casp6, Usp48, Sgk1, Eltd1, Adm, Pax6, Ublcp1, Mrs2, Crbn, Gcom1, MGC112830, Ets1, Rgs1, Tcp11l2, Otc, Tyms, Pebp1, Inhbe, Larp1b, Igf1, Csk, Ubfd1, Dcun1d5, Prkab2, Acsm5, Plcb4, Cpox, Arrb2, Fgfbp1, Nat15, Mare, IL10, Grb7, Galt, Nphs1, Aldob
>rno-miR-214-3p	ACAGCAGGCACAGACAGGCAG	—	<0.1	−0.28	Zbtb20, Luzp1, Atp2a3, Camta1, Smarcd1, Naa15, Endod1, Sec24c, P2rx6, Ldoc1, Pla2g3, Cpsf4, Psmd11, Ammecr1l, Myo18a, Zfand3, Numa1, Ccdc167, Tspan9, Sema4d, Phf6, Zbtb39, Ppih, Itpk1, Atp11c, Pter, Slc8a1, Wnt2, Diaph1, Naa50, Qsox1, Nfatc2, Plekhj1, Ezh1, Hr, Ctdsp1, Fbln5, Nomo1, Slc45a4, Lhx6, Socs7
>rno-miR-33-5p	GUGCAUUGUAGUUGCAUUGGCA	57	<0.1	−0.99	Abca1, Zfp281, Crot, Pdgfra, Ywhah, Hadhb, Sec24c, Arid5b, Scn8a, Cacna1c, Mdm4, Rb1cc1, Slc25a25, Cntln, Fbxw7, Naa15, Tmem33, Tmem86a, Ctnnd1, Map3k3, Dpy19l1, Slltrk5, Gpr158, Zfp286a, Abi1, Tph2, Sgcb, Braf, Txk, Pof1b, Fv1, B3galt2, Enc1, Cdk14, Tbc1d9b, Nfatc2, Plekhj1, Ezh1, Hr, Ctdsp1, Fbln5
>rno-miR-217-5p	UACUGCAUCAGGAACUGACUGG	75	0.20	−1.02	Acer3, Zfp711, Usp6nl, Ppm1d, Atp1b1, Med17, Bai3, RGD1566359, Thoc2, Ezh2, Hivep3, Slc38a2, Ubl3, Kctd9, Dennd6a, Anln, Extl2, Stt3a, Yrpf6, Tmem47, Fn1, Yaf2, Ythdc1, RGD1307830, Appbp2, Gpm6a, Bicd1, Atp11c, Tm9sf3, Zfyve20, Zbtb5, Morf4l2, Tet2, Ptpn21, Foxo1, Kcnh5, Klhl29, Wdr48, Ube4a, Hnrnpa3, Tmed10, Esco1, Slc45a2, Wapal, Ehmt1, Kctd5, RGD1311595, Myef2
>rno-miR-34c-5p	AGGCAGUGUAGUUAGCUGAUUGC	—	0.62	—	Ttc19, Vamp2, Notch1, Rras, E2f5, Tmem79, Arhgap1, Fam76a, Osgin2, Zfhx4, Ctnnd2, Frk, Soga1, Ap1s2, Fam167a, Ubp1, Camta1, Ppp1r11, Gmfb, Slc44a2, Ptpn4, Svop, Mycn, Zfp644, Map1a, Baalc, Pogz, Eml5, Tbl1xr1, Pitpnc1, Shkbp1, Dgkz, Taf5, Ranbp10, Myrf, Plp5k1a, Akap6, Slc35g2, Asic2, Strn3, Zer1, Daam1
>rno-miR-129-5p	CUUUUUGCGGUCUGGGCUUGC	88	<0.1	−1.20	Prkcb, Rbms3, Cib2, Zbtb20, Zfp281, Hmg1l1, Ube2k, Sc5d, Ctdspl2, Rai14, Ccdc43, Eif4g3, Kctd4, Ago3, Mdm4, Sun2, Rims2, Ythdf1, Tnpo1, Ap3b1, Fam63b, Sacm1l, Pdp2, RGD1559786, Ube4b, Klhl32, Nr2c2, Zfp36l1, Gapt, Wdr61, Duoxa1, Ets1, Pou3f1, Tmub2, Ptp4a1, Nrg2, RGD1307621, Paqr5, Add3, Kctd1

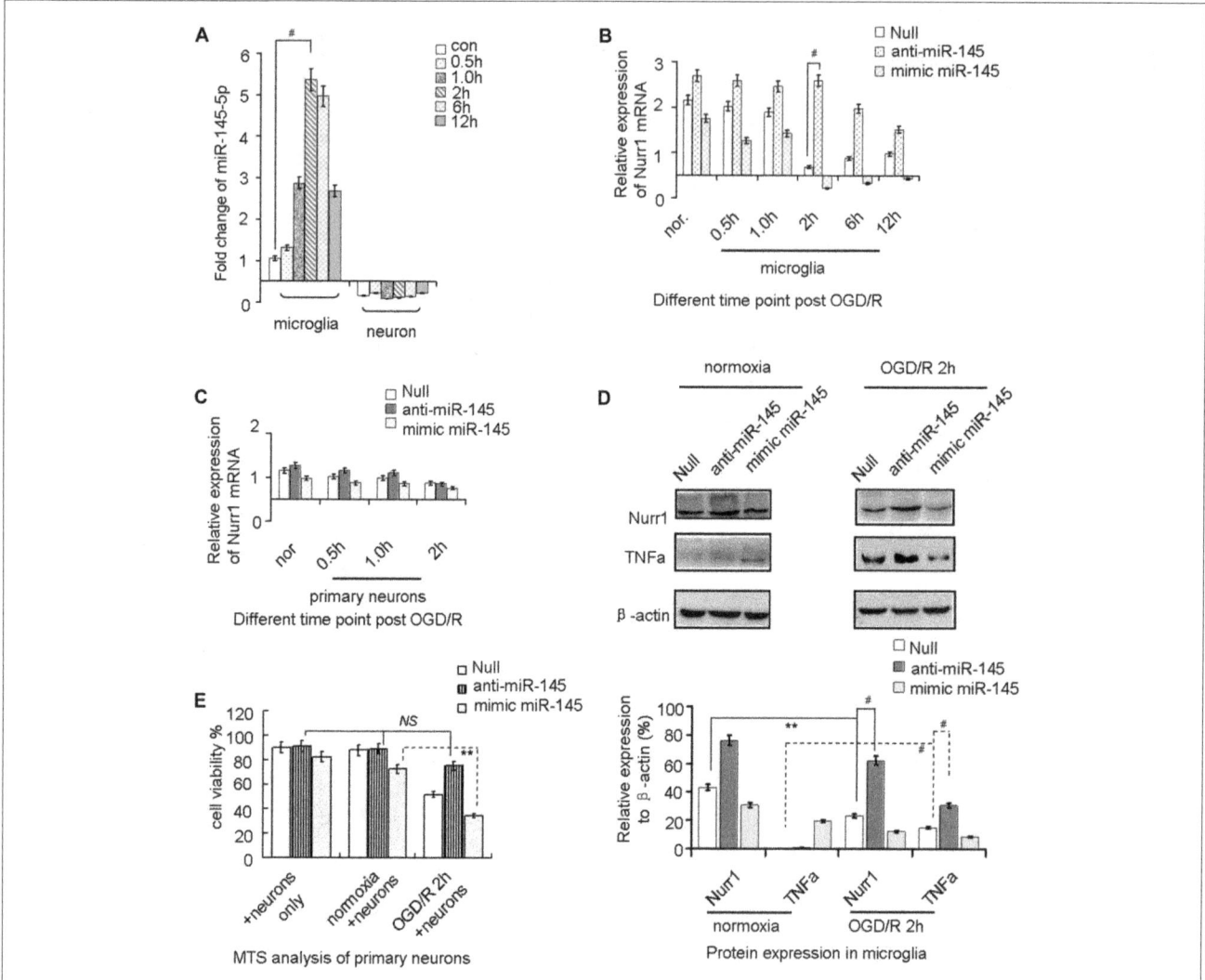

FIGURE 3 | miR-145-5p regulates endogenous Nurr1 levels in microglia. Microglia cultured in normoxia (nor.) and un-transfected (the null group) served as negative controls. **(A)** miR-145-5p expression increased from the onset of OGD/R and peaked at 2 h in microglia (p < 0.001), but did not change significantly in primary neurons. **(B)** Transfection anti-miR-145-5p induced an sharply increasing of Nurr1 mRNA post OGD/R 2 h in microglia (p < 0.001). **(C)** No obvious changes of Nurr1 mRNA were observed in primary neurons subjected to OGD/R, independent of anti-miR-145-5p or miR-145-5p mimic administration. **(D)** By Western blot assay, Nurr1 protein decreased and TNF-α increased significantly after OGD/R 2 h in microglia. miR-145-5p mimic administration significantly reduced Nurr1 protein level after OGD/R 2 h. **(E)** By MTS analysis of neurons, cell viability of primary neurons cocultured with microglia OGD/R 2 h with anti-miR-145-5p transfection did not show significant increase of cell death when compared to microglia normoxia group or neurons only group (NS: non-significance; p > 0.05). Oppositely, administration of miR-145-5p mimic increases neurons death notably in microglia after OGD/R 2 h than those co-cultured with microglia control (p < 0.01). **p < 0.01, #p < 0.001.

Screening for miRNAs that Directly Target the 3′UTR of Nurr1

miRNA expression was measured in our miRNA profiling analysis of rat cortex samples at different time point after MCAO/R (**Figure 2B**). Eleven miRNAs, including miR-145-5p, miR-34c-5p, miR-365-3p, miR-214-3p, miR-151, miR-27a, miR-153-5p, miR-365-3p, miR-33-5p, miR-217-5p and miR-129-5p, were differentially and significantly expressed (P < 0.05; **Figure 2B**). Among them, seven putative relevant miRNAs were screened by bioinformatic analysis of databases to identify miRNAs that

regulate Nurr1 directly through its 3′UTR (**Table 1**). The predicted binding sites of the selected miRNAs are shown in **Figure 2C**.

The 3′UTR of Nurr1 and miRNA binding sites were cloned to construct firefly luciferase reporter plasmids (pMIR-REPORT). Plasmids were cotransfected independently with the respective anti- or mimic miRNAs into HEK293T cells. Results showed that the miR-145-5p mimic clearly decreased the luminescence reporter signal to a greater extent (P < 0.001; **Figure 2D**), as did miR-132 and miR-34c-5p, which were previously shown to directly target Nurr1. However,

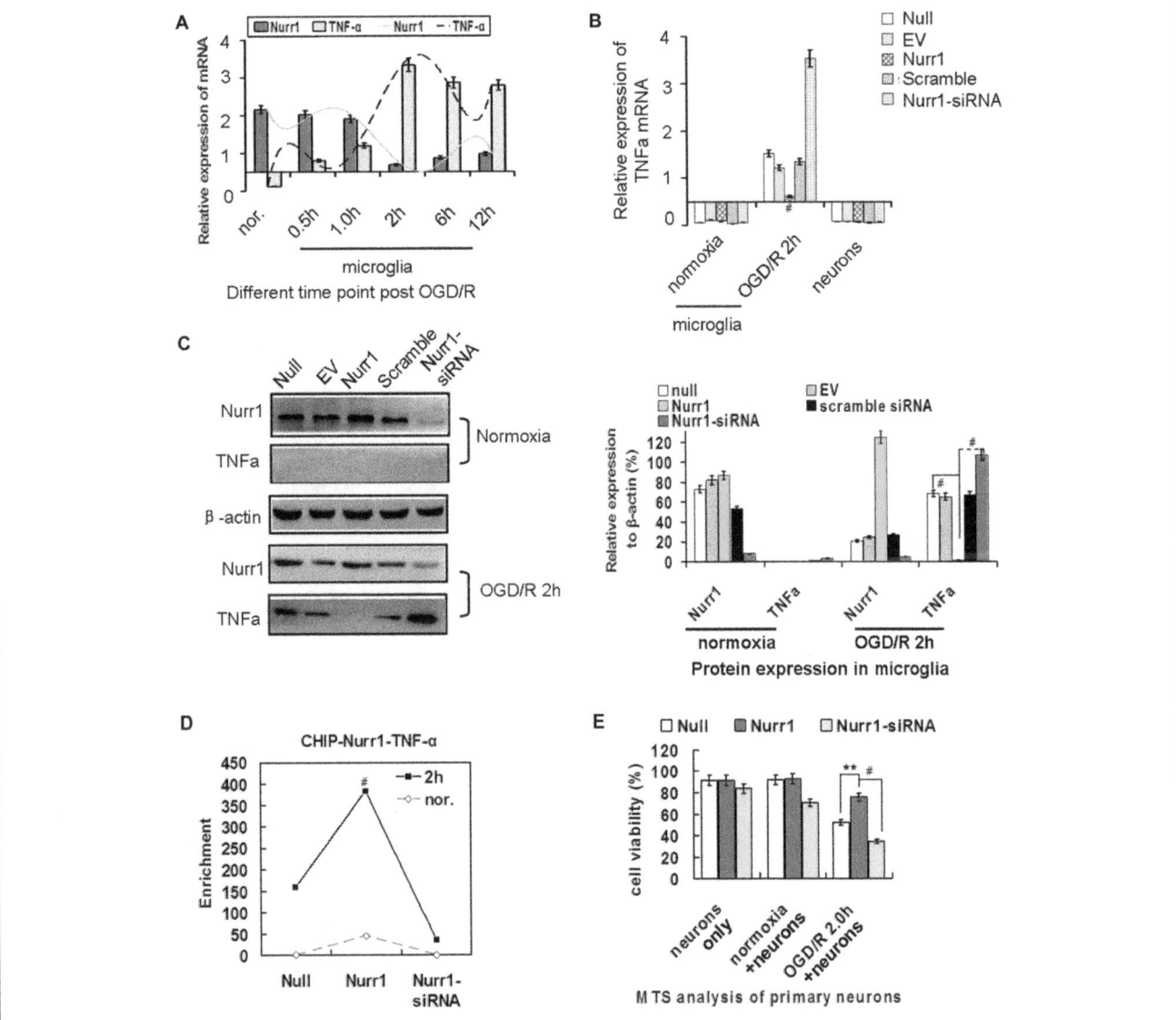

FIGURE 4 | Nurr1 overexpression suppresses TNF-α expression in microglia. **(A)** Quantitation of Nurr1 and TNF-α mRNA in primary microglia which were subjected to OGD/R 0.5 h, 1.0 h, 2 h, 6.0 h and 12 h, respectively. TNF-α mRNA increased from the onset of OGD/R and peaked at OGD/R 2 h in microglia ($p < 0.001$). However, Nurr1 mRNA showed an opposite trend at different time point of OGD/R. **(B)** Overexpression of Nurr1 led to significant attenuation of TNF-α mRNA, otherwise Nurr1-siRNA led to sharply increase of TNF-α expression after OGD/R 2 h in microglia. However, these significant changes of TNF-α mRNA expression were not observed in microglia with normooxia culturing (microglia nor. group) or neurons only group. **(C)** After OGD/R 2 h in microglia, Nurr1 overexpression by pcDNA-Nurr1 transfection significantly inhibited TNF-α protein expression ($p < 0.001$). Oppositely, Nurr1 knockdown by siRNA interference significantly restored TNF-α protein expression ($p < 0.001$). **(D)** ChIP assay of Nurr1 on the TNF-α promoter in response to OGD/R 2 h in microglia cells. Nurr1 overexpression by plasmid transfection increased Nurr1 occupancy at the TNF-α promoter, especially when TNF-α expression reached great amount post OGD/R 2 h. Data are displayed as fold enrichment over control IgG. **(E)** By MTS analysis of neurons, neurons viability co-cultured with microglia OGD/R 2 h which were transfected with Nurr1 overexpression plasmid showed significant lower ratio of cell death than the null ($p < 0.01$) and Nurr1-siRNA group ($p < 0.001$). However, no obvious changes of cell viability were observed in groups of neurons only and neurons+microglia normoxia after overexpression or knockdown Nurr1 expression. **$p < 0.01$, #$p < 0.001$.

miR-217 exhibited a lower significant interaction when compared with miR-145-5p and miR-34c-5p (**Figure 2D**). Mutations of the miR-145-5p and Nurr1 recognition sites in the 3′UTR abolished their interaction (**Figure 2E**). These results suggest that miR-145-5p modulates Nurr1 through its 3′UTR.

miR-145-5p Regulates Endogenous Nurr1 Levels in Microglia

miR-145-5p expression increased from the onset of OGD/R, peaked at 2 h, and then progressively decreased until 12 h in microglia (**Figure 3A**). However, no significant changes of miR-145-5p expression were observed at different time points

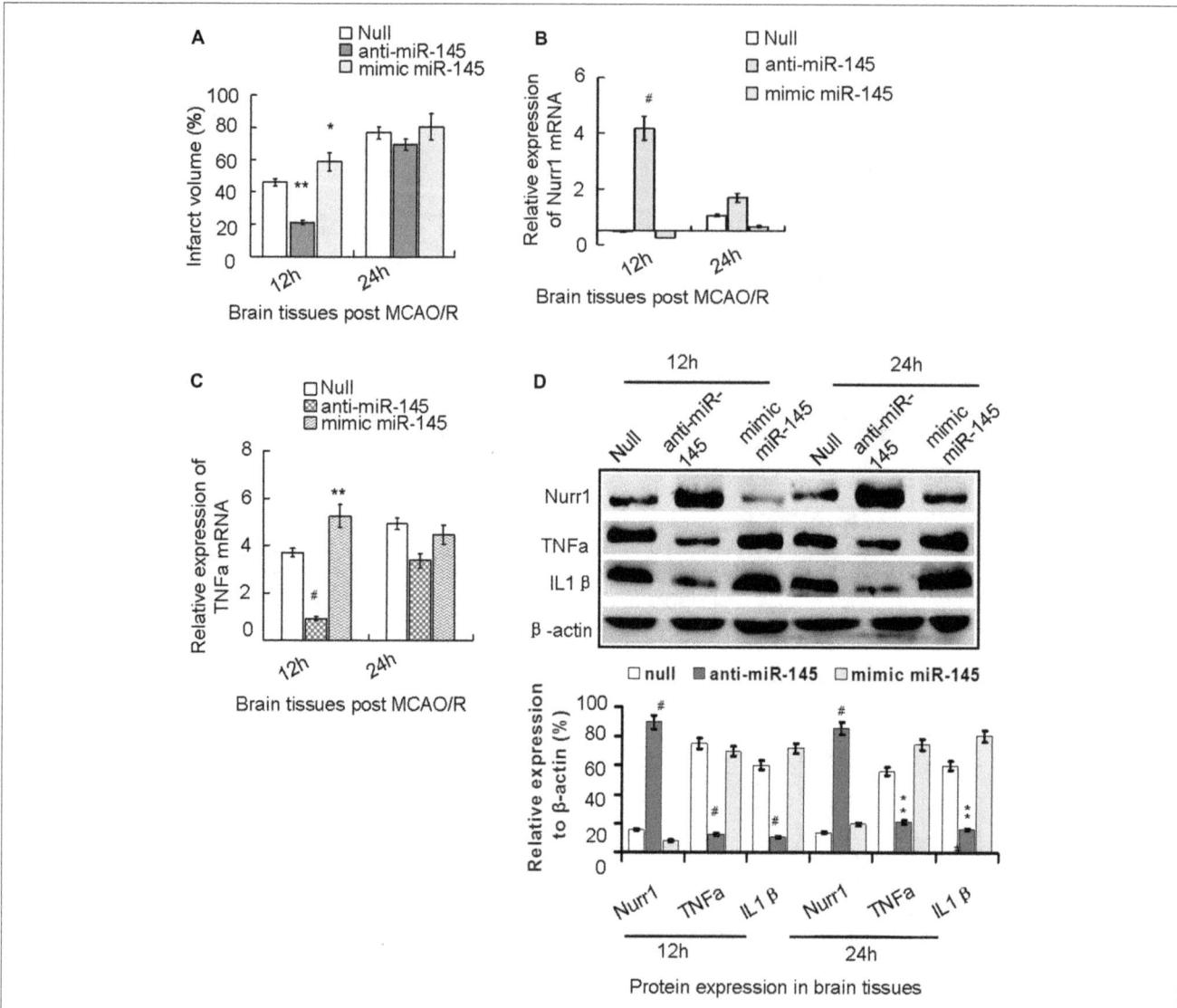

FIGURE 5 | The axis signaling of miR-145-5p-Nurr1-TNF-α in acute *MCAO/R* model of rats by *in vivo* expriments. Infarct volume was plotted as the percentage of ipsilateral cerebral. **(A)** After 12 h of MCAO/R, administration of anti-miR-145-5p *in vivo* via ICV injection to ischemic rats immediately reduced Infarct volume by 39.05% ($p < 0.01$). While administration of miR-145-5p mimic increased infarct volume by 15.05% when compared to the null group ($p < 0.05$). However, these changes of infarct volume were not observed significantly at 24 h. **(B)** By RT-qPCR analysis, Nurr1 mRNA expression significantly increased in anti-miR-145-5p injection samples ($p < 0.001$) whereas decreased after administration of mimic miR-145-5p ($p < 0.05$). **(C)** In contrast, TNF-α mRNA expression significantly decreased in anti-miR-145-5p injection samples post 12 h of MCAO/R ($p < 0.001$) whereas significantly increased when administration of miR-145-5p mimic ($p < 0.01$). However, these significant changes of TNF-α mRNA expression were not observed in samples post 24 h of MCAO/R. **(D)** By western blot analysis, both TNF-α and IL1β expression levels were significantly suppressed by Nurr1 overexpression with administration of anti-miR-145-5p at 12 h and 24 h post-MCAO/R, and vice versa. *$p < 0.05$, **$p < 0.01$, #$p < 0.001$.

of OGD/R in primary neurons (**Figure 3A**). Nurr1 mRNA levels were reduced from the onset of OGD/R and peaked at 2 h in microglia (**Figure 3B**). Administration of anti-miR-145-5p induced a sharp increase of Nurr1 mRNA levels in comparison with the null group at each time point of OGD/R in microglia (**Figure 3B**) but not in primary neurons (**Figure 3C**).

After OGD/R 2 h induction of miR-145-5p overexpression in microglia, the Nurr1 protein was clearly reduced when compared with cells in the normoxia group (**Figure 3D**).

In particular, expression of TNF-α, which was reported to be transrepressed by Nurr1, displayed a pattern of expression opposite that of Nurr1 under treatment with the miR-145-5p mimic or anti-miR-145-5p (**Figure 3D**). Cell viability assessed by MTS analysis showed that primary neuronal cells co-cultured with microglia transfected with anti-miR-145-5p post OGD/R 2 h did not exhibit significant cell death when compared with the normoxia group (**Figure 3E**). Conversely, a noticeable increase in neuronal cell death was observed in those co-cultured with miR-145-5p mimic-transfected

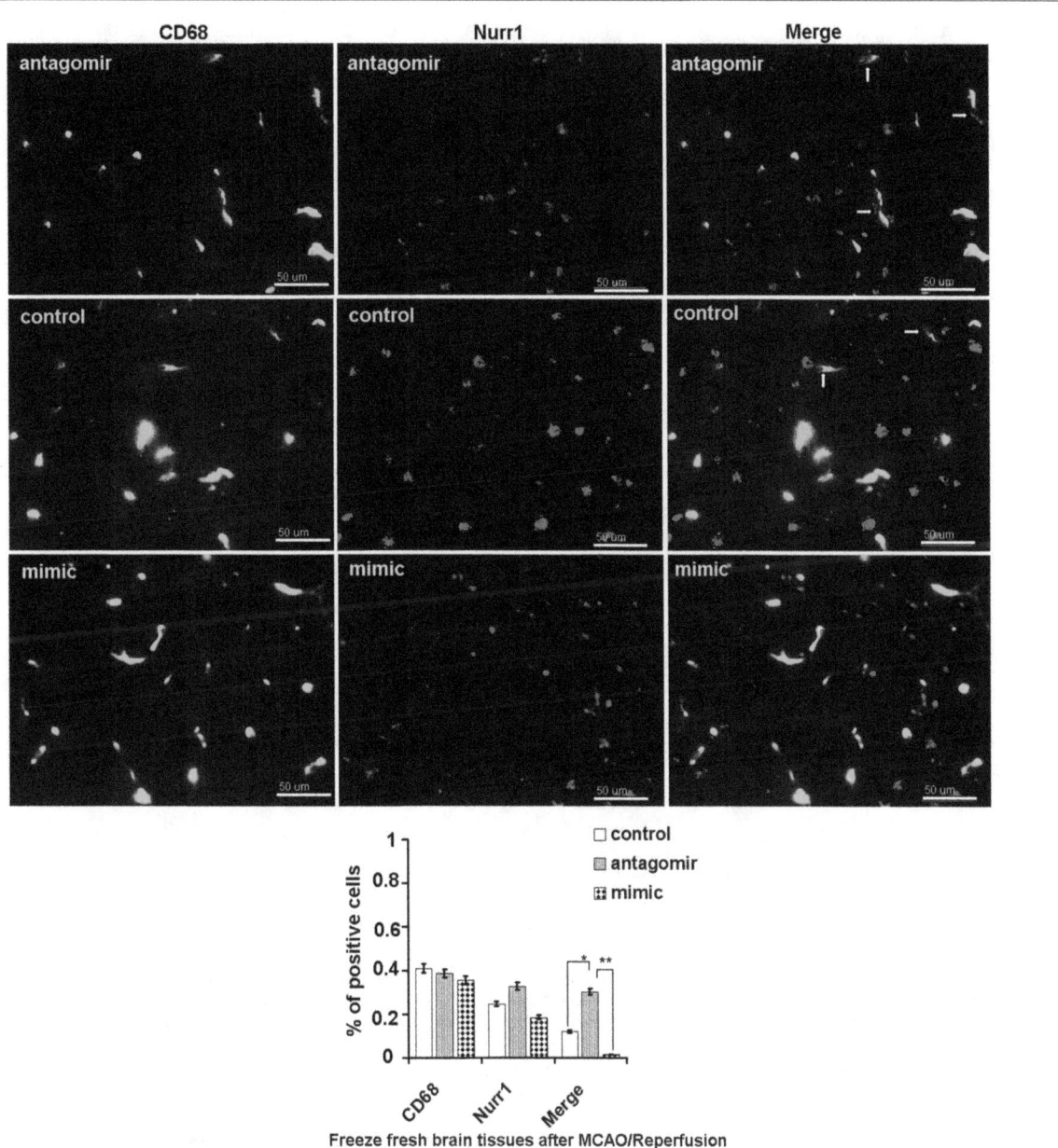

FIGURE 6 | Nurr1 expression in active microglia with administration of miR-145-5p in peri-infarct areas post 12 h of MCAO/R. Double immunofluorescence staining shows more expression of Nurr1 in active microglia with administration of anti-miR-145-5p, and little expression of Nurr1 with administration of miR-145-5p mimic in the peri-infarct areas bordering with intact tissues post 12 h of MCAO/R. Scale bars = 50 μm. Arrows indicate co-localization of CD68 and Nurr1 in active microglia. *$p < 0.05$, **$p < 0.01$.

microglia (**Figure 3E**). These experiments suggest that enhanced levels of miR-145-5p inhibit neuronal viability by regulating expression of Nurr1 and TNF-α in microglia after OGD/R.

Overexpression of Nurr1 Suppresses TNF-α Expression

It was previously reported that Nurr1 induces transrepression of TNF-α by binding to its promoter in Parkinson's disease (Saijo et al., 2009). Herein, we sought to determine the effects

of aberrant expression of Nurr1 on TNF-α expression in microglia and primary neurons subjected to OGD/R at different time. Results of qRT-PCR analysis showed that TNF-α mRNA expression increased and peaked at 2 h of OGD/R in microglia which is opposite to Nurr1 mRNA levels (**Figure 4A**), but not in primary neurons (data not shown). Upon treatment with the Nurr1 activation plasmid, overexpression of Nurr1 significantly attenuated TNF-α mRNA and protein expression. Conversely, an increase in TNF-α expression was observed after treatment with Nurr1-siRNA after OGD/R 2 h in microglia (**Figures 4B,C**).

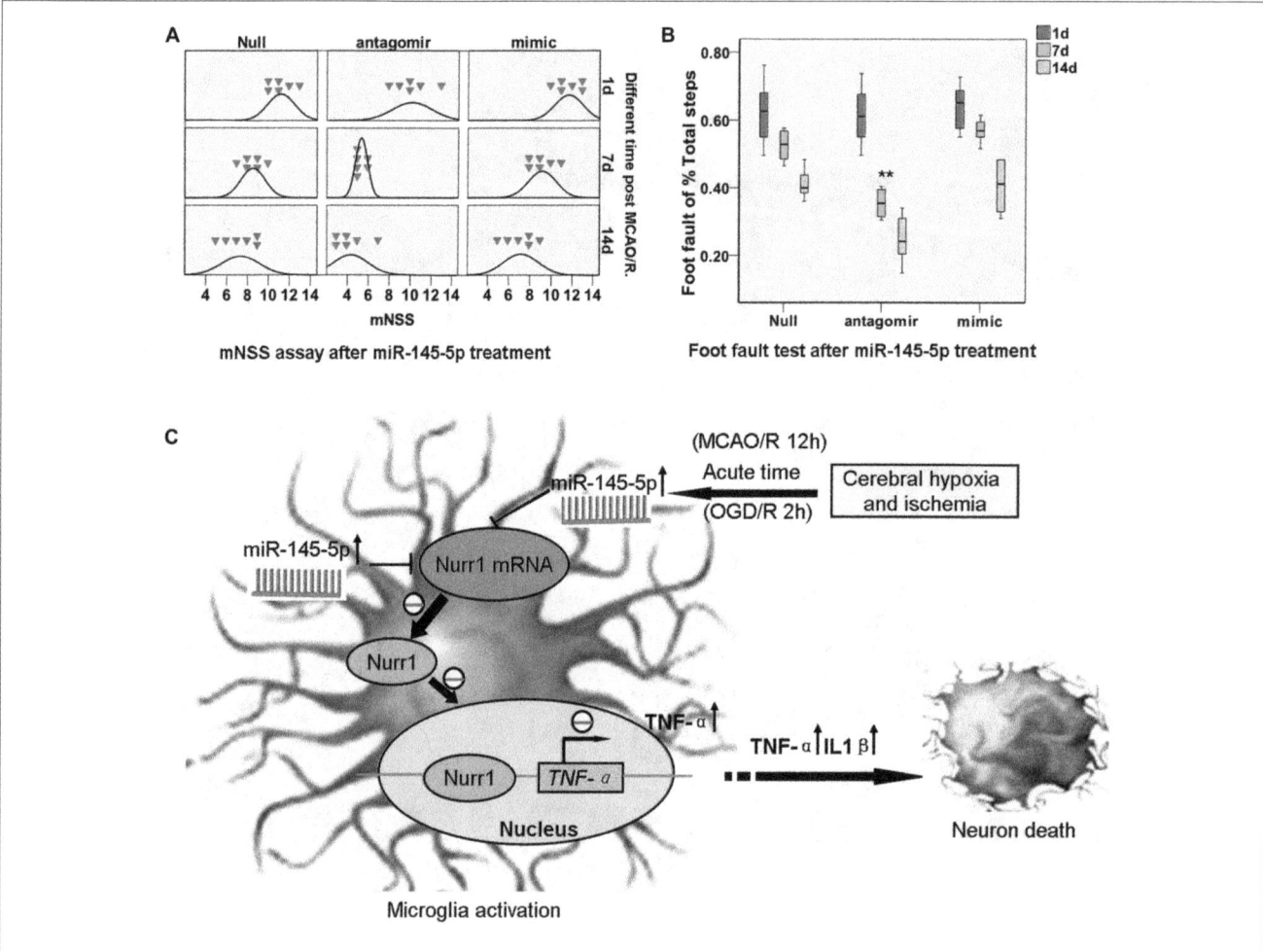

FIGURE 7 | miR-145-5p interruption facilitates neurological outcome of rats post MCAO/R. Modified Neurological Severity Score (mNSS) and foot fault tests were assessed on days 1, 7 and 14 after MCAO/R. **(A)** mNSS in antago-miR-145-5p animals were significantly decreased at both 7d and 14d after MCAO/R ($p < 0.05$). **(B)** Foot fault test, for the front left and hind left limbs were significantly lower in antagomiR-145-5p animals than that of miR-145-5p mimic and null animals. $n = 54$ for the combination group. **(C)** The graphical summary. Here, we describe a novel miR-145-5p regulatory mechanism of Nurr1 that can act downstream of TNF-α activation. In acute cerebral ischemia (MCAO/R 12 h) of rats and OGD/R 2 h of microglia, Nurr1 inhibits TNF-α expression by binding promoter of *TNF-α* gene. This regulatory effect is inhibited by Nurr1 protein decline that induced by miR-145-5p overexpression. Blocking the abnormal activation of miR-145-5p-Nurr1-TNF-α axis signaling can relieve neurons death upon MCAO/R of rats in acute time. **$^{**}p < 0.01$.

However, no significant changes were observed in primary neurons (**Figure 4B**).

To assess whether Nurr1 binds to the *TNF-α* gene promoter, we performed a chromatin immunoprecipitation (ChIP) assay in microglia. As shown in **Figure 4D**, Nurr1 plasmid transfection increased Nurr1 occupancy at the TNF-α promoter, especially when TNF-α expression reached a high level after OGD/R 2 h. Furthermore, neuronal cells co-cultured with microglia after OGD/R 2 h with Nurr1-siRNA transfection showed significant cell death and could be rescued by Nurr1 overexpression after transfection with the Nurr1 plasmid (**Figure 4E**). In contrast, knockdown or overexpression of Nurr1 did not alter cell viability independent of OGD/R in the primary neuron culture only group (**Figure 4E**). These data demonstrate that lower expression of Nurr1 relieves transrepression of TNF-α in microglia and inhibits cell growth in primary neurons after OGD/R 2 h.

Abnormal Inactivation of Nurr1-Mediated Transrepression on TNF-α by miR-145-5p Overexpression Accelerates Inflammatory Injury in Acute MCAO/R of Rats

We demonstrated that increasing expression of miR-145-5p peaked at 12 h in our rat MCAO/R model (**Figure 2A**). We next administered the miR-145-5p mimic and anti-miR-145-5p *in vivo* via ICV injection into ischemic rats immediately after MCAO. Mean infarct volumes were measured at 12 h and 24 h post-MCAO/R. Administration of anti-miR-145-5p reduced infarct volume by 24.05% at 12 h, whereas administration of the miR-145-5p mimic increased infarct volume by 11.05% (**Figure 5A**). Administration of anti-miR-145-5p significantly increased Nurr1 mRNA expression (**Figure 5B**) and reduced TNF-α mRNA expression at 12 h post-MCAO/R (**Figure 5C**) but not at 24 h post-MCAO/R.

Accordingly, Nurr1 protein expression increased 4.53- and 3.42-fold upon treatment with anti-miR-145-5p at 12 h and 24 h post-MCAO/R, respectively (**Figure 5D**). Both TNF-α and IL-1β protein levels were significantly up-regulated at both 12 h and 24 h post-MCAO/R, and they were successfully suppressed by Nurr1 overexpression induced by administration of anti-miR-145-5p (**Figure 5D**). Accordingly, Nurr1 expression increased significantly upon anti-miR-145-5p treatment vs. miR-145-5p mimic treatment in active microglia in peri-infarct areas post 12 h of MCAO/R using immunofluorescence (**Figure 6**). These data reveal that lower expression of Nurr1 induced by miR-145-5p upregulation increases infarct volume at an early stage of cerebral ischemia of rats by activating TNF-α and IL-1β proinflammatory signals.

miR-145-5p Interruption Facilitates Neurological Outcome of Rats Post MCAO/R

Significant functional deficits were observed in the mNSS for animals subjected to mimic miR-145-5p administration compared to anti-miR-145-5p animals at both 7d and 14d after injury ($p < 0.05$; **Figure 7A**). For foot fault tests, the percentages of front left, hind left, total left, and total foot faults at 7d and 14d after injury after anti-miR-145-5p was significantly lower, compared to that of null and mimic miR-145-5p animals ($p < 0.05$; **Figure 7B**). However, in the mNSS and foot fault tests were not significantly different between the null and miR-145-5p mimic administration animals ($p > 0.05$).

DISCUSSION

The present study utilized a coculture model system in which neurotoxicity was induced by activated microglia post OGD/R 2 h. We showed several lines of evidence that Nurr1 plays an important role in protecting neurons from inflammation-induced neurotoxicity by inhibiting expression of inflammatory genes in microglia. Firstly, clear reductions in both Nurr1 mRNA and protein expression were detected in microglia post OGD/R 2 h and accompanied by a substantial elevation in TNF-α gene expression (**Figures 4A,C**). Furthermore, Nurr1 overexpression induced by Nurr1 plasmid indeed leaded to inhibition of TNF-α mRNA and protein expression in microglia post OGD/R 2 h (**Figures 4B,C**). Secondly, reduction of Nurr1 expression induced by OGD/R 2 h in microglia resulted in increased neuronal cell death that could be rescued by Nurr1 plasmid transfection (**Figure 4E**). Thirdly, ChIP assay demonstrated that Nurr1 was recruited to the TNF-α promoter following OGD/R 2 h treatment and was further enhanced by Nurr1 overexpression (**Figure 4D**), indicating that it was acting locally to repress transcription. This finding is consistent with previous studies showing that reduced Nurr1 expression results in death of tyrosine hydroxylase-expressing neurons in Parkinson's disease by targeting the TNF-α gene promoter in microglia (Saijo et al., 2009; Kim et al., 2015). Finally, Nurr1 declined sharply in brain tissues at 12 h following

MCAO/R, and lower Nurr1 expression was partly correlated with infarct volume 12–24 h post MCAO/R by *in vivo* assays (**Figure 2A**).

Results of our study strongly indicate transcript factor Nurr1 protects neurons from ischemia-induced inflammation injury, at least in part, through transrepression of TNF-α in activated microglia post OGD/R. These are supported by several recent studies: (1) up-regulation of VEGF and Nurr1 strongly promote DAergic neuronal survival after Cystatin C treatment, which might be involved in p-PKC-α/p-ERK1/2-Nurr1 signaling and autophagy (Zou et al., 2017). (2) The interaction between Nurr1 and Foxa2 protects midbrain dopamine neurons against various toxic insults, while their expression are absent during aging and degenerative processes (Oh et al., 2015). (3) The Glial cell line-derived neurotrophic factor (GDNF) was found to protect neurons in cerebral ischemia, and Nurr1 was up-regulated by GDNF in the process of dendritic and electrical maturation of neuron cells (Cortés et al., 2017).

The important roles of changes in miRNA expression in ischemic brain injury have been discovered recently using miRNA profiling techniques in a rat MCAO/R model (Dharap et al., 2009; Di et al., 2014). In the present study, miR-145-5p mimic clearly decreased Nurr1 expression in HEK293T ($P < 0.001$), as did miR-132 and miR-34c-5p, which was consistent with the previous study that miR-34c-5p directly regulated Nurr1 in HCT116 cells (Beard et al., 2016) and miR-132 targeted Nurr1 in differentiation of dopamine neurons (Yang et al., 2012). However, miR-217 exhibited a lower significant interaction when compared with miR-145-5p and miR-34c-5p (**Figure 2D**). Further mutation of the miR-145-5p and Nurr1 recognition sites in the 3′UTR abolished their interaction in HEK293T cells as detected by luciferase reporter assay (**Figure 2E**). However, Nurr1 expression is not regulated by miR-365-3p. These data demonstrate that it was essential to elucidate the specific relation between miR-145-5p and Nurr1 in rat model of cerebral I/R injury.

miR-145-5p has been shown to be up-regulated in the pathological process of vascular neointimal lesion formation (Saugstad, 2010; Li et al., 2012), cardiomyocyte survival (Chen et al., 2015), and H_2O_2-induced neuronal injury (Gan et al., 2012). Here, we observed that miR-145-5p expression sharply increased in the cortex of rats 12 h post MCAO/R (**Figure 2B**) and in isolated primary microglia post OGD/R 2 h (**Figure 3A**). Modulation of miR-145-5p expression affects cell viability under *in vitro* ischemic conditions, and this occurs via regulation of Nurr1 (**Figures 3D,E**). Administration of miR-145-5p mimic in primary microglia post OGD/R significantly reduced TNF-α expression and IL-1β by enhancing Nurr1 expression, which subsequently improved neuronal cell viability (**Figures 3E, 4E**). As expected, anti-miR-145-5p administration via ICV injection reduced infarct volume by 24.05% at 12 h post MCAO/R in rats (**Figure 5A**). Whatever, more researches must be performed to elucidate the specific effects of miR-145-5p on cerebral I/R injury in human blood or cerebrospinal fluid.

CD68 is specifically expressed in microglia in ischemic stroke brain (Szalay et al., 2016). By immunofluorescence assay, massive

Nurr1 expression was observed in active microglia after anti-miR-145-5p treatment (**Figure 6**). Our findings are in agreement with a previous report, which demonstrated that antagomir-145 infusion resulted in a decreased area of infarction at 1 day of reperfusion by increasing SOD2 protein expression (Gan et al., 2012). It is noteworthy that 24 h after MCAO/R in rats, miR-145-5p inhibition could not reduce infarct volume despite the significant increase in Nurr1 (**Figure 5A**). Furthermore, significant functional improvement were observed in the mNSS and foot fault tests for animals subjected to mimic miR-145-5p administration compared to anti-miR-145-5p animals after cerebral I/R injury (**Figures 7A,B**). This indicates that the supression of miR-145-5p, which decreased neuron cell death by increasing TNF-α-mediated inflammation, might be insufficient to provide protective effects under these conditions. Therefore, it is still worth exploring function and mechanism(s) of miR-145-5p/Nurr1/TNF-α Signaling in human stoke in the near future.

CONCLUSION

In summary, our study identified and confirmed a novel regulation of Nurr1 by miR-145-5p in both *in vitro* and *in vivo* cerebral ischemic conditions. The hypothetic mechanisms were examined in isolated primary microglia and neurons post OGD/R by *in vitro* study (**Figure 7C**). I/R-induced miR-145-5p overexpression suppresses Nurr1 protein expression and attenuates Nurr1 transrepression of the *TNF-α* promoter in microglia (OGD/R 2 h), which then causes TNF-α-related neuronal injury. Additional *in vivo* studies have shown that administration of anti-miR-145-5p could increase Nurr1 expression and reduce subsequent infarct volume in acute cerebral ischemia (MCAO/R 12 h). It may be an effective therapeutic strategy of reducing neuronal injury following MCAO/R in rats by blocking abnormal activation of miR-145-5p/Nurr1/TNF-α axis signaling in the acute phase.

AUTHOR CONTRIBUTIONS

XX and YoZ conceived and designed the experiments. XX, LP and YaZ conducted the experiments. XX and LL analyzed the results. LP, YC, JZ, SY and XX contributed materials and analysis tools. XX wrote the article. YoZ is the corresponding author. All authors reviewed the manuscript.

ACKNOWLEDGMENTS

This study was supported by the Natural Science Foundation of China, grant 81271460; the Natural Science Foundation of Chongqing Education Committee, grant KJ1500230; and the Science and Technology Innovation Programs for Postgraduates of Chongqing, grant CYB16103.

REFERENCES

Al-Mufti, F., Amuluru, K., Roth, W., Nuoman, R., El-Ghanem, M., and Meyers, P. M. (2017). Cerebral ischemic reperfusion injury following recanalization of large vessel occlusions. *Neurosurgery* doi: 10.1093/neuros/nyx341 [Epub ahead of print].

Beard, J. A., Tenga, A., Hills, J., Hoyer, J. D., Cherian, M. T., Wang, Y. D., et al. (2016). The orphan nuclear receptor NR4A2 is part of a p53-microRNA-34 network. *Sci. Rep.* 6:25108. doi: 10.1038/srep25108

Chen, R., Chen, S., Liao, J., Chen, X., and Xu, X. (2015). MiR-145 facilitates proliferation and migration of endothelial progenitor cells and recanalization of arterial thrombosis in cerebral infarction mice via JNK signal pathway. *Int. J. Clin. Exp. Pathol.* 8, 13770–13776.

Chen, X., Liu, Y., Zhu, J., Lei, S., Dong, Y., Li, L., et al. (2016). GSK-3β downregulates Nrf2 in cultured cortical neurons and in a rat model of cerebral ischemia-reperfusion. *Sci. Rep.* 6:20196. doi: 10.1038/srep20196

Chen, Y., Wu, X., Yu, S., Lin, X., Wu, J., Li, L., et al. (2012). Neuroprotection of tanshinone IIA against cerebral ischemia/reperfusion injury through inhibition of macrophage migration inhibitory factor in rats. *PLoS One* 7:e40165. doi: 10.1371/journal.pone.0040165

Cortés, D., Carballo-Molina, O. A., Castellanos-Montiel, M. J., and Velasco, I. (2017). The non-survival effects of glial cell line-derived neurotrophic factor on neural cells. *Front. Mol. Neurosci.* 10:258. doi: 10.3389/fnmol.2017.00258

Dharap, A., Bowen, K., Place, R., Li, L. C., and Vemuganti, R. (2009). Transient focal ischemia induces extensive temporal changes in rat cerebral microRNAome. *J. Cereb. Blood Flow Metab.* 29, 675–687. doi: 10.1038/jcbfm.2008.157

Di, Y., Lei, Y., Yu, F., Changfeng, F., Song, W., and Xuming, M. (2014). MicroRNAs expression and function in cerebral ischemia reperfusion injury. *J. Mol. Neurosci.* 53, 242–250. doi: 10.1007/s12031-014-0293-8

Fumagalli, S., Perego, C., Pischiutta, F., Zanier, E. R., and De Simoni, M. G. (2015). The ischemic environment drives microglia and macrophage function. *Front. Neurol.* 6:81. doi: 10.3389/fneur.2015.00081

Gan, C. S., Wang, C. W., and Tan, K. S. (2012). Circulatory microRNA-145 expression is increased in cerebral ischemia. *Genet. Mol. Res.* 11, 147–152. doi: 10.4238/2012.January.27.1

Gao, H., Chen, Z., Fu, Y., Yang, X., Weng, R., Wang, R., et al. (2016). Nur77 exacerbates PC12 cellular injury *in vitro* by aggravating mitochondrial impairment and endoplasmic reticulum stress. *Sci. Rep.* 6:34403. doi: 10.1038/srep34403

Gullo, F., Ceriani, M., D'Aloia, A., Wanke, E., Constanti, A., Costa, B., et al. (2017). Plant polyphenols and exendin-4 prevent hyperactivity and TNF-α release in LPS-treated *in vitro* neuron/astrocyte/microglial networks. *Front. Neurosci.* 11:500. doi: 10.3389/fnins.2017.00500

Jeyaseelan, K., Lim, K. Y., and Armugam, A. (2008). MicroRNA expression in the blood and brain of rats subjected to transient focal ischemia by middle cerebral artery occlusion. *Stroke* 39, 959–966. doi: 10.1161/STROKEAHA.107.500736

Jose, S., Tan, S. W., Ooi, Y. Y., Ramasamy, R., and Vidyadaran, S. (2014). Mesenchymal stem cells exert anti-proliferative effect on lipopolysaccharide-stimulated BV2 microglia by reducing tumour necrosis factor-α levels. *J. Neuroinflammation* 11:149. doi: 10.1186/s12974-014-0149-8

Karra, D., and Dahm, R. (2010). Transfection techniques for neuronal cells. *J. Neurosci.* 30, 6171–6177. doi: 10.1523/JNEUROSCI.0183-10.2010

Kim, C. H., Han, B. S., Moon, J., Kim, D. J., Shin, J., Rajan, S., et al. (2015). Nuclear receptor Nurr1 agonists enhance its dual functions and improve behavioral deficits in an animal model of Parkinson's disease. *Proc. Natl. Acad. Sci. U S A* 112, 8756–8761. doi: 10.1073/pnas.1509742112

Li, R., Yan, G., Li, Q., Sun, H., Hu, Y., Sun, J., et al. (2012). MicroRNA-145 protects cardiomyocytes against hydrogen peroxide (H_2O_2)-induced apoptosis through targeting the mitochondria apoptotic pathway. *PLoS One* 7:e44907. doi: 10.1371/journal.pone.0044907

Liu, F. J., Kaur, P., Karolina, D. S., Sepramaniam, S., Armugam, A., Wong, P. T., et al. (2015). MiR-335 regulates hif-1α to reduce cell death in both mouse cell line and rat ischemic models. *PLoS One* 10:e0128432. doi: 10.1371/journal. pone.0128432

Liu, C., Zhao, L., Han, S., Li, J., and Li, D. (2015). Identification and functional analysis of microRNAs in mice following focal cerebral ischemia injury. *Int. J. Mol. Sci.* 16, 24302–24318. doi: 10.3390/ijms161024302

Lourbopoulos, A., Karacostas, D., Artemis, N., Milonas, I., and Grigoriadis, N. (2008). Effectiveness of a new modified intraluminal suture for temporary middle cerebral artery occlusion in rats of various weight. *J. Neurosci. Methods* 173, 225–234. doi: 10.1016/j.jneumeth.2008.06.018

McCarter, K. D., Li, C., Jiang, Z., Lu, W., Smith, H. C., Xu, G., et al. (2017). Effect of low- dose alcohol consumption on inflammation following transient focal cerebral ischemia in rats. *Sci. Rep.* 7:12547. doi: 10.1038/s41598-017-12720-w

Minhas, G., Mathur, D., Ragavendrasamy, B., Sharma, N. K., Paanu, V., and Anand, A. (2017). Hypoxia in CNS pathologies: emerging role of mirna-based neurotherapeutics and yoga based alternative therapies. *Front. Neurosci.* 11:386. doi: 10.3389/fnins.2017.00386

Oh, S. M., Chang, M. Y., Song, J. J., Rhee, Y. H., Joe, E. H., Lee, H. S., et al. (2015). Combined Nurr1 and Foxa2 roles in the therapy of Parkinson's disease. *EMBO Mol. Med.* 7, 510–525. doi: 10.15252/emmm.201404610

Ouyang, Y. B., Stary, C. M., White, R. E., and Giffard, R. G. (2015). The use of microRNAs to modulate redox and immune response to stroke. *Antioxid. Redox Signal.* 22, 187–202. doi: 10.1089/ars.2013.5757

Roqué, P. J., Dao, K., and Costa, L. G. (2016). Microglia mediate diesel exhaust particle-induced cerebellar neuronal toxicity through neuroinflammatory mechanisms. *Neurotoxicology* 56, 204–214. doi: 10.1016/j.neuro.2016.08.006

Saijo, K., Winner, B., Carson, C. T., Collier, J. G., Boyer, L., Rosenfeld, M. G., et al. (2009). A Nurr1/CoREST pathway in microglia and astrocytes protects dopaminergic neurons from inflammation-induced death. *Cell* 137, 47–59. doi: 10.1016/j.cell.2009.01.038

Saugstad, J. A. (2010). MicroRNAs as effectors of brain function with roles in ischemia and injury, neuroprotection, and neurodegeneration. *J. Cereb. Blood Flow Metab.* 30, 1564–1576. doi: 10.1038/jcbfm.2010.101

Sepramaniam, S., Armugam, A., Lim, K. Y., Karolina, D. S., Swaminathan, P., Tan, J. R., et al. (2010). MicroRNA 320a functions as a novel endogenous modulator of aquaporins 1 and 4 as well as a potential therapeutic target

in cerebral ischemia. *J. Biol. Chem.* 285, 29223–29230. doi: 10.1074/Jbc. M110.144576

Szalay, G., Martinecz, B., Lénárt, N., Környei, Z., Orsolits, B., Judák, L., et al. (2016). Microglia protect against brain injury and their selective elimination dysregulates neuronal network activity after stroke. *Nat. Commun.* 7:11499. doi: 10.1038/ncomms11499

Tauskela, J. S., Brunette, E., Monette, R., Comas, T., and Morley, P. (2003). Preconditioning of cortical neurons by oxygen-glucose deprivation: tolerance induction through abbreviated neurotoxic signaling. *Am. J. Physiol. Cell Physiol.* 285, C899–C911. doi: 10.1152/ajpcell.00110.2003

Wang, X., Fan, X., Yu, Z., Liao, Z., Zhao, J., Mandeville, E., et al. (2014). Effects of tissue plasminogen activator and annexin A2 combination therapy on long-term neurological outcomes of rat focal embolic stroke. *Stroke* 45, 619–622. doi: 10.1161/STROKEAHA.113.003823

Wei, X., Gao, H., Zou, J., Liu, X., Chen, D., Liao, J., et al. (2016). Contra-directional coupling ofNur77 and nurr1in neurodegeneration: a novel mechanism for memantine-induced anti-inflammation and anti-mitochondrial impairment. *Mol. Neurobiol.* 53, 5876–5892. doi: 10.1007/s12035-015-9477-7

Wu, S., Sun, H., Zhang, Q., Jiang, Y., Fang, T., Cui, I., et al. (2015). MicroRNA-132 promotes estradiol synthesis in ovarian granulosa cells via translational repression of Nurr1. *Reprod. Biol. Endocrinol.* 13:94. doi: 10.1186/s12958-015-0095-z

Yang, D., Li, T., Wang, Y., Tang, Y., Cui, H., Tang, Y., et al. (2012). miR-132 regulates the differentiation of dopamine neurons by directly targeting Nurr1 expression. *J. Cell Sci.* 125, 1673–1682. doi: 10.1242/jcs.086421

Yuan, Y., Zha, H., Rangarajan, P., Ling, E. A., and Wu, C. (2014). Anti-inflammatory effects of Edaravone and Scutellarin in activated microglia in experimentally induced ischemia injury in rats and in BV-2 microglia. *BMC Neurosci.* 15:125. doi: 10.1186/s12868-014-0125-3

Zhou, Y., Zhou, Y., Yu, S., Wu, J., Chen, Y., and Zhao, Y. (2015). Sulfiredoxin-1 exerts anti-apoptotic and neuroprotective effects against oxidative stress-induced injury in rat cortical astrocytes following exposure to oxygen-glucose deprivation and hydrogen peroxide. *Int. J. Mol. Med.* 36, 43–52. doi: 10.3892/ijmm.2015.2205

Zou, J., Chen, Z., Wei, X., Chen, Z., Fu, Y., Yang, X., et al. (2017). Cystatin C as a potential therapeutic mediator against Parkinson's disease via VEGF-induced angiogenesis and enhanced neuronal autophagy in neurovascular units. *Cell Death Dis.* 8:e2854. doi: 10.1038/cddis.2017.240

15

Apoptosis of Endothelial Cells Contributes to Brain Vessel Pruning of Zebrafish During Development

Yu Zhang [1,2†], Bing Xu [1,2†], Qi Chen [1], Yong Yan [1,3], Jiulin Du [1,2,3]* and Xufei Du [1,2]*

[1]Institute of Neuroscience, State Key Laboratory of Neuroscience, Center for Excellence in Brain Science and Intelligence Technology, Chinese Academy of Sciences, Shanghai, China, [2]School of Future Technology, University of Chinese Academy of Sciences, Beijing, China, [3]School of Life Science and Technology, ShanghaiTech University, Shanghai, China

*Correspondence:
Jiulin Du
forestdu@ion.ac.cn
Xufei Du
xufeidu@ion.ac.cn

†These authors have contributed
equally to this work.

During development, immature blood vessel networks remodel to form a simplified and efficient vasculature to meet the demand for oxygen and nutrients, and this remodeling process is mainly achieved via the pruning of existing vessels. It has already known that the migration of vascular endothelial cells (ECs) is one of the mechanisms underlying vessel pruning. However, the role of EC apoptosis in vessel pruning remains under debate, especially in the brain. Here, we reported that EC apoptosis makes a significant contribution to vessel pruning in the brain of larval zebrafish. Using in vivo long-term time-lapse confocal imaging of the brain vasculature in zebrafish larvae, we found that EC apoptosis was always accompanied with brain vessel pruning and about 15% of vessel pruning events were resulted from EC apoptosis. In comparison with brain vessels undergoing EC migration-associated pruning, EC apoptosis-accompanied pruned vessels were longer and showed higher probability that the nuclei of neighboring vessels' ECs occupied their both ends. Furthermore, we found that microglia were responsible for the clearance of apoptotic ECs accompanying vessel pruning, though microglia themselves were dispensable for the occurrence of vessel pruning. Thus, our study demonstrates that EC apoptosis contributes to vessel pruning in the brain during development in a microglial cell-independent manner.

Keywords: apoptosis, endothelial cells, vessel pruning, microglia, zebrafish

INTRODUCTION

During development, highly ramified immature blood vascular networks undergo extensive remodeling, including refined vessel pruning of selected vessel branches and complete regression of vascular networks, to form an efficient mature vasculature to meet their physiological functions (Adams and Alitalo, 2007; Herbert and Stainier, 2011; Korn and Augustin, 2015; Betz et al., 2016). It is generally thought that vessel pruning is mainly achieved through blood flow fluctuation-induced lateral migration of endothelial cells (ECs) to adjacent unpruned vascular segments (Chen et al., 2012; Kochhan et al., 2013; Franco et al., 2015; Lenard et al., 2015). However, EC apoptosis has been found during the pruning of the cranial division of internal carotid artery (CrDI) in zebrafish (Kochhan et al., 2013) and of the retinal vasculature in mice (Franco et al., 2015). In particular, macrophage-induced EC apoptosis is responsible for the complete regression of the hyaloid vasculature in developing eyes (Lobov et al., 2005). Therefore, the contribution of EC apoptosis to vessel pruning, especially in the brain vasculature, remains unclear.

Microglia, the resident immune cells in the central nervous system (CNS), are specialized macrophages that play crucial roles in mediating immune-related functions (Nayak et al., 2014). Besides being immune mediators, microglia are emerging as important contributors to normal CNS development and function (Paolicelli et al., 2011; Li et al., 2012; Wu et al., 2015). Interestingly, microglia are also shown to be important for the behaviors of vascular tip cells (Fantin et al., 2010; Tammela et al., 2011; Arnold and Betsholtz, 2013), the repair of brain vascular rupture (Liu et al., 2016), and the closure of injured blood-brain barrier (Lou et al., 2016). However, it is still unknown whether microglia participate in brain vascular remodeling during development.

To address whether EC apoptosis and microglia are involved in brain vessel pruning, we performed *in vivo* long-term time-lapse confocal imaging of the brain vasculature in larval zebrafish, and first found that about 15% of brain vessel pruning events were accompanied by EC apoptosis. Furthermore, we demonstrated that microglia were unnecessary for brain vessel pruning but responsible for the clearance of pruned vessel-associated apoptotic ECs. Thus, our findings reveal that microglial cell-independent EC apoptosis is involved in brain vessel pruning of larval zebrafish.

MATERIALS AND METHODS

Zebrafish Husbandry

Adult zebrafish (*Dario rerio*) were maintained in an automatic fish housing system at 28°C following standard protocols (Chen et al., 2012). The *Tg(kdrl:EGFP)s843* (Jin et al., 2005), *Tg(fli1a.ep:DsRedEx)um13* (Covassin et al., 2009), *Tg(fli1a:nEGFP)y7* (Roman et al., 2002) and *Tg(coro1a:DsRedEx)hkz011t* (Li et al., 2012) zebrafish lines were described previously. Zebrafish embryos were raised under a 14 h–10 h light-dark cycle in 10% Hanks' solution that consisted of 140 mM NaCl, 5.4 mM KCl, 0.25 mM Na_2HPO_4, 0.44 mM KH_2PO_4, 1.3 mM $CaCl_2$, 1.0 mM $MgSO_4$ and 4.2 mM $NaHCO_3$ (pH = 7.2). The 0.003% 1-phenyl-2-thiourea (PTU; Sigma, P7629) was added into the Hanks' solution to prevent pigment formation of zebrafish embryos. The zebrafish chorion was removed at 1 day post-fertilization (dpf) with the treatment of 2 mg/ml pronase (Calbiochem, 53702), which was diluted in the Hanks' solution. All animal use and handling procedures were approved by the Institute of Neuroscience, Chinese Academy of Sciences.

In Vivo Confocal Imaging

Imaging was performed on 3–3.5 dpf zebrafish larvae at room temperature (26–28°C). Larvae were embedded in 1% low-melting agarose (Sigma) for imaging without anesthetics. Imaging was carried out under a 40× water immersion objective (N.A., 0.80) with an Olympus Fluoview 1000 confocal microscope (Tokyo, Japan). The z-step of imaging was 3 μm. To trace the fate of each vessel segment during development, long-term time-lapse imaging of the brain vasculature of the same larvae was performed with 1-h interval.

Image Analysis

All the images were analyzed by ImageJ (NIH). The pruned vessels were quantified in the whole zebrafish brain. For the quantification of vessel pruning, the vessel segment which initially displayed lumen morphology but with a collapsed shape and disappeared completely later was counted as a pruned vessel. For the quantification of EC apoptosis, the ECs exhibiting typical apoptotic features, such as plasma membrane blebbing and formation of apoptotic bodies, were counted as apoptotic ECs. For the quantification of microglial density, we counted the number of microglia in the whole brain region of the microglial transgenic zebrafish *Tg(coro1a:DsRedEx)* microinjected with control or *pu.1* MO. The microglial engulfment of apoptotic ECs was defined as a process that the microglia migrated to and engulfed the apoptotic ECs and then the apoptotic ECs faded away.

Morpholino Oligonucleotides and Microinjection

Morpholino oligonucleotides (MOs) were purchased from Gene Tools (Philomath, OR, USA). Lyophilized MOs were dissolved in nuclease-free water. The 4 ng *pu.1* MO (5′-GATATACTGATACTCCATTGGTGGT-3′) or equivalent control MO (5′-CCTCTTACCTCAGTTACAATTTATA-3′) were microinjected into one-cell-stage zebrafish embryos.

DAPI Injection

The DAPI (10 nl, 1 mg/ml) was microinjected into the blood circulation system of 3-dpf *Tg(kdrl:EGFP)* larvae through the common cardinal vein. Immediately after the injection, time-lapse imaging was performed to trace pruned vessels with apoptotic ECs.

Immunofluorescence

Whole zebrafish embryos at 3 dpf were fixed in 4% paraformaldehyde for 2 h at room temperature, washed twice in PBS, and then incubated in a blocking solution (5% normal goat serum in PBS 0.2% Triton) overnight at 4°C. The embryos were then incubated in the blocking solution with a rabbit anti-Caspase-3 primary antibody (1:500, ab13847) for 72 h at 4°C. After washing with PBS 0.2% Triton, the embryos were incubated in the blocking solution with a secondary antibody (1:500) and DAPI (1:1000) for 48 h at 4°C. After washing with PBS 0.2% Triton, images were taken with an Olympus Fluoview 1000 confocal microscope (Tokyo, Japan). The dying ECs showing overlapped anti-Caspase-3 staining and DAPI staining are considered as the apoptotic ECs.

Statistical Analysis

Statistical analyses were performed by using GraphPad Prism v6.0 software. The significance of the difference between two groups was determined by using two-tailed unpaired Student's t-test. Data were represented as mean ± SEM, and $p < 0.05$ was considered to be statistically significant.

RESULTS

Brain Vessel Pruning Is Accompanied With EC Apoptosis in Larval Zebrafish

To examine whether the apoptosis of ECs contributes to brain vessel pruning, we monitored the development of the brain vasculature in individual larval zebrafish with 1-h interval during 3–3.5 dpf, when most brain vessel pruning events occur (Chen et al., 2012). The brain vasculature was visualized by using the double transgenic line *Tg(fli1a:nEGFP);Tg(fli1a.ep:DsRedEx)*, in which the nucleus and cellular morphology of ECs were labeled by enhanced green fluorescent protein (EGFP) and DsRed, respectively (Chen et al., 2012). Besides the previously reported EC lateral migration-associated vessel pruning, we also found EC death-accompanied vessel pruning, during which the dying EC showed typical morphological features of cell apoptosis, including plasma membrane blebbing and formation of apoptotic bodies (**Figure 1A**). In the total of 55 vessel pruning events observed in 17 larvae, 15% of them (8 out of 55) were accompanied with EC apoptosis (**Figure 1B**). Furthermore, by microinjection of DAPI into the circulation system of *Tg(kdrl:EGFP)* larvae to label the nucleus of ECs, we found that the nucleus of apoptotic ECs condensed and fragmented during vessel pruning (**Figure 1C**).

To demonstrate that the dying ECs with typical apoptotic morphology during vessel pruning are indeed undergoing apoptosis, we first screened out the vessel pruning event accompanied with dying ECs by *in vivo* imaging, and then fixed the embryo immediately to examine the expression of the active Caspase-3, a marker of cell apoptosis (Franco et al., 2015). We found that the dying EC indeed expressed active Caspase-3 and showed nuclear condensation and formation of apoptotic bodies, indicating the existence of EC apoptosis (**Figure 1D**). Taken together, these results demonstrate that EC apoptosis is indeed involved in the vessel pruning of the zebrafish brain vasculature during development.

EC Apoptosis-Accompanied Pruned Vessels Are Longer and Show Higher Probability That Adjacent Vessels' EC Nuclei Occupy Their Both Ends

Compared with pruned vessels with EC lateral migration, we found that EC apoptosis-accompanied pruned vessels showed a higher probability that the nuclei of neighboring vessels' ECs located near the both ends of the pruned vessels (**Figures 2A,B**; pruning with EC apoptosis, 7 out of 9; pruning with EC migration, 3 out of 11). Moreover, EC apoptosis-accompanied pruned vessels were much longer (**Figure 2C**; pruning with EC apoptosis, 85 ± 8 μm; pruning with EC migration, 56 ± 5 μm; $p < 0.01$, two-tailed unpaired Student's *t*-test). These results suggest that the vessel segments with longer length and occupied by the nuclei of neighboring vessels' ECs at their both ends are more likely to be pruned via EC apoptosis rather than EC migration.

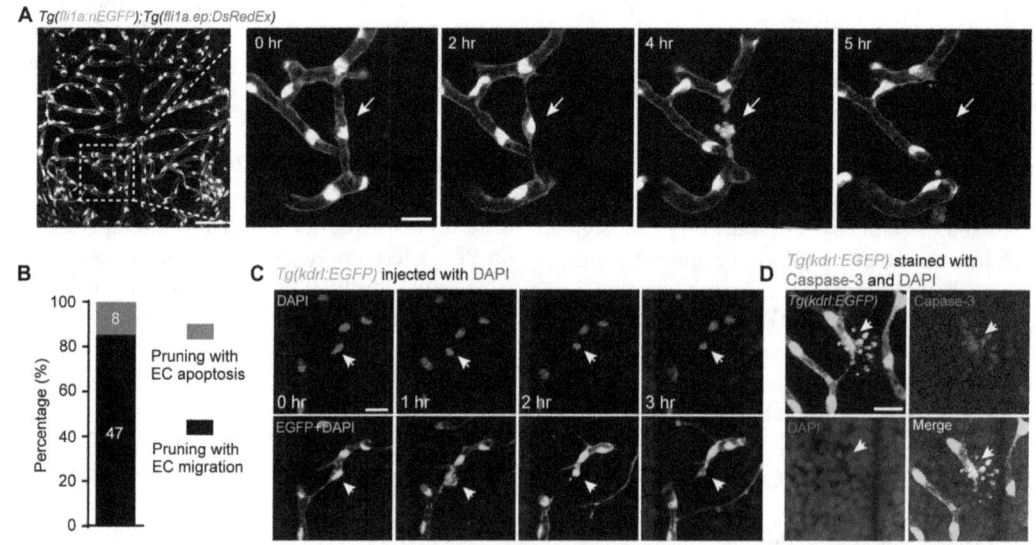

FIGURE 1 | Vessel pruning is accompanied with endothelial cell (EC) apoptosis in the brain of larval zebrafish. **(A)** *In vivo* time-lapse confocal images showing that an EC (arrows) underwent apoptosis on a pruned brain vessel in a *Tg(fli1a:nEGFP);Tg(fli1a.ep:DsRedEx)* larva at 3–3.5 days post-fertilization (dpf), in which EC nuclei were labeled by both enhanced green fluorescent protein (EGFP) and DsRed (yellow). Left, projected confocal image of the whole brain vasculature; Right, time-lapse confocal images of the dashed outlined area in the left. **(B)** Summary of the percentages of EC apoptosis- and migration-accompanied brain vessel pruning (n = 17 larvae). **(C)** *In vivo* time-lapse confocal images showing that an EC (arrows) underwent apoptosis on a brain pruned vessel in a DAPI-injected *Tg(kdrl:EGFP)* larva at 3–3.5 dpf, in which EC nuclei were labeled by DAPI (blue). **(D)** Immunofluorescence images showing that, on a pruned vessel, an EC with typical apoptotic morphology (arrows) expressed Caspase-3. The numbers on the bars **(B)** represent the number of pruned vessels examined. Scales: 50 μm (left in **(A)**, 15 μm (right in **(A)**, 50 μm **(C)** and 15 μm **(D)**.

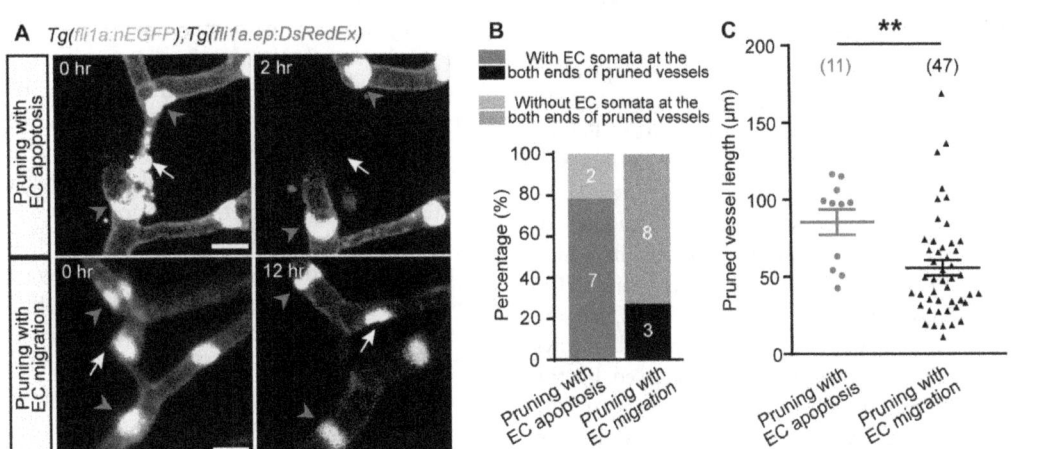

FIGURE 2 | Characterization of EC apoptosis-accompanied pruned vessels. **(A)** *In vivo* time-lapse confocal images showing the relative spatial locations of EC nuclei in pruned and adjacent brain vessels. Top: for an EC apoptosis-accompanied pruned brain vessel, the nuclei of two neighboring vessels' ECs (red arrowheads) located at the both ends of the pruned vessel, respectively (white arrow). Bottom: for an EC migration-accompanied pruned brain vessel, the nuclei of neighboring vessels' ECs (red arrowheads) did not occupy the ends of the pruned vessel (white arrow). **(B)** Summary of data showing that EC apoptosis-accompanied pruning vessels have a higher probability that neighboring vessels' EC nuclei occupy their both ends (7 out of 9, $n = 7$ larvae) than EC migration-accompanied pruning vessels (3 out of 11, $n = 6$ larvae). **(C)** Length of pruned vessels with EC apoptosis ($n = 9$ larvae) or EC migration ($n = 13$ larvae). The numbers on the bars **(B)** or in the brackets **(C)** represent the number of pruned vessels examined. Data are shown as mean $\pm$ SEM. **$p < 0.01$ (two-tailed unpaired Student's t-test). Scale bar: 10 μm **(A)**.

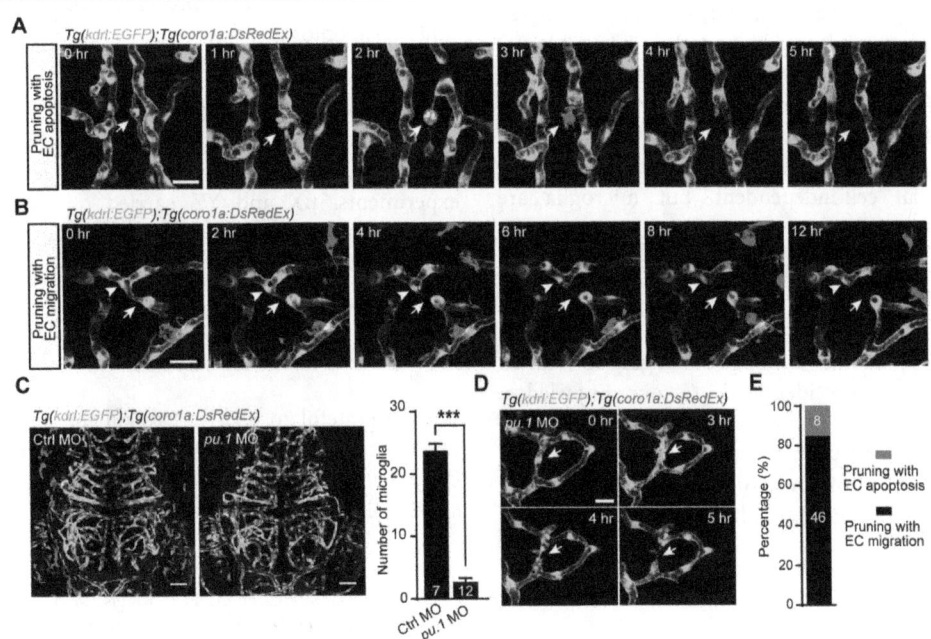

FIGURE 3 | Role of microglia in EC apoptosis-accompanied brain vessel pruning. **(A)** *In vivo* time-lapse confocal images showing that, during EC apoptosis-accompanied vessel pruning, a microglial cell (red) migrated to, engulfed and cleaned the apoptotic EC (white arrow). The *Tg(coro1a:DsRedEx); Tg(kdrl:EGFP)* larvae at 3–3.5 dpf were used. **(B)** *In vivo* time-lapse confocal images showing that, during EC migration-accompanied vessel pruning (white arrow), there was no obvious interaction between microglia (red) and the migrating EC (white arrowhead). **(C)** Representative projected confocal images (left) and summary data (right) showing that knockdown of *pu.1* significantly diminished the number of microglia (red) in the brain at 3 dpf. **(D)** *In vivo* time-lapse confocal images showing that EC apoptosis-accompanied brain vessel pruning (arrow) still occurred in *pu.1* morphants. **(E)** Summary of the percentages of EC apoptosis- and migration-accompanied brain vessel pruning from *pu.1* MO-injected *Tg(coro1a:DsRedEx);Tg(kdrl:EGFP)* ($n = 30$ larvae). The numbers on the bars represent the number of animals **(C)** or pruned vessels **(E)** examined. Data are shown as mean $\pm$ SEM. ***$p < 0.001$ (two-tailed unpaired Student's t-test). Scales: 20 μm **(A,B)**, 50 μm **(C)** and 15 μm **(D)**.

Microglia Engulf Apoptotic ECs but Are Dispensable for Vessel Pruning

As previous studies showed that macrophage-dependent EC apoptosis is responsible for the complete regression of the hyaloid vessel network (Kochhan et al., 2013; Franco et al., 2015), we hypothesized that microglia may be involved in EC apoptosis-accompanied brain vessel pruning. Thus, we performed *in vivo* time-lapse confocal imaging of *Tg(coro1a:DsRedEx);Tg(kdrl:EGFP)* zebrafish larvae, in which DsRed and EGFP were expressed in microglia and ECs, respectively. Interestingly, we found that after the EC on the pruning vessel went into apoptosis, a microglial cell always migrated to and engulfed the apoptotic EC, followed by the completion of vessel pruning (**Figure 3A**). Whereas, in the EC lateral migration-accompanied vessel pruning, we did not find the interaction between microglia and pruned vessels (**Figure 3B**). The fact that the migration of microglia toward pruning vessel is after the EC apoptosis suggests that EC apoptosis is microglial cell-independent. However, microglia contribute to the clearance of the apoptotic ECs.

To further examine whether microglia are essential for EC apoptosis-accompanied vessel pruning, we downregulated the expression of the *pu.1* transcription factor, which is required for macrophage differentiation, through *pu.1* morpholino oligonucleotide (MO; Chen et al., 2012). In the *pu.1* MO-injected *Tg(coro1a:DsRedEx);Tg(kdrl:EGFP)* larvae, microglia were largely diminished (**Figure 3C**; $p < 0.001$, two-tailed unpaired Student's *t*-test). However, in accordance with in normal larvae (see **Figure 1B**), the percentage of EC apoptosis-accompanied brain vessel pruning in microglial cell-deficient fish was also about 15% (**Figures 3D,E**, 8 out of 54). These results indicate that the occurrence of EC apoptosis-accompanied brain vessel pruning is microglial cell-independent, but microglia are responsible for the clearance of apoptotic ECs.

DISCUSSION

Taking advantage of *in vivo* time-lapse imaging on larval zebrafish, in this study we identified, to our knowledge, for the first time that EC apoptosis makes a significant contribution to brain vessel pruning during development in a microglial cell-independent manner. Although unnecessary for EC apoptosis-accompanied vessel pruning, microglia can engulf and clear apoptotic ECs, a behavior that may be important for protecting the brain microenvironment from inflammation.

A previous study showed that the pruning of a defined blood vessel CrDI, which locates along the anterior margin of the eye in zebrafish and contains 3–4 ECs, was accompanied by the death of 1–2 ECs (Kochhan et al., 2013), suggesting that the cell apoptosis are caused by the redundancy of ECs. Franco et al. (2015) reported the association of apoptotic events with regression of long vessels in the retina and hypothesized that cell apoptosis are associated with failure of ECs to integrate into adjacent vessels. Consistently, for the developing brain vasculature, we showed that EC apoptosis-accompanied pruned vessels were longer and more likely to have the nuclei of neighboring vessels' ECs occupying at their both ends. These characteristics may prevent ECs in pruning vessels migrating to and integrating into adjacent vessels, thus leading to EC apoptosis.

In our previous study, we found that vessel pruning in the brain of larval zebrafish occurs preferentially at loop-forming vessels, which usually show inefficient blood flow and are functionally redundant (Chen et al., 2012). Here, EC apoptosis-accompanied vessel pruning also occurred at loop-forming vessels. We speculated that hemodynamic changes in loop-forming vessels will usually trigger the ECs in pruning vessels to migrate and integrate into adjacent unpruned vessels, but if the loop-forming vessels are long in length and occupied by the nuclei of neighboring vessels' ECs at their both ends, the ECs in pruning vessels may be prevented from migration and instead undergo apoptosis.

AUTHOR CONTRIBUTIONS

BX, YZ, JD and XD conceived the project and designed the experiments. BX and YZ carried out the experiments and analyzed the data with QC's and YY's help. BX, YZ, JD and XD wrote the manuscript.

ACKNOWLEDGMENTS

We are grateful to Dr. N. Lawson for providing the *Tg(kdrl: EGFP)* line, Dr. Z. L. Wen for providing the *Tg(coro1a: DsRedEx)* line, and Dr. Q. Jin for providing the *Tg(fli1a: nEGFP)* line.

REFERENCES

Adams, R. H., and Alitalo, K. (2007). Molecular regulation of angiogenesis and lymphangiogenesis. *Nat. Rev. Mol. Cell Biol.* 8, 464–478. doi: 10.1038/nrm2183

Arnold, T., and Betsholtz, C. (2013). The importance of microglia in the development of the vasculature in the central nervous system. *Vasc. Cell* 5:4. doi: 10.1186/2045-824x-5-4

Betz, C., Lenard, A., Belting, H. G., and Affolter, M. (2016). Cell behaviors and dynamics during angiogenesis. *Development* 143, 2249–2260. doi: 10.1242/dev.135616

Chen, Q., Jiang, L., Li, C., Hu, D., Bu, J. W., Cai, D., et al. (2012). Haemodynamics-driven developmental pruning of brain vasculature in zebrafish. *PLoS Biol.* 10:e1001374. doi: 10.1371/journal.pbio.1001374

Covassin, L. D., Siekmann, A. F., Kacergis, M. C., Laver, E., Moore, J. C., Villefranc, J. A., et al. (2009). A genetic screen for vascular mutants in zebrafish reveals dynamic roles for Vegf/Plcg1 signaling during artery development. *Dev. Biol.* 329, 212–226. doi: 10.1016/j.ydbio.2009.02.031

Fantin, A., Vieira, J. M., Gestri, G., Denti, L., Schwarz, Q., Prykhozhij, S., et al. (2010). Tissue macrophages act as cellular chaperones for vascular anastomosis downstream of VEGF-mediated endothelial tip cell induction. *Blood* 116, 829–840. doi: 10.1182/blood-2009-12-257832

Franco, C. A., Jones, M. L., Bernabeu, M. O., Geudens, I., Mathivet, T., Rosa, A., et al. (2015). Dynamic endothelial cell rearrangements drive developmental vessel regression. *PLoS Biol.* 13:e1002125. doi: 10.1371/journal.pbio.1002125

Herbert, S. P., and Stainier, D. Y. (2011). Molecular control of endothelial cell behaviour during blood vessel morphogenesis. *Nat. Rev. Mol. Cell Biol.* 12, 551–564. doi: 10.1038/nrm3176

Jin, S. W., Beis, D., Mitchell, T., Chen, J. N., and Stainier, D. Y. (2005). Cellular and molecular analyses of vascular tube and lumen formation in zebrafish. *Development* 132, 5199–5209. doi: 10.1242/dev.02087

Kochhan, E., Lenard, A., Ellertsdottir, E., Herwig, L., Affolter, M., Belting, H. G., et al. (2013). Blood flow changes coincide with cellular rearrangements during blood vessel pruning in zebrafish embryos. *PLoS One* 8:e75060. doi: 10.1371/journal.pone.0075060

Korn, C., and Augustin, H. G. (2015). Mechanisms of vessel pruning and regression. *Dev. Cell* 34, 5–17. doi: 10.1016/j.devcel.2015.06.004

Lenard, A., Daetwyler, S., Betz, C., Ellertsdottir, E., Belting, H. G., Huisken, J., et al. (2015). Endothelial cell self-fusion during vascular pruning. *PLoS Biol.* 13:e1002126. doi: 10.1371/journal.pbio.1002126

Li, Y., Du, X. F., Liu, C. S., Wen, Z. L., and Du, J. L. (2012). Reciprocal regulation between resting microglial dynamics and neuronal activity *in vivo*. *Dev. Cell* 23, 1189–1202. doi: 10.1016/j.devcel.2012.10.027

Liu, C., Wu, C., Yang, Q., Gao, J., Li, L., Yang, D., et al. (2016). Macrophages mediate the repair of brain vascular rupture through direct physical adhesion and mechanical traction. *Immunity* 44, 1162–1176. doi: 10.1016/j.immuni.2016.03.008

Lobov, I. B., Rao, S., Carroll, T. J., Vallance, J. E., Ito, M., Ondr, J. K., et al. (2005). WNT7b mediates macrophage-induced programmed cell death in patterning of the vasculature. *Nature* 437, 417–421. doi: 10.1038/nature03928

Lou, N., Takano, T., Pei, Y., Xavier, A. L., Goldman, S. A., and Nedergaard, M. (2016). Purinergic receptor P2RY12-dependent microglial closure of the injured blood-brain barrier. *Proc. Natl. Acad. Sci. U S A* 113, 1074–1079. doi: 10.1073/pnas.1520398113

Nayak, D., Roth, T. L., and McGavern, D. B. (2014). Microglia development and function. *Annu. Rev. Immunol.* 32, 367–402. doi: 10.1146/annurev-immunol-032713-120240

Paolicelli, R. C., Bolasco, G., Pagani, F., Maggi, L., Scianni, M., Panzanelli, P., et al. (2011). Synaptic pruning by microglia is necessary for normal brain development. *Science* 333, 1456–1458. doi: 10.1126/science.1202529

Roman, B. L., Pham, V. N., Lawson, N. D., Kulik, M., Childs, S., Lekven, A. C., et al. (2002). Disruption of acvrl1 increases endothelial cell number in zebrafish cranial vessels. *Development* 129, 3009–3019.

Tammela, T., Zarkada, G., Nurmi, H., Jakobsson, L., Heinolainen, K., Tvorogov, D., et al. (2011). VEGFR-3 controls tip to stalk conversion at vessel fusion sites by reinforcing Notch signalling. *Nat. Cell Biol.* 13, 1202–1213. doi: 10.1038/ncb2331

Wu, Y., Dissing-Olesen, L., MacVicar, B. A., and Stevens, B. (2015). Microglia: dynamic mediators of synapse development and plasticity. *Trends Immunol.* 36, 605–613. doi: 10.1016/j.it.2015.08.008

16

Oxytocin Rapidly Changes Astrocytic GFAP Plasticity by Differentially Modulating the Expressions of pERK 1/2 and Protein Kinase A

Ping Wang[1], Danian Qin[2] and Yu-Feng Wang[1]*

[1]*School of Basic Medical Sciences, Harbin Medical University, Harbin, China,* [2]*Department of Physiology, Shantou University, Shantou, China*

*Correspondence:
Yu-Feng Wang
yufengwang@ems.hrbmu.edu.cn

The importance of astrocytes to normal brain functions and neurological diseases has been extensively recognized; however, cellular mechanisms underlying functional and structural plasticities of astrocytes remain poorly understood. Oxytocin (OT) is a neuropeptide that can rapidly change astrocytic plasticity in association with lactation, as indicated in the expression of glial fibrillary acidic protein (GFAP) in the supraoptic nucleus (SON). Here, we used OT-evoked changes in GFAP expression in astrocytes of male rat SON as a model to explore the cellular mechanisms underlying GFAP plasticity. The results showed that OT significantly reduced the expression of GFAP filaments and proteins in SON astrocytes in brain slices. In lysates of the SON, OT receptors (OTRs) were co-immunoprecipitated with GFAP; vasopressin (VP), a neuropeptide structurally similar to OT, did not significantly change GFAP protein level; OT-evoked depolarization of astrocyte membrane potential was sensitive to a selective OTR antagonist (OTRA) but not to tetanus toxin, a blocker of synaptic transmission. The effects of OT on GFAP expression and on astrocyte uptake of Bauer-Peptide, an astrocyte-specific dye, were mimicked by isoproterenol (IPT; β-adrenoceptor agonist), U0126 or PD98059, inhibitors of extracellular signal-regulated protein kinase (ERK) 1/2 kinase and blocked by the OTRA or KT5720, a protein kinase A (PKA) inhibitor. The effect of OT on GFAP expressions and its association with these kinases were simulated by mSIRK, an activator of Gβγ subunits. Finally, suckling increased astrocytic expression of the catalytic subunit of PKA (cPKA) at astrocytic processes while increasing the molecular associations of GFAP with cPKA and phosphorylated ERK (pERK) 1/2. Upon the occurrence of the milk-ejection reflex, spatial co-localization of the cPKA with GFAP filaments further increased, which was accompanied with increased molecular association of GFAP with pERK 1/2 but not with cPKA. Thus, OT-elicited GFAP plasticity is achieved by sequential activation of ERK 1/2 and PKA via OTR signaling pathway in an antagonistic but coordinated manner.

Keywords: astrocyte, glial fibrillary acidic protein, phosphorylated extracellular signal-regulated protein kinase1/2, oxytocin, protein kinase A, supraoptic nucleus

INTRODUCTION

Astrocytes are extensively involved in normal brain functions (Vasile et al., 2017) and neurological diseases (Pekny et al., 2016; Verkhratsky et al., 2016). However, cellular mechanisms underlying functional and structural plasticities of astrocytes are largely unknown, which requires further investigation. In astrocytes, functional roles of glial fibrillary acidic protein (GFAP), a major cytoskeletal element, have been identified in acute astrocyte plasticity (Langle et al., 2003; Sun et al., 2016). GFAP largely determines astrocytic morphology, which in turn changes astrocytic absorption of glutamate and ions, synaptic innervation and interactions between neighboring neurons (Hatton, 2004; Theodosis et al., 2008).

GFAP molecules are posttranslationally modified by a series of protein kinase. For example, protein kinase A (PKA) phosphorylated and destabilized GFAP *in vitro* (Inagaki et al., 1990; Heimfarth et al., 2016), and caused retraction of astrocytic processes (Hatton et al., 1991) from the surrounding of neurons. Increases in phosphorylated extracellular signal-regulated protein kinase (pERK) 1/2 elevated GFAP levels (Li D. et al., 2017), an effect similar to that of chronic cAMP/PKA activation (Hsiao et al., 2007). These findings thus highlight the dependence of GFAP metabolism on interactions between different protein kinases in astrocytes. However, these studies are mainly based on cell cultures or cell lines in a chronic time course, and it remains unclear how these kinases *in vivo* interact with each other in acute modulation of GFAP plasticity that is essential for both physiological regulation and pathogenesis of diseases (Wang and Hamilton, 2009; Wang and Parpura, 2016). Thus, determining the interaction between GFAP and these protein kinases in acute physiological processes is important to understand the cellular mechanisms that regulate GFAP/astrocytic plasticity.

In the study on glial neuronal interaction in the supraoptic nucleus (SON), GFAP plastic change occurred in concert with the change in oxytocin (OT) neuronal activity in the SON of lactating rats (Wang and Hatton, 2009). In the SON, OT can increase pERK1/2 expression; the distribution of pERK 1/2 was spatiotemporally associated with astrocyte/GFAP morphology (Wang and Hatton, 2007b), which is opposite to the effect of increased intracellular cyclic AMP (Hatton et al., 1991). Activation of OT receptors (OTRs) can trigger both pERK 1/2 and PKA signaling, and the functions of two kinases are mutually antagonistic in uterus (Zhong et al., 2003), which could also occur in the SON. Thus, we hypothesized that both pERK 1/2 and PKA are involved in GFAP plasticity while they function differently in OT-elicited astrocytic plasticity in the SON.

To test this hypothesis, we used OT-evoked GFAP plasticity in male rats first to study the contribution of these two signaling pathways to GFAP plasticity. We found that astrocytes in the SON expressed OTRs and OT could depolarize astrocyte membrane potential while reducing GFAP expression via OTRs. Effects of OT on GFAP plasticity and on the absorption of an astrocyte-specific peptide were differentially modulated by PKA and pERK 1/2. The effect of OT on

GFAP expressions and its association with pERK 1/2 and PKA were simulated by an activator of $G\beta\gamma$ subunits. In lactating rats, suckling of pups increased astrocytic expression of PKA at astrocytic processes, which was accompanied with increased molecular associations between GFAP with PKA and pERK 1/2. When the milk-ejection reflex occurred, catalytic subunit of PKA (cPKA) co-localized with GFAP filaments further increased, and the molecular association of GFAP with pERK 1/2 but not with cPKA was also increased. Along with findings of pERK 1/2 involvement in GFAP plasticity in the SON (Wang and Hatton, 2009), the present results highlight that OT-elicited GFAP plasticity is associated with sequential activation of ERK 1/2 and PKA downstream to OTR signaling in an antagonistic but coordinated manner.

MATERIALS AND METHODS

Experiments were performed using adult male (42–60 days old) and lactating female Sprague-Dawley rats. This study was carried out in accordance with the recommendations of NIH guidelines. The protocol was approved by Institutional Animal Care and Use Committees of the University of California-Riverside and Harbin Medical University, respectively.

Drugs, Reagents and Antibodies

OT, β-Mercapto-β, [β-cyclopentamethylene-propionyl^1, O-Me-Tyr^2, Orn^8]-OT (OTR antagonist, OTRA), vasopressin (VP), isoproterenol (IPT, activator of β-adrenoceptor and PKA), KT5720 (inhibitor of PKA), PD98059 and U0126 (inhibitors of ERK 1/2 kinase), U73122 (inhibitor of the coupling of G protein-phospholipase C activation), tetanus toxin and others were from Sigma except as otherwise noted. Myristoylated G-protein $\beta\gamma$-binding peptide [myristoyl-SIRKALNILGYPDYD (mSIRK)] was from EMD Biosciences. Reagents for Western blots were from GE Healthcare. Primary antibodies were from Santa Cruz Biotechnology except for mouse anti-pERK 1/2 (Cellular Signaling), mouse anti-OT neurophysin (NP) and VP-NP (Dr. H. Gainer, NIH, Bethesda, MD, USA). Bauer peptide (β-Ala-Lys-N$_\varepsilon$-AMCA) was provided by Dr. K. Bauer (Max-Planck-Institut für experimentelle Endokrinologie, Hannover, Germany). Secondary antibodies were from Thermo Fisher Scientific.

Slice Preparation

Rats were decapitated; the brain was quickly removed and immersed in ice-cold slicing solution that was oxygenated through bubbling with a compressed gas mixture of 95% O_2/5% CO_2. The slicing solution contained 1/3 of 10% sucrose and 2/3 regular artificial CSF. Then, coronal slices (200 μm thick) were obtained from the SON as previously described (Wang and Hatton, 2009). Slices were pre-incubated at room temperature (21–23°C) for 1 h in oxygenated regular artificial CSF before drug treatment or application of other procedures. The regular artificial CSF contained (in mM): 126 NaCl, 3 KCl, 1.3 $MgSO_4$, 2.4 $CaCl_2$, 1.3 NaH_2PO_4, 26 $NaHCO_3$, 10 Glucose, 0.2 ascorbic acid, pH 7.4 and 305 mOsm/kg, oxygenated with 95% O_2/5% CO_2. The slice was then randomly assigned to groups for immunostaining or protein analyses.

In suckling experiments, lactating rats with 4 h separation from pups were allowed to suckling of 10 pups for 0–30 min as previously described (Wang and Hatton, 2009). According to the occurrence of the milk-ejection reflex, brains were collected in three groups, i.e., non-suckling, suckling (for 5–10 min before the first milk ejection), and milk-ejection reflex (suckling until the occurrence of the third or fourth milk ejections). Brain were fixed immediately without pre-incubation or homogenized to obtain protein lysates.

Immunocytochemistry and Confocal Microscopy

Immunostaining was performed based on our earlier reports (Wang and Hatton, 2009) with minor modifications. In brief, slices were permeated with 0.3% Triton X-100 in 0.1 M PBS for 30 min, and non-specific binding was blocked with 0.3% gelatin-PBS. The slices were then incubated overnight at 4°C with primary antibodies against goat or mouse OT-NP (1:400) and VP-NP (1:400), goat or mouse GFAP (1:300), mouse pERK 1/2 (1:1000), rabbit cPKA (1:250) and goat OTRs (1:250). After rinsing, the slice was incubated with species-matched fluorescent donkey anti-goat/mouse/rabbit antibodies (Alexa Fluor® 647/555/488, 1:1000) for 1.5 h at room temperature (22–24°C). Finally, Hoechst stain (0.5 μg/ml for 15 min) was used to label nuclei.

For each treatment, 6–12 pieces of slices from the middle part of the SON of 3–6 rats were imaged at high magnification (630×), 10–20 μm from the surface using a laser scanning confocal microscope (Leica TCP SP2 or Zeiss LSM510). Multiple fluorophores were imaged sequentially, and distribution pattern and colocalization of different molecules were analyzed. To avoid false positive or negative results of immunostaining, serial dilutions of the primary antibody, staining with pre-absorbed (immune-neutralization) primary antibody, no-primary and no secondary antibody controls were applied.

Western Blots and Co-Immunoprecipitation (Co-IP)

Methods for protein analysis were the same as previously reported (Wang et al., 2013a,b). In brief, hypothalamic slices from three to six rats were obtained as described above. SONs were punched out and then lysed. The lysates were centrifuged to remove insoluble components before protein levels were quantitated using a plate reader. Protein aliquots (60 μg) were loaded and separated on 10% SDS-PAGE gels, and then transferred onto polyvinylidene difluoride membranes. After blocking with 5% milk solids (or 1% gelatin for detecting primary antibodies from goat) for 1 h at room temperature, membranes were incubated with mouse or goat anti-GFAP, rabbit or goat anti-OTRs, rabbit anti-cPKA and mouse anti-pERK 1/2 (all 1:500) overnight at 4°C. To calibrate protein levels, mouse anti-tubulin (1:300), rabbit anti-actin (1:500) or

anti-total ERK2 (tERK2, 1:1000) were also detected (1 h at room temperature). Bands were visualized using horseradish peroxidase-conjugated secondary antibodies and an enhanced chemiluminescence system (Tanon 5200, Shanghai). Data are reported for 3–6 replicates.

For Co-IP experiments, SON lysates were precleared with protein G agarose beads to reduce nonspecific binding. Mouse anti-GFAP (1.5 μg/7.5 μl) was then added to the lysates (1500 μg/500 μl protein) to form an immunocomplex and incubated overnight at 4°C. Immunocomplexes were captured by adding 50 μl of protein G agarose bead slurry and gently rocking for 2 h at 4°C. They were then collected, washed and resuspended in 50 μl 2× Western blotting sample buffer, and boiled for 10 min to dissociate proteins from the beads. Target proteins were then detected using Western blotting.

Patch-Clamp Recordings

Patch-clamp recording procedures for SON were similar to those described previously (Wang and Hatton, 2004). Briefly, after pre-incubation in the regular artificial CSF, slices were incubated in artificial CSF containing Bauer peptide that is fluorescent and can be selectively taken up by astrocytes via peptide transporter PepT2 (Dieck et al., 1999) at 35°C for 2–4 h, with or without tetanus toxin (1 nM) to block synaptic vesicle release. Whole cell patch-clamp recordings were obtained from fluorescent cells visualized using an epifluorescence microscope using an Axopatch 200B amplifier or Multiclamp 700B amplifier (Molecular Devices). The pipette solution for recording SON astrocytes contained (in mM): 145 K-gluconate, 10 KCl, 1MgCl₂, 10 HEPES, 1 EGTA, 0.01 CaCl₂, 2 Mg-ATP, 0.5 Na₂-GTP, pH 7.3, adjusted with KOH. Signals that were filtered, sampled at 5 kHz, and analyzed offline using Clampfit 10 software (Molecular Devices). Exemplary cells were also tested their voltage-current relationship and examined in immunohistochemistry after drug tests to verify their astrocytic nature.

Real Time Imaging of Astrocytes

To link modulating effects of OT on GFAP plasticity to astrocytic functions, slices were incubated with the artificial CSF containing 0.1 μM Bauer peptide as described above. Then, astrocytic somata in the ventral glial lamina (VGL) of the SON were focused and their fluorescence intensity was captured through a Microfire Camera in single frame following a brief (20–30 s) exposure to UV light. Drugs were bath-applied for 15 min, and images were captured immediately before, 5 min and 15 min after drug application, respectively. Preparations were fixed in situ at the end of observations for further analyzing images by using confocal microscopy as previously reported (Wang and Hatton, 2009).

Data Analysis

Methods for analyzing data of immunocytochemistry, Western blots and patch-clamp recordings were modified from our previous experiment (Wang and Hatton, 2009). To evaluate GFAP levels, the fluorescence intensity in each channel was normalized to a standard curve (1–256) to allow for comparison between different experiments. The background fluorescence level was set as 1 through minimum baseline correction using Leica LCS Lite or ZeissLSM software and maximum intensity was set at 256. To assay GFAP expression in single scan-based confocal image, whole frame of the image was compared on the same background level of fluorescence intensity. The efficiency of this single sectioned image in reflecting GFAP plasticity had been validated by Z-stack scanning (0.5 μm/section) and fluorescent microscopy as described previously (Wang and Hatton, 2009). The increase or decrease in the expression level of GFAP was defined as a change more than 20% from the control. In analyzing the expression of GFAP filaments, they were first distinguished from the cell body by its extending from but not surrounding "astrocytic nucleus" and appearing thread-like rather than circular morphology. The diameter of GFAP filaments was measured at five sub-sections including those near each of the four corners and the center in a square frame and their average was used to represent the diameter of a section, which was applicable in 95% of the images.

In determining different components of GFAP protein, the monomers were identified by the appearance of a single 50 KDa band and the fragments included the diffuse bands below 50 KDa and above 35 KDa. In patch-clamp recordings, the membrane potential was an average level of 1 min before and 1 min (9th–10th min) after drug application. In analyzing the fluorescent intensity of Bauer peptide-loaded astrocytes, background fluorescence was subtracted from the dorsal portion of the SON that did not show clear astrocyte soma; changes in fluorescence were calculated by comparing the intensity after drugs with those before drugs.

Student's t-test and ANOVA were used for statistical analyses where appropriate, as instructed by SigmaStat 12 program, and $p < 0.05$ was considered significant. When abnormal distribution or large variances appeared, square-root transformations (for some Co-IP studies with $n = 3$) were applied to minimize the influence of individual data points on the evaluation of whole significance level. All measures were expressed as mean $\pm$ SEM in percentage of control values, or as otherwise noted.

RESULTS

In this study, we first examined the features of OT-evoked GFAP plasticity of astrocytes in the SON of male rats and its dependence on OTR-G$\beta\gamma$ signaling. Next, we explored the roles of pERK 1/2 and PKA in the GFAP plasticity and their association with astrocytic absorption of fluorescent peptide. Lastly, we linked the roles of pERK 1/2 and PKA to suckling-evoked GFAP plasticity of the SON in lactating rats to verify the applicability of identified features of kinase modulation of acute GFAP plasticity.

OT Changed GFAP Expression in the SON of Male Rats

To analyze the potential modulatory effects of signaling molecules on GFAP plasticity in the SON, we first examined effects of OT on GFAP expression in brain slices from six male rats. The results showed that OT, at 10 pM, 1 nM and 0.1 μM for 30 min, concentration-dependently reduced GFAP levels in the SON in confocal microscopy. As shown in **Figure 1A**, before OT treatment (Control), GFAP staining was less compact in perinuclear areas of astrocytes while GFAP filaments were in clear and rich bundles. Accompanying with the general reduction of GFAP, the length and diameter of GFAP filaments were reduced significantly by OT, whereas GFAP staining was increased at the somata (**Figure 1Aa**). Pretreatment of the slices with [β-Mercapto-β, β-cyclopentamethylene-propionyl[1], O-Me-Tyr[2], Orn[8]]-OT blocked OT-elicited GFAP reduction (**Figure 1Ab**). In Western blots, different concentrations of OT differentially influenced the expressions of different components of GFAP protein (**Figure 1B**). OT at 10 pM for 30 min significantly decreased 50 kDa GFAP protein and smaller GFAP fragments. At 1 nM and 0.1 μM, OT still decreased 50 kDa GFAP but significantly increased small GFAP fragments. The concentration-dependent effects of OT on GFAP plasticity were exhibited as the increases in GFAP fragments rather than 50 KDa components. Moreover, pretreatment of slices with the OTRA blocked this OT effect (**Figure 1Ba**). These results summarized in **Figures 1Ac,Bb** are consistent with the effect of OT on GFAP plasticity in lactating rats (Wang and Hatton, 2009).

In the SON, there are two major forms of astrocytes, radial glia-like morphology in the VGL, and stellate morphology in the somatic area (Israel et al., 2003). Here, we analyzed OT effects on GFAP expression in these two forms of astrocytes by observing GFAP expression with nuclear staining. The result showed that OT-reduced GFAP staining was significantly stronger ($n = 6$, $P < 0.05$ by paired t-test) in the dorsal SON (34.8 $\pm$ 7.4% of control) than in the ventral SON (72.7 $\pm$ 8.6% of the control by paired t-test). This is in agreement with the finding that osmotic stimulation causes glial retraction around dorsally located OT neurons but not ventrally located VP neurons in the SON (Chapman et al., 1986).

Specificity of OT Actions on GFAP Expression in Astrocytes

To verify the specificity of OT actions on astrocytes, we performed Co-IP of GFAP with OTRs in three rats. The result showed a clear molecular association between GFAP and OTRs (**Figure 2A**). This result is consistent with our previous finding in immunocytochemistry that OTRs were present in GFAP-positive astrocytes in the SON (Wang and Hatton, 2006).

Next, we tested effects of VP, a nonapeptide in the SON structurally similar to OT, on GFAP protein levels. In contrast to the strong effects of OT on GFAP level in the SON, VP (0.1 nM, 30 min) did not significantly influence the expression of GFAP (101.0 $\pm$ 17.3% of control, $n = 5$, $P > 0.05$ by pared t-test) in Western blots (**Figure 2B**). Moreover, in Bauer peptide (20 μM)-loaded astrocytes in brain slices (**Figure 2Ca1**) that possessed linear voltage-current relationship (**Figures 2Ca2,Ca3**) we found that OT (0.1 nM, 5–10 min) significantly depolarized astrocytic membrane potentials (67.4 $\pm$ 2.3 mV vs. 63.5 $\pm$ 1.8 mV at 10 min, $n = 7$, $P < 0.01$ by paired *t*-test; **Figure 2Cb1**), which was blocked by pretreatment of slices with OTRA (67.1 ± 3.0 mV vs. 67.0 ± 3.0 mV at 10 min, *n* = 7, *P* > 0.05 by paired *t*-test; **Figure 2Cb2**). By contrast, pretreatment of astrocytes in three slices with tetanus toxin (10 nM, 2–4 h) did not block the depolarizing effects of OT (66.0 ± 3.0 mV vs. 62.3 ± 3.3 mV at 10 min, *n* = 6, *P* < 0.01 by paired *t*-test; **Figure 2Cb3**) while VP had no significantly effect on the membrane potential (67.0 ± 2.4 mV vs. 67.1 ± 2.2 mV at 10 min, *n* = 6, *P* > 0.05 by paired *t*-test; **Figure 2Cb4**). These results are consistent with our previous finding in lactating rats (Wang and Hatton, 2009) that the presence of tetanus toxin did not influence OT-elicited GFAP reduction.

Signaling Cascades Mediating OT-Evoked GFAP Plasticity

Gβγ signaling cascade was the major approach mediating OT modulation of OT neuronal activity (Wang and Hatton, 2007a,b) and could also be the mediator of OT modulation of GFAP plasticity. To clarify this issue, we first examined GFAP expression after treatment of slices with mSIRK, an activator of Gβγ subunits. As shown in **Figure 3A**, mSIRK

(0.5 μM) dually changed GFAP expression significantly (*n* = 6, *P* < 0.05 by ANOVA) in a time-dependent manner. At 5 min after mSIRK treatment, GFAP filaments were increased significantly (*n* = 6, *P* < 0.05), which then significantly reduced at 30 min (*n* = 6, *P* < 0.05). In Western blots, mSIRK (0.5 μM, 30 min) significantly reduced 50 kDa GFAP bands but increased GFAP fragments (**Figure 3B**), an effect similar to that of OT. These findings are in agreement with the time-dependent effect of OT on GFAP plasticity in the SON of lactating rats (Wang and Hatton, 2009).

Next, we observed effects of blocking downstream signaling of Gβγ subunits on mSIRK-evoked GFAP plasticity. As shown in **Figure 4A**, mSIRK (0.5 μM, 30 min)-evoked reduction of GFAP and pERK 1/2 immunostaining was differentially influenced by pretreatments of the slices (*n* = 6) with OT (1 nM, 30 min), KT5720 (10 μM), PD98059 (20 μM) and U73122 (10 μM). Compared to the effect of mSIRK only on GFAP filaments, addition of OT, PD98059 or U73122 did not significantly influence GFAP levels; however, the addition of KT5720 significantly increased both GFAP and pERK

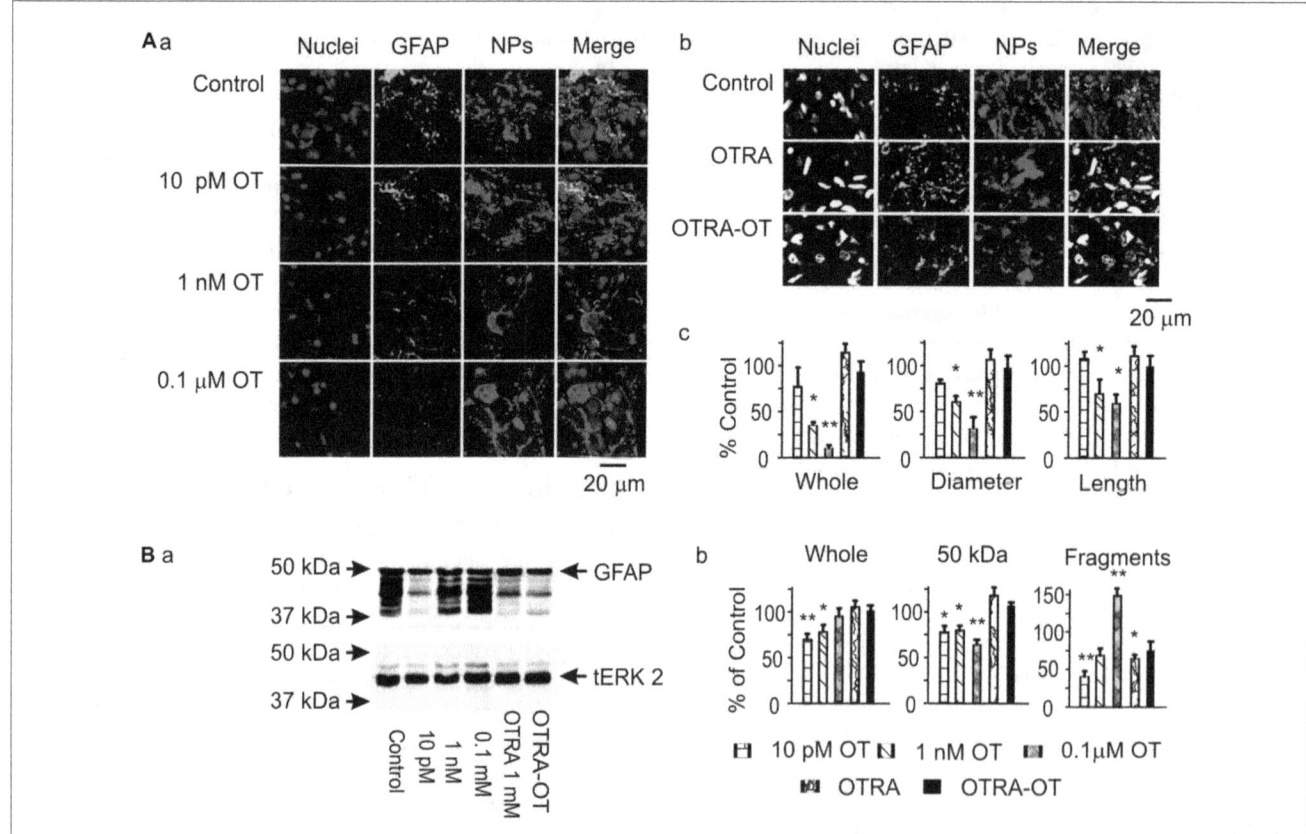

FIGURE 1 | Effects of oxytocin (OT) on glial fibrillary acidic protein (GFAP) plasticity in the supraoptic nucleus (SON) in brain slices from male rats. **(A)** Effects of OT (10 pM, 1 nM and 0.1 μM for 30 min) and OT receptor antagonist (OTRA, 1 μM for 30 min) on GFAP expression in confocal images. **(Aa,Ab)** Staining of nuclei, GFAP, neurophysins (NPs) and their merges (left to right). **(Ac)** Summary graphs showing the intensity of GFAP filaments relative to the control in whole panel, diameters and lengths of the filaments. Note that: *P < 0.05 and **P < 0.01 compared to the control (*n* = 6) in ANOVA with Holm-Sidak comparisons. **(B)** GFAP expression in Western blots. **(Ba)** Bands (left to right) show GFAP proteins (top) for control, 10 pM OT, 1 nM OT, 0.1 μM OT, 1 μM OTRA and 0.1 nM OT after OTRA for 30 min, respectively. Bands at the bottom are total extracellular signal regulated protein kinase 2 (tERK 2), serving as loading controls for corresponding treatments (*n* = 6, compared in ANOVA). **(Bb)** Summary graphs showing GFAP relative to control in whole bands, bands at 50 kDa (full size), and their smaller fragments, respectively.

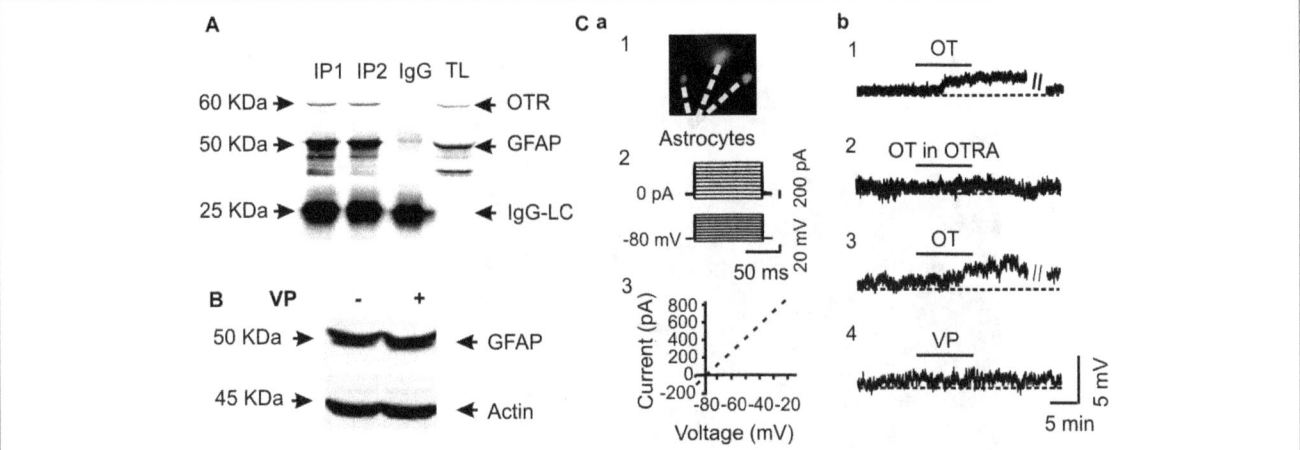

FIGURE 2 | Specific effects of OT on GFAP plasticity. **(A)** Co-immunoprecipitation (Co-IP) of GFAP with OTRs: Western blotting of OTRs (top panels), immunoprecipitated GFAP (the middle panels and IgG light chains, IgG-LC, bottom) in two duplicates (IP1 and IP2) along with the loading of total lysis (TL). **(B)** Western blots showing that effect of vasopressin (VP, 0.1 nM, 30 min) on the expression of GFAP in SON protein lysates. **(C)** Patch-clamp recordings showing effects of OT on the membrane potential of astrocytes in brain slices. **(Ca)** Identification of astrocytes by Bauer peptide loading **(Ca1)** and by the linear current-voltage relationship **(Ca2,Ca3)**. **(Cb1)** Depolarizing effects of OT (0.1 nM, 10 min) on astrocytes in the SON. **(Cb2)** Effects of OTRA (1 μM, 10 min before OT) on OT-evoked depolarization. **(Cb3,Cb4)** Effects of OT **(Cb3)** and VP **(Cb4)** at 0.1 nM for 10 min on the membrane potentials after pretreatment with tetanus toxin (10 nM, 2–4 h). Other annotations refer to **Figure 1**.

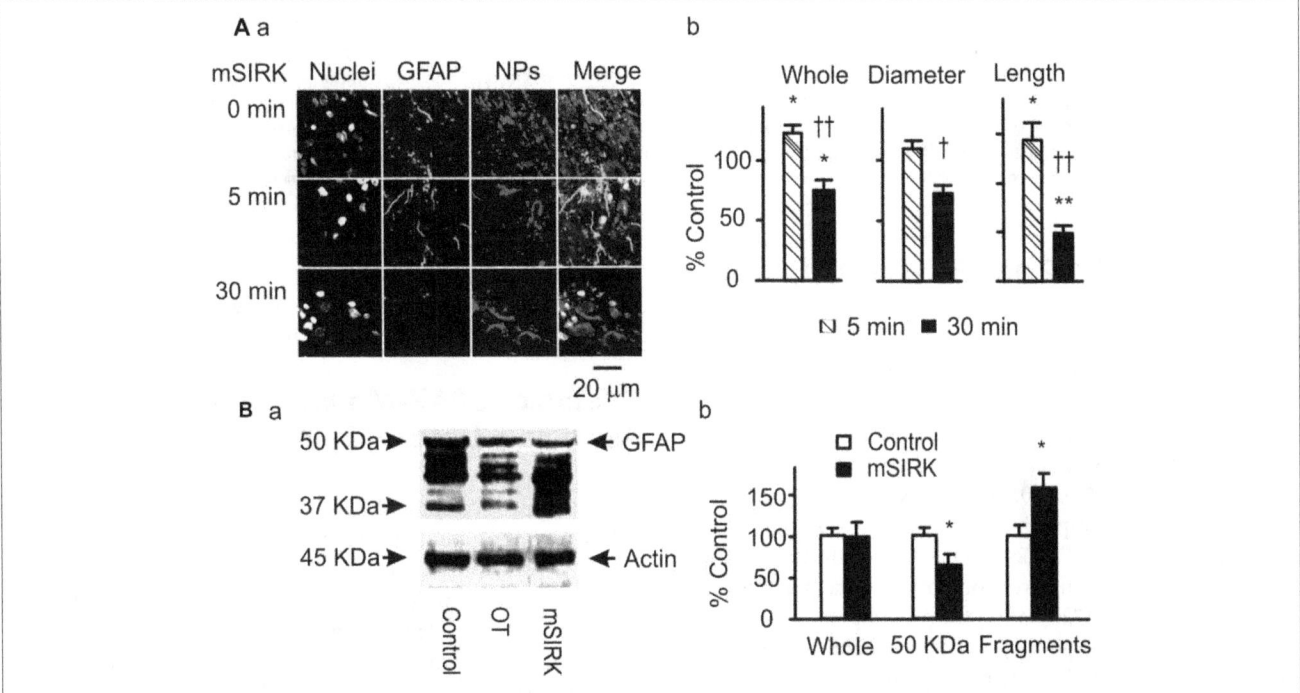

FIGURE 3 | Roles of Gβγ subunits in OT-elicited GFAP plasticity. **(A)** Exemplary confocal images showing (left to right) nuclei, GFAP, OT plus VP NPs and their merges at 0 min, 5 min and 30 min after treatments with myristoyl-SIRKALNILGYPDYD (mSIRK; 0.5 μM), respectively **(Aa)**. **(Ab)** Summary graphs showing mSIRK effects on different components of GFAP. *$P < 0.05$ and **$P < 0.01$ compared to 0 min controls; $^{†}P < 0.05$ and $^{††}P < 0.01$ compared to mSIRK at 5 min. **(B)** Western blot showing effects of mSIRK on different bands of GFAP proteins **(Ba**, relative to the effect of OT) and the summary graphs of the mSIRK effect **(Bb)**. Other annotations refer to **Figure 1**.

1/2 expressions. This finding is in agreement with previous finding that Gβγ signaling cascade including pERK 1/2 was the major mediator of OT effect in the SON (Wang and Hatton, 2007b) and that PKA antagonized pERK 1/2 signaling (Zhong et al., 2003) and promoted GFAP depolymerization (Inagaki et al., 1990; Heimfarth et al., 2016).

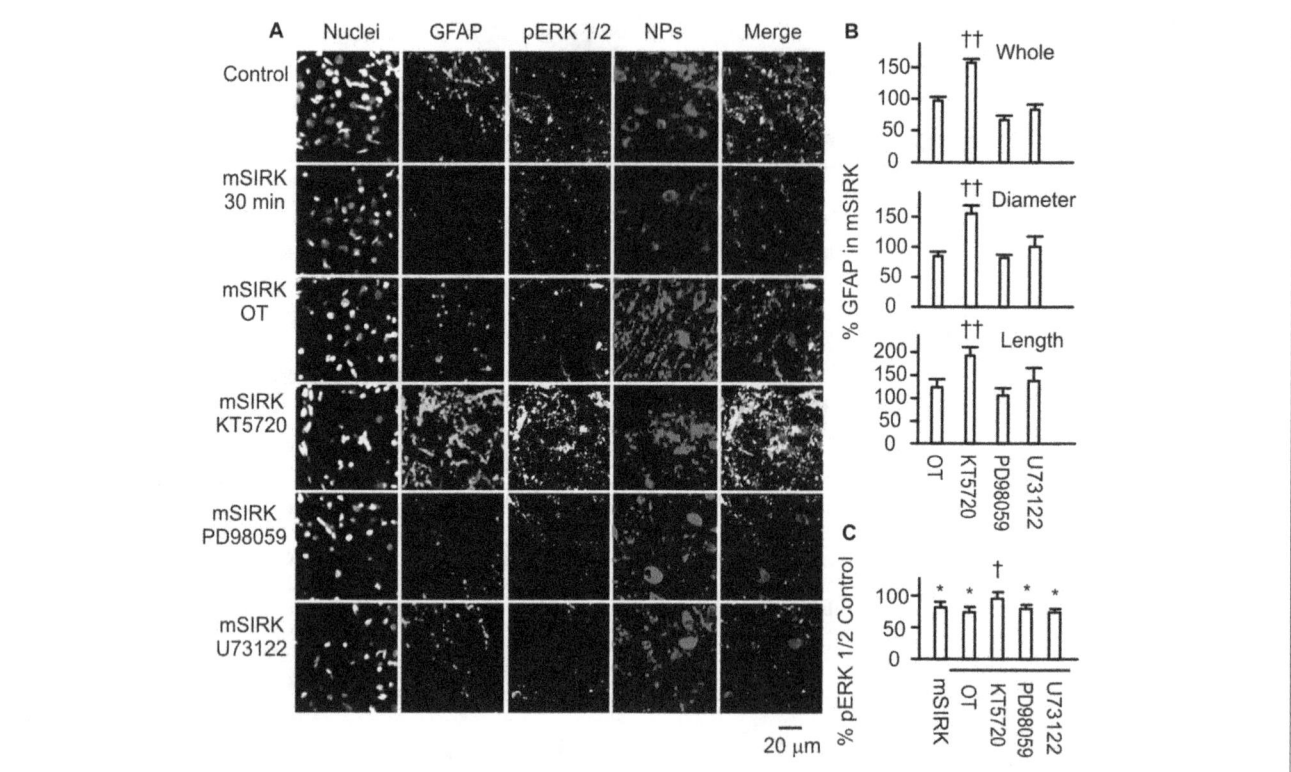

FIGURE 4 | Roles of OTR-associated signals in Gβγ subunit-modulated expressions of GFAP and phosphorylated extracellular signal-regulated protein kinase (pERK) 1/2. **(A)** Confocal images (left to right) show staining for nuclei, GFAP, pERK 1/2, NPs and their merges. From top to the bottom, images were acquired in control, mSIRK (0.5 μM, 30 min) and mSIRK following 30 min pretreatment with OT, KT5720 (10 μM), PD98059 (20 μM), or U73122 (10 μM), respectively. **(B)** Summary graphs of GFAP expression in whole field view, the diameter, or length of GFAP filaments relative to mSIRK alone. **(C)** Summary graphs of pERK 1/2 under different conditions. Note that, $^{*}P < 0.05$ compared to the control; $^{†}P < 0.05$ and $^{††}P < 0.01$ compared with mSIRK alone. Other annotations refer to **Figures 1, 3**.

Roles of pERK 1/2 and PKA in OT-Evoked GFAP Plasticity

To further test if and how pERK 1/2 and PKA were involved in OT-elicited GFAP plasticity, we observed effects of OT (1 nM) on GFAP plasticity after changing the activity of pERK 1/2 or PKA 30 min before OT application. Immunostaining and confocal microscopy (**Figure 5Aa**) showed that the staining of GFAP was substantially ($p < 0.05$ or 0.01, $n \geq 6$) reduced by IPT (10 μM) and U0126 (1 μM) respectively, and slightly but significantly ($P < 0.05$) reduced by KT5720 (10 μM). The effects of IPT and U0126 on GFAP filaments were on both somata and processes, while the effect of KT5720 was on the processes only. Moreover, addition of OT (**Figure 5Ab**) did not significantly influence the actions of IPT or U0126, but reversed the effect of KT5720. The effects of these agents on OT-evoked GFAP plasticity are summarized in **Figure 5Ac** and are consistent with their effects on mSIRK-evoked GFAP plasticity.

Following the immunohistochemical experiments, Western blots were performed to detect GFAP protein after KT5720 and U0126 treatments with and without the presence of OT in brain slices from 6 rats. As shown in **Figure 5Ba**, KT5720 significantly decreased the 50 kDa GFAP bands; addition of OT strongly increased both 50 kDa and the fragments of GFAP proteins. By contrast, U0126 significantly reduced small GFAP bands but

enhanced 50 kDa bands, which were not significantly changed by addition of OT (see summary graph in **Figure 5Bb**). These results are in agreement with the finding in confocal microscopy.

Effects of GFAP-Modulating Agents on Astrocytic Uptake/Retention of Bauer Peptide

To link the GFAP-modulating effects of pERK 1/2 and PKA with astrocytic functions, we observed effects of OT, PD98059 and IPT on the fluorescence intensity of astrocytes after loading Bauer peptide in brain slices ($n = 6$) from three rats. As shown in **Figure 6A**, OT significantly reduced the fluorescence intensity after 15 min treatment compared to the time-matched controls, effects of which were mimicked by using PD98059 and IPT, and blocked by pretreatment of the slices with OTRA from 5 min before these agents (see **Figure 6B** for summary).

Effects of Suckling Stimulation on cPKA Expression and the Molecular Associations of GFAP with pERK 1/2 and PKA

To link the differential GFAP-modulating effects of the kinases to astrocytic functions, we further examined cPKA expressions

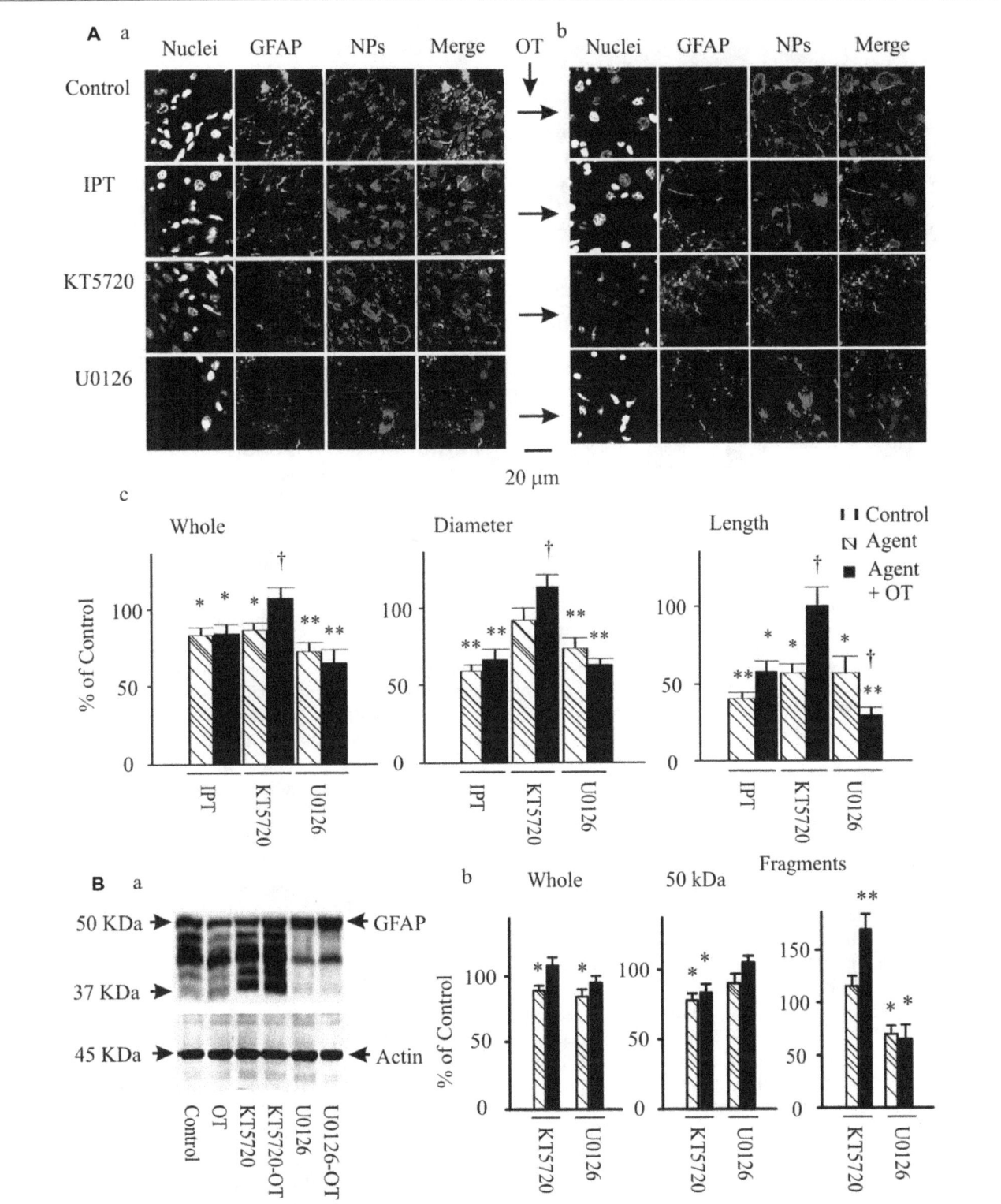

FIGURE 5 | Effects of modulating protein phosphorylation on OT-evoked GFAP reduction. **(A)** Exemplary confocal images showing (left to right) nuclei, GFAP, NPs and their merges before **(Aa)** and after **(Ab)** application of OT (1 nM, 30 min). Note: IPT, isoproterenol (10 μM), KT5720 and U0126 (1 μM). OT was added 30 min after these agents. **(Ac)** Summary graphs showing relative intensity of GFAP staining under different conditions. *$P < 0.05$ and **$P < 0.01$ compared to the control; †$P < 0.05$ compared with agents only. **(B)** GFAP expressions in Western blots. **(Ba)** Bands (left to right) showing GFAP proteins (top) in control, 1 nM OT, KT5720, KT5720 plus OT, U0126 and U0126 plus OT, respectively. Bands at bottom are actin, serving as loading controls for corresponding treatments ($n = 6$).
(Bb) Summary graphs showing different components of GFAP relative to controls. Other annotations refer to **Figures 1**, **4**.

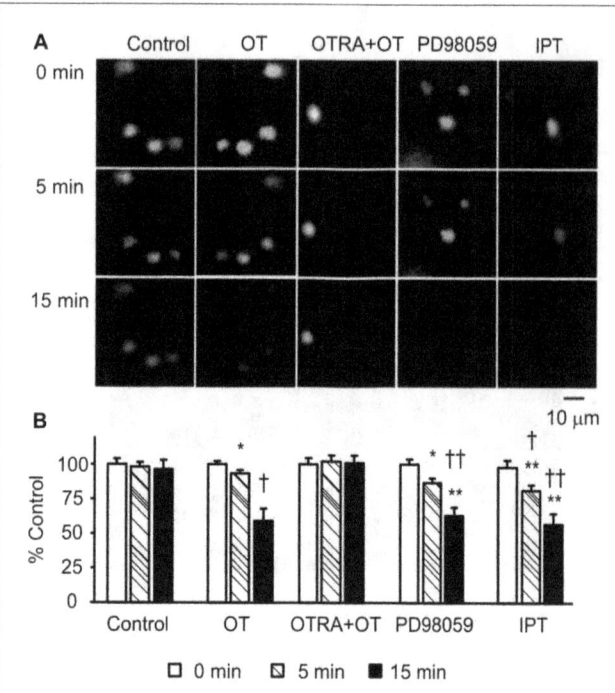

FIGURE 6 | Effects of GFAP-modulating agents on astrocytic uptake of Bauer peptide. **(A)** Exemplary fluorescence images showing temporal changes of fluorescence intensity of astrocytes that were preloaded with Bauer peptide (20 µM, 2–4 h at 35°C) in the SON. From left to the right the columns were control, OT (1 nM), OTRA (1 µM, from 5 min before OT) plus OT, PD98059 (20 µM) and IPT (10 µM), respectively. **(B)** Summary graphs. Note that, *$P < 0.05$, **$P < 0.01$ compared to the controls; †$P < 0.05$ and ††$P < 0.01$ compared with those at 5 min. Other annotations refer to **Figures 1**, **4**.

at different stages of suckling in immunocytochemistry in three sets of lactating rats, based on previous observation of pERK 1/2 (Wang and Hatton, 2009). As shown in **Figure 7A**, suckling significantly increased cPKA expression in whole SON. In astrocyte profiles, cPKA was markedly increased at the processes as indicated by its heavy overlapping with GFAP. The occurrence of the milk-ejection reflex further increased the co-localization of cPKA with GFAP filaments, but reduced cPKA expression at neuronal profiles from the peak at the initial stage of suckling. Further examination of the molecular association of cPKA and pERK 1/2 with GFAP in Co-IP experiments (**Figure 7B**) revealed that molecular association of GFAP with pERK 1/2 ($204.0 \pm 23.6\%$ of control, $n = 3$, $P < 0.05$) increased significantly ($n = 3$, $P < 0.05$) at the initial stage of suckling; the Co-IP of cPKA with GFAP was increased in all the three cases although the increase in the average level ($178.2 \pm 30.8\%$ of control, $P = 0.053$) did not reach statistical significance due to the relatively low power. Upon the occurrence of the milk-ejection reflex, the association of GFAP with pERK 1/2 ($431.4 \pm 66.0\%$ of control, $n = 3$, $P < 0.05$ compared to those of the control and during suckling) was further increased (**Figure 7Ba**), whereas the association between GFAP and cPKA was decreased significantly from that during suckling ($104.5 \pm 54.9\%$ of the control, $n = 3$,

$P < 0.05$ compared to that during suckling; **Figure 7Bb**). These results are in agreement with the findings in pharmacological manipulation in the males.

DISCUSSION

In the present study, we found that OT-evoked reduction of GFAP filaments depends on an antagonistic but coordinated pERK 1/2 and PKA interaction, downstream to OTR-Gβγ signaling. Upregulation of pERK 1/2 facilitates GFAP polymerization and PKA suppresses this polymerization while providing a basal condition for pERK 1/2 to function. Moreover, the two kinases work in coordination to localize GFAP at different compartments of astrocytes. This cellular signaling process is common between male and lactating female rats, likely contributing to astrocytic modulation of OT neuronal activity under diverse physiological conditions.

Methodological Consideration

To study astrocytic plasticity, classically electron microscopy is the first choice. However, at this level, GFAP is often absent from fine astrocytic processes around synapses and neuronal somata (Theodosis et al., 2008). Thus, despite the power of electron microscopy in revealing detailed morphological features of astrocytic processes and their association with particular neuronal and astrocytic populations, it is not more helpful in studying GFAP plasticity than confocal microscopy plus proteomic analyses. *In situ* observation of GFAP plasticity using rats that express luciferase under GFAP promoters to image GFAP can provide a real-time image of GFAP filaments. However, *in vivo* imaging of fluorescent GFAP in the SON is limited by the anatomical complexity in exposing the SON surgically. In the present study, combining confocal microscopy with protein analysis, patch-clamp recordings, and fluorescence detection of astrocyte-specific peptide in brain slices could not only reveal OT-elicited GFAP reduction and the underlying molecular mechanisms, but also establish a functional association of GFAP with astrocytic plasticity as discussed below.

OTR Mediation of OT-Evoked GFAP Reduction

GFAP monomers and filaments were reduced upon exposing to OT in males via activating OTRs on astrocytes in the SON. This finding is in agreement with previous reports that suckling acutely reduced GFAP expression via activation of OTRs (Wang and Hatton, 2009) and that a global reduction of GFAP expression occurred at the late pregnancy and during lactation in the SON (Perlmutter et al., 1984), the periods with enhanced OT actions (Hatton and Wang, 2008). The present study unambiguously verified that OT-reduced GFAP is directly mediated by astrocytic OTRs. The supportive evidences include that: (1) astrocytes in the SON express OTRs and OT but not VP elicited GFAP reduction under the same condition; (2) the effect of OT on GFAP did not require neuron-mediation since tetanus toxin could not block this OT action; and (3) OT-evoked

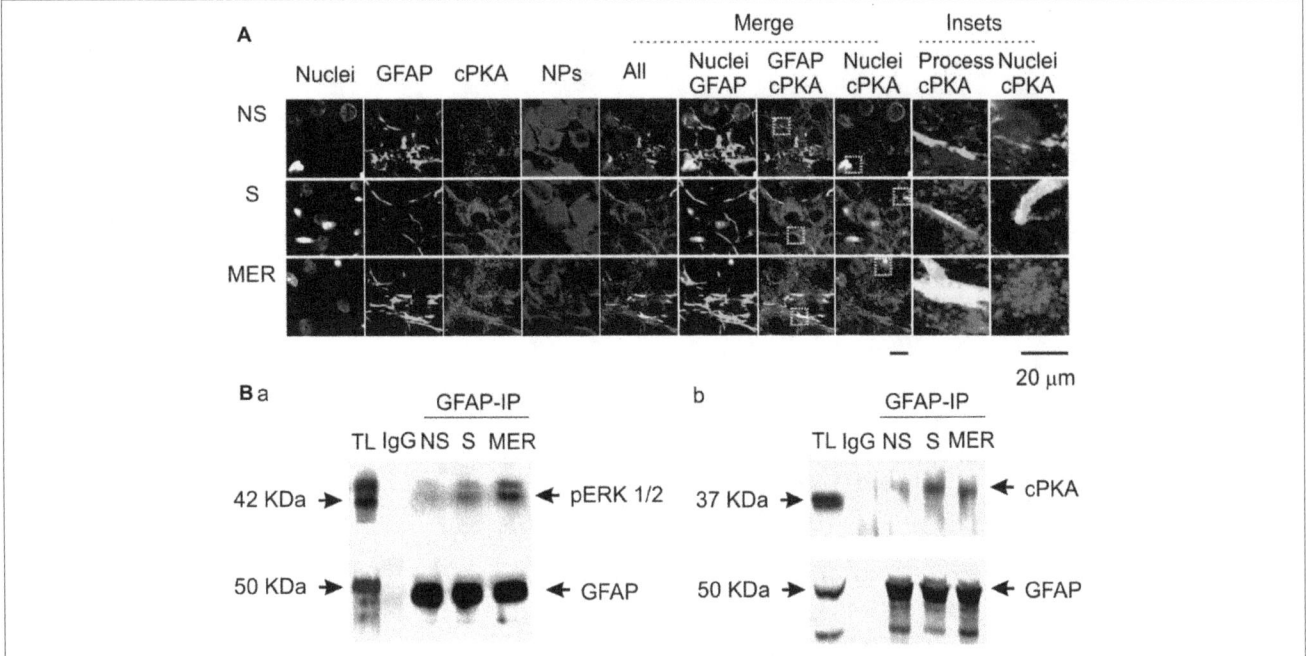

FIGURE 7 | Spatial and molecular associations of GFAP with protein kinases. **(A)** Exemplary confocal images showing effects of suckling at different stages on the expression of catalytic subunit of protein kinase A (cPKA). Images from left to the right showing nuclei, GFAP, cPKA, NPs, the merges of all channels, GFAP with nuclei, GFAP with cPKA, nuclei with cPKA, and the insets expanded from white dashed squares in corresponding merged channels. **(B)** Molecular association of GFAP with pERK 1/2 **(Ba)** and with cPKA **(Bb)** at different stages of suckling as indicated. Note that, NS, non-suckling; S, suckling; MER, immediately after the milk-ejection reflex. Other annotations refer to **Figures 1, 4**.

depolarization of astrocytic membrane potential was blocked by the OTRA, even in the present of tetanus toxin; and most importantly, there was molecular association between GFAP and OTRs. One argument could be that tetanus toxin does not block the release of lipophilic compounds and gases, e.g., endocannabinoids (Hirasawa et al., 2004) and nitric oxide (Luckman et al., 1997), which may mediate OT actions. However, endocannabinoids (Aguado et al., 2006) and nitric oxide (Guo et al., 2007) increase but does not decrease GFAP expression. If their release is not blocked by tetanus toxin, OT-reduced GFAP expression (Wang and Hatton, 2009) should be weakened, even blocked but not increased. Thus, we conclude that OT can directly act on astrocytes and reduce GFAP expression via OTRs.

OT Modulation of GFAP Metabolism and its Concentration- and Time-Dependence

Similar to those found in lactating rats (Wang and Hatton, 2009), OT concentration dependently reduced GFAP filaments in male rats. Interestingly, different concentrations of OT reduced GFAP likely via different metabolic processes. At lower levels, OT reduced both GFAP monomers at 50 kDa and their fragments, illustrating an increase in decomposition of GFAP molecules (Wang and Hatton, 2009). At higher levels, OT increased GFAP fragments while maintaining lower levels of GFAP monomers. The increased GFAP fragments did not contribute to the polymerization of GFAP filaments since GFAP polymerization is based on monomers but not their fragments.

Moreover, effects of activating OTRs on GFAP levels are related to the time course of OT actions. In previous study, we showed that OT reduced GFAP protein and filaments in brain slices from lactating rats in a time-dependent manner (Wang and Hatton, 2009). This finding is confirmed in the present study that mSIRK-reduced GFAP expression (**Figure 3**) or OT-reduced astrocyte absorption of Bauer peptide (**Figure 6**) were also time-dependent in the SON of male rats. This is in agreement with the feature of activation of G Protein-coupled receptors and the relatively slow metabolic processes of GFAP.

Intracellular Processes Following OTR Activation

In this study, we confirmed not only the expression of OTRs at astrocytes in the SON reported previously (Wang and Hatton, 2006), but also identified the mediation of Gβγ signaling in OT effects on SON functions and the mutually interactive processes between pERK 1/2 and PKA in OT-evoked GFAP plasticity.

In previous study on lactating rats, we found that OT-evoked activation of OT neurons was mainly mediated by Gβγ signaling but not Gα signaling following OTR activation of $G_{q/11}$ type G protein (Wang and Hatton, 2007a). The present study extends this finding by showing that mSIRK evoked GFAP reduction (**Figure 3**) as OT did, which was in synergy with the expression of pERK 1/2 (**Figure 4**). Moreover, along with the activation of ERK 1/2 in 5 min

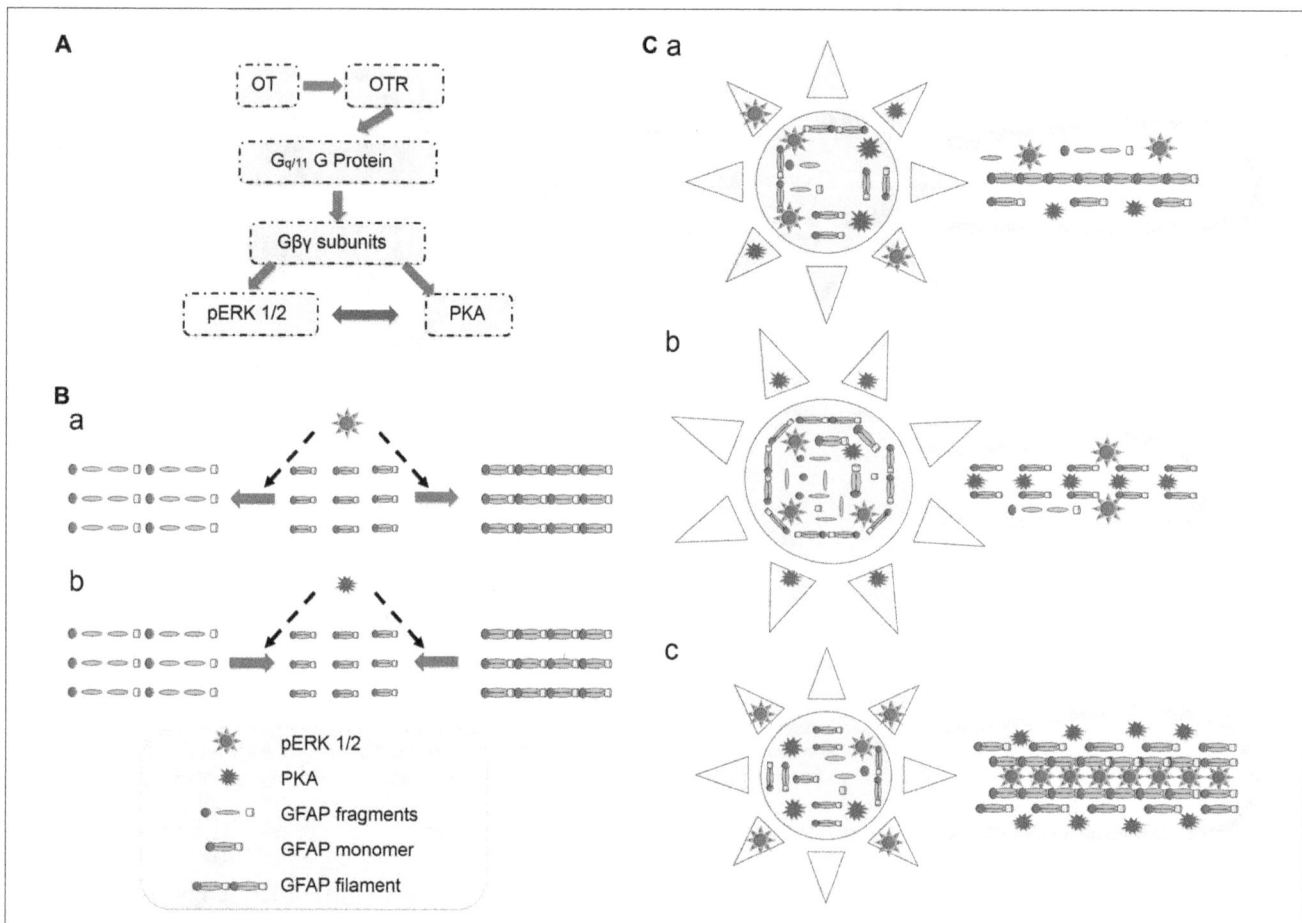

FIGURE 8 | Diagram of hypothetic cellular and molecular mechanisms underlying OT modulation of GFAP plasticity in astrocytes. **(A)** OTR signaling in association with OT-evoked GFAP reduction. The red arrows show the flow of signaling transduction and the bidirectional blue arrow indicates a relationship of mutual inhibition. **(B)** Effects of activation of pERK 1/2 **(a)** and PKA **(b)** on GFAP metabolism/plasticity, respectively. **(C)** GFAP plasticity of astrocytic somata (left panels) and processes (right panels) in the SON at different stages of suckling at its differential associations with pERK 1/2 and PKA. **(Ca)** Before suckling; **(Cb)** during the first 5–10 min of suckling with the occurrence of milk ejection; and **(Cc)** 1 min within the occurrence of the third or fourth milk ejections.

and its reversal at 30 min (Wang and Hatton, 2007b), OT- and/or mSIRK-evoked reduction of GFAP as well as OT-modulated astrocytic absorption of Bauer-peptide became significant after 15–30 min. These processes were modulated not only by pERK 1/2 blockers but also by PKA-modulating agents. Lastly, at different stages of suckling stimulation that are associated with different levels of GFAP and pERK 1/2 expressions (Wang and Hatton, 2007b), cPKA (**Figure 7**) also had different expressions. Lastly, activating PKA with IPT or blocking ERK 1/2 phosphorylation decreased GFAP polymerization; blocking PKA blocked OT-evoked GFAP reduction while blocking ERK 1/2 activation simulated effects of OT on GFAP. Thus, both pERK1/2 and PKA are involved in OT-modulation of GFAP plasticity downstream to OTR-Gβγ signaling pathway.

In addition to these classical signaling processes, our finding also fills in a gap between OT-evoked GFAP plasticity and astrocytic functions by showing the depolarizing effects of OT on astrocyte membrane potentials and the OT-reduced absorption of Bauer Peptide. As shown in **Figure 2**, OT could depolarize

astrocyte membrane potentials in 5–10 min of action. This could be an effect of OTR activation on the mobilization of IP3-sensitive intracellular Ca^{2+} stores (Di Scala-Guenot et al., 1994), one of the major downstream events of OTR activation. The increased intracellular Ca^{2+} levels could account for this initial depolarizing effect of OT because of its neutralization of intracellular negative charges. Moreover, this depolarization could account for the reduced absorption of Bauer Peptide at 30 min of OT treatment (**Figure 6**). Increased intracellular Ca^{2+} levels in astrocytes could cause hyper-phosphorylation of GFAP and its depolymerization (Heimfarth et al., 2016); the disruption of GFAP filaments could influence the expression of peptide transporter PepT2 (Dieck et al., 1999) due to losses of the scaffolding and guiding roles of GFAP (Hou et al., 2016; Wang and Parpura, 2016). As a result, astrocyte absorption of Bauer Peptide and the fluorescence intensity were also reduced. This proposal is supported by previous findings that reduction of GFAP also reduced its molecular association with glutamine synthetase (Wang et al., 2013b) and vesicular GABA transporters (Wang et al., 2013a) in astrocytes although directly assaying intracellular Ca^{2+} levels remains to be performed.

Antagonistic but Coordinated Interaction between pERK 1/2 and PKA in OT-Modulation of GFAP Plasticity

Both pERK 1/2 and PKA are involved in OT-elicited GFAP reduction but play different roles. Activation of PKA can mediate OT-reduced GFAP filaments while certain level of basal PKA activity is necessary to maintain basal GFAP expression. In the presence of OT, KT5720 did not only increase GFAP filaments and 50 kDa GFAP, but also increased GFAP fragments. However, KT5720 itself reduced basal GFAP filaments. Thus, combining with Western blotting results, we propose that at resting conditions, PKA promotes GFAP polymerization by reducing dissembling or decomposition of GFAP monomers in the presence of pERK 1/2; when strongly activated, PKA causes GFAP depolymerization. The increased GFAP filaments by KT5720 to OT stimulation can be attributable to a simultaneously increased dissembling and polymerization of GFAP monomers by OT-increased pERK 1/2 (Wang and Hatton, 2007b).

It is clear that pERK 1/2 can promote dissembling of GFAP monomers since U0126 reduced GFAP fragments with or without OT. Thus, reduction of GFAP monomers by OT-activated pERK 1/2 may counterbalance, even overrule, PKA-increased GFAP monomers, accounting for the increases in GFAP fragments in response to KT5720 plus OT. Moreover, pERK 1/2 can promote GFAP polymerization despite its increasing the dissembling of GFAP monomers. This proposal is supported by the fact that inhibition of ERK 1/2 phosphorylation significantly reduced basal GFAP filaments and increased 50 kDa GFAP proteins. Alternatively, OT-elicited activation of PKA in the presence of U0126 can explain the decrease in GFAP fragments while increasing GFAP monomers. However, without pERK 1/2, even OT could still activate PKA, GFAP monomers failed to be polymerized into GFAP filaments. As a whole, the actions of these two protein kinases are antagonistic but coordinated in OT-modulated GFAP plasticity. This conclusion is in agreement with that PKA and pERK 1/2 play antagonistic roles in astrocytic proliferation (Bayatti and Engele, 2001) and in the OTR signaling (Zhong et al., 2003).

Further analysis highlights the possibility that there are spatiotemporally coordinative actions between the two kinases in OT-elicited GFAP reduction. It is clear, simply increasing cAMP or reducing pERK 1/2 levels did not elicit the spatial feature of GFAP plasticity observed in lactating rats. By contrast, using confocal microscopy, we found critical clues for a microdomain-specific regulation of the two kinases. During suckling, pERK 1/2 is translocated into somata (Wang and Hatton, 2007b), which indicates a separation of pERK 1/2 from the majority of GFAP in astrocytic processes. Following a reversal increase in GFAP upon occurrence of milk-ejection, pERK 1/2 was also expressed in astrocytic proximal processes (Wang and Hatton, 2007b).

A possible role for the increased PKA levels at the milk-ejection reflex is to provide GFAP monomers for pERK 1/2 to build new GFAP filaments at astrocytic processes. The expression of pERK 1/2 at somata (Wang and Hatton, 2007b) and activation of PKA at the processes could account for the GFAP increases in astrocytic somata and decreases in the processes at the initial stages of suckling (**Figure 7**), explaining the inhibitory effect of OT on GFAP expression. Noteworthy is that following the occurrence of the milk-ejection reflex, cPKA expression in the processes further increased from the elevated levels during suckling, which was accompanied with increased molecular association between GFAP and pERK 1/2 but decreased molecular association between GFAP and cPKA. It is likely that the increased cPKA provided GFAP monomers that are essential for pERK 1/2-evoked GFAP polymerization and filament formation; however, due to the reduced direct interaction of cPKA with GFAP, its depolymerization role gave the way to the polymerization role of pERK 1/2, and resulted in the increase in GFAP filaments and its ensuing expansion of astrocytic processes. Since GFAP filament is positively correlated to the extension of astrocyte processes (Pekny et al., 2007) and OT neuronal activity (Wang and Hamilton, 2009), pERK 1/2 and PKA modulated GFAP plasticity thus could determine OT neuronal activity under a variety of physiological and pathological conditions. **Figure 8** is a schematic drawing of OTR signaling in OT-elicited GFAP plasticity.

CONCLUSION

The present study revealed that OT-elicited GFAP plasticity is associated with sequential activation of ERK 1/2 and PKA via OTR signaling pathway in an antagonistic but coordinated manner with microdomain-specific features. This finding not only confirms that OT can efficiently facilitate neuronal activity by eliciting retraction of astrocyte processes (Wang and Hatton, 2009), but also provide a novel approach to alter astrocyte functions by directly changing GFAP plasticity through modulating the activity of pERK 1/2 and PKA. Since GFAP-associated astrocytic plasticity is extensively involved in physiological processes, such as OT secretion in lactation (Liu X. et al., 2016), reproduction (Liu X. Y. et al., 2016), and immunological activity (Wang et al., 2015; Li T. et al., 2017), as well as VP secretion in hyponatremia (Wang et al., 2011; Jiao et al., 2017) and ischemic stroke (Jia et al., 2016; Wang and Parpura, 2016) in addition to psychiatric disorders (Verkhratsky et al., 2016), neurodegenerative diseases (Vardjan et al., 2017) and many others (Yang et al., 2013; Pekny et al., 2016), further exploration of signaling process regulating GFAP plasticity is warranted.

AUTHOR CONTRIBUTIONS

PW and Y-FW performed the experiment and data analysis; PW wrote the first draft; DQ participated in discussion and revision; Y-FW designed the study and made the final revision.

ACKNOWLEDGMENTS

We thank late Dr. Glenn I. Hatton for initial support and advices, Dr. Kathryn A. Hamilton for discussion and critical reading, Dr. Harold Gainer for neurophysin antibodies and Dr. Karl Bauer for Bauer peptide. This work was supported by National Institutes of Health NS009140 (GIH), the National Natural Science Foundation of China (grant No. 31471113, Y-FW), and Strong Province of Higher Education and University Quality Engineering Fund of Heilongjiang (grant No. 002000154, Y-FW). Initial part of this study was performed in the Department of Cell Biology and Neuroscience, University of California, Riverside.

REFERENCES

Aguado, T., Palazuelos, J., Monory, K., Stella, N., Cravatt, B., Lutz, B., et al. (2006). The endocannabinoid system promotes astroglial differentiation by acting on neural progenitor cells. *J. Neurosci.* 26, 1551–1561. doi: 10.1523/jneurosci.3101-05.2006

Bayatti, N., and Engele, J. (2001). Cyclic AMP modulates the response of central nervous system glia to fibroblast growth factor-2 by redirecting signalling pathways. *J. Neurochem.* 78, 972–980. doi: 10.1046/j.1471-4159.2001.00464.x

Chapman, D. B., Theodosis, D. T., Montagnese, C., Poulain, D. A., and Morris, J. F. (1986). Osmotic stimulation causes structural plasticity of neurone-glia relationships of the oxytocin but not vasopressin secreting neurones in the hypothalamic supraoptic nucleus. *Neuroscience* 17, 679–686. doi: 10.1016/0306-4522(86)90039-4

Dieck, S. T., Heuer, H., Ehrchen, J., Otto, C., and Bauer, K. (1999). The peptide transporter PepT2 is expressed in rat brain and mediates the accumulation of the fluorescent dipeptide derivative β-Ala-Lys-N$_\epsilon$-AMCA in astrocytes. *Glia* 25, 10–20. doi: 10.1002/(SICI)1098-1136(19990101)25:1<10::AID-GLIA2>3.0.co;2-y

Di Scala-Guenot, D., Mouginot, D., and Strosser, M. T. (1994). Increase of intracellular calcium induced by oxytocin in hypothalamic cultured astrocytes. *Glia* 11, 269–276. doi: 10.1002/glia.440110308

Guo, W., Wang, H., Watanabe, M., Shimizu, K., Zou, S., LaGraize, S. C., et al. (2007). Glial-cytokine-neuronal interactions underlying the mechanisms of persistent pain. *J. Neurosci.* 27, 6006–6018. doi: 10.1523/JNEUROSCI.0176-07.2007

Hatton, G. I. (2004). "Morphological plasticity of astroglial/neuronal interactions: functional implications," in *Glial Neuronal Signaling*, eds G. I. Hatton and V. Parpura (Boston, MA: Kluwer Academic Publishers), 99–124.

Hatton, G. I., Luckman, S. M., and Bicknell, R. J. (1991). Adrenalin activation of β2-adrenoceptors stimulates morphological changes in astrocytes (pituicytes) cultured from adult rat neurohypophyses. *Brain Res. Bull.* 26, 765–769. doi: 10.1016/0361-9230(91)90173-h

Hatton, G. I., and Wang, Y. F. (2008). Neural mechanisms underlying the milk ejection burst and reflex. *Prog. Brain Res.* 170, 155–166. doi: 10.1016/s0079-6123(08)00414-7

Heimfarth, L., da Silva Ferreira, F., Pierozan, P., Loureiro, S. O., Mingori, M. R., Moreira, J. C., et al. (2016). Calcium signaling mechanisms disrupt the cytoskeleton of primary astrocytes and neurons exposed to diphenylditelluride. *Biochim. Biophys Acta* 1860, 2510–2520. doi: 10.1016/j.bbagen.2016.07.023

Hirasawa, M., Schwab, Y., Natah, S., Hillard, C. J., Mackie, K., Sharkey, K. A., et al. (2004). Dendritically released transmitters cooperate via autocrine and retrograde actions to inhibit afferent excitation in rat brain. *J. Physiol.* 559, 611–624. doi: 10.1113/jphysiol.2004.066159

Hou, D., Jin, F., Li, J., Lian, J., Liu, M., Liu, X., et al. (2016). Model roles of the hypothalamo-neurohypophysial system in neuroscience study. *Biochem. Pharmacol. (Los Angel)* 5:3. doi: 10.4172/2167-0501.1000211

Hsiao, H.-Y., Mak, O.-T., Yang, C.-S., Liu, Y.-P., Fang, K.-M., and Tzeng, S. F. (2007). TNF-α/IFN-γ-induced iNOS expression increased by prostaglandin E$_2$ in rat primary astrocytes via EP2-evoked cAMP/PKA and intracellular calcium signaling. *Glia* 55, 214–223. doi: 10.1002/glia.20453

Inagaki, M., Gonda, Y., Nishizawa, K., Kitamura, S., Sato, C., Andog, S., et al. (1990). Phosphorylation sites linked to glial filament disassembly *in vitro* locate in a non-a-helical head domain. *J. Biol. Chem.* 265, 4722–4729.

Israel, J. M., Le Masson, G., Theodosis, D. T., and Poulain, D. A. (2003). Glutamatergic input governs periodicity and synchronization of bursting activity in oxytocin neurons in hypothalamic organotypic cultures. *Eur. J. Neurosci.* 17, 2619–2629. doi: 10.1046/j.1460-9568.2003.02705.x

Jia, S.-W., Liu, X.-Y., Wang, S. C., and Wang, Y.-F. (2016). Vasopressin hypersecretion-associated brain edema formation in ischemic stroke: underlying mechanisms. *J. Stroke Cerebrovasc. Dis.* 25, 1289–1300. doi: 10.1016/j.jstrokecerebrovasdis.2016.02.002

Jiao, R., Cui, D., Wang, S. C., Li, D., and Wang, Y. F. (2017). Interactions of the mechanosensitive channels with extracellular matrix, integrins, and cytoskeletal network in osmosensation. *Front. Mol. Neurosci.* 10:96. doi: 10.3389/fnmol.2017.00096

Langle, S. L., Poulain, D. A., and Theodosis, D. T. (2003). Induction of rapid, activity-dependent neuronal-glial remodelling in the adult rat hypothalamus *in vitro*. *Eur. J. Neurosci.* 18, 206–214. doi: 10.1046/j.1460-9568.2003.02741.x

Li, D., Liu, N., Zhao, H.-H., Zhang, X., Kawano, H., Liu, L., et al. (2017). Interactions between Sirt1 and MAPKs regulate astrocyte activation induced by brain injury *in vitro* and *in vivo*. *J. Neuroinflammation* 14:67. doi: 10.1186/s12974-017-0841-6

Li, T., Wang, P., Wang, S. C., and Wang, Y.-F. (2017). Approaches mediating oxytocin regulation of the immune system. *Front. Immunol.* 7:693. doi: 10.3389/fimmu.2016.00693

Liu, X. Y., Hou, D., Wang, J., Lv, C., Jia, S., Zhang, Y., et al. (2016). Expression of glial fibrillary acidic protein in astrocytes of rat supraoptic nucleus throughout estrous cycle. *Neuro. Endocrinol. Lett.* 37, 41–45.

Liu, X., Jia, S., Zhang, Y., and Wang, Y.-F. (2016). Pulsatile but not tonic secretion of oxytocin plays the role of anti-precancerous lesions of the mammary glands in rat dams separated from the pups during lactation. *M. J. Neurol.* 1:002.

Luckman, S. M., Huckett, L., Bicknell, R. J., Voisin, D. L., and Herbison, A. E. (1997). Up-regulation of nitric oxide synthase messenger RNA in an integrated forebrain circuit involved in oxytocin secretion. *Neuroscience* 77, 37–48. doi: 10.1016/s0306-4522(96)00498-8

Pekny, M., Pekna, M., Messing, A., Steinhäuser, C., Lee, J. M., Parpura, V., et al. (2016). Astrocytes: a central element in neurological diseases. *Acta Neuropathol.* 131, 323–345. doi: 10.1007/s00401-015-1513-1

Pekny, M., Wilhelmsson, U., Bogestål, Y. R., and Pekna, M. (2007). The role of astrocytes and complement system in neural plasticity. *Int. Rev. Neurobiol.* 82, 95–111. doi: 10.1016/s0074-7742(07)82005-8

Perlmutter, L. S., Tweedle, C. D., and Hatton, G. I. (1984). Neuronal/glial plasticity in the supraoptic dendritic zone: dendritic bundling and double synapse formation at parturition. *Neuroscience* 13, 769–779. doi: 10.1016/0306-4522(84)90095-2

Sun, H., Li, R., Xu, S., Liu, Z., and Ma, X. (2016). Hypothalamic astrocytes respond to gastric mucosal damage induced by restraint water-immersion stress in rat. *Front. Behav. Neurosci.* 10:210. doi: 10.3389/fnbeh.2016.00210

Theodosis, D. T., Poulain, D. A., and Oliet, S. H. (2008). Activity-dependent structural and functional plasticity of astrocyte-neuron interactions. *Physiol. Rev.* 88, 983–1008. doi: 10.1152/physrev.00036.2007

Vardjan, N., Verkhratsky, A., and Zorec, R. (2017). Astrocytic pathological calcium homeostasis and impaired vesicle trafficking in neurodegeneration. *Int. J. Mol. Sci.* 18:E358. doi: 10.3390/ijms18020358

Vasile, F., Dossi, E., and Rouach, N. (2017). Human astrocytes: structure and functions in the healthy brain. *Brain Struct. Funct.* 222, 2017–2029. doi: 10.1007/s00429-017-1383-5

Verkhratsky, A., Steardo, L., Peng, L., and Parpura, V. (2016). Astroglia, glutamatergic transmission and psychiatric diseases. *Adv. Neurobiol.* 13, 307–326. doi: 10.1007/978-3-319-45096-4_12

Wang, Y.-F., and Hamilton, K. (2009). Chronic vs. acute interactions between supraoptic oxytocin neurons and astrocytes during lactation: role of glial fibrillary acidic protein plasticity. *ScientificWorldJournal* 9, 1308–1320. doi: 10.1100/tsw.2009.148

Wang, Y. F., and Hatton, G. I. (2004). Milk ejection burst-like electrical activity evoked in supraoptic oxytocin neurons in slices from lactating rats. *J. Neurophysiol.* 91, 2312–2321. doi: 10.1152/jn.00697.2003

Wang, Y. F., and Hatton, G. I. (2006). Mechanisms underlying oxytocin-induced excitation of supraoptic neurons: prostaglandin mediation of actin polymerization. *J. Neurophysiol.* 95, 3933–3947. doi: 10.1152/jn.01 267.2005

Wang, Y. F., and Hatton, G. I. (2007a). Dominant role of βγ subunits of G-proteins in oxytocin-evoked burst firing. *J. Neurosci.* 27, 1902–1912. doi: 10.1523/jneurosci.5346-06.2007

Wang, Y. F., and Hatton, G. I. (2007b). Interaction of extracellular signal-regulated protein kinase 1/2 with actin cytoskeleton in supraoptic oxytocin neurons and astrocytes: role in burst firing. *J. Neurosci.* 27, 13822–13834. doi: 10.1523/jneurosci.4119-07.2007

Wang, Y. F., and Hatton, G. I. (2009). Astrocytic plasticity and patterned oxytocin neuronal activity: dynamic interactions. *J. Neurosci.* 29, 1743–1754. doi: 10.1523/jneurosci.4669-08.2009

Wang, Y.-F., Liu, L.-X., and Yang, H.-P. (2011). Neurophysiological involvement in hypervolemic hyponatremia-evoked by hypersecretion of vasopressin. *Transl. Biomed.* 2:3. doi: 10:3823/425

Wang, Y.-F., and Parpura, V. (2016). Central role of maladapted astrocytic plasticity in ischemic brain edema formation. *Front. Cell. Neurosci.* 10:129. doi: 10.3389/fncel.2016.00129

Wang, Y.-F., Sun, M.-Y., Hou, Q., and Hamilton, K. A. (2013a). GABAergic inhibition through synergistic astrocytic neuronal interaction transiently decreases vasopressin neuronal activity during hypoosmotic challenge. *Eur. J. Neurosci.* 37, 1260–1269. doi: 10.1111/ejn.12137

Wang, Y.-F., Sun, M. Y., Hou, Q., and Parpura, V. (2013b). Hyposmolality differentially and spatiotemporally modulates levels of glutamine synthetase and serine racemase in rat supraoptic nucleus. *Glia* 61, 529–538. doi: 10.1002/glia.22453

Wang, P., Yang, H. P., Tian, S., Wang, L., Wang, S. C., Zhang, F., et al. (2015). Oxytocin-secreting system: a major part of the neuroendocrine center regulating immunologic activity. *J. Neuroimmunol.* 289, 152–161. doi: 10.1016/j.jneuroim.2015.11.001

Yang, H. P., Wang, L., Han, L., and Wang, S. C. (2013). Nonsocial functions of hypothalamic oxytocin. *ISRN Neurosci.* 2013:179272. doi: 10.1155/2013/179272

Zhong, M., Yang, M., and Sanborn, B. M. (2003). Extracellular signal-regulated kinase 1/2 activation by myometrial oxytocin receptor involves GαqGβγ and epidermal growth factor receptor tyrosine kinase activation. *Endocrinology* 144, 2947–2956. doi: 10.1210/en.2002-221039

17

A Brief History of Microglial Ultrastructure: Distinctive Features, Phenotypes and Functions Discovered Over the Past 60 Years by Electron Microscopy

Julie C. Savage[1,2]*, Katherine Picard[1,2], Fernando González-Ibáñez[1,2] and Marie-Ève Tremblay[1,2]*

Axe neurosciences, Centre de Recherche du CHU de Québec – Université Laval, Québec City, QC, Canada, [2] Département de médecine moléculaire, Université Laval, Québec City, QC, Canada

*Correspondence:
Julie C. Savage
julie.savage.1@ulaval.ca;
Marie-Ève Tremblay
tremblay.marie-eve@
crchudequebec.ulaval.ca

The first electron microscope was constructed in 1931. Several decades later, techniques were developed to allow the first ultrastructural analysis of microglia by transmission electron microscopy (EM). In the 50 years that followed, important roles of microglia have been identified, specifically due to the ultrastructural resolution currently available only with EM. In particular, the addition of electron-dense staining using immunohistochemical EM methods has allowed the identification of microglial cell bodies, as well as processes, which are difficult to recognize in EM, and to uncover their complex interactions with neurons and synapses. The ability to recognize neuronal, astrocytic, and oligodendrocytic compartments in the neuropil without any staining is another invaluable advantage of EM over light microscopy for studying intimate cell–cell contacts. The technique has been essential in defining microglial interactions with neurons and synapses, thus providing, among other discoveries, important insights into their roles in synaptic stripping and pruning *via* phagocytosis of extraneous synapses. Recent technological advances in EM including serial block-face imaging and focused-ion beam scanning EM have also facilitated automated acquisition of large tissue volumes required to reconstruct neuronal circuits in 3D at nanometer-resolution. These cutting-edge techniques which are now becoming increasingly available will further revolutionize the study of microglia across stages of the lifespan, brain regions, and contexts of health and disease. In this mini-review, we will focus on defining the distinctive ultrastructural features of microglia and the unique insights into their function that were provided by EM.

Keywords: microglia, ultrastructure, electron microscopy, correlative light and electron microscopy, 3D ultrastructure

INTRODUCTION

Microglia are the only immune cells that permanently reside in the brain. Originally believed to mediate inflammatory responses to infection (1), trauma (2), ischemia (3), or neurodegenerative disease (4), recent studies identified microglia as crucial actors in the proper development and maintenance of neuronal circuits (5). del Río-Hortega provided the original morphological description

---

Abbreviations: 2p, two-photon; EM, electron microscopy; TEM, transmission electron microscope; SEM, scanning electron microscope; DAB, diaminobenzidine; Iba1, ionized calcium-binding adapter molecule 1; CLEM, correlative light and electron microscopy; AD, Alzheimer's disease; SBEM, serial block-face scanning electron microscope; FIB-SEM, focused-ion beam coupled with scanning electron microscope.

of microglia at the turn of the twentieth century, having modified Golgi's silver stain to identify microglia (6). His manuscripts have recently been translated into English and annotated (7). Early research into microglial physiology prompted researchers to posit hypotheses that still hold true: microglia are phagocytic; they are capable of generating inflammation in response to infection; they may be responsible for some aspects of neurodegenerative disease; they originate outside of the brain and colonize it early in development (8). Between the early twentieth and twenty-first centuries microglia remained mainly uninvestigated as a cell type [reviewed in Ref. (9)], until Davalos et al. and Nimmerjahn et al. uncovered their incredibly dynamic processes in the adult brain under physiological conditions using two-photon (2p) microscopy (10, 11). Following this discovery and with the development of genetic tools to specifically identify microglia and their progeny (12–14), high throughput gene-expression analysis (15–18), and investigation into expression of cell surface receptors (19, 20), researchers have completed a whirlwind of studies in an attempt to unravel microglial roles in a myriad of healthy and disease processes (21). Recent developments in super-resolution and 2p microscopy have provided insight into microglial interaction with dendritic spines (22–24). However, genetic manipulations required for marker expression in neurons and microglia can induce cellular stress and impair normal functions (25–27). Electron microscopy (EM) can be used to investigate the unique ultrastructure of microglia and their relationship with synapses, and identify their phagocytic cargo without any immunohistochemical or genetic labeling. While super-resolution microscopy has surpassed the diffraction limit of light microscopy, its resolution is still insufficient to discern samples smaller than 50 nm, especially in the z-dimension, and requires specific labeling probes to prevent steric hindrance from

influencing the resulting image (28). In this review, we will focus on the use of EM to unravel structural and functional mysteries of microglia and their interaction with healthy and diseased brain tissue.

HISTORY AND DEVELOPMENT OF EM

Electron microscopy utilizes focused electron beams to illuminate the subject of interest. Since an electron's wavelength is up to 100,000 times shorter than a photon's, EM is capable of resolving atomic structures, while most light microscopes are diffraction-limited to 500 nm resolution.

Hans Busch, a pioneer in the field of electron optics, laid the theoretical groundwork for EM by determining the motion of electrons in a magnetic field and the potential to focus electron beams (29). The first EM was invented by Knoll and Ruska in 1932, based on the Bush's published theories (30). The first transmission electron microscope (TEM) functioned by projecting electrons through a thin sample and onto film, and investigating the regions of the sample that were electron-permissive versus electron-dense. Shortly after TEMs were developed, the first scanning electron microscope (SEM) was invented in 1940 (31). SEM differs from TEM as it visualizes electrons that are scattered off the surface of the specimen instead of electrons that pass through the specimen.

Over the following three decades, scientists perfected multiple ways to process and preserve biological samples in order to garner useful images of *in situ* tissue preparations (**Figure 1**). Aldehyde fixation cross-links proteins in tissues (32, 33), while osmium tetroxide fixation mainly preserves lipids and renders membranes electron-dense (34). The development of transcardiac perfusions provided fast delivery of fixatives to deep regions of the brain and

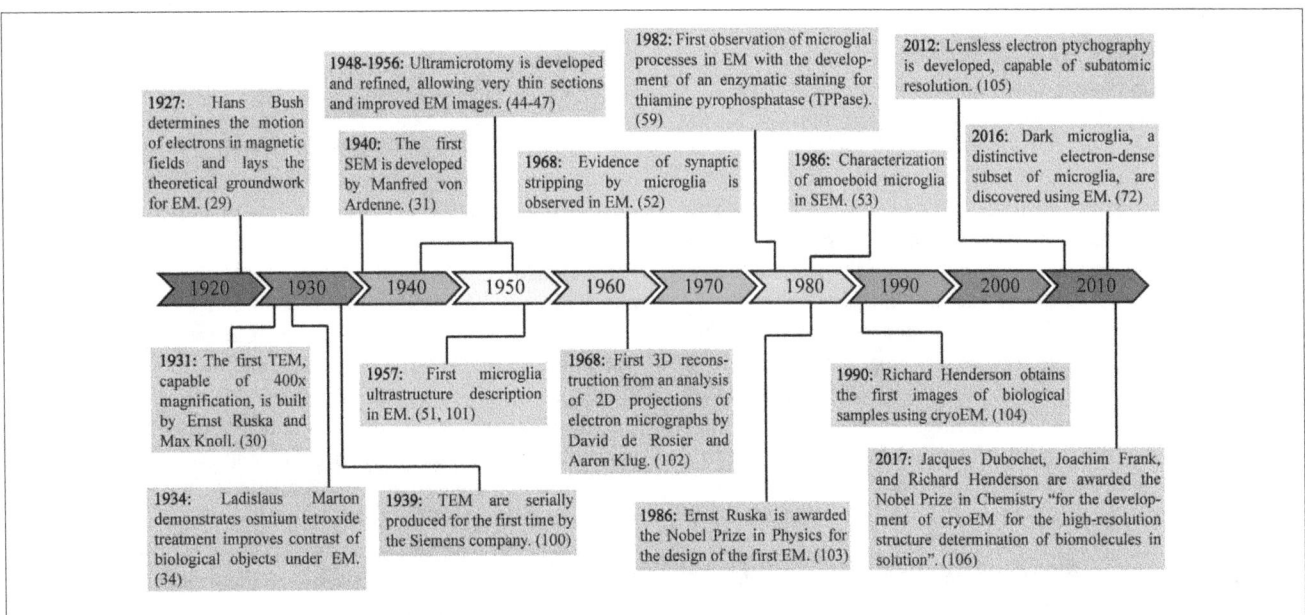

FIGURE 1 | Milestones in electron microscopy (EM) engineering and discovery. This timeline highlights the major theoretical and experimental advances in EM, from the invention of the first electron microscope to the 2017 Nobel Prize in Chemistry for the discoveries leading to cryoEM. Purple frames contain information about the development of technology required for EM, while orange frames contain information about microglial discoveries made possible through the use of EM.

other biological tissues, arresting any possible degradation that may have occurred in diffusion-dependent fixation techniques (35–37). However, using aldehydes or other fixatives results in tissue shrinkage and loss of extracellular space. This can be avoided by freeze substitution, a type of cryoEM: flash-freezing the tissue of interest followed by fixation performed at very low temperatures (38). Fixing the specimen in buffers that match the osmolarity of the tissue of interest can preserve extracellular space (39). Alternatively, if the specimen and chamber of the EM are kept below −140°C, samples can be visualized without any fixation (40, 41). Cell viability assays and staining for cell surface markers can be performed on live cells in suspension prior to deposition onto TEM grids and flash-freezing (42, 43).

Particularly important for TEM imaging was the development of ultramicrotomy, which allowed ultrathin (50–80 nm) sections to be cut from larger specimens, thus improving resolution and focus (44–47). These ultrathin sections allowed researchers to visualize ultrastructural images of various biological samples by capturing the transmitted electrons after they passed through the specimen onto films. The conventional protocol to prepare biological tissue for TEM is well explained by several groups (48–50).

EM AND MICROGLIA

In 1957, the first ultrastructural image of microglia in the rat parietal cortex was published (51), and in 1968, TEM images showed microglia physically separating presynaptic terminals from postsynaptic dendrites or neuronal cell bodies, a term defined as synaptic stripping (52). The first TEM images of microglia uncovered clues to the dynamic nature of these cells, decades before 2p microscopy discovered their movements to survey the brain parenchyma in real-time. Cultured microglia investigated using SEM identified many tiny processes projecting directly from cell somas, and draw stark attention to the two-dimensional stressors placed on cells in culture (53). Pioneering studies in EM identified many unique characteristics of microglial cell bodies, before any cell-specific immunological studies were developed.

Microglial cell bodies can be discerned from those of other cell types by their small size (3–6 µm), electron-dense cytoplasm, and characteristically bean-shaped nuclei. They also display a distinct heterochromatin pattern. A thick, dark band of electron-dense heterochromatin is located near the nuclear envelope, with pockets of compact heterochromatin nets throughout the nucleus. These nets are often visualized as small islands of dark heterochromatin within a sea of more loosely packed, lighter euchromatin within the central part of the nucleus (54, 55). Microglial cell bodies have a very thin ring of cytoplasm separating their nuclei from their cell membranes, and contain few organelles within a single ultrathin section, but those visible are mostly mitochondria, long stretches of endoplasmic reticulum, Golgi saccules, and lysosomes (54, 56, 57). They are often phagocytic and contain lipidic inclusions, especially in older animals (58) (**Figures 2A,D**).

The development of microglial-specific stains compatible with EM has been a major aid in determining their functions *in situ*. Labeling microglial membranes and cytoplasm, originally with enzymatic reactions and more recently with immunoEM, allowed researchers to investigate microglial processes in animal models

and human postmortem tissue (57, 59, 60). Current immunoEM studies utilize either diaminobenzidine or gold-conjugated antibodies (or colabeling using both) to deposit electron-dense precipitate and identify proteins of interest (22, 61). Ionized calcium-binding adapter molecule 1 (Iba1) is often used to identify microglia/macrophages within the brain (62). After much study using immunoEM to identify their main characteristics (22), trained researchers can identify microglial processes based solely on their unique ultrastructure. Microglia's ramified projections are long, thin, and almost never contiguous with their cell bodies in ultrathin sections examined by TEM (**Figures 2B,C**). They are often in close, direct contact with neuronal cell bodies, or separated only by a very thin astrocytic process (22, 57, 59, 63). A single microglial process can contact multiple synaptic elements, and interacts with axon terminals, dendritic spines, perisynaptic astrocytic processes, and encircles parts of synaptic clefts (22, 63). Their processes often perform extracellular degradation, visible as pockets of extracellular space sometimes containing pinpoints of membrane degradation. They frequently contain vacuoles or multivesicular bodies, long stretches of endoplasmic reticulum, and phagocytic inclusions (**Figure 2C**).

Microglia promote proper neuronal wiring and activity, and EM studies were vital for discovering their role in development and maintenance of functional neuronal connections (21). Elegant EM studies demonstrated that glia (performing the functions of microglia) in *Drosophila* (64), macrophages and microglia in zebrafish (65), as well as microglia in rodents (66–68) phagocytose degenerating axonal tracts, axon terminal fragments, and dendritic spines during development of the thalamus, cerebral cortex, and hippocampus. Interestingly, no phagocytic interactions between microglia and synapses were identified in TEM studies of a mouse model of prion disease (69), although immunoEM was not performed and microglial processes may have been overlooked. Microglia also phagocytose putative neuronal debris following saponin-induced cholinergic cell death in rats (70). Sequential EM images are required to verify phagocytic cargo is fully enclosed within a microglial process and has been demonstrated for phagocytosis of synaptic elements by both microglia and astrocytes (22, 67, 71). Automation of sequential EM using knives or focused-ion beams inside SEM chambers can provide nanometer-scale resolution images of microglia in 3D.

Recent TEM studies have uncovered a new microglial phenotype, named dark microglia. These dark microglia share many ultrastructural characteristics (including cell size, immunohistochemical markers, and phagocytic phenotype) with healthy microglia, yet appear strikingly different under TEM. Their cell bodies can be quickly identified by their condensed, electron-dense cytoplasm that makes them appear as dark as mitochondria. Dark microglia display many signs of cellular stress, including nuclear and chromatin condensation and dilation of their endoplasmic reticulum (**Figure 2F**). Additionally, they are present in greater numbers in pathological contexts often associated with neuronal dystrophy and distress. They have been identified in the APP-PS1 mouse model of Alzheimer's disease (AD), aged mice, animals subjected to social defeat stress, fractalkine receptor-deficient mice, and mouse models of schizophrenia (72, 73). They show reduced expression levels of some microglial markers, including

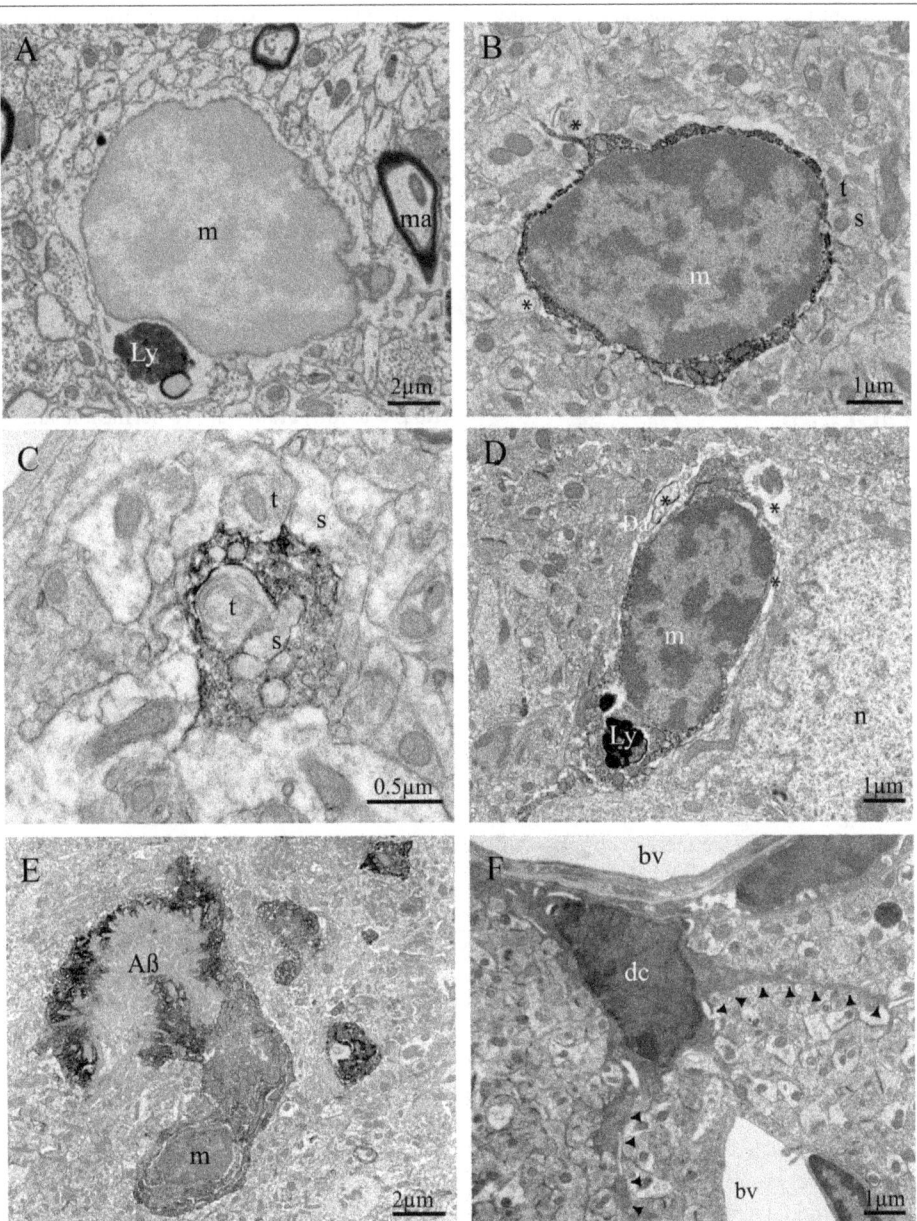

FIGURE 2 | Ultrastructural features of murine brain microglia in health and disease. Example of microglia imaged using a focused-ion beam coupled with scanning electron microscope without any immunostaining **(A)**, containing lipofuscin granules (Ly) and a lipid body (Lb). Diaminobenzidine staining against ionized calcium-binding adapter molecule 1 (Iba1) creates a dark immunoprecipitate in the cytoplasm as shown by transmission electron microscopy (TEM) **(B–E)**. Iba1 staining allows identification of microglial processes in fractalkine receptor-knockout mice, for instance, allowing researchers to investigate their contacts with synaptic terminals and study phagocytic inclusions. **(B)** A microglial cell body in an APP-PS1 mouse is contacting a synapse between two axon terminals and a dendritic spine, as well as juxtaposing cellular debris. **(C)** A microglial process in a C57Bl/6 mouse contains several inclusions, notably an axon terminal making a synaptic contact on a dendritic spine. **(D)** A microglial cell body in a mouse model of Werner syndrome juxtaposes myelin debris and contains lipofuscin granules. **(E)** A microglial cell body in an APP-PS1 mouse is found in intimate contact with an amyloid beta plaque. **(F)** Example of dark microglia observed by TEM in a stressed fractalkine receptor-deficient mouse, characterized by its dark cytoplasm and thin processes projecting from the cell body (black arrowheads). Symbols and abbreviations: m, microglia; n, neuron; dc, dark microglia; t, axon terminal; s, dendritic spine; bv, blood vessel; Ly, lipofuscin; Da, degenerated axon; ma, myelinated axon; AB, amyloid-beta plaque. Asterisk (*) denotes evidence of cellular debris undergoing digestion in the extracellular space. Pseudocolor code: phagocytic inclusions = purple, examples of dilated endoplasmic reticulum = blue, examples of mitochondria = orange, amyloid-beta plaque = green, lipid bodies = red.

Iba1, but are strongly immunopositive for others, including complement receptor subunit CD11b and microglia-specific antibody 4D4. Most dark microglia located near amyloid plaques in APP-PS1 mouse model express TREM2, though dark microglia in other disease models do not. While normal microglia rarely have contiguous processes attached to their somas in ultrathin sections, dark microglia show many long, thin processes encircling dystrophic neurons, wrapping around synaptic structures, investing themselves deep into amyloid beta plaques, and interacting with synapses in regions of high synaptic turnover (72). Although they display many signs of cellular stress, they have never been found expressing apoptotic or necrotic cell markers. Dark microglia are often located near blood vessels, which could imply a possible peripheral origin or perhaps a route of egress for the stressed cell to leave the brain parenchyma (**Figure 2F**). As there is not yet a definitive marker of dark microglia, they can only be investigated with EM, highlighting its relevance in modern microglial biology studies.

CORRELATIVE LIGHT AND ELECTRON MICROSCOPY (CLEM)

The combination of both light microscopy and EM can be used to uncover more information than either technique individually. CLEM was first used in 1969. Silver staining originally described by Río-Hortega was used to identify and investigate microglia in light microscopy. After confirming microglial-specific staining, researchers investigated ultrathin sections under TEM and published the first description of microglial ultrastructure (54).

Electron microscopy is currently used to unravel details and variations in ultrastructure that cannot be investigated with light microscopy (**Table 1**). Light microscopy is often used to detect changes in microglial density and morphology in health and disease (74), and to identify particular regions of interest. After identifying a region of interest, such as one affected by hypoxia in stroke or amyloid-beta positive plaque-containing tissue in AD, EM can delve further into specific changes in microglial ultrastructure, cellular viability and stress, all without requiring further immunostaining markers (75, 76). EM can also reveal structures which are not otherwise visible, and discern subcellular localization of proteins and mRNA using immunostaining, *in situ* hybridization, or *in situ* RT-PCR (57, 77, 78). EM was recently used to clarify microglial process fragmentation observed with light microscopy in postmortem human tissue from an individual suffering from AD. Unexpectedly, EM studies revealed no fragmentation as the two parts of the microglial process were linked by a cytoplasmic bridge, thus invalidating the original hypothesis (79).

Light microscopy can also be performed on living cells prior to investigation with EM to tie temporal information to ultrastructural resolution. The technique used is a specific type of CLEM. Live imaging using 2p microscopy studies cellular relationships, interactions with the surrounding environment, and intracellular dynamics in real-time; but lacks complex ultrastructural information. These imaging techniques are also limited to genetically encoded or virally introduced cell-specific fluorescent markers, which may introduce phenotypic changes on their own (26). EM can uncover structural information, but the specimen must be fixed (or flash-frozen for cryoEM), and can only be investigated as a snapshot moment in time. CLEM integrates imaging of fluorescent proteins in live cells with the ultrastructural resolution of EM. After live-cell imaging is performed, various fixation and staining techniques can be employed to investigate ultrastructure in the same tissues. van Rijnsoever and colleagues used CLEM to study the endolysosomal system by confocal microscopy

followed by cryoEM to image protein structures with nanometer resolution (85). This technique could be used to obtain insight into microglial proteins, phagocytic machinery, and organelle biogenesis.

It is also possible to combine EM with 2p microscopy. While 2p studies allow investigation of live microglial interactions with nearby neurons, EM performed afterward can study the intimate contacts between microglia and synaptic elements, and their changes in response to various behavioral experiences and pathologies (22, 63, 86–88). Light microscopy also informs EM studies, making it much easier to solve needle-in-haystack problems and identify rare events within the neuropil. For example, an Alzheimer's study injected animals with methoxy-X04, a blood–brain barrier permeable amyloid-beta fluorescent marker. Researchers then selected sections containing the region of interest known to contain amyloid-beta, thus increasing the likelihood of finding plaque-associated microglia in ultrathin sections (**Figure 2E**) (50). Similarly, fluorescent microscopy can be used to target a specific microglial population to be analyzed in EM. Bechmann and Nitsch fluorescently labeled axons prior to performing entorhinal lesions and traced the clearance of degenerating tissue by identifying the fluorescent compound within nearby microglia. By focusing EM studies on regions containing fluorescently labeled microglia, they were able to investigate the subpopulation of microglial cells which had phagocytosed degenerating axons (89).

Another CLEM technique is the use of light microscopy in correlation with cryoEM. A study utilizing both techniques recently discovered the native folding of herpes simplex virus as it moved throughout axons. 3D visualization permits analysis of vesicle fusion and actin bond formation (90). CryoEM was also recently used to image Golgi apparati in two different conformations within neurons (91) and investigate minor changes in ultrastructure following intracerebral injections (a common technique used to introduce vectors into mouse models) (92). Cryo-fixation preserves extracellular space, especially notable at synapses and blood vessels (93). This method could be used to determine native folding and unfolding of proteins within microglia, to better understand their morphological and functional changes in various disease conditions.

THE FUTURE OF EM AND MICROGLIA

The past 15 years have seen a whirlwind in EM development. Previously, when investigating 3D ultrastructure, serial ultrathin sections were manually cut and collected at the ultramicrotome, imaged individually onto film under a TEM, and painstakingly reoriented and collated prior to analysis (94). As digital imaging improved, TEMs were outfitted with digital cameras allowing for faster imaging, but the electron beam of the TEM could still deform ultrathin sections, making perfect alignment of sequential sections almost impossible. Developments in SEM opened the door for array tomography studies on ribbons of serial ultrathin sections, allowing CLEM on the same tissue, and solving the problem of deformation introduced when using TEM (81).

The first major revolution in 3D ultrastructure imaging came when Denk and Horstmann engineered a working diamond knife

TABLE 1 | Types of EM.

Type of EM	Typical sample preparation	Maximal resolution	Advantages	Disadvantages
Transmission electron microscopy (TEM) (49)	– Fixation with aldehydes and plastic resin embedding – Manually cut ultramicrotomy (thin sections of 50–80 nm stored on metal grids)	Nanometer resolution in x, y Resolution in z limited by section thickness	– Tissue can be archived and reimaged – Block of tissue may be saved and recut – Highest resolution and magnification – Osmium fixation is not required	– Biological specimens must be fixed with gluteraldehyde or acrolein – Low throughput – Electron beam can cause deformation of ultrathin tissue sections – Smaller magnification range (680× to greater than 30,000×)
Scanning transmission electron microscopy (STEM) (49)	– Fixation with aldehydes, strong post-fixation with osmium (OTO), and plastic resin embedding – Manually cut ultramicrotomy (thin sections of 50–80 nm stored on metal grids)	Nanometer resolution in x, y Resolution in z limited by section thickness	– Tissue can be archived and reimaged – Block of tissue may be saved and recut – Faster imaging throughput than traditional TEM – Large magnification range (20× to greater than 30,000×)	– Biological specimens must be fixed with gluteraldehyde or acrolein – Stronger osmium fixation required than traditional TEM – Electron beam can cause deformation of ultrathin tissue sections – Risk of tissue destruction is higher than with traditional TEM
Scanning electron microscopy (SEM) (80)	– Dehydration – Strong post-fixation with osmium (OTO) if material contrast imaging is desired – Entire specimen (entire insect, dissected organ, etc.) mounted on a stub of metal with adhesive – Coated with a conductive metal	Nanometer resolution in x, y, and z for surface topography	– Tissue can be archived and reimaged – Large magnification range (20× to greater than 30,000×) – Can create images of up to several cm³, which provides a good representation of the 3D shape of the specimen – Secondary electron detector measures surface topography – Backscatter electron detector measures material contrast (i.e., cell membrane versus cytoplasm)	– Biological specimens must be fixed with gluteraldehyde or acrolein – Image is created using scattered electrons and limited to the surface of the specimen
Scanning electron microscopy with array tomography (81)	– Fixation with aldehydes, strong post-fixation with osmium (OTO), and plastic resin embedding – Manually or automatically cut serial sections ultramicrotomy (thin sections of 50–80 nm stored on silicon chips or magnetic tape)	Nanometer resolution in x, y Resolution in z limited by section thickness	– Tissue can be archived and reimaged – Image large and serial sections – Large magnification range (20× to greater than 30,000×) – Compatible with correlative light-EM imaging – No deformation of tissue, making serial reconstruction simpler	– Serial section cutting and collecting is technically challenging – Stronger fixation required for proper material contrast
Focused-ion beam–scanning electron microscopy (FIB–SEM) (82)	– Fixation with aldehydes, strong post-fixation with osmium (OTO), and plastic resin embedding – Prepared tissue specimen (3–10 mm² wide × 3–10 mm² tall × 50–75 μm thick) mounted on a stub of metal with adhesive – Coated with a conductive metal	Nanometer resolution in x, y Up to 5 nm resolution in z (83)	– Nanometer resolution (less than 5 nm per pixel) in all three dimensions – Simplest serial image reconstruction	– The entire tissue block must be mounted and cannot be resectioned – Limited to a very small area, usually less than 15 μm × 15 μm – Smaller magnification range (400× to greater than 30,000×) – The sample is destroyed as it is imaged and cannot be reimaged
CryoTEM (84) CryoSEM (84)	– High-pressure freezing – Manually or automatically cut sections using cryo-ultramicrotomy (40–100 nm thick)	Nanometer resolution in x, y Resolution in z limited to section thickness	– No fixation required – Allows imaging of specimens in a native-like state	– Technically challenging – The sample must be flash-frozen to preserve native protein folding – The sample must remain frozen through entire process

Table of the major types of electron microscopy (EM) described in this mini-review, highlighting sample preparation, maximal resolution, magnification power, and advantages and disadvantages to each technique. Typical sample preparation is provided for each method, but fixation with aldehydes can be avoided if the researchers instead flash-freeze samples and perform freeze-substitution following sample collection.

into the chamber of a SEM to perform serial block-face scanning electron microscope (SBEM) (95). With this approach, the block face is imaged, a 50-nm section is cut away, followed by another image. Peddie and Collinson recently reviewed the many types of 3D EM and its applications to biological tissues (96). A decade after SBEM was invented, it allowed researchers to confirm microglial synaptic stripping (97). Research in a mouse model of multiple sclerosis used SBEM to unravel different roles of microglia *versus* infiltrating monocytes very early in the disease. The authors performed SBEM and differentiated resident microglia from invading myeloid cells by their ultrastructural differences (changes in mitochondrial makeup, nuclear shape, and presence of osmiophilic granules) in order to determine that demyelination in experimental autoimmune encephalitis is initially performed by invading monocytes, while resident microglia did not contribute to the early stages of inflammation (98). More recently, focused-ion beam coupled with SEM (FIB-SEM) has improved resolution from 50 nm to less than 10 nm in the z-dimension (99). By employing a focused-ion beam to atomize a very thin layer from a small (usually less than 500 μm^2) area, researchers may image at 5 nm resolution in x, y, and z (82).

Both SBEM and FIB-SEM are capable of investigating neuronal ultrastructure, and can follow a single process through several microns of neuropil, but FIB-SEM is also capable of resolving synaptic vesicles, lysosomes, and phagosomes in three dimensions. If the FIB-SEM process began within a microglial cell body, researchers could trace fine microglial processes through several microns of neuropil, without having to perform immunoEM. This offers a better chance to investigate lipidic inclusions and other pathological changes in organelles obscured by electron-dense precipitates used in immunoEM.

In addition to technological advances in both SEM and TEM, the rapid development of cryoEM techniques described here could uncover native protein structures within microglia.

It could additionally pave the way for discoveries into the snapshot of microglial–neuron and microglia–glia interactions without requiring fixatives, and without the corresponding tissue deformation that occurs with rapid fixation currently required to preserve ultrastructure. While fixatives and ultrathin sections required for EM are not compatible with post-imaging analysis of RNA or proteins, future iterations of CLEM (perhaps cryoCLEM) and advances in single-cell mRNA isolation may be able to isolate subcellular tissue fractions for further analysis. Armed with these new tools, biologists may investigate the complex interactions between glia and neurons in a number of diseases. The unique nature of EM allows researchers to characterize unique ultrastructural characteristics of microglia and other immune cells, and uncover possible paths for therapeutic intervention.

AUTHOR CONTRIBUTIONS

JS and MET conceived the ideas and drafted the manuscript to which all authors contributed. KP and FG-I created the figures. All authors read and approved the final version of the manuscript.

ACKNOWLEDGMENTS

The authors would like to thank Kanchan Bisht, Kaushik Sharma, Marie-Kim St.-Pierre, and Hassan El-Hajj for providing EM images of microglia, as well as Nathalie Vernoux, Cynthia Lecours, and Julie-Christine Levesque for technical assistance. MET is a recipient of a Canada Research Chair (Tier 2) in *Neuroimmune Plasticity in Health and Therapy*, JS is a recipient of a Fonds de recherche du Québec-Santé (FRQS) fellowship, KP is the recipient of a Fondation du CHU de Québec scholarship, and FG-I is the recipient of a scholarship from Consejo Nacional de Ciencia y Tecnologia (Conacyt).

REFERENCES

Mariani MM, Kielian T. Microglia in infectious diseases of the central nervous system. *J Neuroimmune Pharmacol* (2009) 4:448–61. doi:10.4049/jimmunol.178.9.5753

Witcher KG, Eiferman DS, Godbout JP. Priming the inflammatory pump of the CNS after traumatic brain injury. *Trends Neurosci* (2015) 38:609–20. doi:10.1016/j.tins.2015.08.002

Zarruk JG, Greenhalgh AD, David S. Microglia and macrophages differ in their inflammatory profile after permanent brain ischemia. *Exp Neurol* (2018) 301:120–32. doi:10.1016/j.expneurol.2017.08.011

Ransohoff RM. How neuroinflammation contributes to neurodegeneration. *Science* (2016) 353:777–83. doi:10.1126/science.aag2590

Wu Y, Dissing-Olesen L, MacVicar BA, Stevens B. Microglia: dynamic mediators of synapse development and plasticity. *Trends Immunol* (2015) 36:605–13. doi:10.1016/j.it.2015.08.008

del Rio-Hortega P. El tercer elemento de los centros nerviosos. I. La microglia en estado normal. II. Intervencion de la microglia en los procesos patologi- cos. III. Naturaleza probable de la microglia. *Bol de la Soc Esp de Biol* (1919) 8:69–120.

Sierra A, de Castro F, Del Río-Hortega J, Rafael Iglesias-Rozas J, Garrosa M, Kettenmann H. The "Big-Bang" for modern glial biology: translation and comments on Pío del Río-Hortega 1919 series of papers on microglia. *Glia* (2016) 64:1801–40. doi:10.1111/ejn.13256

Rezaie P, Hanisch U-K. Historical context. In: Tremblay M-È, Sierra A, editors. *Microglia in Health and Disease*. New York, NY: Springer New York (2014). p. 7–46.

Tremblay M-È, Lecours C, Samson L, Sánchez-Zafra V, Sierra A. From the

Cajal alumni Achúcarro and Río-Hortega to the rediscovery of never-resting microglia. *Front Neuroanat* (2015) 9:45. doi:10.3389/ fnana.2015.00045

Davalos D, Grutzendler J, Yang G, Kim JV, Zuo Y, Jung S, et al. ATP medi- ates rapid microglial response to local brain injury in vivo. *Nat Neurosci* (2005) 8:752–8. doi:10.1038/nn1472

Nimmerjahn A, Kirchhoff F, Helmchen F. Resting microglial cells are highly dynamic surveillants of brain parenchyma in vivo. *Science* (2005) 308:1314–8. doi:10.1126/science.1110647

Ginhoux F, Greter M, Leboeuf M, Nandi S, See P, Gokhan S, et al. Fate map- ping analysis reveals that adult microglia derive from primitive macrophages. *Science* (2010) 330:841–5. doi:10.1126/science.1194637

Goldmann T, Wieghofer P, Müller PF, Wolf Y, Varol D, Yona S, et al. A new type of microglia gene targeting shows TAK1 to be pivotal in CNS autoimmune inflammation. *Nat Neurosci* (2013) 16:1618–26. doi:10.1038/nn.3531

Parkhurst CN, Yang G, Ninan I, Savas JN, Yates JR, Lafaille JJ, et al. Microglia promote learning-dependent synapse formation through brain-derived neu- rotrophic factor. *Cell* (2013) 155:1596–609. doi:10.1016/j.cell.2013.11.030

Butovsky O, Siddiqui S, Gabriely G, Lanser AJ, Dake B, Murugaiyan G, et al. Modulating inflammatory monocytes with a unique microRNA gene signa- ture ameliorates murine ALS. *J Clin Invest* (2012) 122:3063–87. doi:10.1172/ JCI62636

Hickman SE, Kingery ND, Ohsumi TK, Borowsky ML, Wang L-C, Means TK, et al. The microglial sensome revealed by direct RNA sequencing. *Nat Neurosci* (2013) 16:1896–905. doi:10.1038/nn.3554

Chiu IM, Morimoto ETA, Goodarzi H, Liao JT, O'Keeffe S, Phatnani HP, et al. A neurodegeneration-specific gene-expression signature of acutely isolated microglia from an amyotrophic lateral sclerosis mouse model. *Cell Rep* (2013) 4:385–401. doi:10.1016/j.celrep.2013.06.018

Zhang Y, Chen K, Sloan SA, Bennett ML, Scholze AR, O'Keeffe S, et al. An RNA-sequencing transcriptome and splicing database of glia, neurons, and vascular cells of the cerebral cortex. *J Neurosci* (2014) 34:11929–47. doi:10.1523/JNEUROSCI.1860-14.2014

Pocock JM, Kettenmann H. Neurotransmitter receptors on microglia. *Trends Neurosci* (2007) 30:527–35. doi:10.1016/j.tins.2007.07.007

Domercq M, Vázquez-Villoldo N, Matute C. Neurotransmitter signaling in the pathophysiology of microglia. *Front Cell Neurosci* (2013) 7:49. doi:10.3389/fncel.2013.00049

Tay TL, Savage J, Hui C-W, Bisht K, Tremblay M-È. Microglia across the lifes- pan: from origin to function in brain development, plasticity and cognition. *J Physiol* (2017) 595:1929–45. doi:10.1113/JP272134

Tremblay M-È, Lowery RL, Majewska AK. Microglial interactions with synapses are modulated by visual experience. *PLoS Biol* (2010) 8:e1000527. doi:10.1371/journal.pbio.1000527

Bethge P, Chéreau R, Avignone E, Marsicano G, Nägerl UV. Two-photon excitation STED microscopy in two colors in acute brain slices. *Biophys J* (2013) 104:778–85. doi:10.1016/j.bpj.2012.12.054

Szalay G, Martinecz B, Lénárt N, Környei Z, Orsolits B, Judák L, et al. Microglia protect against brain injury and their selective elimination dysreg- ulates neuronal network activity after stroke. *Nat Commun* (2016) 7:1–13. doi:10.1038/ncomms11499

Porrero C, Rubio-Garrido P, Avendaño C, Clascá F. Mapping of fluorescent protein-expressing neurons and axon pathways in adult and developing Thy1-eYFP-H transgenic mice. *Brain Res* (2010) 1345:59–72. doi:10.1016/j.brainres.2010.05.061

Lee S, Varvel NH, Konerth ME, Xu G, Cardona AE, Ransohoff RM, et al. CX3CR1 deficiency alters microglial activation and reduces beta-amyloid deposition in two Alzheimer's disease mouse models. *Am J Pathol* (2010) 177:2549–62. doi:10.2353/ajpath.2010.100265

Comley LH, Wishart TM, Baxter B, Murray LM, Nimmo A, Thomson D, et al. Induction of cell stress in neurons from transgenic mice expressing yellow fluorescent protein: implications for neurodegeneration research. *PLoS One* (2011) 6:e17639. doi:10.1371/journal.pone.0017639.t001

MacDonald L, Baldini G, Storrie B. Does super-resolution fluorescence microscopy obsolete previous microscopic approaches to protein co-localiza- tion? In: Tang B, editor. *Methods in Molecular Biology. Membrane Trafficking*. New York, NY: Springer New York (2015). p. 255–75.

Busch H. Über die Wirkungsweise der Konzentrierungsspule bei der Braunschen Röhre. *Archiv für Elektrotechnik* (1927) 18:583–94. doi:10.1007/BF01656203

Knoll M, Ruska E. Das Elektronenmikroskop. *Zeitschrift für Physik* (1932) 78:318–39. doi:10.1007/BF01342199

Ardenne Von M, Beischer D. Untersuchung von Metalloxyd-Rauchen mit dem Universal-Elektronenmikroskop. *Berichte der Bunsengesellschaft für physikalische Chemie* (1940) 46:270–7. doi:10.1002/bbpc.19400460406

Luft JH. The use of acrolein as a fixative for light and electron microscopy. *Anat Record* (1959) 133:305.

Sabatini DD, Bensch K, Barrnett RJ. Cytochemistry and electron microscopy. The preservation of cellular ultrastructure and enzymatic activity by aldehyde fixation. *J Cell Biol* (1963) 17:19–58. doi:10.1083/jcb.17.1.19

Marton L. Electron microscopy of biological objects. *Nature* (1934) 133:911. doi:10.1038/133911b0

Palay SL, McGee-Russell SM, Gordon S, Grillo MA. Fixation of neural tissues for electron microscopy by perfusion with solutions of osmium tetroxide. *J Cell Biol* (1962) 12:385–410. doi:10.1083/jcb.12.2.385

Webster H, Collins GH. Comparison of osmium tetroxide and glutaralde- hyde perfusion fixation for the electron microscopic stufy of the normal rat peripheral nervous system. *J Neuropathol Exp Neurol* (1964) 23:109–26. doi:10.1093/jnen/23.1.109

Williams TH, Jew JY. An improved method for perfusion fixation of neural tissues for electron microscopy. *Tissue Cell* (1975) 7:407–18. doi:10.1016/0040-8166(75)90015-4

Van Harreveld A, Crowell J, Malhotra SK. A study of extracellular space in central nervous tissue by freeze-substitution. *J Cell Biol* (1965) 25:117–37. doi:10.1083/jcb.25.1.117

Pallotto M, Watkins PV, Fubara B, Singer JH, Briggman KL. Extracellular space preservation aids the connectomic analysis of neural circuits. *Elife* (2015) 4. doi:10.7554/eLife.08206

Dubochet J, Adrian M, Chang JJ, Homo JC, Lepault J, McDowall AW, et al. Cryo-electron microscopy of vitrified specimens. *Q Rev Biophys* (1988) 21:129–228. doi:10.1017/S0033583500004297

Al-Amoudi A, Norlen LPO, Dubochet J. Cryo-electron microscopy of vitreous sections of native biological cells and tissues. *J Struct Biol* (2004) 148:131–5. doi:10.1016/j.jsb.2004.03.010

Cloutier N, Tan S, Boudreau LH, Cramb C, Subbaiah R, Lahey L, et al. The exposure of autoantigens by microparticles underlies the formation of potent inflammatory components: the microparticle-associated immune complexes. *EMBO Mol Med* (2012) 5:235–49. doi:10.1016/j.autrev.2005.03.005

Milasan A, Tessandier N, Tan S, Brisson A, Boilard E, Martel C. Extracellular vesicles are present in mouse lymph and their level differs in atherosclerosis. *J Extracell Vesicles* (2016) 5:31427. doi:10.1038/srep27862

Pease DC, Baker RF. Sectioning techniques for electron microscopy using a conventional microtome. *Exp Biol Med* (1948) 67:470–4. doi:10.3181/ 00379727-67-16344

Fernández-Morán H. A diamond knife for ultrathin sectioning. *Exp Cell Res* (1953) 5:255–6. doi:10.1016/0014-4827(53)90112-8

Sjöstrand FS. A new microtome for ultrathin sectioning for high res- olution electron microscopy. *Experientia* (1953) 9:114–5. doi:10.1007/ BF02178346

Fernández-Morán H. Applications of a diamond knife for ultrathin section- ing to the study of the fine structure of biological tissues and metals. *J Biophys Biochem Cytol* (1956) 2:29–30. doi:10.1083/jcb.2.4.29

Tremblay M-È, Riad M, Majewska A. Preparation of mouse brain tissue for immunoelectron microscopy. *J Vis Exp* (2010) (41):e2021. doi:10.3791/2021

Winey M, Meehl JB, O'Toole ET, Giddings TH. Conventional transmis- sion electron microscopy. *Mol Biol Cell* (2014) 25:319–23. doi:10.1083/ jcb.148.4.635

Bisht K, Hajj El H, Savage JC, Sánchez MG, Tremblay M-È. Correlative light and electron microscopy to study microglial interactions with β-amyloid plaques. *J Vis Exp* (2016) (112):e54060. doi:10.3791/54060

Schultz RL, MAYNARD EA, Pease DC. Electron microscopy of neurons and neuroglia of cerebral cortex and corpus callosum. *Am J Anat* (1957) 100:369–407. doi:10.1002/aja.1001000305

Blinzinger K, Kreutzberg G. Displacement of synaptic terminals from regen- erating motoneurons by microglial cells. *Z Zellforsch Mikrosk Anat* (1968) 85:145–57. doi:10.1007/BF00325030

Giulian D, Baker TJ. Characterization of ameboid microglia isolated from developing mammalian brain. *J Neurosci* (1986) 6:2163–78. doi:10.1523/ JNEUROSCI.06-08-02163.1986

Mori S, Leblond CP. Identification of microglia in light and electron micros- copy. *J Comp Neurol* (1969) 135:57–80. doi:10.1002/cne.901350104

García-Cabezas MÁ, John YJ, Barbas H, Zikopoulos B. Distinction of neurons, glia and endothelial cells in the cerebral cortex: an algorithm based on cytological features. *Front Neuroanat* (2016) 10:107. doi:10.3389/ fnana.2016.00107

Herndon RM. The fine structure of the rat cerebellum. II The stellate neurons, granule cells, and glia. *J Cell Biol* (1964) 23:277–93. doi:10.1083/jcb.23.2.277

Shapiro LA, Perez ZD, Foresti ML, Arisi GM, Ribak CE. Morphological and ultrastructural features of Iba1-immunolabeled microglial cells in the hippocampal dentate gyrus. *Brain Res* (2009) 1266:29–36. doi:10.1016/j.brainres.2009.02.031

Peinado MA, Quesada A, Pedrosa JA, Torres MI, Martinez M, Esteban FJ, et al. Quantitative and ultrastructural changes in glia and pericytes in the parietal cortex of the aging rat. *Microsc Res Tech* (1998) 43:34–42. doi:10.1002/ (SICI)1097-0029(19981001)43:1<34:AID-JEMT6>3.0.CO;2-G

Murabe Y, Sano Y. Morphological studies on neuroglia. V. Microglial cells in the cerebral cortex of the rat, with special reference to their possible involve- ment in synaptic function. *Cell Tissue Res* (1982) 223:493–506.

Almolda B, González B, Castellano B. Microglia detection by enzymatic histochemistry. In: Joseph B, Venero JL, editors. *Microglia: Methods and Protocols*. Totowa, NJ: Humana Press (2013). p. 243–59.

Tremblay M-È, Zhang I, Bisht K, Savage JC, Lecours C, Parent M, et al. Remodeling of lipid bodies by docosahexaenoic acid in activated microglial cells. *J Neuroinflammation* (2016) 13:116. doi:10.1186/s12974-016-0580-0

Ito D, Imai Y, Ohsawa K, Nakajima K, Fukuuchi Y, Kohsaka S. Microglia- specific localisation of a novel calcium binding protein, Iba1. *Brain Res Mol Brain Res* (1998) 57:1–9. doi:10.1016/S0169-328X(98)00040-0

Wake H, Moorhouse AJ, Jinno S, Kohsaka S, Nabekura J. Resting microglia directly monitor the functional state of synapses in vivo and determine the fate of ischemic terminals. *J Neurosci* (2009) 29:3974–80. doi:10.1523/JNEUROSCI.4363-08.2009

Watts RJ, Schuldiner O, Perrino J, Larsen C, Luo L. Glia engulf degenerating axons during developmental axon pruning. *Curr Biol* (2004) 14:678–84. doi:10.1016/j.cub.2004.03.035

van Ham TJ, Brady CA, Kalicharan RD, Oosterhof N, Kuipers J, Veenstra- Algra A, et al. Intravital correlated microscopy reveals differential macro- phage and microglial dynamics during resolution of neuroinflammation. *Dis Model Mech* (2014) 7:857–69. doi:10.1242/dmm.014886

Paolicelli RC, Bolasco G, Pagani F, Maggi L, Scianni M, Panzanelli P, et al. Synaptic pruning by microglia is necessary for normal brain development. *Science* (2011) 333:1456–8. doi:10.1126/science.1202529

Schafer DP, Lehrman EK, Kautzman AG, Koyama R, Mardinly AR, Yamasaki R,   et al. Microglia sculpt postnatal neural circuits in an activity and complement-dependent manner. *Neuron* (2012) 74:691–705. doi:10.1016/j.neuron.2012.03.026

Squarzoni P, Oller G, Hoeffel G, Pont-Lezica L, Rostaing P, Low D, et al. Microglia modulate wiring of the embryonic forebrain. *Cell Rep* (2014) 8:1271–9. doi:10.1016/j.celrep.2014.07.042

Sisková Z, Page A, O'Connor V, Perry VH. Degenerating synaptic boutons in prion disease. *Am J Pathol* (2009) 175:1610–21. doi:10.2353/ ajpath.2009.090372

Seeger G, Härtig W, Rossner S, Schliebs R, Brückner G, Bigl V, et al. Electron microscopic evidence for microglial phagocytic activity and cholinergic cell death after administration of the immunotoxin 192IgG-saporin in rat. *J Neurosci Res* (1997) 48:465–76. doi:10.1002/(SICI)1097-4547(19970601) 48:5<465::AID-JNR7>3.0.CO;2-C

Chung W-S, Clarke LE, Wang GX, Stafford BK, Sher A, Chakraborty C, et al. Astrocytes mediate synapse elimination through MEGF10 and MERTK pathways. *Nature* (2013) 504:394–400. doi:10.1038/nature12776

Bisht K, Sharma KP, Lecours C, Gabriela Sánchez M, Hajj El H, Milior G, et al. Dark microglia: a new phenotype predominantly associated with pathological states. *Glia* (2016) 64:826–39. doi:10.1002/glia.22966

Hui C-W, St-Pierre A, Hajj El H, Remy Y, Hebert SS, Luheshi G, et al. Prenatal immune challenge in mice leads to partly sex-dependent behavioral, microg- lial, and molecular abnormalities associated with schizophrenia. *Front Mol Neurosci* (2018) 11:13. doi:10.3389/fnmol.2018.00013

Streit WJ, Xue Q-S, Tischer J, Bechmann I. Microglial pathology. *Acta Neuropathol Commun* (2014) 2:142. doi:10.1186/s40478-014-0142-6

Milior G, Lecours C, Samson L, Bisht K, Poggini S, Pagani F, et al. Fractalkine receptor deficiency impairs microglia and neuronal responsiveness to chronic stress. *Brain Behav Immun* (2016) 55:114–25. doi:10.1016/j.bbi.2015.07.024

Tseng CY, Ling EA, Wong WC. Light and electron microscopic and cyto- chemical identification of amoeboid microglial cells in the brain of prenatal rats. *J Anat* (1983) 136:837–49.

Le Guellec D. Ultrastructural in situ hybridization: a review of technical aspects. *Biol Cell* (1998) 90:297–306. doi:10.1111/j.1768-322X.1998.tb01040.x

Cubas-Nuñez L, Duran-Moreno M, Castillo-Villalba J, Fuentes-Maestre J, Casanova B, García-Verdugo JM, et al. In situ RT-PCR optimized for electron microscopy allows description of subcellular morphology of target mRNA-expressing cells in the brain. *Front Cell Neurosci* (2017) 11:141. doi:10.3389/fncel.2017.00141

Tischer J, Krueger M, Mueller W, Staszewski O, Prinz M, Streit WJ, et al. Inhomogeneous distribution of Iba-1 characterizes microglial pathology in Alzheimer's disease. *Glia* (2016) 64:1562–72. doi:10.1002/glia.23024

Vernon-Parry KD. Scanning electron microscopy: an introduction. *III-Vs Review* (2000) 13:40–4. doi:10.1016/S0961-1290(00)80006 X

Micheva KD, Smith SJ. Array tomography: a new tool for imaging the molecular architecture and ultrastructure of neural circuits. *Neuron* (2007) 55:25–36. doi:10.1016/j.neuron.2007.06.014

Knott G, Marchman H, Wall D, Lich B. Serial section scanning electron microscopy of adult brain tissue using focused ion beam milling. *J Neurosci* (2008) 28:2959–64. doi:10.1523/JNEUROSCI.3189-07.2008

Maco B, Cantoni M, Holtmaat A, Kreshuk A, Hamprecht FA, Knott GW. Semiautomated correlative 3D electron microscopy of in vivo-im- aged axons and dendrites. *Nat Protoc* (2014) 9:1354–66. doi:10.1038/ nprot.2014.101

Thompson RF, Walker M, Siebert CA, Muench SP, Ranson NA. An intro- duction to sample preparation and imaging by cryo-electron microscopy for structural biology. *Methods* (2016) 100:3–15. doi:10.1016/j.ymeth. 2016.02.017

van Rijnsoever C, Oorschot V, Klumperman J. Correlative light-electron microscopy (CLEM) combining live-cell imaging and immunolabeling of ultrathin cryosections. *Nat Meth* (2008) 5:973–80. doi:10.1038/nmeth.1263

Knott GW, Holtmaat A, Wilbrecht L, Welker E, Svoboda K. Spine growth precedes synapse formation in the adult neocortex in vivo. *Nat Neurosci* (2006) 9:1117–24. doi:10.1038/nn1747

Majewska AK, Lamantia CE, Kelly EA, Sipe GO, Tremblay M-E, Lowery RL. Microglial P2Y12 is necessary for synaptic plasticity in mouse visual cortex. *Nat Commun* (2016) 7:10905. doi:10.1038/ncomms10905

Vasek MJ, Garber C, Dorsey D, Durrant DM, Bollman B, Soung A, et al. A complement–microglial axis drives synapse loss during virus-induced memory impairment. *Nature* (2016) 534:538–43. doi:10.1038/nature18283

Bechmann I, Nitsch R. Astrocytes and microglial cells incorporate degenerating fibers following entorhinal lesion: a light, confocal, and electron microscopical study using a phagocytosis-dependent labeling technique. *Glia* (1997) 20:145–54. doi:10.1002/(SICI)1098-1136(199706) 20:2<145:AID-GLIA6>3.0.CO;2-8

Ibiricu I, Huiskonen JT, Döhner K, Bradke F, Sodeik B, Grünewald K. Cryo electron tomography of herpes simplex virus during axonal transport and secondary envelopment in primary neurons. *PLoS Pathog* (2011) 7:e1002406. doi:10.1371/journal.ppat.1002406.s002

Fernandez-Fernandez MR, Ruiz-Garcia D, Martin-Solana E, Chichon FJ, Carrascosa JL, Fernandez J-J. 3D electron tomography of brain tissue unveils distinct Golgi structures that sequester cytoplasmic contents in neurons. *J Cell Sci* (2017) 130:83–9. doi:10.1242/jcs.188060

Weil M-T, Ruhwedel T, Möbius W, Simons M. Intracerebral injections and ultrastructural analysis of high-pressure frozen brain tissue. *Curr Protoc Neurosci* (2017) 78:2.27.1–2.27.18. doi:10.1002/cpns.22

Korogod N, Petersen CCH, Knott GW. Ultrastructural analysis of adult mouse neocortex comparing aldehyde perfusion with cryo fixation. *Elife* (2015) 4. doi:10.7554/eLife.05793

Fiala JC, Harris KM. Extending unbiased stereology of brain ultrastructure to three-dimensional volumes. *J Am Med Inform Assoc* (2001) 8:1–16. doi:10.1136/jamia.2001.0080001

Denk W, Horstmann H. Serial block-face scanning electron microscopy to reconstruct three-dimensional tissue nanostructure. *PLoS Biol* (2004) 2:e329. doi:10.1371/journal.pbio.0020329.s002

Peddie CJ, Collinson LM. Exploring the third dimension: volume elec- tron microscopy comes of age. *Micron* (2014) 61:9–19. doi:10.1016/j.micron.2014.01.009

Chen Z, Jalabi W, Hu W, Park H-J, Gale JT, Kidd GJ, et al. Microglial displace- ment of inhibitory synapses provides neuroprotection in the adult brain. *Nat Commun* (2014) 5:4486. doi:10.1038/ncomms5486

Yamasaki R, Lu H, Butovsky O, Ohno N, Rietsch AM, Cialic R, et al. Differential roles of microglia and monocytes in the inflamed central nervous system. *J Exp Med* (2014) 211:1533–49. doi:10.1182/blood-2009-02-200543

Heymann JAW, Hayles M, Gestmann I, Giannuzzi LA, Lich B, Subramaniam S. Site-specific 3D imaging of cells and tissues with a dual beam microscope. *J Struct Biol* (2006) 155:63–73. doi:10.1016/j.jsb.2006.03.006

Wolpers C. Electron microscopy in Berlin 1928–1945. *Adv Electron Electron Phys* (1991) 81:211–29. doi:10.1016/S0065-2539(08)60866-5

Farquhar MG, Hartmann JF. Neuroglial structure and relationships as revealed by electron microscopy. *J Neuropathol Exp Neurol* (1957) 16:18–39. doi:10.1097/00005072-195701000-00003

De Rosier DJ, Klug A. Reconstruction of three dimensional structures from electron micrographs. *Nature* (1968) 217:130. doi:10.1038/217130a0

Nobelprize.org. *Press Release: The 1986 Nobel Prize in Physics*. Nobel Media AB (2014). Available from: http://www.nobelprize.org/nobel_prizes/physics/laureates/1986/press.html (Accessed: April 6, 2018).

Henderson R, Baldwin JM, Ceska TA, Zemlin F, Beckmann E, Downing KH. Model for the structure of bacteriorhodopsin based on high-resolution electron cryo-microscopy. *J Mol Biol* (1990) 213:899–929. doi:10.1016/S0022-2836(05)80271-2

Humphry MJ, Kraus B, Hurst AC, Maiden AM, Rodenburg JM. Ptychographic electron microscopy using high-angle dark-field scattering for sub-nanometre resolution imaging. *Nat Commun* (2012) 3:730. doi:10.1038/ ncomms1733

Nobelprize.org. *The 2017 Nobel Prize in Chemistry – Press Release*. Nobel Media AB (2014). Available from: http://www.nobelprize.org/nobel_prizes/ chemistry/laureates/2017/press.html (Accessed: April 6, 2018).

18

Enhancement of Astroglial Aerobic Glycolysis by Extracellular Lactate-Mediated Increase in cAMP

Nina Vardjan[1,2]*, Helena H. Chowdhury[1,2], Anemari Horvat[1], Jelena Velebit[1,2],
Maja Malnar[2], Marko Muhič[1], Marko Kreft[1,2,3], Špela G. Krivec[1,2], Saša T. Bobnar[1,2],
Katarina Miš[4], Sergej Pirkmajer[4], Stefan Offermanns[5], Gjermund Henriksen[6,7],
Jon Storm-Mathisen[8], Linda H. Bergersen[9,10] and Robert Zorec[1,2]*

[1] Laboratory of Neuroendocrinology - Molecular Cell Physiology, Institute of Pathophysiology, Faculty of Medicine, University of Ljubljana, Ljubljana, Slovenia, [2] Laboratory of Cell Engineering, Celica Biomedical, Ljubljana, Slovenia, [3] Department of Biology, Biotechnical Faculty, University of Ljubljana, Ljubljana, Slovenia, [4] Laboratory for Molecular Neurobiology, Institute of Pathophysiology, Faculty of Medicine, University of Ljubljana, Ljubljana, Slovenia, [5] Department of Pharmacology, Max Planck Institute for Heart and Lung Research, Bad Nauheim, Germany, [6] Nuclear and Energy Physics, Department of Physics, The Faculty of Mathematics and Natural Sciences, University of Oslo, Oslo, Norway, [7] Norwegian Medical Cyclotron Centre Ltd., Oslo, Norway, [8] Division of Anatomy, Department of Molecular Medicine, CMBN/SERTA Healthy Brain Ageing Centre, Institute of Basic Medical Sciences, Faculty of Medicine, University of Oslo, Oslo, Norway, [9] Institute of Oral Biology, Faculty of Dentistry, University of Oslo, Oslo, Norway, [10] Center for Healthy Aging, Faculty of Health and Medical Sciences, University of Copenhagen, Copenhagen, Denmark

*Correspondence:
Nina Vardjan
nina.vardjan@mf.uni-lj.si
Robert Zorec
robert.zorec@mf.uni-lj.si

Besides being a neuronal fuel, L-lactate is also a signal in the brain. Whether extracellular L-lactate affects brain metabolism, in particular astrocytes, abundant neuroglial cells, which produce L-lactate in aerobic glycolysis, is unclear. Recent studies suggested that astrocytes express low levels of the L-lactate GPR81 receptor ($EC_{50} \approx 5$ mM) that is in fat cells part of an autocrine loop, in which the G_i-protein mediates reduction of cytosolic cyclic adenosine monophosphate (cAMP). To study whether a similar signaling loop is present in astrocytes, affecting aerobic glycolysis, we measured the cytosolic levels of cAMP, D-glucose and L-lactate in single astrocytes using fluorescence resonance energy transfer (FRET)-based nanosensors. In contrast to the situation in fat cells, stimulation by extracellular L-lactate and the selective GPR81 agonists, 3-chloro-5-hydroxybenzoic acid (3Cl-5OH-BA) or 4-methyl-N-(5-(2-(4-methylpiperazin-1-yl)-2-oxoethyl)-4-(2-thienyl)-1,3-thiazol-2-yl)cyclohexanecarboxamide (Compound 2), like adrenergic stimulation, elevated intracellular cAMP and L-lactate in astrocytes, which was reduced by the inhibition of adenylate cyclase. Surprisingly, 3Cl-5OH-BA and Compound 2 increased cytosolic cAMP also in GPR81-knock out astrocytes, indicating that the effect is GPR81-independent and mediated by a novel, yet unidentified, excitatory L-lactate receptor-like mechanism in astrocytes that enhances aerobic glycolysis and L-lactate production via a positive feedback mechanism.

Keywords: astrocytes, aerobic glycolysis, L-lactate receptor, cAMP, L-lactate

INTRODUCTION

Aerobic glycolysis, non-oxidative metabolism of glucose despite the presence of adequate levels of oxygen, a phenomenon termed "the Warburg effect," is an inefficient way to generate energy in the form of ATP. The advantage of this process, however, appears to be in providing intermediates for the biosynthesis of lipids, nucleic acids, and amino acids (Vander Heiden et al., 2009).

These are needed for making new cells during division, such as in cancer cells and in developing normal cells, and for cells engaged in morphological reshaping, such as neural cells in the central nervous system (CNS; Goyal et al., 2014).

At least in the brain, aerobic glycolysis appears to be regulated. For example, during alerting, sensory stimulation, exercise, and pathophysiological conditions, L-lactate production and release are upregulated. Although it is still unclear how this takes place at the cellular level, the process likely involves noradrenergic neurons from *locus coeruleus* (LC; Dienel and Cruz, 2016). L-lactate can be produced from glycogen stored in astrocytes, an abundant glial cell type in the CNS, and can be used by neurons and oligodendrocytes, which seem to require L-lactate in addition to D-glucose for their optimal function, including memory formation (Cowan et al., 2010; Barros, 2013; Gao et al., 2016; Dong et al., 2017), myelin production, and the sustenance of long axons (Lee et al., 2012; Rinholm and Bergersen, 2012). In addition, L-lactate is considered to be neuroprotective against various types of brain damage (Castillo et al., 2015) and is required for cancer cell survival (Roland et al., 2014).

These effects suggest that L-lactate not only acts as a fuel but also has extracellular signaling roles (Chen et al., 2005; Barros, 2013; Roland et al., 2014). L-lactate modulates the activity of primary cortical neurons through a receptor-mediated pathway (Bozzo et al., 2013), and the activation of astrocytes by LC neurons results in the release of L-lactate, which back-excites LC neurons and stimulates the further release of noradrenaline (NA; Tang et al., 2014). These effects are supported by the observation that the monocarboxylate transporter 2 (MCT2), transporting L-lactate, is selectively co-located with glutamate receptors at the postsynaptic membranes of fast-acting excitatory synapses (Bergersen et al., 2001). Interestingly, L-lactate appears to promote gene expression that mediates N-methyl-D-aspartate (NMDA)-related neuronal plasticity (Yang et al., 2014) and the expression of membrane metabolite receptors at the plasma membrane (Castillo et al., 2015). L-lactate is known to mediate cerebral vasodilatation, causing increased brain blood flow (Gordon et al., 2007, 2008).

The notion of multiple signaling roles of L-lactate and its widespread diffusion in tissues led to the concept of L-lactate being a "volume transmitter" of metabolic information (Chen et al., 2005) and perhaps also a gliotransmitter (Tang et al., 2014) in the brain.

L-lactate signaling may occur through several mechanisms, including the modulation of prostaglandin action (Gordon et al., 2008), redox regulation (Brooks, 2009), and the activation of the L-lactate-sensitive receptors, such as G_i-protein coupled receptors GPR81 (Lauritzen et al., 2014) or the yet unidentified plasma membrane receptors (Tang et al., 2014).

The L-lactate selectivity of the GPR81 receptor ($EC_{50} \approx 5$ mM for rat GPR81; Liu et al., 2009), also known as hydroxycarboxylic acid receptor 1 (HCA$_1$ or HCAR1), was discovered in adipose tissues (reported EC_{50} range from 1 to 5 mM; Cai et al., 2008; Liu et al., 2009), where GPR81 is highly expressed and down-regulates the formation of cytosolic cyclic adenosine monophosphate ($[cAMP]_i$) by coupling to G_i-protein and inhibiting the cAMP producing enzyme adenylate cyclase (AC),

thereby in an autocrine loop inhibiting lipolysis and promoting energy storage (Ahmed et al., 2010). Whether a similar signaling loop is present in astrocytes is not known and was explored in this study. Quantitative RT-PCR analysis of human, rat, and mouse brain tissue revealed the presence of mRNA GPR81 in the brain, although at very low levels compared to adipose tissue (Liu et al., 2009). Consistent with RT-PCR analysis, RNA sequencing transcriptome databases revealed the presence of GPR81 mRNA in individual types of mouse brain cells, including astrocytes (Zhang et al., 2014; Sharma et al., 2015). Immunohistochemical studies on mouse brain tissue slices also suggested the presence of the GPR81 receptor in neurons, endothelial cells, and at low density of expression in astrocytes, in particular in membranes of perivascular astrocytic processes and not so much in perisynaptic processes of astrocytes (Lauritzen et al., 2014; Morland et al., 2017). The mechanism of how L-lactate via activation of a GPR81 receptor would modulate $[cAMP]_i$, cytosolic levels of D-glucose ($[glucose]_i$) and L-lactate ($[lactate]_i$) in brain, in particular in astrocytes that are actively involved in the regulation of brain metabolism and produce L-lactate, is currently unknown.

In contrast to the situation in adipocytes, where the GPR81 receptor agonist decreases $[cAMP]_i$ (Ahmed et al., 2010), we show here by fluorescence resonance energy transfer (FRET) microscopy on single astrocytes expressing FRET-nanosensors for cAMP and L-lactate that very high levels of extracellular L-lactate and the agonist for the L-lactate GPR81 receptor 3-chloro-5-hydroxybenzoic acid (3Cl-5OH-BA; Dvorak et al., 2012) elevate $[cAMP]_i$ and $[lactate]_i$ in astrocytes, as does the activation of adrenergic receptors (ARs). The 3Cl-5OH-BA-dependent elevation in $[lactate]_i$ and the extracellular L-lactate-mediated rise in $[cAMP]_i$, both act through the activation of AC in astrocytes, as demonstrated by the use of an AC inhibitor. Interestingly, in astrocytes from the L-lactate specific GPR81 receptor knock-out (KO) mice, 3Cl-5OH-BA still elevated $[cAMP]_i$, indicating that the supposedly selective GPR81 agonist also activates a second, yet unidentified excitatory L-lactate receptor-like mechanism. Pretreatment of rat astrocytes with the sub-effective doses of 3Cl-5OH-BA reduced the L-lactate-induced elevation in $[cAMP]_i$, suggesting that 3Cl-5OH-BA and L-lactate at least to some extent bind to the same yet unidentified receptor. A new generation GPR81 selective high affinity agonist lead compound, Compound 2 (4-methyl-N-(5-(2-(4-methylpiperazin-1-yl)-2-oxoethyl)-4-(2-thienyl)-1,3-thiazol-2-yl)cyclohexanecarboxamide; Sakurai et al., 2014) reproduced the cAMP enhancing effects of L-lactate and 3Cl-5OH-BA, further supporting the existence of an unidentified L-lactate receptor. The new excitatory L-lactate receptor-mediated mechanism ("metabolic excitability") may participate in maintaining high $[lactate]_i$ in cells exhibiting aerobic glycolysis, such as in astrocytes (Mächler et al., 2016), contributing to the elevated levels of extracellular L-lactate in comparison to the plasma levels (Abi-Saab et al., 2002). While generating metabolic intermediates required for cell division and morphological plasticity, this regulation presumably facilitates the exit of L-lactate into the extracellular space, where it can become an autocrine and paracrine signal. Since relatively high concentrations of L-lactate (20 mM, but not 2 mM) are required for the increase in second messenger cAMP (the predicted brain physiological concentrations of L-lactate are up to 2 mM), the putative novel facilitatory L-lactate receptor-like mechanism may have a role under conditions of very high extracellular L-lactate that may occur during extreme exercise (Osnes and Hermansen, 1972; Matsui et al., 2017) or neuronal activity (for example during seizures; During et al., 1994). Such a scenario may also be relevant if local fluctuations of extracellular L-lactate concentration exist in brain microdomains (Morland et al., 2015; Mosienko et al., 2015).

MATERIALS AND METHODS

Cell Culture, Transfection, and Reagents

Primary cultures of astrocytes were prepared from cortices of 2–3 and 1–4 days old rat and wild type (WT) or GPR81 KO mice pups, respectively, as described previously (Schwartz and Wilson, 1992), and grown in high-glucose (25 mM) Dulbecco's Modified Eagle's Medium containing 10% fetal bovine serum, 1 mM sodium pyruvate, 2 mM L-glutamine, and 25 µg/ml penicillin-streptomycin at 37°C in 95% air–5% CO_2 until reaching 70–80% confluency. Confluent cultures were shaken at 225 rpm overnight, and the medium was changed the next morning; this was repeated three times. After the third overnight shaking, the cells were trypsinized and put in flat tissue culture tubes with 10 cm^2 growth area. After reaching confluency, the cells were trypsinized and plated on 22-mm diameter glass cover slips coated with poly- L-lysine. By using quantitative PCR, we verified that astrocytes from the GPR81 KO animals were devoid of the GPR81 RNA transcript. The purity of rat astrocytes was determined immunocytochemically using antibodies against the astrocytic marker GFAP (Abcam, Cambridge, United Kingdom) and >94% of imaged cultured cells were GFAP-positive (Stenovec et al., 2016). 3T3-L1 fibroblasts (ATCC-LGC Standards, VA, United States) were grown in high-glucose Dulbecco's Modified Eagle's Medium containing 10% fetal bovine serum and 2 mM L-glutamine. BT474 cancer cells (BT-474 Clone 5; ATCC® CRL-3247™; ATCC-LGC Standards) were grown in Hybri-Care medium (ATCC® 46-X™; ATCC-LGC Standards) supplemented with 1.5 g/L sodium bicarbonate and 10% fetal bovine serum. 3T3-L1 and BT474 cells were grown in culture flasks with 25 cm^2 growth area at 37°C in 95% air–5% CO_2. After reaching 70–80% confluency, cells were trypsinized and plated on 22-mm diameter glass cover slips coated with poly- L-lysine.

After 24–72 h, transfection with plasmids carrying FRET-based nanosensors was performed with FuGENE® 6 transfection reagent according to manufacturer's instructions (Promega Co., Madison, WI, United States). Transfection medium contained no serum or antibiotics.

GPR81 KO and WT mice (Ahmed et al., 2010) in C57Bl/6N background were bred in Oslo from founder mice obtained from Stefan Offermanns, Max-Planck-Institute for Heart and Lung Research, Department of Pharmacology, D-61231 Bad Nauheim, Germany. The experimental animals were cared for in accordance with the International Guiding Principles for Biomedical Research Involving Animals developed by the Council for International Organizations of Medical Sciences and the Animal Protection Act (Official Gazette RS, No. 38/13). The experimental protocol was approved by The Administration of the Republic of Slovenia for Food Safety, Veterinary and Plant Protection (Republic of Slovenia, Ministry of Agriculture, Forestry and Food, Ljubljana), Document No. U34401-47/2014/7, signed by Barbara Tomše, DVM. All experiments were performed on rat astrocytes isolated from at least two different animals. All chemicals were from Sigma-Aldrich (St. Louis, MO, USA) unless otherwise noted.

FRET Measurements of [cAMP]$_i$ and PKA Activity

Cells expressing the FRET-based cAMP nanosensor Epac1-camps (Nikolaev et al., 2004) or the PKA nanosensor AKAR2 (Zhang et al., 2005) were examined 24–48 h after transfection with a Plan NeoFluoar 40×/1.3 NA oil differential interference contrast (DIC) immersion objective (Carl Zeiss, Jena, Germany) using a Zeiss LSM510 META confocal microscope (Carl Zeiss, Jena, Germany). Cells were excited at 458 nm, and images (512 × 512) were acquired every 3.5 s using lambda stack acquisition. Emission spectra were collected from the META detector in 8 channels (lambda stack) ranging from 470 nm to 545 nm, each with a 10.7-nm width. Two-channel [cyan fluorescent protein (CFP) and yellow fluorescent protein (YFP)] images were generated using the analytical software Extract channels (Zeiss LSM510 META, Carl Zeiss, Jena, Germany). Channels with emission spectra at 470 and 481 nm and emission spectra at 513, 524, and 534 nm were extracted to the CFP and YFP channels, respectively. YFP and CFP fluorescence intensities were quantified within a region of interest (ROI) selected for individual cells expressing Epac1-camps or AKAR2 using the LSM510 META software.

In some experiments, astrocytes expressing the Epac1-camps FRET nanosensor, were examined 24 h after transfection with a fluorescence microscope Zeiss Axio Observer.A1 (Zeiss, Oberkochen, Germany) with a CCD camera and monochromator Polychrome V (Till Photonics, Graefelfing, Germany) as a monochromatic source of light with a wavelength 436 nm/10 nm. Dual emission intensity ratios were recorded using an image splitter (Optical Insights, Tucson, AZ, United States) and two emissions filters (465/30 nm for CFP and 535/30 nm for YFP). Images were acquired every 3.5 s with an exposure time of 0.1 s. CFP and YFP fluorescence intensities were obtained from the integration of ROI over the entire cell using Life Acquisition software (Till Photonics, Graefelfing, Germany).

In the graphs, the FRET ratio signal was reported as the ratio of the CFP/YFP (Epac1-camps) and YFP/CFP (AKAR2) fluorescence signals after subtracting the background fluorescence from the signals using Excel (Microsoft, Seattle, WA, United States). The values of the FRET signals were normalized to 1.0. An increase in the FRET ratio signal reflects an increase in the [cAMP]$_i$ and the PKA activity.

Initially, astrocytes were kept in extracellular solution (10 mM Hepes/NaOH, pH 7.2, 10 mM D-glucose, 131.8 mM NaCl, 1.8 mM $CaCl_2$, 2 mM $MgCl_2$, and 5 mM KCl) or extracellular solution with sodium bicarbonate (10 mM Hepes/NaOH, pH 7.2, 10 mM D-glucose, 131.8 mM NaCl, 1.8 mM CaCl₂, 2 mM MgCl₂, 5 mM KCl, 0.5 mM NaH₂PO₄ × H₂O, and 5 mM NaHCO₃) and were then treated with various reagents following a 100-s baseline: 1 μM NA, 10 μM isoprenaline (ISO), 2 or 20 mM L-lactate (osmolality was adjusted by lowering NaCl in extracellular solution containing sodium bicarbonate) and 0.5 mM 3Cl-5OH-BA. In some experiments cell were after initial a 100-s baseline pretreated with 100 μM 2',5'-dideoxyadenosine (DDA) or 50 μM 3Cl-5OH-BA for 450 s and then stimulated with 20 mM L-lactate in the presence of DDA or 3Cl-5OH-BA, respectively. The 4-methyl-N-(5-(2-(4-methylpiperazin-1-yl)-2-oxoethyl)-4-(2-thienyl)-1,3-thiazol-2-yl)cyclohexanecarboxamide (Compound 2; Sakurai et al., 2014) was custom synthesized (ABX advanced biochemical compounds, D-01454 Radeberg, Germany). After the initial 90–100-s baseline, WT and GPR81 KO astrocytes were treated with GPR81 receptor agonist 3Cl-5OH-BA (0.5 mM) or Compound 2 (50 nM) and recorded for another 300 s. In control experiments, astrocytes were treated only with extracellular solution (Vehicle). Extracellular solution osmolality was ~300 mOsm, measured with a freezing point osmometer (Osmomat030, Gonotech GmbH, Germany).

FRET Measurements of [glucose]$_i$ and [lactate]$_i$

Astrocytes, 3T3-L1 embryonic preadipocyte fibroblast cells and BT474 cancer cells expressing the FRET-based glucose nanosensor FLII[12]Pglu-700 μδ6[1] (Takanaga et al., 2008; Prebil et al., 2011) or the FRET-based lactate nanosensor Laconic (San Martin et al., 2013) were examined 16–48 h after transfection with a fluorescence microscope (Zeiss Axio Observer.AI, Zeiss, Oberkochen, Germany), with a CCD camera and monochromator Polychrome V (Till Photonics, Graefelfing, Germany) as a monochromatic source of light with a wavelength 436 nm/10 nm. Dual emission intensity ratios were recorded using an image splitter (Optical Insights, Tucson, AZ, United States) and two emission filters; 465/30 nm for ECFP or mTFP and 535/30 nm for EYFP or Venus. Images were acquired every 10 s with an exposure time of 0.1 s. The background fluorescence was subtracted from individual EYFP or Venus and ECFP or mTFP fluorescence signals. The FRET ratio signals, EYFP/ECFP (FLII[12]Pglu-700 μδ6) and mTFP/Venus (Laconic), were obtained from the integration of the ratio signal over the entire cell using Life Acquisition software (Till Photonics, Graefelfing, Germany). The values of the FRET signals were normalized to 1.0. An increase in the FRET ratio signal reflects increases in the [glucose]$_i$ and [lactate]$_i$.

Initially, cells were kept in extracellular solution with sodium bicarbonate (10 mM Hepes/NaOH, pH 7.2, 3 mM D-glucose, 135.3 mM NaCl, 1.8 mM CaCl₂, 2 mM MgCl₂, 5 mM KCl, 0.5 mM NaH₂PO₄ × H₂O, and 5 mM NaHCO₃), and were then treated with various reagents following a 200-s baseline: 200 μM ISO, 200 μM NA, 2 or 20 mM L-lactate, 0.5 mM 3Cl-5OH-BA, and 6 mM α-cyano-4-hydroxycinnamate (CHC). In some experiments, cells were, after an initial 200

---

[1] www.addgene.org

s-baseline, pretreated with 100 μM DDA for 450 s and then stimulated with 0.5 mM 3Cl-5OH-BA in the presence of DDA. Extracellular solution osmolality was ~300 mOsm, measured with a freezing point osmometer (Osmomat030, Gonotech GmbH, Germany).

Extraction of mRNA and Quantitative Real-Time PCR (qPCR)

Rat, WT mouse, and GPR81 KO mouse astrocytes were cultured in six-well plates. Total RNA was extracted from cultured astrocytes with the RNeasy Mini Plus Kit (Qiagen, Hilden, Germany). cDNA was synthesized from total RNA using the High-Capacity cDNA Reverse Transcription Kit (Applied Biosystems, Thermo Fisher Scientific, Vilnius, Lithuania). qPCR was performed on ABI PRISM SDS 7500 (Applied Biosystems, Thermo Fisher Scientific) in a 96-well format using TaqMan Universal PCR Master Mix (Applied Biosystems, Thermo Fisher Scientific, Foster City, United States) and gene expression assays for GPR81 (Rn03037047_sH) and 18S rRNA (TaqMan Endogenous Control). Standard quality controls were performed in line with the MIQE Guidelines (Bustin et al., 2009). Expression level of GPR81 mRNA was calculated as gene expression ratio (GPR81 mRNA/18S rRNA) according to the equation: $E_{18S\ rRNA}^{Ct,18S\ rRNA} / E_{GPR81}^{Ct,GPR81}$, where E is the PCR efficiency and C_t is the threshold cycle for the reference (18S rRNA) or the target (GPR81) gene (Ruijter et al., 2009; Tuomi et al., 2010). PCR efficiency was estimated using the LinRegPCR program (Ruijter et al., 2009; Tuomi et al., 2010).

Statistical Analysis

Single-exponential increase to maximum functions [$F = F_0 + c \times (1 - \exp(-t/\tau))$] were fitted to the diagrams with FRET ratio signals using SigmaPlot. The time constant (τ) and the FRET ratio signal amplitudes (c) were determined from the fitted curves. F is the FRET ratio signal at time t, F_0 is the baseline FRET ratio signal, c is FRET ratio signal amplitude of $F - F_0$, and τ is the time constant of the individual exponential component. In some experiments, the initial rate of the FRET signal increase (ΔFRET/Δtime) after the addition of various reagents was calculated as the slope of the linear regression function [ΔFRET (%) = slope (%/min) × Δtime (min)] fitting the initial FRET signal decrease or increase. In these experiments, the amplitude of the ΔFRET (%) was determined by subtracting the mean FRET ratio signal of the first 100 s from the last 100 s upon stimulation or *vice versa*, if the FRET signal increased.

The average traces of the predominant responses are presented in the figures for individual stimuli; all other responses are listed in **Table 1**. Unless stated otherwise, the Student's *t*-test was performed to determine statistical significance. $P < 0.05$ was considered to be significant.

RESULTS

We studied here how extracellular L-lactate and agonists of the L-lactate GPR81 receptor affect cAMP signaling and

TABLE 1 | Responsiveness of astrocytes to adrenergic and L-lactate receptor activation.

FRET nanosensor		Stimulus	n (%) increase	n (%) decrease	n (%) transient decrease	n (%) unresponsive	n all
Rat astrocytes							
cAMP		ISO (10 μM)	8 (100%)	0 (0%)	0 (0%)	0 (0%)	8
		NA (1 μM)	16 (100%)	0 (0%)	0 (0%)	0 (0%)	16
		L-lactate (20 mM)	7 (36.8%)	3 (15.8%)	0 (0%)	9 (47.4%)	19
		3Cl-5OH-BA (0.5 mM)	9 (42.9%)	2 (9.5%)	0 (0%)	10 (47.6%)	21
PKA		L-lactate (2 mM)	0 (0%)	0 (0%)	0 (0%)	7 (100%)	7
		L-lactate (20 mM)	8 (57.1%)	1 (7.1%)	0 (0%)	5 (35.7%)	14
		3Cl-5OH-BA (0.5 mM)	9 (56.3%)	2 (12.5%)	0 (0%)	5 (31.3%)	16
Glucose		ISO (200 μM)	5 (16.1%)	0 (%)	0 (0%)	26 (83.9%)	31
		NA (200 μM)	8 (40%)	0 (%)	0 (0%)	12 (60%)	20
		L-lactate (2 mM)	0 (0%)	0 (0%)	0 (0%)	11 (100%)	11
		L-lactate (20 mM)	3 (15%)	4 (20.0%)	13 (65.0%)	0 (0%)	20
		3Cl-5OH-BA (0.5 mM)	1 (5.3%)	13 (68.4%)	1 (5.3%)	4 (21.1%)	19
Lactate		ISO (200 μM)	10 (63%)	0 (0%)	0 (0%)	6 (27%)	16
		NA (200 μM)	8 (88.9%)	0 (0%)	0 (0%)	1 (11.1%)	9
		L-lactate (20 mM)	14 (93.3%)	0 (0%)	0 (0%)	1 (6.7%)	15
		3Cl-5OH-BA (0.5 mM)	14 (100%)	0 (0%)	0 (0%)	0 (0%)	14
Mouse astrocytes							
cAMP	WT	3Cl-5OH-BA (0.5 mM)	11 (68.8)	0 (0)	0 (0)	5 (31.2)	16
		Compound 2 (50 nM)	10 (66.7%)	0 (0)	0 (0)	5 (33.3%)	15
	KO GPR81	3Cl-5OH-BA (0.5 mM)	11 (73.3)	0 (0)	0 (0)	4 (26.7)	15
		Compound 2 (50 nM)	11 (100%)	0 (0)	0 (0)	0 (0)	11

ISO, isoprenaline; NA, noradrenaline; 3Cl-5OH-BA, 3-chloro-5-hydroxybenzoic acid; Compound 2, 4-methyl-N-(5-(2-(4-methylpiperazin-1-yl)-2-oxoethyl)-4-(2-thienyl)-1,3-thiazol-2-yl)cyclohexanecarboxamide; n, number of cells studied. The increase or decrease in the FRET ratio signal was set by determining a threshold of 3 standard deviations of the average signal measured prior to the application of agents (baseline), as described (Horvat et al., 2016). The frequency of unresponsive cells did not differ in KO group compared to WT group, when treated with 3Cl-5OH-BA (P = 0.78). The frequency of unresponsive cells was significantly lower in KO group compared to WT group, when treated with Compound 2 (P = 0.03; chi-square statistic).

changes in the intracellular levels of metabolites in isolated astrocytes, which can express the GPR81 receptor (Lauritzen et al., 2014; Sharma et al., 2015) and mRNA for GPR81 can be measured in these cells (Supplementary Figure S1). Real-time FRET imaging of cultured cortical astrocytes, expressing FRET-nanosensors for cAMP, glucose, and lactate was performed.

Extracellular L-Lactate and GPR81 Lactate Receptor Agonists Increase [cAMP]$_i$ in Astrocytes

We performed real-time monitoring of [cAMP]$_i$ while extracellular L-lactate or the GPR81 lactate receptor agonist 3Cl-5OH-BA (Dvorak et al., 2012; Liu et al., 2012) were applied. Using the FRET-nanosensor for cAMP Epac1-camps, in 7 (37%) out of 19 cells an increase in [cAMP]$_i$ was measured when a high concentration of L-lactate (20 mM) was applied (**Figure 1A**). Decreased [cAMP]$_i$ was measured only in 3 (16%) cells out of 19 (**Table 1**). Although GPR81 receptor in adipocytes is coupled to G$_i$-proteins decreasing [cAMP]$_i$ (Ahmed et al., 2010), the observed L-lactate-mediated elevation in [cAMP]$_i$ in 37% of cells studied might be due to the activation of a G$_s$-protein coupled to L-lactate GPR81 receptor, as a similar rise in [cAMP]$_i$ was recorded in astrocytes by

adding the selective GPR81 receptor agonist 3Cl-5OH-BA (0.5 mM, **Figure 1B**); 9 (43%) of 21 cells responded with an increase (**Figure 1B**) and 2 (10%) with a small decrease in [cAMP]$_i$ (**Table 1**). Inhibition of AC activity by 100 μM DDA (Vardjan et al., 2014) reduced the 20 mM extracellular L-lactate-induced increase in [cAMP]$_i$ by ~50 % (**Figure 3A**), indicating that in the majority of astrocytes, extracellular L-lactate activates the cAMP pathways via binding to receptors that activate AC.

To verify whether the 3Cl-5OH-BA-induced increase in [cAMP]$_i$ is mediated via the GPR81 lactate receptor, we used isolated astrocytes from the GPR81 KO mice (Ahmed et al., 2010). Interestingly, even in GPR81 KO astrocytes, the application of 3Cl-5OH-BA (0.5 mM), like in WT astrocytes, elicited an increase in [cAMP]$_i$ (**Figure 2** and **Table 1**). Similar results were obtained by using a much higher affinity GPR81 receptor agonist Compound 2 (Sakurai et al., 2014; histograms in **Figures 2C,D**; **Table 1**), suggesting that these agonists activate in astrocytes a second, yet unidentified receptor-like mechanism, thus resembling the unidentified L-lactate receptor in neurons, coupled to G$_s$-protein increasing the production of cAMP (Tang et al., 2014).

To see if L-lactate and GPR81 agonist 3Cl-5OH-BA target the same, yet unidentified receptor-like mechanism, sharing the binding on the receptor, we pretreated rat astrocytes with

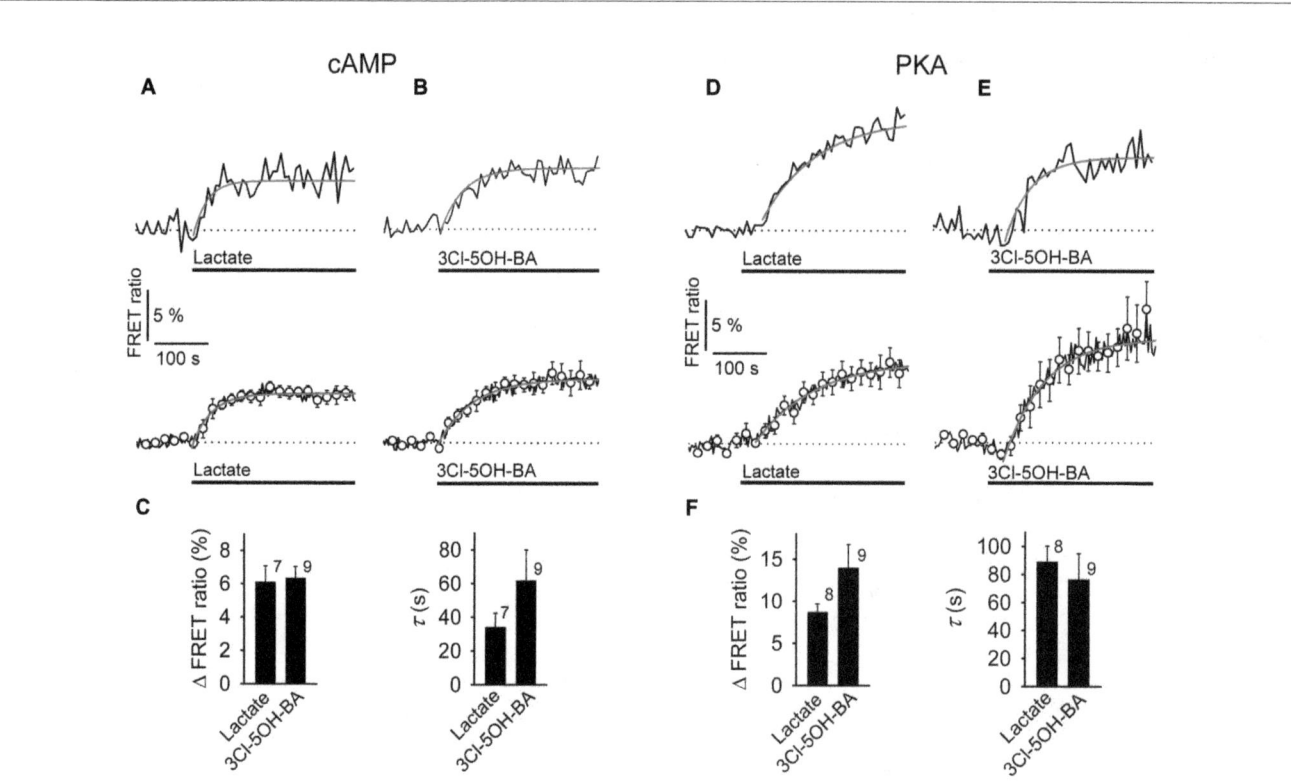

FIGURE 1 | Application of L-lactate and 3Cl-5OH-BA increases [cAMP]$_i$ and the intracellular PKA activity in rat astrocytes. **(A,B)** Representative (above) and mean time-course (below) of the Epac1-camps FRET ratio signal upon the addition of **(A)** L-lactate (20 mM) and **(B)** 3Cl-5OH-BA (0.5 mM), a selective agonist of the GPR81 receptor. Data are expressed as percentages of the inverse FRET ratio signal (CFP/YFP) relative to the baseline values. Single exponential rise functions were fitted to the curves: (**A**, above) FRET ratio = [0.98 ± 0.01] + [0.06 ± 0.01] × (1 − exp(−t/[25.00 ± 5.63 s])), (**A**, below) FRET ratio = [1.00 ± 0.00] + [0.05 ± 0.00] × (1 − exp(−t/[35.71 ± 2.55 s])), and (**B**, above) FRET ratio = [0.94 ± 0.00] + [0.06 ± 0.01] × (1 − exp(−t/[41.49 ± 5.85 s])), (**B**, below) FRET ratio = [1.00 ± 0.00] + [0.07 ± 0.00] × (1 − exp(t/[66.67 ± 4.42 s])). Note that the addition of L-lactate and 3Cl-5OH-BA increased the FRET ratios, indicating increases in [cAMP]$_i$. Each data point represents the mean ± s.e.m. **(C)** Mean changes in the FRET ratio (ΔFRET ratio) and mean time-constants (τ) upon L-lactate and 3Cl-5OH-BA stimulation. Changes in FRET ratio are expressed as percentages relative to the initial values. The numbers by the bars depict the number of cells analyzed. Data are shown as the means ± s.e.m. **(D,E)** Representative (above) and mean time-course (below) of the AKAR2 FRET ratio signal upon addition of **(D)** L-lactate (20 mM) and **(E)** 3Cl-5OH-BA (0.5 mM). Data are expressed as percentages of the FRET ratio signals (YFP/CFP) relative to the baseline ratio values. Single exponential rise functions were fitted to the curves: (**A**, above) FRET ratio = [1.01 ± 0.00] + [0.10 ± 0.00] × (1 − exp(−t/[90.91 ± 8.26 s])), (**A**, below) FRET ratio = [1.00 ± 0.00] + [0.08 ± 0.00] × (1 − exp(−t/[111.11 ± 9.91 s])), and (**B**, above) FRET ratio = [0.97 ± 0.01] + [0.10 ± 0.01] × (1 − exp(−t/[47.62 ± 4.54 s])), (**B**, below) FRET ratio = [0.98 ± 0.00] + [0.12 ± 0.00] × (1 − exp(−t/[71.43 ± 5.10 s])). Note that the addition of L-lactate and 3Cl-5OH-BA increased the FRET ratios, indicating increased PKA activity. Each data point represents the mean ± s.e.m. **(F)** Mean changes in the FRET ratio (ΔFRET ratio) and mean time-constants (τ) upon L-lactate and 3Cl-5OH-BA stimulation. The numbers by the bars depict the number of cells analyzed. Data shown are in the format of the mean ± s.e.m. Data for every set of experiment was acquired from at least two different animals.

the sub-effective dose of GPR81 receptor agonist 3Cl-5OH-BA (50 μM). The application of 50 μM 3Cl-5OH-BA is ineffective in increasing [cAMP]$_i$ in astrocytes, since the application of 3Cl-5OH-BA results in a FRET ratio signal change (1.63 ± 0.38 %, n = 13), not significantly different to the vehicle-induced response in controls (1.01 ± 0.46 %, n = 13, P = 0.31). However, the sub-effective dose of 3Cl-5OH-BA (50 μM) reduced the L-lactate-induced elevation in [cAMP]$_i$ (Supplementary Figure S4), indicating that in addition to interacting with the GPR81 lactate receptor, this supposedly selective GPR81 receptor agonist (Dvorak et al., 2012), binds like L-lactate to the yet unidentified receptor in astrocytes.

Furthermore, to evaluate whether the L-lactate receptor-like mechanism elevates [cAMP]$_i$ and consequently also activates the cAMP-effector protein kinase A (PKA) we used the AKAR2,

PKA activation nanosensor (Zhang et al., 2005). Significant increase in PKA activity was recorded (**Figures 1D,E**) when astrocytes were exposed to 20 mM L-lactate in 57% of 14 cells or to 0.5 mM 3Cl-5OH-BA in 56% of 16 cells (**Table 1**). The differences in the responsiveness of cells between Epac1-camps and AKAR2 nanosensors (∼40% vs. ∼60%, respectively) may be due to a higher sensitivity of the AKAR2 nanosensor, since Epac1-camps can only detect [cAMP]$_i$ that is >100 nM (Börner et al., 2011). No changes in PKA activity were observed, when a 10-fold lower concentration of L-lactate (2 mM) was applied to cells (Supplementary Figure S5). The time-course and the extent of increase in the PKA activity were similar for both types of stimuli, L-lactate and 3Cl-5OH-BA (**Figure 1F**). We observed a delay between the addition of L-lactate or 3Cl-5OH-BA and subsequent increase in AKAR2 FRET ratio signal,

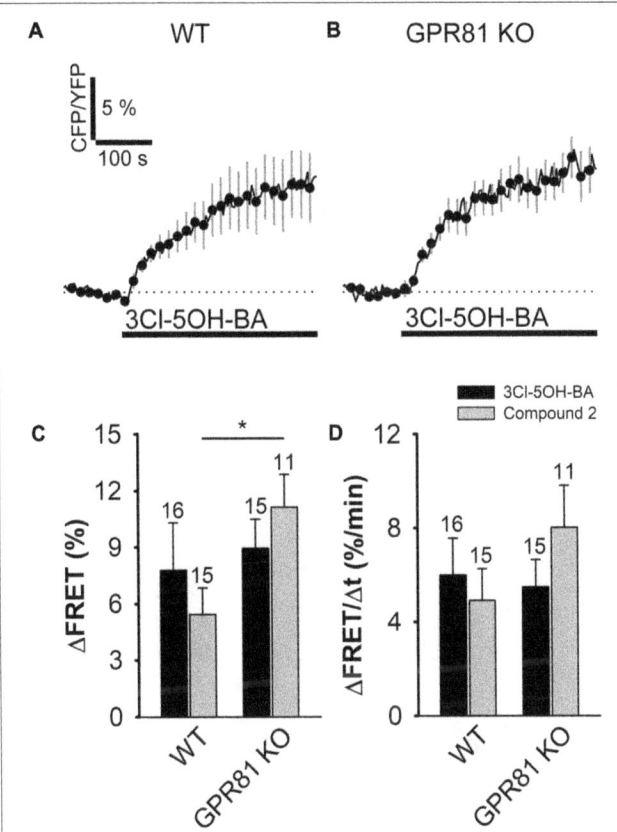

FIGURE 2 | Stimulation with 3Cl-5OH-BA and Compound 2 elicits persistent increase in [cAMP]$_i$ in WT and GPR81 KO mice astrocytes expressing the FRET nanosensor Epac1-camps. **(A,B)** Mean normalized time-courses of the Epac1-camps FRET ratio signal after stimulation with 3Cl-5OH-BA (0.5 mM) in **(A)** WT (n = 16) and **(B)** GPR81 KO (n = 15) cultured mouse astrocytes at t = 100 s (black lines). Changes in the FRET ratio signal are expressed as percentages of the inverse FRET ratio signal (CFP/YFP) relative to the baseline values. The monophasic exponential increase in the FRET signal (represented by the CFP/YFP ratio) after 3Cl-5OH-BA stimulation reflects an increase in [cAMP]$_i$. Each data point represents mean ± s.e.m. **(C,D)** Mean changes in the Epac1-camps FRET ratio signal (ΔFRET; **C**) and initial rates of the FRET ratio signal increase (ΔFRET/Δt; **D**) after the addition of 3Cl-5OH-BA (black bars) and Compound 2 (gray bars) in WT and GPR81 KO astrocytes. Numbers by the error bars depict the number of cells analyzed. (*P < 0.05; Mann Whitney U test) Data for every set of experiment was acquired from at least two different animals except for Compound 2.

determined from the intersection of the reference baseline with the exponential curve (time of delay was 35 ± 7 s (20 mM L-lactate) vs. 49 ± 7 s (0.5 mM 3Cl-5OH-BA); P = 0.19), consistent with PKA activation occurring downstream of cAMP production.

The observation that a 10-fold lower concentrations of 3Cl-5OH-BA (50 μM vs. 500 μM) and L-lactate (2 mM vs. 20 mM) does not affect cAMP levels and PKA activity in astrocytes, respectively, indicates that astrocytes respond to extracellular 3Cl-5OH-BA and L-lactate in a concentration-dependent manner, further suggesting that these two agonists in astrocytes cause a rise in cAMP via activation of an excitatory receptor-like mechanism in the plasma membrane.

Extracellular L-Lactate and GPR81 Receptor Agonist Trigger Similar Increases in [lactate]$_i$ in Astrocytes and Cancer Cells as Adrenergic Stimulation via Adenylate Cyclase Activation

Brain-based aerobic glycolysis (Goyal et al., 2014) is upregulated by noradrenergic stimuli affecting astrocyte metabolism (Dienel and Cruz, 2016). It has been reported that AR stimulation elevates astrocytic [cAMP]$_i$ (Vardjan et al., 2014; Horvat et al., 2016) and also intracellular concentration in free glucose ([glucose]$_i$; Prebil et al., 2011). We show here using lactate FRET nanosensor Laconic (San Martin et al., 2013) that astrocytes respond to AR activation also with an increase in intracellular L-lactate (Supplementary Figure S2) from its basal to the new steady state levels.

Although we have observed that in all studied astrocytes exposed to ISO (n = 8) and NA (n = 16) the [cAMP]$_i$ increased, the elevations in [glucose]$_i$ were only rarely observed, when cells were treated with selective β-AR agonist ISO [5 (16%) of the 31 cells]. The percentage of responsive cells, however, increased by twofold, when cells were exposed to NA, nonselective AR agonist [8 (40%) of the 20 cells; **Table 1**] and when cells were treated with selective α$_1$-AR agonist (data not shown), implying the role of Ca^{2+} in the elevation of [glucose]$_i$. In contrast to measurements of [glucose]$_i$, the majority of astrocytes (63%) responded to ISO application with an elevation in [lactate]$_i$ (10 out of 16 cells); 89% of astrocytes (8 out of 9 cells) responded with enhanced [lactate]$_i$ to stimulation by NA (Supplementary Figure S2C, **Table 1**). The time-course of the measured [cAMP]$_i$ increase was faster compared to that of [glucose]$_i$ and [lactate]$_i$ (Supplementary Figure S3).

Since L-lactate and GPR81 agonist 3Cl-5OH-BA (**Figure 1**), like β-AR agonists (Supplementary Figure S2; Vardjan et al., 2014; Horvat et al., 2016) increase [cAMP]$_i$ and PKA activity in astrocytes, we further examined whether extracellular L-lactate or the GPR81 agonist 3Cl-5OH-BA affect [glucose]$_i$ and [lactate]$_i$ in astrocytes. **Figure 4** shows that both extracellular L-lactate (20 mM) and 3Cl-5OH-BA (0.5 mM) decreased [glucose]$_i$ in the majority of studied cells. In the former case, the decrease was transient (65% of 20 cells responded), but the change was persistent in the latter case, where 68% of 19 cells responded. Interestingly, these treatments elevated [lactate]$_i$ in 93% of 15 cells (L-lactate) and in all 14 (100%) cells in case of 3Cl-5OH-BA (**Figures 3B** (left panel), **5**). The increase in [lactate]$_i$ was faster when the cells were exposed to extracellular L-lactate than 3Cl-5OH-BA, likely due to the entry of L-lactate into the cytosol through the plasma membrane MCTs. The extent of [lactate]$_i$ elevation was more than twofold higher upon the addition of L-lactate vs. 3Cl-5OH-BA (**Figure 5C**), which was expected due to L-lactate entry into the cytoplasm via the MCTs and possibly channels (Sotelo-Hitschfeld et al., 2015).

A few cells responded with a [glucose]$_i$ increase to L-lactate or 3Cl-5OH-BA stimulation [3 (15%) and 1 (5%) cells, respectively], but no cell responded with a decrease in [lactate]$_i$ to these stimuli.

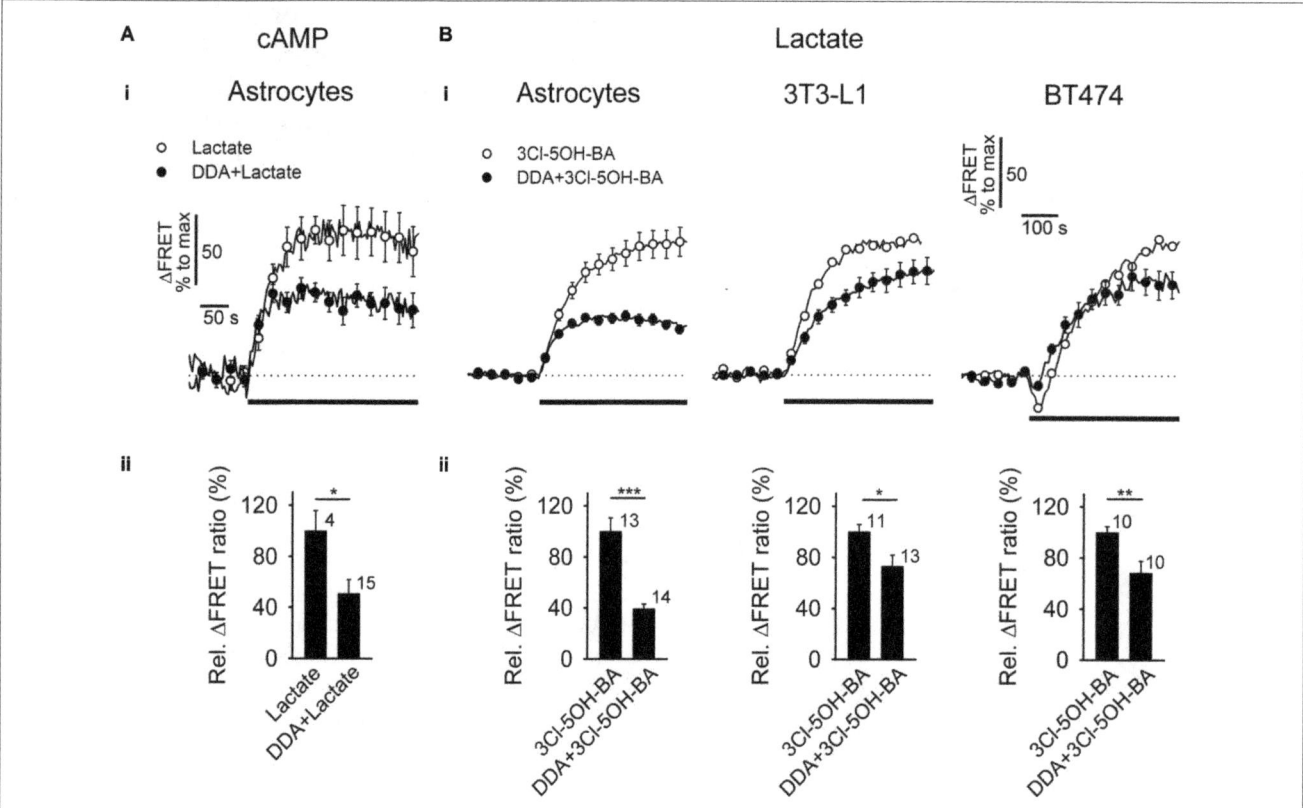

FIGURE 3 | Inhibition of AC reduces the L-lactate- and 3Cl-5OH-BA-induced increase in [cAMP]$_i$ and [lactate]$_i$ **(A,B,** panels **i)** Mean time-courses of FRET ratio signal changes normalized to the maximum signal change for **(A,i)** Epac1-camps upon addition of 20 mM L-lactate and **(B,i)** Laconic upon addition of 0.5 mM 3Cl-5OH-BA (black lines) in the absence (white circles) and presence (black circles) of 100 μM DDA, an inhibitor of AC. Each data point represents mean ± s.e.m. **(A,B,** panels **ii)** Mean relative changes in FRET ratio (Rel. ΔFRET ratio) upon L-lactate **(A)** and **(B)** 3Cl-5OH-BA stimulation in the absence and presence of DDA. Relative ΔFRET values (%) were calculated by dividing individual ΔFRET values with the average ΔFRET value upon L-lactate or 3Cl-5OH-BA stimulation. Note that the inhibition of AC by DDA causes ~50 % reduction in L-lactate-induced increase in [cAMP]$_i$ in astrocytes and a ~30–60% reduction in 3Cl-5OH-BA-induced increase in [lactate]$_i$ in astrocytes, 3T3-L1 and BT474 cells. In BT474 cells the application of 3Cl-5OH-BA initiated a transient reduction in [lactate]$_i$ that was diminished in the presence of DDA. Numbers adjacent to black bars represent number of cells. Data are in the format of the mean ± s.e.m (*P < 0.05, **P < 0.01, ***P < 0.001). Data for every set of experiment was acquired from at least two different animals.

As in the case of the cAMP-dependent PKA activity ($n = 7$), the addition of 2 mM L-lactate had no significant effect on [glucose]$_i$ ($n = 11$; Supplementary Figure S5), further indicating that astrocytes respond to extracellular L-lactate in a concentration-dependent manner.

Taken together these results show that in astrocytes, extracellular L-lactate and 3Cl-5OH-BA trigger an elevation in [lactate]$_i$, likely via the cAMP-mediated activation of glucose consumption, as decreased [glucose]$_i$ was recorded in the majority of the 3Cl-5OH-BA-treated cells. Moreover, inhibition of AC by 100 μM DDA (Vardjan et al., 2014), reduced the 3Cl-5OH-BA-induced increase in [lactate]$_i$ in astrocytes. We have observed inhibition of 3Cl-5OH-BA-induced increase in [lactate]$_i$ with DDA also in 3T3-L1 embryonic murine cells with genetic trait as seen in human ectodermal cancers (Leibiger et al., 2013) and in human mammary ductal carcinoma BT474 cells **(Figure 3B)**. Both cell lines were shown to be positive for the GPR81 receptor (Liu et al., 2009; Staubert et al., 2015), indicating that AC activation by a membrane receptor is responsible for the increase in [lactate]$_i$ in cells that exhibit aerobic

glycolysis such as astrocytes and cancer cells (Supplementary Table S1).

Receptor-Regulated Increase in the Rate of Aerobic Glycolysis in Astrocytes

Aerobic glycolysis in the brain is likely mediated by astrocytes, since these cells strongly favor L-lactate as the end glycolytic product whether or not oxygen is present (Halim et al., 2010). To measure the glycolytic rate in these cells, we estimated the maximal initial rate of [lactate]$_i$ increase in the presence of CHC (6 mM), a non-specific inhibitor of membrane MCTs (MCTs 1-4). In the absence of L-lactate exchange through the plasma membrane, the predominant pathway available for cytosolic L-lactate accumulation is considered to be glycolytic L-lactate production. Although cytosolic L-lactate may be transported to mitochondria for oxidation in astrocytes (Passarella et al., 2008), in the presence of MCT blockers the initial rate of cytosolic L-lactate accumulation, measured as ΔFRET/Δt, is comparable to that of the initial rate of aerobic glycolysis. **Figure 6A,** ii

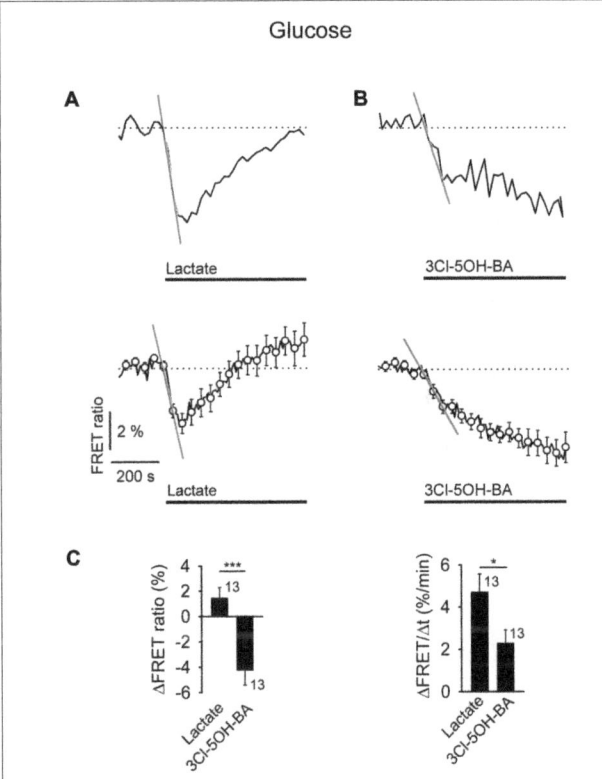

FIGURE 4 | Application of L-lactate and 3Cl-5OH-BA decreases [glucose]$_i$ in astrocytes. **(A,B)** Representative (above) and mean time-course (below) of FLII^{12}Pglu-700$\mu\delta$6 FRET ratio signal, reporting [glucose]$_i$, upon the addition of **(A)** L-lactate (20 mM) and **(B)** 3Cl-5OH-BA (0.5 mM). Data are expressed as the percentages of the FRET ratio signals (EYFP/ECFP) relative to the baseline ratio values. The initial rate of change in [glucose]$_i$ was determined by fitting the regression lines to the signal decreases; ΔFRET (%) = $-k$ (%/min) $\times$ Δt (min), where t is the time and the k–slope is the initial ΔFRET decline: $k_{(A,above)}$ = 4.94 $\pm$ 0.28 %/min, $k_{(A,below)}$ = 2.90 $\pm$ 0.36 %/min, and $k_{(B,above)}$ = 2.32 $\pm$ 0.29 %/min, $k_{(B,below)}$ = 1.05 $\pm$ 0.11 %/min. Note that the addition of L-lactate transiently decreased the FRET ratio, while the 3Cl-5OH-BA persistently decreased the FRET ratio. Each data point represents the mean $\pm$ s.e.m. **(C)** Mean changes in the FRET ratio (ΔFRET ratio; left) and initial rates in the FRET ratio decrease (ΔFRET/Δt; right) upon L-lactate and 3Cl-5OH-BA stimulation. Numbers adjacent to the bars depict the number of cells analyzed. Data shown are in the format of the mean $\pm$ s.e.m (*$P < 0.05$, ***$P < 0.001$). Data for every set of experiment was acquired from at least two different animals.

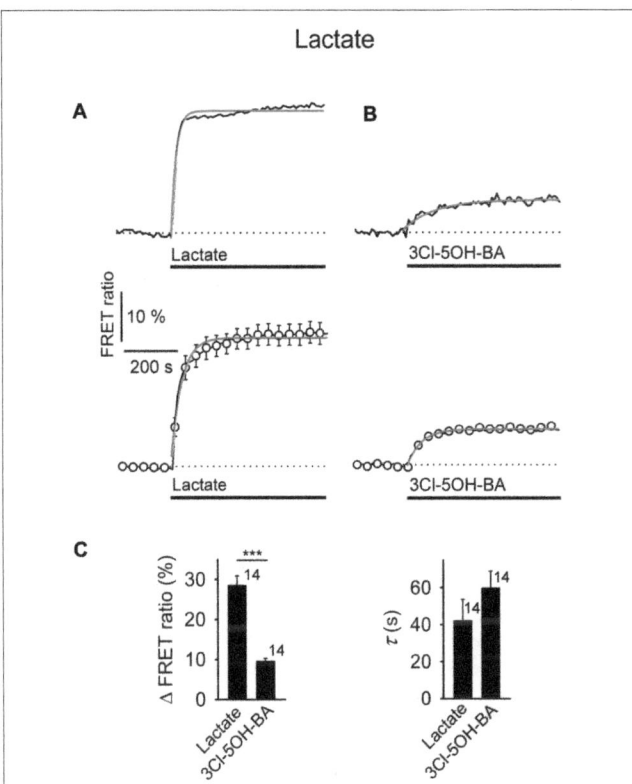

FIGURE 5 | Comparison of increases in [lactate]$_i$ by extracellular L-lactate and 3Cl-5OH-BA in astrocytes. **(A,B)** Representative (above) and mean time-course (below) of the Laconic FRET ratio signal upon the addition of **(A)** L-lactate (20 mM) and **(B)** 3Cl-5OH-BA (0.5 mM). Data are expressed as the percentages of FRET ratio signals (mTFP/Venus) relative to the baseline ratio values. Single exponential rise functions were fitted to the curves: **(A**, above) FRET ratio = [0.81 $\pm$ 0.02] + [0.44 $\pm$ 0.02] $\times$ (1 $-$ exp ($-t$/[18.80 $\pm$ 1.10 s])), **(A**, below) FRET ratio = [1.02 $\pm$ 0.01] + [0.24 $\pm$ 0.01] $\times$ (1 $-$ exp($-t$ / [39.06 $\pm$ 2.44 s])), and **(B**, above) FRET ratio = [1.00 $\pm$ 0.00] + [0.06 $\pm$ 0.00] $\times$ (1 $-$ exp($-t$/[116.28 $\pm$ 14.87 s])), **(B**, below) FRET ratio = [0.98 $\pm$ 0.00] + [0.09 $\pm$ 0.00] $\times$ (1 $-$ exp($-t$/[54.05 $\pm$ 2.34 s])). Note that the addition of L-lactate and 3Cl-5OH-BA increased the FRET ratios, indicating an increase in [lactate]$_i$. Each data point represents the mean $\pm$ s.e.m. **(C)** Mean changes in the FRET ratio (ΔFRET ratio; left) and mean time-constants (τ; right) upon L-lactate and 3Cl-5OH-BA stimulation. Numbers by the error bars depict the number of cells analyzed. Data shown are in the format of the mean $\pm$ s.e.m (***$P < 0.001$). Data for every set of experiment was acquired from at least two different animals.

shows that upon the addition of CHC, the initial maximal rate of the [lactate]$_i$ increase was 3.2 $\pm$ 0.6 %/min ($P < 0.001$), which was consistent with previous results (Sotelo-Hitschfeld et al., 2015). It appears that this effect is not an artifact, as vehicle addition did not affect the [lactate]$_i$ (**Figure 6B**; $P = 0.18$). With the addition of 3Cl-5OH-BA (0.5 mM), the rate of [lactate]$_i$ increase was approximately twofold higher than the resting rate of glycolysis. NA (200 μM) appeared as a weaker stimulus of [lactate]$_i$ increase in comparison to 3Cl-5OH-BA (0.5 mM) (**Figure 6B**) at the concentrations of respective agonists used. The strongest effect on the measured rate in the [lactate]$_i$ increase was recorded when extracellular L-lactate (20 mM) was applied. This was expected, as L-lactate can enter the cytoplasm via the MCTs. Consistent with this

finding, when L-lactate was added in the presence of the MCT blocker CHC, the increase in [lactate]$_i$ was strongly attenuated in comparison to the conditions in the absence of CHC (data not shown).

Taken together, the results indicate that in astrocytes, extracellular L-lactate mediates an increase in [cAMP]$_i$ that stimulates aerobic glycolysis and elevates [lactate]$_i$ via a receptor-mediated pathway involving AC.

DISCUSSION

Here we studied whether extracellular L-lactate affects aerobic glycolysis via G-protein coupled L-lactate receptors, such as

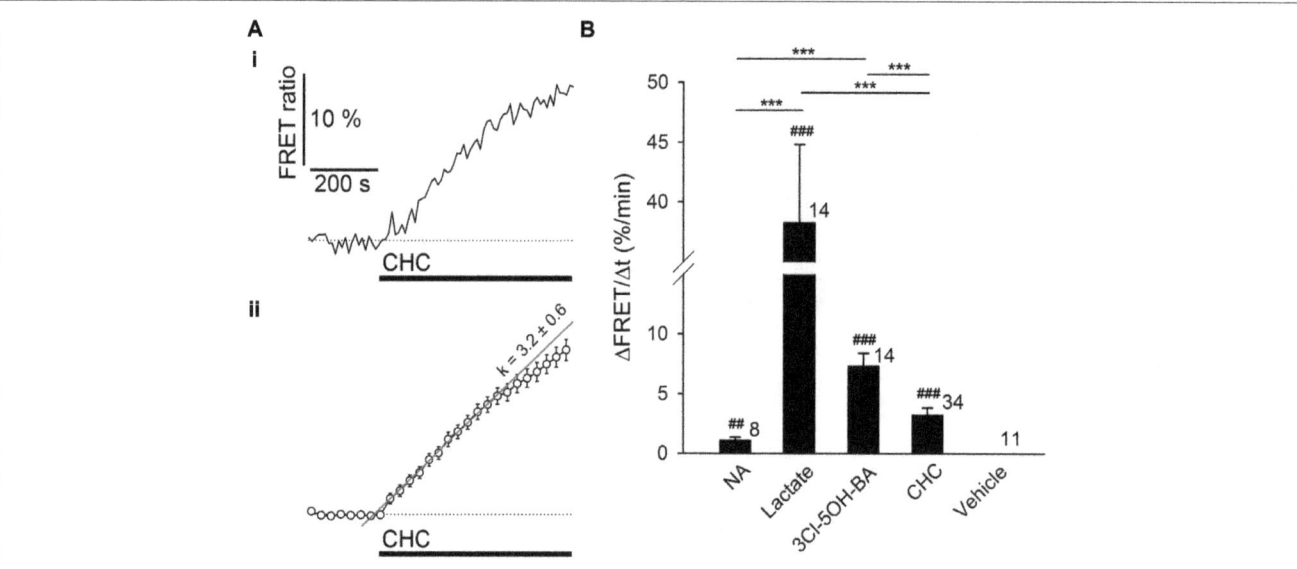

FIGURE 6 | The resting rate of glycolysis and lactate production is modulated by noradrenaline and 3Cl-5OH-BA in astrocytes. **(A)** Representative **(i)** and mean **(ii)** time-course of Laconic FRET ratio signal upon addition of 6 mM CHC. Data are expressed as the percentage of the FRET ratio signal (mTFP/Venus) relative to the baseline ratio values. Note that the increased FRET ratio indicates an increase in [lactate]$_i$. Each data point in **(ii)** represents mean ± s.e.m. The initial rate of change in [lactate]$_i$ was determined by fitting the regression lines to the initial rise of the curves. The regression line in **(ii)** has the form of: ΔFRET (%) = k (%/min) × Δt (min) + ΔFRET$_0$, where t is time, ΔFRET$_0$ is ΔFRET at the time of stimulus, and the k-slope is the initial ΔFRET rise: k = 3.2 ± 0.6 %/min. **(B)** Comparison of mean maximal initial rates in FRET ratio changes (ΔFRET/Δt) upon the addition of NA (200 μM), L-lactate (20 mM), 3Cl-5OH-BA (0.5 mM), CHC (6 mM), and vehicle (control) in astrocytes. Maximal initial rates are expressed as percent change of FRET ratio per minute. Numbers by the error bars depict the number of cells analyzed. ***P < 0.001 one-way ANOVA comparison between different types of stimuli; ##P < 0.01, ###P < 0.001 one-sample t-test. Data for every set of experiment was acquired from at least two different animals.

the GPR81 receptor in astrocytes (Lauritzen et al., 2014; Sharma et al., 2015). The main finding of this study is that extracellular L-lactate in astrocytes activates AC, elevates [cAMP]$_i$, and accelerates aerobic glycolysis. Interestingly, by using the selective agonists for the GPR81 receptor, such as 3Cl-5OH-BA (Dvorak et al., 2012; Liu et al., 2012) or Compound 2 (Sakurai et al., 2014), the results revealed that even in the absence of the GPR81 receptor expression in astrocytes from GPR81 KO mice, elevations in [cAMP]$_i$ were still recorded, indicating that in addition to the GPR81 receptors, these agonists activate also a yet unidentified L-lactate receptor-like mechanism.

L-Lactate and Adrenergic Receptor Stimulation Increases [lactate]$_i$ in Astrocytes

Genes associated with aerobic glycolysis are mainly expressed in neocortical areas where neuronal cell plasticity is taking place (Goyal et al., 2014). In these areas, astrocytes are considered the primary site of aerobic glycolysis, and the primary source of L-lactate release. The results of this study demonstrate that the activation of astrocytic receptors by extracellular L-lactate (20 mM) or the GPR81 lactate receptor agonist 3Cl-5OH-BA (0.5 mM) (Dvorak et al., 2012) increases [lactate]$_i$ within 1 min to a relatively high and stable level.

If the L-lactate-mediated increase in [cAMP]$_i$ is present *in vivo* it may contribute to the maintenance of the L-lactate gradient

between astrocytes and neurons (Mächler et al., 2016) and also the gradient between the extracellular L-lactate in the brain and in the plasma (Abi-Saab et al., 2002). A positive feedback mechanism involving L-lactate as an extracellular signal that controls L-lactate production and subsequent L-lactate release from astrocytes (**Figure 7**) may be important during neuronal activity, when [lactate]$_i$ may rapidly decline in astrocytes due to its facilitated exit through lactate MCTs and putative ion channels (Sotelo-Hitschfeld et al., 2015) and/or its diffusion through the gap junctions in astroglial syncytia (Hertz et al., 2014). Thus, in the absence of a positive feedback mechanism, the L-lactate concentration gradient may dissipate, reducing the availability of L-lactate as a metabolic fuel for neurons (Barros, 2013; Mosienko et al., 2015) and thereby limiting the support for neural network activity.

Experiments in slices have revealed that optical activation of astrocytes in LC triggers the release of L-lactate from astrocytes, which then excites LC neurons and triggers the widespread release of NA, likely in a receptor-dependent manner involving AC and PKA activation (Tang et al., 2014). It has been proposed that astrocytes also release L-lactate at the axonal varicosities of the noradrenergic neurons, where it facilitates the release of NA from the varicosities (Tang et al., 2014). However, whether NA released from the LC neurons affects astrocyte L-lactate production has not been studied directly.

The real-time [lactate]$_i$ monitoring in this study revealed that AR activation by NA can elicit a sustained increase in

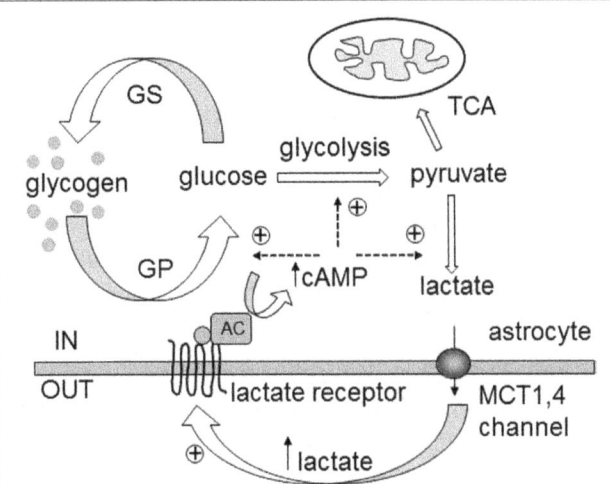

FIGURE 7 | Extracellular L-lactate controls cytosolic lactate synthesis via yet unidentified receptors coupled to adenylate cyclase activity and cytosolic cAMP increase in astrocytes. In the brain, L-lactate is formed in the cytoplasm of astrocytes (IN) and is released through monocarboxylate transporters (MCTs) 1,4 and/or lactate-permeable channels. Extracellularly (OUT), L-lactate can be transported to neighboring cells as a fuel. However, it can also act as a signaling molecule by binding to the L-lactate receptors of neighboring cells, stimulating adenylate cyclase activity (AC) and an increase in cAMP synthesis. Elevated cytosolic cAMP levels facilitate glycogen degradation by activating glycogen phosphorylase (GP) and glycolysis with L-lactate as the end product. In the absence of the L-lactate positive feedback mechanism ("metabolic excitability"), the L-lactate tissue concentration gradient could be dissipated, reducing the availability of L-lactate as a metabolic fuel, when local energy demands, especially in the brain, are high. This model shares similarities with the 'autocrine lactate loop' acting (oppositely) on [cAMP]$_i$ through GPR81 receptor in adipocytes (Ahmed et al., 2010). GS, glycogen synthase; TCA, tricarboxylic acid cycle. Glucose denotes phosphorylated and free glucose.

[lactate]$_i$ in astrocytes, with an average time constant of ~100 s (Supplementary Figure S2). The increase in [lactate]$_i$ is predominantly a consequence of increased L-lactate production in aerobic glycolysis, although changes in L-lactate fluxes across the plasma membrane or astrocytic consumption of L-lactate in oxidative metabolic pathway could also partly contribute to the new L-lactate steady state levels (San Martin et al., 2013).

The increased [lactate]$_i$ likely leads to L-lactate exiting astrocytes, which, *in vivo*, may back-excite LC neurons and further stimulate NA release (Tang et al., 2014). Such bidirectional communication between astrocytes and neurons may be part of a tissue-coordinated astrocyte activity, which can be studied by monitoring widespread LC-mediated Ca^{2+} signaling in astrocytes in awake-behaving mice *in vivo* (Ding et al., 2013; Paukert et al., 2014). Consistent with this possibility, cultured primary cortical neurons respond to extracellular L-lactate and the GPR81 lactate receptor agonists with a decrease in spontaneous Ca^{2+}-spiking activity that is concentration-dependent, with 50% inhibitory concentration (IC$_{50}$) of ~4 mM L-lactate (Bozzo et al., 2013), which is close to the sensitivity of the GPR81 receptor for L-lactate (Liu et al., 2012).

Extracellular L-Lactate-Induced Increase in [lactate]$_i$ Involves Receptor-Like Mediated Adenylate Cyclase Activation and cAMP Production

In astrocytes, similar to the activation of β-ARs (Vardjan et al., 2014; Horvat et al., 2016), extracellular L-lactate (20 mM) and GPR81 lactate receptor agonist 3Cl-5OH-BA (0.5 mM) elevate [cAMP]$_i$ (**Figures 1A,B**), as independently confirmed by monitoring the increase in activity of cAMP effector PKA (**Figures 1D,E**), but do not elevate [Ca^{2+}]$_i$ (data not shown). Thus, extracellular L-lactate in astrocytes likely stimulates a plasma membrane L-lactate sensitive receptor, such as the recently identified GPR81 receptor in brain astrocytes, which was linked to downregulation of cAMP synthesis (Lauritzen et al., 2014). To exclude the possibility that the addition of extracellular L-lactate acidifies the cytoplasm, thus affecting the fluorescence of fluorophores in a FRET nanosensor, especially the fluorescence of YFP (Patterson et al., 2001), which declines with acidification, producing an artifact that can be read as an increase in cAMP, we used two different types of FRET nanosensors to monitor cAMP activity; Epac1-camps (Nikolaev et al., 2004), and AKAR2 (Zhang et al., 2005), which report cAMP activity in opposite directions, as decrease and increase in YFP/CFP ratio, respectively. While in Epac1-camps transfected cells the L-lactate application induced a reduction in the YFP/CFP ratio, AKAR2-transfected cells responded to L-lactate with the increase in YFP/CFP ratio. Since the readouts from the Epac1-camps and AKAR2 YFP/CFP signals are in opposite directions it is highly unlikely that the observed YFP/CFP responses are due to acidification of the cytoplasm and not a consequence of cAMP activity. Furthermore, if the responses in cAMP and PKA activity would simply be due to a pH-dependent artifact, then we would have recorded signals that would be in phase with the application of L-lactate. This was not the case. As the PKA response was significantly delayed, it is more likely that delayed PKA activity is due to an elevation of cAMP, followed by the cAMP-mediated activation of PKA.

Interestingly, in both mouse WT and GPR81 KO astrocytes an increase in [cAMP]$_i$ was detected (**Table 1**), suggesting that the observed effects of L-lactate and 3Cl-5OH-BA on cAMP signaling and metabolism in astrocytes are GPR81 independent. These results indicate the existence of a yet unidentified receptor-like mechanism of L-lactate production, activated with L-lactate as well as with the GPR81 selective agonists, 3Cl-5OH-BA (**Figure 1**) and Compound 2 (**Figure 2**; Sakurai et al., 2014). The unidentified receptor-like mechanism exhibits a relatively low 3Cl-5OH-BA affinity (we could not observe any response using 50 μM concentration) in comparison with the GPR81 lactate receptor (EC$_{50}$ = 17 μM for 3Cl-5OH-BA) (Dvorak et al., 2012). Pretreatment of rat astrocytes with the 50 μM (subeffective) dose of 3Cl-5OH-BA reduced the L-lactate-induced elevations in [cAMP]$_i$, indicating that this GPR81 receptor agonist may share with L-lactate the binding site on a yet unidentified receptor stimulating L-lactate production in astrocytes.

The synthesis of cAMP upon receptor activation is likely achieved via activation of AC, as was suggested for LC neurons, where extracellular L-lactate has been considered to activate AC and PKA, although the receptor triggering cAMP elevation in LC neurons is unknown (Tang et al., 2014; Mosienko et al., 2015). The application of DDA, an AC inhibitor, reduced the L-lactate-mediated increase in [cAMP]$_i$ and also the 3Cl-5OH-BA-induced increase in [lactate]$_i$ (**Figure 3**). An AC-dependent activity by the 3Cl-5OH-BA-induced increase in [lactate]$_i$ appears to be taking place also in 3T3-L1 and BT474 cells, indicating that the Warburg effect-bearing cells regulate L-lactate synthesis via receptors (Supplementary Table S1).

A 10-fold lower L-lactate concentration (2 mM) did not affect [cAMP]$_i$ or [glucose]$_i$, implying a role of L-lactate receptor-like mechanism *in vivo* only at relatively elevated extracellular L-lactate levels, as likely takes place during exercise (Matsui et al., 2017). During abnormal conditions (e.g., hypoxia, hyperglycemia, seizures) the local resting extracellular L-lactate concentration of 0.1–2 mM in the narrow brain interstices (in humans 5 mM; Abi-Saab et al., 2002) can increase ∼10- to 20-fold to values >10 mM (Barros, 2013; Mosienko et al., 2015). Moreover, L-lactate production in the normal brain might occur in microdomains, which could create higher-than-average local concentrations (Morland et al., 2015; Mosienko et al., 2015).

In a few rat astrocytes (up to 16%, **Table 1**), the addition of 3Cl-5OH-BA, but not AR agonists, resulted in small sustained decreases in [cAMP]$_i$ and PKA activity (**Table 1**), indicating that in these cells, L-lactate preferentially activates receptors coupled to the G$_i$-proteins to downregulate cAMP, i.e., the coupling originally identified for GPR81 receptor (Ahmed et al., 2010). In mouse WT and GPR81 KO astrocytes, however, the decrease in [cAMP]$_i$ was not detected (**Table 1**). The observed heterogeneity in the recorded L-lactate receptor-mediated cAMP responses may be species specific and/or due to molecular and functional heterogeneity of the astrocytes (Zhang and Barres, 2010), determined by the neuron-specific circuits (Chai et al., 2017).

The time-constants of the increases in [cAMP]$_i$ and [lactate]$_i$ were similar (58 s vs. 60 s, respectively; $P = 0.90$) upon 3Cl-5OH-BA application. However, upon AR activation, the increase in [cAMP]$_i$ was approximately fivefold faster than that for the [lactate]$_i$ increase (20 s vs. 105 s, respectively; $P < 0.001$; Supplementary Figure S2). The observed difference could be due to the distinct molecular coupling mechanisms between the respective receptors and the cAMP pathway generating distinct cAMP pools inside cells (compartmentalized cAMP signaling; Pidoux and Taskén, 2010), but altogether, the results indicate that lactate production depends on cAMP and occurs downstream of cAMP synthesis.

Taken together, these results show that the activation of not only ARs but also L-lactate receptor-like mechanism can accelerate L-lactate production in astrocytes via cAMP signaling, suggesting the existence of a yet unidentified excitatory L-lactate receptor in astrocytes. These findings diverge from the previously reported results in CHO-K1 cells (Cai et al., 2008), primary cortical neurons (Bozzo et al., 2013), homogenized mouse adipose tissue slices (Ahmed et al., 2010), and rat hippocampal slices (Lauritzen et al., 2014). In these tissues L-lactate (presumably via activation of the L-lactate GPR81 receptor) was considered to inhibit cAMP and subsequent L-lactate production.

L-Lactate Receptor-Like Mechanisms Increase Intracellular L-Lactate More Potently Than Adrenergic Receptors

In astrocytes the NA-mediated glycogen breakdown and [glucose]$_i$ increase (Prebil et al., 2011) may regulate the extent of aerobic glycolysis, a metabolic process favored in astrocytes (Gandhi et al., 2009; Barros, 2013). Intracellular [glucose]$_i$ is (i) a function of glucose uptake from the extracellular space, (ii) is affected by glycogen degradation (Prebil et al., 2011), and (iii) involves the activity of glucose-6-phosphatase, present in astrocytes (Ghosh et al., 2005; Sharma et al., 2015), which converts glucose-6-phosphate to free glucose.

Glycogenolysis in astroglial cells is considered to be mainly regulated by β-ARs, although α$_2$-ARs may also enhance it (Subbarao and Hertz, 1990; Hertz et al., 2010). Consistent with this possibility, the responsiveness of astrocytes to β-AR stimulation was lower than to α-/β-AR agonist NA when [lactate]$_i$ was measured [63% (ISO) and 90% (NA), respectively, **Table 1**]. Upon the addition of 3Cl-5OH-BA, in contrast to adrenergic stimulation, a sustained decrease in [glucose]$_i$ was recorded, indicating that D-glucose is rapidly consumed upon 3Cl-5OH-BA-mediated L-lactate receptor-like mechanism of activation (**Figure 4**). However, extracellular L-lactate triggered only a transient decrease in D-glucose, likely due to its interference with cytoplasmic L-lactate-sensitive enzymes (Costa Leite et al., 2007).

To estimate the extent by which the ARs and L-lactate receptors modulate the rate of aerobic glycolysis, we monitored [lactate]$_i$ at rest in the presence of blocked L-lactate membrane transport. The addition of CHC, an inhibitor of several MCTs, resulted in a persistent increase in [lactate]$_i$ (**Figure 6**). If it is assumed that L-lactate cannot exit cells under these conditions, the putative L-lactate-permeable channels are inactive (Sotelo-Hitschfeld et al., 2015), and L-lactate is not substantially metabolized (Passarella et al., 2008) then the rate of change in [lactate]$_i$ reflects the resting rate of aerobic glycolysis in astrocytes. The addition of NA accelerated the rate of glycolysis to only ∼30% of the resting rate, whereas in the presence of 3Cl-5OH-BA, the glycolytic rate was increased by >600% relative to the NA-induced response (**Figure 6**). Although the effectiveness of the agonists used and their affinities for respective receptors may not be easily compared, these results suggest that L-lactate via receptor-like mechanism activates glycolysis more potently vs. ARs in astrocytes.

The results of this work bring new insights that in astrocytes L-lactate receptor-like mechanisms increase the rate of aerobic glycolysis, which manifests itself in increased [lactate]$_i$. This metabolite can exit astrocytes to further accelerate L-lactate signaling at autocrine and paracrine sites. Hence, designating this process as "metabolic excitability" appears appropriate, as it may provide the means for maintaining a high and stable source of L-lactate levels in astrocytes, as measured *in vivo* (Mächler et al., 2016) and likely contributes to the difference between the brain extracellular fluid and plasma levels of L-lactate (Abi-Saab et al., 2002).

AUTHOR CONTRIBUTIONS

NV conceived and co-directed the study, performed experiments, analyzed data, and wrote the manuscript. HHC, AH, JV, MMu, STB, KM, and SP performed experiments and analyzed data. MMa and ŠGK performed experiments. MK analyzed data. GH designed and provided custom synthesized Compound 2. LHB

and JS-M provided GPR81 KO mouse pups (founder mice from SO's lab). SO generated the GPR81 KO mouse line. RZ conceived and directed the study and wrote the manuscript. All authors read and contributed to the completion of the draft manuscript.

ACKNOWLEDGMENTS

The authors thank Drs. W. B. Frommer for providing FLII^{12}Pglu-700 μδ6 (Addgene plasmid # 17866), M. Lohse for providing Epac1-camps, Roger Y. Tsien for providing AKAR2, and L. F. Barros for providing Laconic.

REFERENCES

Abi-Saab, W. M., Maggs, D. G., Jones, T., Jacob, R., Srihari, V., Thompson, J., et al. (2002). Striking differences in glucose and lactate levels between brain extracellular fluid and plasma in conscious human subjects: effects of hyperglycemia and hypoglycemia. *J. Cereb. Blood Flow Metab.* 22, 271–279. doi: 10.1097/00004647-200203000-00004

Ahmed, K., Tunaru, S., Tang, C., Muller, M., Gille, A., Sassmann, A., et al. (2010). An autocrine lactate loop mediates insulin-dependent inhibition of lipolysis through GPR81. *Cell Metab.* 11, 311–319. doi: 10.1016/j.cmet.2010.02.012

Barros, L. F. (2013). Metabolic signaling by lactate in the brain. *Trends Neurosci.* 36, 396–404. doi: 10.1016/j.tins.2013.04.002

Bergersen, L., Waerhaug, O., Helm, J., Thomas, M., Laake, P., Davies, A. J., et al. (2001). A novel postsynaptic density protein: the monocarboxylate transporter MCT2 is co-localized with delta-glutamate receptors in postsynaptic densities of parallel fiber-Purkinje cell synapses. *Exp. Brain Res.* 136, 523–534. doi: 10.1007/s002210000600

Börner, S., Schwede, F., Schlipp, A., Berisha, F., Calebiro, D., Lohse, M. J., et al. (2011). FRET measurements of intracellular cAMP concentrations and cAMP analog permeability in intact cells. *Nat. Protoc.* 6, 427–438. doi: 10.1038/nprot.2010.198

Bozzo, L., Puyal, J., and Chatton, J. Y. (2013). Lactate modulates the activity of primary cortical neurons through a receptor-mediated pathway. *PLoS One* 8:e71721. doi: 10.1371/journal.pone.0071721

Brooks, G. A. (2009). Cell-cell and intracellular lactate shuttles. *J. Physiol.* 587, 5591–5600. doi: 10.1113/jphysiol.2009.178350

Bustin, S. A., Benes, V., Garson, J. A., Hellemans, J., Huggett, J., Kubista, M., et al. (2009). The MIQE guidelines: minimum information for publication of quantitative real-time PCR experiments. *Clin. Chem.* 55, 611–622. doi: 10.1373/clinchem.2008.112797

Cai, T. Q., Ren, N., Jin, L., Cheng, K., Kash, S., Chen, R., et al. (2008). Role of GPR81 in lactate-mediated reduction of adipose lipolysis. *Biochem. Biophys. Res. Commun.* 377, 987–991. doi: 10.1016/j.bbrc.2008.10.088

Castillo, X., Rosafio, K., Wyss, M. T., Drandarov, K., Buck, A., Pellerin, L., et al. (2015). A probable dual mode of action for both L- and D-lactate neuroprotection in cerebral ischemia. *J. Cereb. Blood Flow Metab.* 35, 1561–1569. doi: 10.1038/jcbfm.2015.115

Chai, H., Diaz-Castro, B., Shigetomi, E., Monte, E., Octeau, J. C., Yu, X., et al. (2017). Neural circuit-specialized astrocytes: transcriptomic, proteomic, morphological, and functional evidence. *Neuron* 95, 531–549.e9. doi: 10.1016/j.neuron.2017.06.029

Chen, X., Wang, L., Zhou, Y., Zheng, L. H., and Zhou, Z. (2005). "Kiss-and-run" glutamate secretion in cultured and freshly isolated rat hippocampal astrocytes. *J. Neurosci.* 25, 9236–9243. doi: 10.1523/JNEUROSCI.1640-05.2005

Costa Leite, T., Da Silva, D., Guimarães Coelho, R., Zancan, P., and Sola-Penna, M. (2007). Lactate favours the dissociation of skeletal muscle 6-phosphofructo-1-kinase tetramers down-regulating the enzyme and muscle glycolysis. *Biochem. J.* 408, 123–130. doi: 10.1042/BJ20070687

Cowan, C. M., Shepherd, D., and Mudher, A. (2010). Insights from *Drosophila* models of Alzheimer's disease. *Biochem. Soc. Trans.* 38, 988–992. doi: 10.1042/BST0380988

Dienel, G. A., and Cruz, N. F. (2016). Aerobic glycolysis during brain activation: adrenergic regulation and influence of norepinephrine on astrocytic metabolism. *J. Neurochem.* 138, 14–52. doi: 10.1111/jnc.13630

Ding, F., O'donnell, J., Thrane, A. S., Zeppenfeld, D., Kang, H., Xie, L., et al. (2013). α1-Adrenergic receptors mediate coordinated Ca2+ signaling of cortical astrocytes in awake, behaving mice. *Cell Calcium* 54, 387–394. doi: 10.1016/j.ceca.2013.09.001

Dong, J. H., Wang, Y. J., Cui, M., Wang, X. J., Zheng, W. S., Ma, M. L., et al. (2017). Adaptive activation of a stress response pathway improves learning and memory through Gs and beta-arrestin-1-regulated lactate metabolism. *Biol. Psychiatry* 81, 654–670. doi: 10.1016/j.biopsych.2016.09.025

During, M. J., Fried, I., Leone, P., Katz, A., and Spencer, D. D. (1994). Direct measurement of extracellular lactate in the human hippocampus during spontaneous seizures. *J. Neurochem.* 62, 2356–2361. doi: 10.1046/j.1471-4159.1994.62062356.x

Dvorak, C. A., Liu, C., Shelton, J., Kuei, C., Sutton, S. W., Lovenberg, T. W., et al. (2012). Identification of hydroxybenzoic acids as selective lactate receptor (GPR81) agonists with antipolytic effects. *ACS Med. Chem. Lett.* 3, 637–639. doi: 10.1021/ml3000676

Gandhi, G. K., Cruz, N. F., Ball, K. K., and Dienel, G. A. (2009). Astrocytes are poised for lactate trafficking and release from activated brain and for supply of glucose to neurons. *J. Neurochem.* 111, 522–536. doi: 10.1111/j.1471-4159.2009.06333.x

Gao, V., Suzuki, A., Magistretti, P. J., Lengacher, S., Pollonini, G., Steinman, M. Q., et al. (2016). Astrocytic β2-adrenergic receptors mediate hippocampal long-term memory consolidation. *Proc. Natl. Acad. Sci. U.S.A.* 113, 8526–8531. doi: 10.1073/pnas.1605063113

Ghosh, A., Cheung, Y. Y., Mansfield, B. C., and Chou, J. Y. (2005). Brain contains a functional glucose-6-phosphatase complex capable of endogenous glucose production. *J. Biol. Chem.* 280, 11114–11119. doi: 10.1074/jbc.M410894200

Gordon, G., Mulligan, S., and Macvicar, B. (2007). Astrocyte control of the cerebrovasculature. *Glia* 55, 1214–1221. doi: 10.1002/glia.20543

Gordon, G. R., Choi, H. B., Rungta, R. L., Ellis-Davies, G. C., and Macvicar, B. A. (2008). Brain metabolism dictates the polarity of astrocyte control over arterioles. *Nature* 456, 745–749. doi: 10.1038/nature07525

Goyal, M. S., Hawrylycz, M., Miller, J. A., Snyder, A. Z., and Raichle, M. E. (2014). Aerobic glycolysis in the human brain is associated with development and neotenous gene expression. *Cell Metab.* 19, 49–57. doi: 10.1016/j.cmet.2013.11.020

Halim, N. D., Mcfate, T., Mohyeldin, A., Okagaki, P., Korotchkina, L. G., Patel, M. S., et al. (2010). Phosphorylation status of pyruvate dehydrogenase distinguishes metabolic phenotypes of cultured rat brain astrocytes and neurons. *Glia* 58, 1168–1176. doi: 10.1002/glia.20996

Hertz, L., Gibbs, M. E., and Dienel, G. A. (2014). Fluxes of lactate into, from, and among gap junction-coupled astrocytes and their interaction with noradrenaline. *Front. Neurosci.* 8:261. doi: 10.3389/fnins.2014.00261

Hertz, L., Lovatt, D., Goldman, S. A., and Nedergaard, M. (2010). Adrenoceptors in brain: cellular gene expression and effects on astrocytic metabolism and [Ca(2+)]i. *Neurochem. Int.* 57, 411–420. doi: 10.1016/j.neuint.2010.03.019

Horvat, A., Zorec, R., and Vardjan, N. (2016). Adrenergic stimulation of single rat astrocytes results in distinct temporal changes in intracellular Ca(2+) and cAMP-dependent PKA responses. *Cell Calcium* 59, 156–163. doi: 10.1016/j.ceca.2016.01.002

Lauritzen, K. H., Morland, C., Puchades, M., Holm-Hansen, S., Hagelin, E. M., Lauritzen, F., et al. (2014). Lactate receptor sites link neurotransmission, neurovascular coupling, and brain energy metabolism. *Cereb. Cortex* 24, 2784–2795. doi: 10.1093/cercor/bht136

Lee, Y., Morrison, B. M., Li, Y., Lengacher, S., Farah, M. H., Hoffman, P. N., et al. (2012). Oligodendroglia metabolically support axons and contribute to neurodegeneration. *Nature* 487, 443–448. doi: 10.1038/nature11314

Leibiger, C., Kosyakova, N., Mkrtchyan, H., Glei, M., Trifonov, V., and Liehr, T. (2013). First molecular cytogenetic high resolution characterization of the NIH 3T3 cell line by murine multicolor banding. *J. Histochem. Cytochem.* 61, 306–312. doi: 10.1369/0022155413476868

Liu, C., Kuei, C., Zhu, J., Yu, J., Zhang, L., Shih, A., et al. (2012). 3,5-Dihydroxybenzoic acid, a specific agonist for hydroxycarboxylic acid 1, inhibits lipolysis in adipocytes. *J. Pharmacol. Exp. Ther.* 341, 794–801. doi: 10.1124/jpet.112.192799

Liu, C., Wu, J., Zhu, J., Kuei, C., Yu, J., Shelton, J., et al. (2009). Lactate inhibits lipolysis in fat cells through activation of an orphan G-protein-coupled receptor, GPR81. *J. Biol. Chem.* 284, 2811–2822. doi: 10.1074/jbc.M806409200

Mächler, P., Wyss, M. T., Elsayed, M., Stobart, J., Gutierrez, R., Von Faber-Castell, A., et al. (2016). In Vivo evidence for a lactate gradient from astrocytes to neurons. *Cell Metab.* 23, 94–102. doi: 10.1016/j.cmet.2015.10.010

Matsui, T., Omuro, H., Liu, Y. F., Soya, M., Shima, T., Mcewen, B. S., et al. (2017). Astrocytic glycogen-derived lactate fuels the brain during exhaustive exercise to maintain endurance capacity. *Proc. Natl. Acad. Sci. U.S.A.* 114, 6358–6363. doi: 10.1073/pnas.1702739114

Morland, C., Andersson, K. A., Haugen, O. P., Hadzic, A., Kleppa, L., Gille, A., et al. (2017). Exercise induces cerebral VEGF and angiogenesis via the lactate receptor HCAR1. *Nat. Commun.* 8:15557. doi: 10.1038/ncomms15557

Morland, C., Lauritzen, K. H., Puchades, M., Holm-Hansen, S., Andersson, K., Gjedde, A., et al. (2015). The lactate receptor, G-protein-coupled receptor 81/hydroxycarboxylic acid receptor 1: expression and action in brain. *J. Neurosci. Res.* 93, 1045–1055. doi: 10.1002/jnr.23593

Mosienko, V., Teschemacher, A. G., and Kasparov, S. (2015). Is L-lactate a novel signaling molecule in the brain? *J. Cereb. Blood Flow Metab.* 35, 1069–1075. doi: 10.1038/jcbfm.2015.77

Nikolaev, V. O., Bünemann, M., Hein, L., Hannawacker, A., and Lohse, M. J. (2004). Novel single chain cAMP sensors for receptor-induced signal propagation. *J. Biol. Chem.* 279, 37215–37218. doi: 10.1074/jbc.C400302200

Osnes, J. B., and Hermansen, L. (1972). Acid-base balance after maximal exercise of short duration. *J. Appl. Physiol.* 32, 59–63. doi: 10.1152/jappl.1972.32.1.59

Passarella, S., De Bari, L., Valenti, D., Pizzuto, R., Paventi, G., and Atlante, A. (2008). Mitochondria and L-lactate metabolism. *FEBS Lett.* 582, 3569–3576. doi: 10.1016/j.febslet.2008.09.042

Patterson, G., Day, R. N., and Piston, D. (2001). Fluorescent protein spectra. *J. Cell Sci.* 114, 837–838.

Paukert, M., Agarwal, A., Cha, J., Doze, V. A., Kang, J. U., and Bergles, D. E. (2014). Norepinephrine controls astroglial responsiveness to local circuit activity. *Neuron* 82, 1263–1270. doi: 10.1016/j.neuron.2014.04.038

Pidoux, G., and Taskén, K. (2010). Specificity and spatial dynamics of protein kinase A signaling organized by A-kinase-anchoring proteins. *J. Mol. Endocrinol.* 44, 271–284. doi: 10.1677/JME-10-0010

Prebil, M., Vardjan, N., Jensen, J., Zorec, R., and Kreft, M. (2011). Dynamic monitoring of cytosolic glucose in single astrocytes. *Glia* 59, 903–913. doi: 10.1002/glia.21161

Rinholm, J. E., and Bergersen, L. H. (2012). Neuroscience: the wrap that feeds neurons. *Nature* 487, 435–436. doi: 10.1038/487435a

Roland, C. L., Arumugam, T., Deng, D., Liu, S. H., Philip, B., Gomez, S., et al. (2014). Cell surface lactate receptor GPR81 is crucial for cancer cell survival. *Cancer Res.* 74, 5301–5310. doi: 10.1158/0008-5472.CAN-14-0319

Ruijter, J. M., Ramakers, C., Hoogaars, W. M., Karlen, Y., Bakker, O., Van Den Hoff, M. J., et al. (2009). Amplification efficiency: linking baseline and bias in the analysis of quantitative PCR data. *Nucleic Acids Res.* 37:e45. doi: 10.1093/nar/gkp045

Sakurai, T., Davenport, R., Stafford, S., Grosse, J., Ogawa, K., Cameron, J., et al. (2014). Identification of a novel GPR81-selective agonist that suppresses lipolysis in mice without cutaneous flushing. *Eur. J. Pharmacol.* 727, 1–7. doi: 10.1016/j.ejphar.2014.01.029

San Martin, A., Ceballo, S., Ruminot, I., Lerchundi, R., Frommer, W. B., and Barros, L. F. (2013). A genetically encoded FRET lactate sensor and its use to detect the Warburg effect in single cancer cells. *PLoS One* 8:e57712. doi: 10.1371/journal.pone.0057712

Schwartz, J., and Wilson, D. (1992). Preparation and characterization of type 1 astrocytes cultured from adult rat cortex, cerebellum, and striatum. *Glia* 5, 75–80. doi: 10.1002/glia.440050111

Sharma, K., Schmitt, S., Bergner, C. G., Tyanova, S., Kannaiyan, N., Manrique-Hoyos, N., et al. (2015). Cell type- and brain region-resolved mouse brain proteome. *Nat. Neurosci.* 18, 1819–1831. doi: 10.1038/nn.4160

Sotelo-Hitschfeld, T., Niemeyer, M. I., Mächler, P., Ruminot, I., Lerchundi, R., Wyss, M. T., et al. (2015). Channel-mediated lactate release by K^+-stimulated astrocytes. *J. Neurosci.* 35, 4168–4178. doi: 10.1523/JNEUROSCI.5036-14.2015

Staubert, C., Broom, O. J., and Nordstrom, A. (2015). Hydroxycarboxylic acid receptors are essential for breast cancer cells to control their lipid/fatty acid metabolism. *Oncotarget* 6, 19706–19720. doi: 10.18632/oncotarget.3565

Stenovec, M., Trkov, S., Lasiè, E., Terzieva, S., Kreft, M., Rodríguez Arellano, J. J., et al. (2016). Expression of familial Alzheimer disease presenilin 1 gene attenuates vesicle traffic and reduces peptide secretion in cultured astrocytes devoid of pathologic tissue environment. *Glia* 64, 317–329. doi: 10.1002/glia.22931

Subbarao, K. V., and Hertz, L. (1990). Effect of adrenergic agonists on glycogenolysis in primary cultures of astrocytes. *Brain Res.* 536, 220–226. doi: 10.1016/0006-8993(90)90028-A

Takanaga, H., Chaudhuri, B., and Frommer, W. B. (2008). GLUT1 and GLUT9 as major contributors to glucose influx in HepG2 cells identified by a high sensitivity intramolecular FRET glucose sensor. *Biochim. Biophys. Acta* 1778, 1091–1099. doi: 10.1016/j.bbamem.2007.11.015

Tang, F., Lane, S., Korsak, A., Paton, J. F., Gourine, A. V., Kasparov, S., et al. (2014). Lactate-mediated glia-neuronal signalling in the mammalian brain. *Nat. Commun.* 5:3284. doi: 10.1038/ncomms4284

Tuomi, J. M., Voorbraak, F., Jones, D. L., and Ruijter, J. M. (2010). Bias in the Cq value observed with hydrolysis probe based quantitative PCR can be corrected with the estimated PCR efficiency value. *Methods* 50, 313–322. doi: 10.1016/j.ymeth.2010.02.003

Vander Heiden, M. G., Cantley, L. C., and Thompson, C. B. (2009). Understanding the Warburg effect: the metabolic requirements of cell proliferation. *Science* 324, 1029–1033. doi: 10.1126/science.1160809

Vardjan, N., Kreft, M., and Zorec, R. (2014). Dynamics of β-adrenergic/cAMP signaling and morphological changes in cultured astrocytes. *Glia* 62, 566–579. doi: 10.1002/glia.22626

Yang, J., Ruchti, E., Petit, J. M., Jourdain, P., Grenningloh, G., Allaman, I., et al. (2014). Lactate promotes plasticity gene expression by potentiating NMDA signaling in neurons. *Proc. Natl. Acad. Sci. U.S.A.* 111, 12228–12233. doi: 10.1073/pnas.1322912111

Zhang, J., Hupfeld, C. J., Taylor, S. S., Olefsky, J. M., and Tsien, R. Y. (2005). Insulin disrupts beta-adrenergic signalling to protein kinase A in adipocytes. *Nature* 437, 569–573. doi: 10.1038/nature04140

Zhang, Y., and Barres, B. A. (2010). Astrocyte heterogeneity: an underappreciated topic in neurobiology. *Curr. Opin. Neurobiol.* 20, 588–594. doi: 10.1016/j.conb.2010.06.005

Zhang, Y., Chen, K., Sloan, S. A., Bennett, M. L., Scholze, A. R., O'keeffe, S., et al. (2014). An RNA-sequencing transcriptome and splicing database of glia, neurons, and vascular cells of the cerebral cortex. *J. Neurosci.* 34, 11929–11947. doi: 10.1523/JNEUROSCI.1860-14.2014

19

EGF Enhances Oligodendrogenesis from Glial Progenitor Cells

Junlin Yang[1]*, Xuejun Cheng[1], Jiajun Qi[1], Binghua Xie[1], Xiaofeng Zhao[1], Kang Zheng[1], Zunyi Zhang[1] and Mengsheng Qiu[1,2]*

[1]The Institute of Developmental and Regenerative Biology, Zhejiang Key Laboratory of Organ Development and Regeneration, College of Life and Environment Sciences, Hangzhou Normal University, Hangzhou, China, [2]Department of Anatomical Sciences and Neurobiology, University of Louisville, Louisville, KY, USA

*Correspondence:
Junlin Yang
yjl8121@yahoo.com
Mengsheng Qiu
m0qiu001@yahoo.com

Emerging evidence indicates that epidermal growth factor (EGF) signaling plays a positive role in myelin development and repair, but little is known about its biological effects on the early generation and differentiation of oligodendrocyte (OL) lineage cells. In this study, we investigated the role of EGF in early OL development with isolated glial restricted precursor (GRP) cells. It was found that EGF collaborated with Platelet Derived Growth Factor-AA (PDGFaa) to promote the survival and self-renewal of GRP cells, but predisposed GRP cells to develop into O4$^-$ early-stage oligodendrocyte precursor cells (OPCs) in the absence of or PDGFaa. In OPCs, EGF synergized with PDGFaa to maintain their O4 negative antigenic phenotype. Upon PDGFaa withdrawal, EGF promoted the terminal differentiation of OPCs by reducing apoptosis and increasing the number of mature OLs. Together, these data revealed that EGF is an important mitogen to enhance oligodendroglial development.

Keywords: GRP cell, OPC, oligodendrocyte lineage, self-renewal, synergistic effect

INTRODUCTION

Oligodendrocytes (OLs) elaborate insulating myelin sheaths that wrap around axons to ensure the rapid conduction of nerve impulses and axonal survival (Qiu, 2013; Zhang et al., 2013; Blank and Prinz, 2014; Alizadeh et al., 2015; Rao and Pearse, 2016). Tripotential glial-restricted precursor (GRP) cells were initially found in embryonic spinal tissues (Herrera et al., 2001; Gregori et al., 2002; Wu et al., 2002; Cao et al., 2005; Phillips et al., 2012), as they can generate OLs and type I and type II astrocytes *in vitro* (Rao et al., 1998; Gregori et al., 2002; Yang et al., 2016) and *in vivo* (Herrera et al., 2001; Hill et al., 2004). The germination of GRP cells from neuroepithelial stem cells was viewed as the beginning of OL generation (Rao et al., 1998; Gregori et al., 2002). Tripotential GRP cells then generate bipotential oligodendrocyte precursor cells (OPCs) which are capable of differentiating into either OLs or type II astrocytes (Morath and Mayer-Pröschel, 2001; Gregori et al., 2002). OPCs proliferate and migrate throughout the CNS during late mouse embryonic development, and later differentiate into mature myelinating OLs (Fernandez et al., 2004; Cai et al., 2010; Chen et al., 2015).

Emerging evidence suggests that epidermal growth factor (EGF) signaling plays an important role in oligodendroglial development (Aguirre et al., 2007; Chong et al., 2008; Hu et al., 2010).

Abbreviations: bFGF, Basic fibroblast growth factor; EGF, Epidermal growth factor; EGFR, Epidermal growth factor receptor; GRP, Glial restricted precursor; OL, Oligodendrocyte; OPC, Oligodendrocyte precursor cells; PDGFaa, Platelet derived growth factor-AA.

Loss-of-function of epidermal growth factor receptor (EGFR) reduced oligodendrogenesis *in vivo* (Aguirre et al., 2007); conversely, intraventricular infusion of EGF induced subventricular zone (SVZ) type B cells to migrate and differentiate into OLs (Gonzalez-Perez et al., 2009). More recently, it was shown that intranasal EGF treatment immediately after brain injury promoted oligodendrogenesis and remyelination (Scafidi et al., 2014). Although the importance of EGF signaling in the development of OL lineage has been established, it remains elusive at what stage of oligodendrogenesis EGF starts to function and how it regulates the development of OL lineage progression.

In this study, we used mouse GRP cells as the starting point to systematically investigate the role of EGF signaling in OL lineage development. It was found that cells of OL lineage were responsive to EGF at all developmental stages. In GRP cells, EGF promoted their proliferation and survival by augmenting their responsiveness to Platelet Derived Growth Factor-AA (PDGFaa) for self-renewal. In the absence of PDGFaa, EGF predisposed GRP cells to differentiate into O4⁻ early-stage OPCs. At OPC stage, EGF collaborated with PDGFaa to enhance OPC self-renewal. Upon PDGFaa withdrawal, OPCs underwent terminal differentiation and EGF functioned to reduce apoptosis and increase the number of mature OLs.

MATERIALS AND METHODS

Isolation and Culture of GRP Cells

GRP cells were isolated from E13.5 C57BL/6 mouse spinal cord by A2B5 immunopanning as described elsewhere (Gregori et al., 2002), all experimental procedures were carried out in accordance with institutional guidelines for the care and use of laboratory animals, and the protocol was approved by the Animal Ethics Committee of Hangzhou Normal University, China. A2B5$^+$ cells were then grown in glial basal medium (DMEM/F12, $1 \times$ N2, $1 \times$ B27, $1 \times$ P/S, and 0.1% w/v BSA, all from Gibco) supplemented with 10 ng/ml PDGFaa and EGF (Peprotech) on fibronectin/laminin coated 12-well plates at 2000 cells/well for mass culture experiments or on coated 24-well plated at 1000 cells/well for immunofluorescence staining. The immunostaining was performed 3 days after seeding using the standard protocols.

Clonal Analysis and Sub-Cloning of GRP Cells

Immunopurified A2B5$^+$ cells from E13.5 spinal tissues were adjusted to a cell density of 10 cells/ml with glial basal medium supplemented with 10 ng/ml EGF and PDGFaa, then the cell suspension was added into fibronectin/laminin-coated 96-well plates at 100 μl/well, and wells with a single cell were marked for further culture. When primary clones were generated, 10 clones were randomly selected and only one was found to be A2B5 negative. Of the nine A2B5$^+$ clones, three were subjected to differentiation potential analysis as indicated in the "Results" Section. The other six A2B5$^+$ clones were used for sub-cloning analysis. Each clone was replated on three separate grid dishes at equal clonal density, and cultured in presence of EGF, PDGFaa or both. After 6 days, the numbers of secondary clones were scored.

Cell Proliferation and Apoptosis Analysis

For cell proliferation and apoptosis analysis, 1×10^4 cells were plated to each fibronectin/laminin-coated 24-well plates. Cell proliferation was analyzed by adding BrdU (Sigma) to a final concentration of 30 ng/ml. Following 24 h of incorporation, cells were fixed in 4% paraformaldehyde at RT for 10 min, and BrdU positive cells were detected by anti-BrdU immunostaining. For apoptosis assays, cells were fixed in 4% paraformaldehyde 3 days after replating and apoptotic cells were detected by TUNEL FITC Apoptosis Detection Kit (Vazyme Biotech). Positive cells were counted from three different areas of each well under fluorescence microscopy. The results were expressed as mean values and standard deviation.

GRP Cell Development

The function of EGF on the development of GRP cells into OPCs was confirmed by single-cell tracking clonal differentiation analysis as described previously (Gregori et al., 2002). Freshly immunopurified GRP cells from E13.5 spinal tissues were adjusted to a cell density of 10 cells/ml with glial basal medium supplemented, then the cell suspension was added into fibronectin/laminin-coated 48-well plates at 100 μl/well in the presence of EGF and PDGFaa (10 ng/ml) for 24 h before being exposed to the factors as indicated in the "Results" Section wells with a single cell were marked for further culture. Since there is no specific immunological marker to distinguish GRP cells from OPCs, one candidate marker for distinguishing them is the O4 monoclonal antibody, which labels late-stage OPCs and OLs (Gard and Pfeiffer, 1990; Bansal et al., 1992). Therefore, the proportion of O4$^+$ cells was used as a standard to estimate the differentiation of GRP cells.

OPC Culture

O4⁻ early-stage OPCs were induced from GRP cells and plated on the fibronectin/laminin-coated plates and fed every other day with glial basal medium supplemented with EGF and PDGFaa. Because of the presence of contact inhibition, OPCs were plated more sparsely than GRP cells, and passaged more frequently. OPCs were plated at 1500 cells/cm^2 for mass culture, and 750 cells/cm^2 for differentiation experiments due to process growth.

Western Immunoblotting

Western blotting was carried out as previously described (Yang et al., 2009). Briefly, cells were lysed in sample buffer plus a cocktail of protease inhibitors (Roche). For each sample, 20 μg of protein was used for electrophoresis in SDS-PAGE gel. Primary antibodies were used as follows; anti-rabbit PDGFRα (1:1000, Santa Cruz), anti-rabbit EGFR (1:200, Abcam), anti-rabbit Olig2 (1:1000, Millipore), anti-rabbit Nestin (1:5000, Covance) and anti-mouse MBP (1:1000, Abcam). Horseradish peroxidase (HRP)-conjugated secondary antibody (Promaga) was used at 1:2500. Chemiluminescent signals were detected by autoradiography using the ECL System (Amersham, Piscataway, NJ, USA).

Immunocytochemical Analysis

Immunocytochemical analysis was carried out as previously described (Cheng et al., 2017). Antibodies used include anti-mouse A2B5 IgM, anti-BrdU IgM, O4 IgM (1:1 dilution in DPBS + 10% goat serum), anti-mouse Olig2 (1:1000, Millipore), anti-mouse MBP (1:500, Abcam), anti- rabbit EGFR (1:200, Abcam), anti-mouse GFAP (1:300, Chemicon), anti-rabbit Nestin (1:2000, Covance), and anti-rabbit neurofilament (1:100, Sigma). The Alexa-488 or Alexa-594 conjugated secondary antibodies were obtained from Molecular Probes (Thermo fisher). The nucleic acid dye 4′,6-diamidino-2-phenylindole (DAPI) was obtained from Roche.

Statistical Analysis

All quantitative data are presented as means ± SD. Statistical significance of the difference was evaluated by Student's t-test. P-value < 0.05 was considered statistically significant.

RESULTS

EGF Enhances the Survival and Extensive Self-Renewal of GRP Cells in Culture

To investigate the role of EGF on OL lineage development, we first immunopurified A2B5$^+$ cells from E13.5 spinal cord tissues (Liu et al., 2002). These A2B5$^+$ cells expressed typical GRP markers including PDGFRα (Rao et al., 1998), Olig2 (Zhao et al., 2009) and Nestin (Yoo and Wrathall, 2007; **Figures 1A–D,I–Q**), but not neuronal marker neurofilament (Tang et al., 2007; Rao and Pearse, 2016) and astrocyte marker GFAP (Sun et al., 2008; Sántha et al., 2016). The GRP cells were also immunoreactive for EGFR (**Figures 1E–H,Q**) suggesting their potential EGF-responsiveness. The effects of EGF on A2B5$^+$ cell proliferation were investigated in culture exposed to different concentrations of EGF (0, 2.5, 5, 10, 20 and 40 ng/ml) for 24 h. BrdU labeling revealed a dose-dependent effect on cell division at the concentration range of 0–10 ng/ml (**Figure 2A**). However, no statistical differences were found in the percentage of BrdU$^+$ cells among the 10, 20 and 40 ng/ml EGF groups (**Figure 2A**), indicating that cell proliferation plateaued at 10 ng/ml. When we grew A2B5$^+$ cells for 7 days, EGF was found to promote the cell proliferation and survival as effectively as basic fibroblast growth factor (bFGF), but the percentages of BrdU$^+$ cells on d3 and d7 in EGF groups were higher than those of bFGF groups (**Figures 2B,C**). When GRP cells were treated with EGF and Erlotinib HCl (an antagonist of EGFR) simultaneously, the biological effects of EGF on promoting the division and survival was neutralized (**Figures 2B,C**), and the antigenic phenotypes were similar to those of the control (**Figure 4B**). T3 did not significantly enhance the proliferation of A2B5$^+$ cells compared to control group (**Figure 2B**), but it significantly reduced cell apoptosis (**Figure 2C**), probably by promoting A2B5$^+$ cell differentiation into OLs (**Figures 3D,E**).

EGF also cooperated with PDGFaa to promote a vibrant proliferation of A2B5$^+$ cells as bFGF + PDGFaa did (**Figures 2B,C**). As a result, EGF + PDGFaa stimulated A2B5$^+$ cells to divide continuously and form clones (**Figures 2B, 3A**). Three randomly chosen primary clones (EGF + PDGFaa treatment) were digested into single cells with trypsin and replated for antigen phenotyping and differentiation potential analysis. All clones from the freshly immunopurified A2B5$^+$ cells expressed the same antigens as described above. Cells grown in the presence of cholinergic neurotrophic factor (CNTF) and bFGF mainly yielded A2B5$^+$/GFAP$^+$ type II astrocytes, but A2B5$^-$/GFAP$^+$ type I astrocytes in the presence of FBS (**Figures 3B,C**). When exposed to thyroid hormone T3 (Rodríguez-Peña, 1999) for 5 days, all clones gave rise to MBP$^+$ mature OLs with multiple interconnecting processes (**Figures 3D,E**; Shaw et al., 1981). No neurofilament$^+$ neurons were generated in these cultures. Thus, these A2B5$^+$ cells are the bona fide tripotential glial progenitor cells (GRP cells), with the potential to generate OLs and two distinct types of astrocytes (Gregori et al., 2002; Dadsetan et al., 2009; Haas et al., 2012).

Recloning experiments showed EGF and PDGFaa have a synergistic effect on the extensive self-renewal of GRP cells. After primary clones were amplified in EGF + PDGFaa for 10 days, randomly selected clones were digested into single cells and evenly plated at clonal density on three separate grid dishes, and then cultured in the presence of EGF, PDGFaa, or both. EGF and PDGFaa treatment yielded an average of 21 and 6 secondary clones, respectively. When these two factors were added together, the number of secondary clones reached an average of 51, far greater than the sum of individual factors (**Figure 2D**), suggesting that EGF synergized with PDGFaa in stimulating the clonal expansion of GRP cells. Secondary clones exhibited an identical pattern of antigen expressions to primary clones and can differentiate into MBP$^+$ OLs, A2B5$^+$ or A2B5$^-$ astrocytes under corresponding environmental cues. Based on these observations, it is concluded that EGF signaling participated in the survival and extensive self-renewal of GRP cells.

EGF Promoted the Generation of Early-Stage OPCs from GRP Clones

Tripotential GRP cells are capable of developing into bipotential OPCs under appropriate signals (Gregori et al., 2002). To investigate how EGF signaling influences this transition process, we grew freshly immunopurified GRP cells at clonal density on 48-well plates for 5 days in the presence or absence of EGF, and then analyzed the formation of OPCs from a single GRP cell by tracking clonal differentiation (**Figure 4A**) as described previously (Gregori et al., 2002). Since common GRP markers such as A2B5, Olig2, PDGFRα and Nestin were also positive for OPCs (Crang et al., 2004; Yang et al., 2016), only O4 monoclonal antibody can be used to define a secondary stage of OPC development (Sommer and Schachner, 1981; Chen et al., 2007; Dincman et al., 2012). In EGF group, 10.7% ± 1.3% of the clones generated O4$^+$/MBP$^-$ cells (**Figures 4A,B**). While in the control group, GRP cells were unable to divide to form clones due to the lack of growth factors, so only a few single

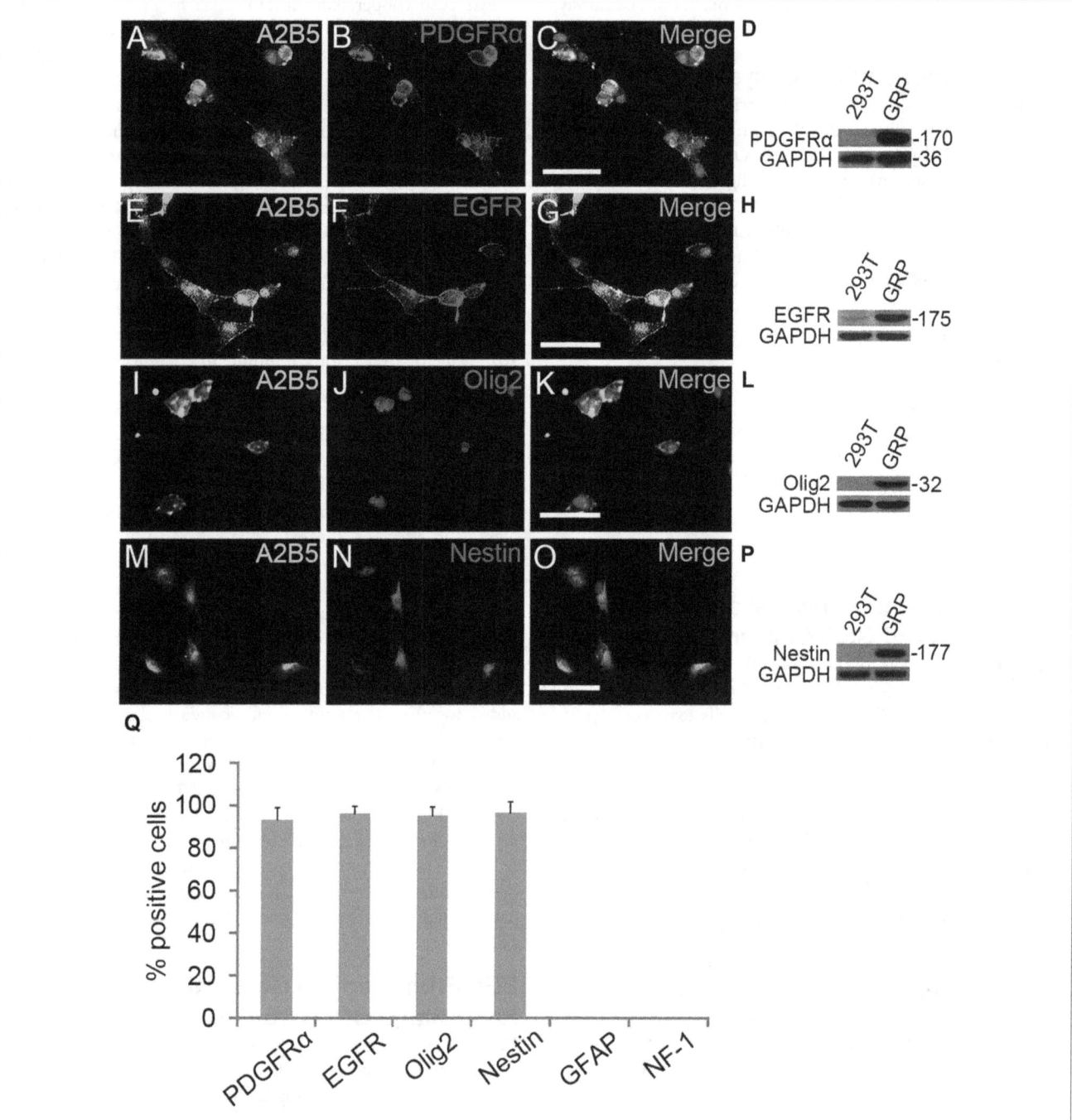

FIGURE 1 | The antigenic phenotypes of E13.5 spinal cord derived A2B5+ cells. (A–P) Immunostaining of A2B5+ cells with PDGFRα, epidermal growth factor receptor (EGFR), Olig2 and Nestin antibodies, respectively, and the antigenic phenotypes were confirmed by Western blotting **(D,H,L,P)**. **(Q)** Quantification of PDGFRα+, EGFR+, Olig2+, Nestin+, GFAP+ and NF-1+ cells in A2B5+ cell cultures, n = 3. Scale bars: 50 μm.

O4+/MBP− cells (1.3% + 0.3%) were observed (**Figure 4B**). The specificity of EGF signaling in promoting oligodendrogenesis was confirmed with Erlotinib HCl. When GRP cells were treated with EGF and Erlotinib HCl simultaneously, the antigenic phenotypes were similar to those of the controls (**Figure 4B**). Based on these results, we postulated that EGF plays a modest role in promoting the development of GRP cells into OPCs.

However, it is plausible that a considerable number of clones have differentiated into O4− early-stage OPCs that could not be detected immunologically. To examine this possibility, we continued to culture these single GRP cell derived clones in glial basal medium for another 5 days in the absence of EGF. In the control group, only 1.5% ± 0.2% of the cells were MBP+ (**Figure 4D**), similar to the percentage of O4+ clones prior to the

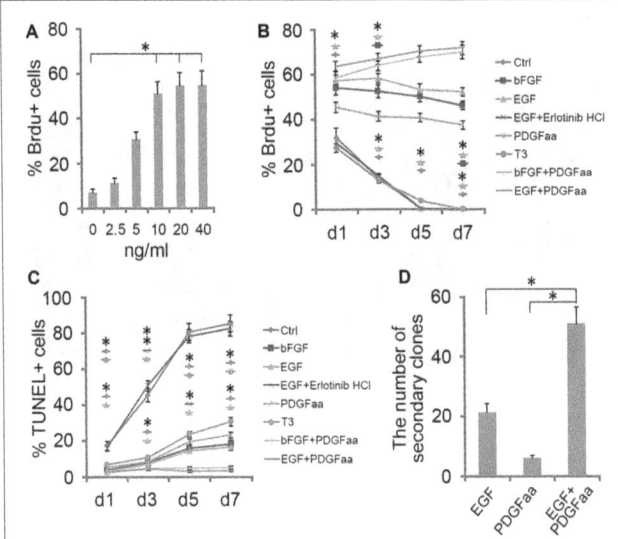

FIGURE 2 | Effect of EGF on A2B5⁺ cells. (A) Histogram of the number of BrdU⁺ cells found in A2B5⁺ cell cultures after exposure to different doses of EGF. *Indicates differences between groups 10, 20 and 40 ng/ml EGF vs. the low-dose groups, P < 0.05. No differences were found among the groups of 10, 20 and 40 ng/ml. n = 3. (B,C) Quantification of BrdU⁺ and TUNEL⁺ cells in A2B5⁺ cell cultures after EGF, EGF + Erlotinib-HCl, basic fibroblast growth factor (bFGF), PDGFaa, T3, bFGF + PDGFaa and EGF + PDGFaa treatments for various time lengths, Ctrl refers to the groups without any supplemented factor, n = 3. (D) Histogram of the number of secondary clones in A2B5⁺ cell cultures at clonal density after exposure to EGF, PDGFaa and EGF + PDGFaa, respectively. n = 3. Statistical analyses are presented as mean ± SD. *P < 0.05, **P < 0.01.

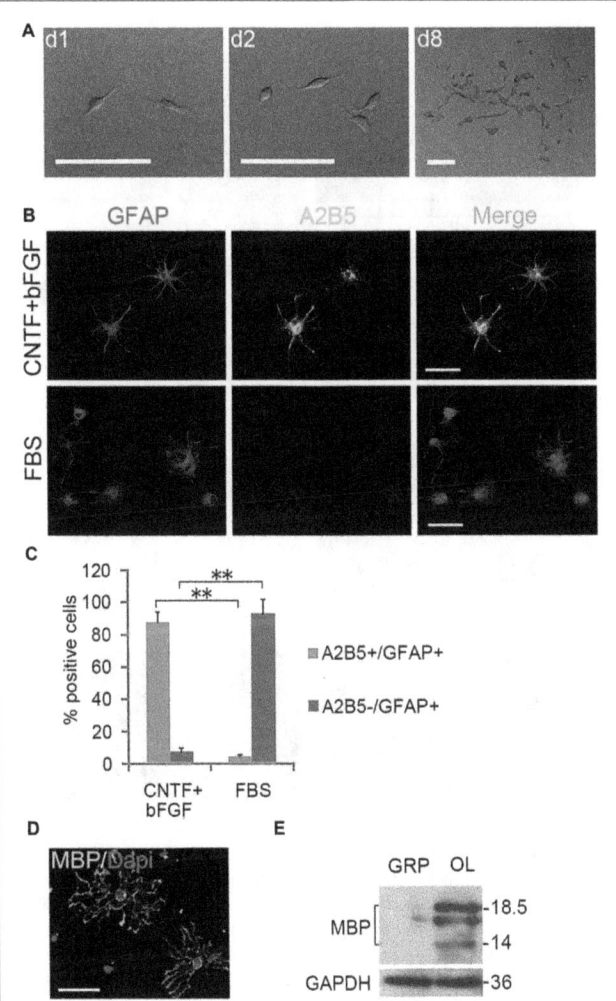

FIGURE 3 | Expanded clones of A2B5⁺ cell expressed typical differentiation phenotype of tripotential glial restricted precursor (GRP) cells. (A) An example of an expanded clone in presence of EGF + PDGFaa. (B) A2B5⁺ clones were digested and replated into separate wells of 24-well plates and induced to differentiate for 5 days in the presence of bFGF + CNTF or FBS, A2B5⁺/GFAP⁺ and A2B5⁻/GFAP⁺ astrocytes were obtained, respectively. (C) Quantification of A2B5⁺/GFAP⁺ and A2B5⁻/GFAP⁺ astrocytes in the differentiation cultures exposed to bFGF + CNTF and FBS, n = 3. (D) A2B5⁺ cells were cultured in T3 for 5 days and MBP⁺ Oligodendrocytes (OLs) can be detected. (E) The culture described in (D) was confirmed by western blotting with MBP antibody. Abbreviation: OL, oligodendrocyte. **P < 0.01. Scale bars: (A) 100 μm; (B,D) 50 μm.

5-day culture (**Figure 4B**). In the EGF group, 32.4% ± 2.7% of the clones generated MBP⁺ cells (**Figures 4C,D**), suggesting that about 21.4% (32.4% MBP⁺–10.7% O4⁺MBP⁻) of the clones in EGF group generated O4⁻ OPCs. Thus, EGF predisposed GRP cells to develop along OPCs.

However, EGF treatment alone was less efficient in fully transforming GRP cells into OPCs and maintaining their O4 negative antigen phenotype. We found that GRP cells exposed to a combination of EGF + PDGFaa + T3 exhibited a significant change from fibroblast morphology to bipolar or tripolar morphology, and remained O4 negative antigen phenotype. Moreover, the vast majority of clones (89.1% ± 3.7%) generated O4⁺/MBP⁺ OLs after growth factor withdrawal. This conjecture was further strengthened by their differentiation into A2B5⁺/GFAP⁺ instead of A2B5⁻/GFAP⁺ astrocytes upon FBS treatment (**Figure 4E**), a bipotential differentiation characteristic of OPCs (Sommer and Schachner, 1981; Barnett et al., 1993).

EGF Enhances the Self-Renewal of OPCs

PDGFaa is an important mitogen for the self-renewal of OPCs (Noble et al., 1988; Raff et al., 1988; Hart et al., 1989; Neman and de Vellis, 2012), but it alone cannot maintain the O4 negative antigenic phenotype in OPCs. When GRP-derived early-stage O4⁻ OPCs from EGF + PDGFaa + T3 treatment were cultured in PDGFaa alone for 5 days, most cells became O4 positive (88.7% ± 5.9%; **Figure 5A**) but MBP negative with few processes.

These O4⁺ late-stage OPCs (Sommer and Schachner, 1981; Gard and Pfeiffer, 1990) continued to cycle (**Figures 5A,C**). When EGF was present, the majority of cells maintained bipolar or tripolar morphology, and the rate of O4⁺ cells decreased substantially (7.4% ± 2.6%) with a significant increase of BrdU⁺ cells (80% ± 3.8%; **Figures 5A,B**). Moreover, these expanded cultures of OPCs maintained by EGF + PDGFaa can differentiate into either mature OLs or A2B5⁺/GFAP⁺ astrocytes under specific culture conditions, indicating the differentiation potential was not compromised by proliferation, nor did they revert to GRP cells. Therefore, EGF have a synergistic effect

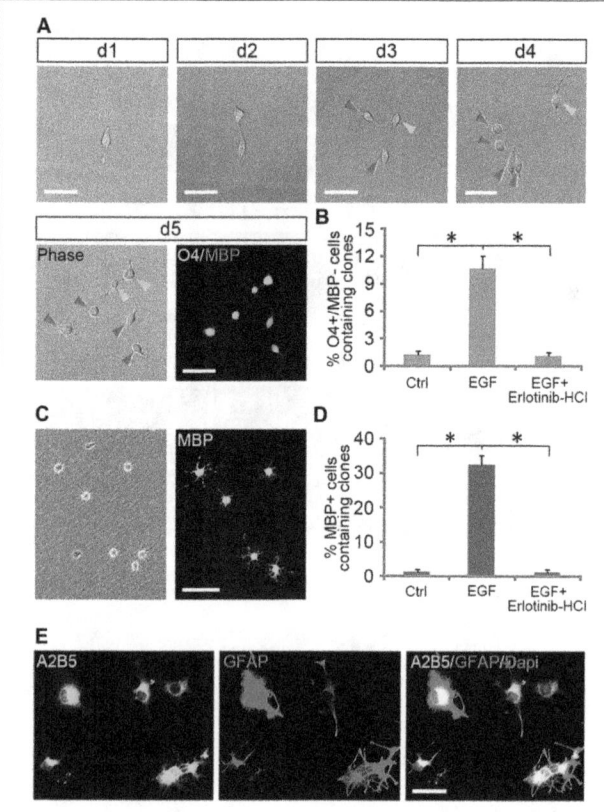

FIGURE 4 | EGF enhanced the formation of OPCs from GRP clones.
(A) The single cell tracking of a GRP cell generating O4+/MBP− daughter cells in EGF for 5 days. With the progression of cell division, a fibroblast-like GRP cell was gradually converted to typical OPCs of bipolar or tripolar morphology, well separated from each other, and expressed O4 antigen but not MBP. Daughter cells of the same progenitor were indicated by arrows of different color. **(B)** Quantification of the clones containing O4+/MBP− cells in the GRP cell cultures exposed to various combinations of factors for 5 days, respectively, n = 3. **(C)** OPC-like cells from **(B)** differentiated into MBP+ cells after 5 days of culture in glial basal medium without supplemented factors. Left: phase image; right: anti-MBP immunostaining. **(D)** Quantification of the clones containing MBP+ cells in OL differentiation cultures described in **(C)**, n = 3. **(E)** OPCs derived from EGF + PDGFaa + T3 treatment differentiated into A2B5+/GFAP+ astrocytes in presence of FBS. *P < 0.05. Scale bars: **(A,C)** 75 μm; **(E)** 50 μm.

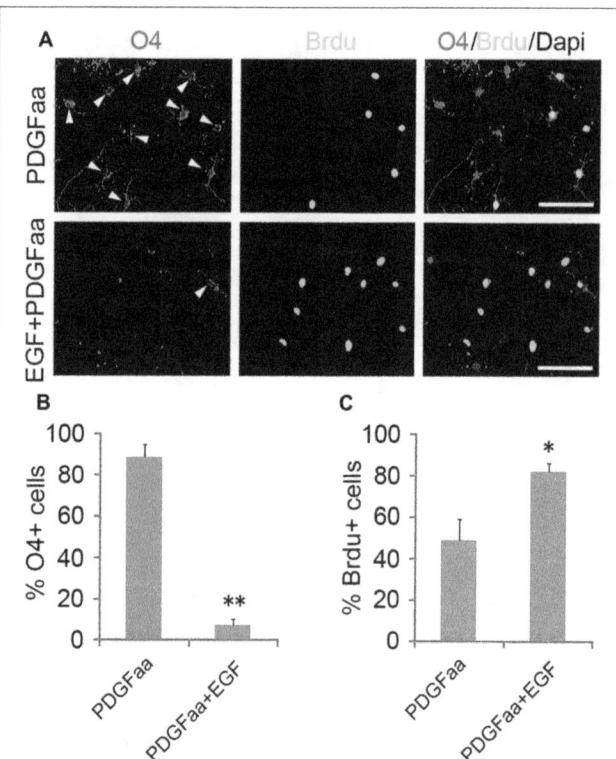

FIGURE 5 | Synergistic effect of PDGFaa and EGF in promoting the self-renewal of O4− OPCs. (A) Representative images of O4+ and/or BrdU+ cells cultured in PDGFaa or PDGFaa + EGF for 5 days, cell proliferation was analyzed by BrdU incorporation for 24 h before fixation. **(B)** Quantification of O4+ cells in PDGFaa and PDGFaa + EGF cultures. **(C)** Quantification of BrdU+ cells in PDGFaa and PDGFaa + EGF cultures. Statistical analyses are presented as mean ± SD, n = 3. *P < 0.05, **P < 0.01, Scale bars: 100 μm.

with PDGFaa in the self-renewal of O4− OPCs. And this cooperative effect of EGF and PDGFaa can be further amplified by other growth factor such as bFGF. When GRP-derived early-stage O4− OPCs were cultured in EGF + PDGFaa + bFGF *in vitro*, a faster cell division (87.3% ± 4.1%) and less O4+ cells (2.6% ± 0.9%) were observed without compromising their differentiation characteristics, suggesting that the proliferation and self-renewal of OPCs are regulated by multiple signaling pathways.

EGF Treatment Increased the Number of Differentiated OLs

When GRP-derived O4− OPCs were cultured in EGF alone, cells were initially active in proliferation, and many more BrdU-positive cells were found in EGF group than in control group. However, the percentage of BrdU+ cells in EGF group decreased with time, and by day 4, less than 3% of cells were proliferative (**Figure 6A**). Consistent with this, the EGFR expression was reduced more than half by day 4 (**Figures 6D,E**), which may partly contribute to the reduced effect of EGF on cell proliferation. It was found that bFGF cannot substitute for the loss of EGFR (data not shown), and the promotion of OPC proliferation by EGF or bFGF were based on the existence of PDGFaa. The reduced cell proliferation was accompanied by the increased proportion of MBP+ cells in culture (**Figure 6B**). Although the rate of OPC differentiation under EGF treatment was slightly lower than that of control and T3 groups at the beginning, it increased rapidly on day 3 and exceeded the other two groups (**Figure 6B**), suggesting that EGF has a significant effect in promoting OL maturation when PDGFaa is not present. A synergistic effect was observed when T3 and EGF was combined, the differentiation was faster and the efficiency was higher than that of T3 alone or EGF alone (**Figure 6B**). TUNEL labeling experiments revealed that the rate of apoptosis with EGF treatment was significantly lower than that of the other two groups (**Figure 6C**). This raised the possibility that EGF treatment may partly increase the number of mature OLs indirectly by promoting their post-mitotic survival.

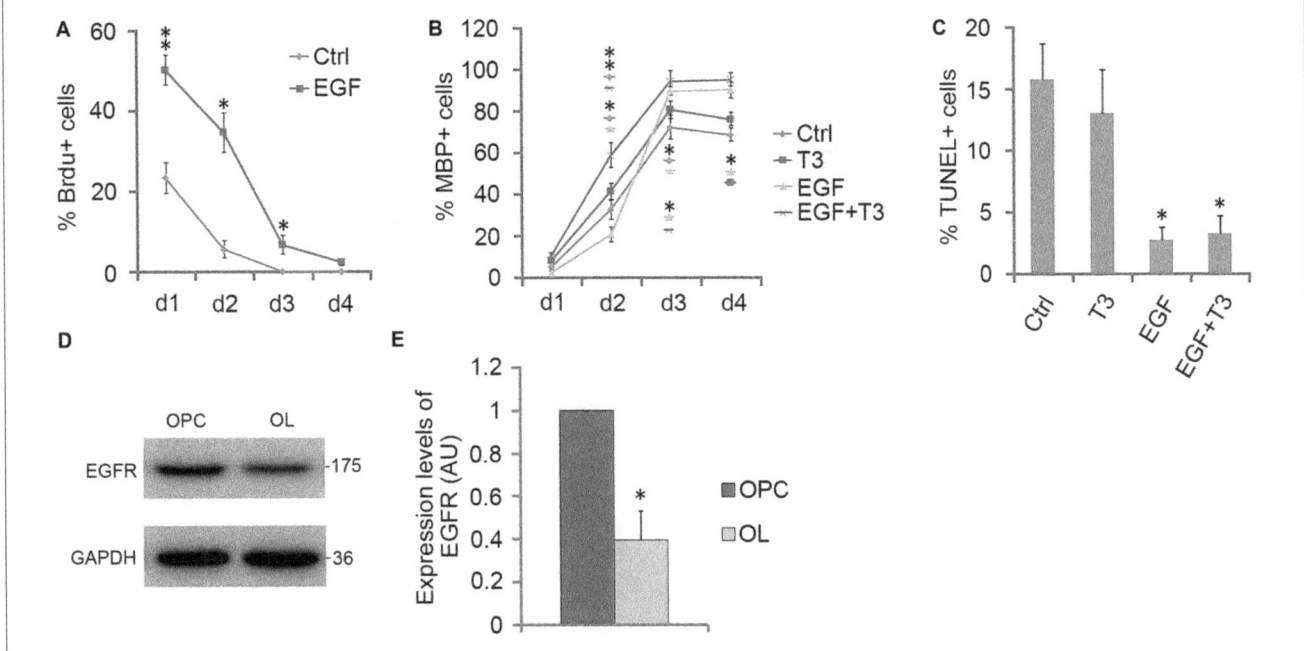

FIGURE 6 | EGF promotes OPC differentiation in cooperation with T3. (A) Quantification of BrdU⁺ cells for OPCs cultures with EGF or not for various time lengths. **(B)** Quantification of MBP⁺ cells in OPCs differentiation cultures treated with T3, EGF or T3 + EGF for various time lengths. **(C)** Quantification of TUNEL⁺ cells in OPCs differentiation cultures on day 4. **(D,E)** Western blotting and quantitative analysis of EGFR protein expression in O4⁻ OPCs and MBP⁺ OLs, histograms express results in arbitrary units, taking GRP cells values as 100%. Statistical analyses are presented as mean ± SD, $n = 3$. *$P < 0.05$, **$P < 0.01$.

DISCUSSION

EGF is an extracellular signal molecule that binds to its specific receptor (EGFR) on the target cell membrane and then stimulates the phosphorylation of the receptor. The EGFR, also known as erbB1, is a glycoprotein that belongs to the related proteins family of c-erbB, there are three other members of this family: erbB2, erbB3 and erbB4 (Galvez-Contreras et al., 2013). The EGFR can generate homodimerization or heterodimerization with all ErbB family members, therefore, activated EGFR can stimulate a large number of downstream signaling molecules, and the complexity of its signal transduction determines the diversity of its biological effects. The diversity of EGF biological effects is reflected in the progression of OL lineage, as it plays distinct roles at different stages of OL development (**Figure 7**). The biological diversity of EGF signaling could be achieved by changing the balance between different signaling pathways (Tzahar et al., 1996).

Role of EGF Signaling in GRP Development

GRP cells are probably the earliest progenitor cells for OL and astrocyte lineages (Rao et al., 1998; Gregori et al., 2002). GRP cells exhibited strong responsiveness to EGF for cell survival and proliferation in chemically defined medium. This result suggests that EGF signaling may be involved in the origin and expansion of GRP cells. Consistent with this idea, EGFR overexpression in postnatal white matter led to diffuse hyperplasia of progenitor cells that possess the same antigen phenotype as GRP cells *in vitro* (Ivkovic et al., 2008).

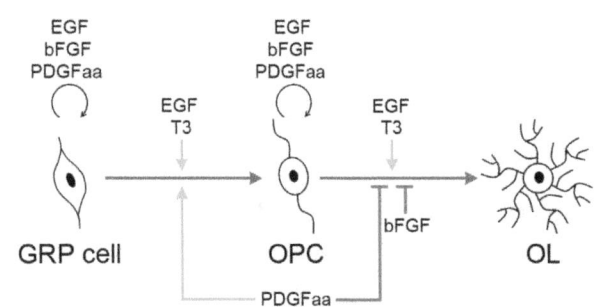

FIGURE 7 | Biological effects of EGF, PDGFaa, bFGF and T3 in the progression of OL lineage. EGF has multiple biological effects on oligodendrogenesis, and its functional output is influenced by other signal molecules, such as PDGFaa and T3. In GRP cells, EGF and bFGF collaborate with PDGFaa to promote the self-renewal of GRP cells. When EGF is present alone, it favors the development of GRP cells to OPCs, and this progress is accelerated by supplementing PDGFaa and T3 simultaneously. In OPCs, EGF and bFGF enhance their responsiveness to PDGFaa and thus maintains their O4 negative phenotype as well as bipolar or tripolar cell morphology. When EGF is present alone, it synergizes with T3 to promote the terminal differentiation of OPCs, whereas bFGF and PDGFaa inhibit this differentiation process.

Moreover, following lysolecithin (LPC)-induced demyelination, a significantly higher number of GRP cells was observed in EGFR-overexpressing transgenic mice than that of the wild-type, leading to faster and more extensive remyelination, and more rapid functional recovery. Conversely, reduced EGFR signaling *in vivo* decreased the generation of OLs in postnatal brains (Aguirre et al., 2007).

EGF has a variety of biological effects on GRP cells, including cell survival and proliferation, and enhanced OL lineage progression. However, its eventual functional output is influenced by other signal molecules. For instance, PDGFaa and EGF work synergistically to maintain the self-renewal, the rate of cell proliferation decreased substantially after PDGFaa withdrawal, and GRP cells progressed into OPCs. The number of secondary clones in EGF group was much lower than that of PDGFaa + EGF group, probably due to the higher differentiation rate in the EGF group, and thus the reduced ability of GRP cells to form clones.

Role of EGF Signaling in OPC Lineage Development

OPCs retained strong responsiveness to EGF, and EGF enhanced the response of early-stage OPCs to PDGFaa in their self-renewal. In the absence of EGF, PDGFaa alone was difficult to prevent OPCs from expressing O4 antigen, and the cells then entered the secondary stage of OPC development with reduced proliferative capacity. The mechanisms underlying this enhanced response is not clear, one possibility is that it may function to maintain a high level of PDGFRα expression or signaling in OPCs (McKinnon et al., 1990). Thus, EGF may "set" a PDGF-driven clock in OPCs by establishing their sensitivity to PDGFaa (Galvez-Contreras et al., 2013). In this way, EGF may enlarge the OPC pool during CNS development.

The biological effect of EGF in promoting oligodendrogenesis was also reflected in the maturation of OPCs into OLs. EGF alone could not maintain the self-renewal of OPCs; instead it appeared to promote the differentiation of OPCs and the survival of mature OLs (**Figure 6**), leading to an increased number of MBP$^+$ mature OLs.

Comparison of EGF and bFGF Effects in Oligodendroglial Lineage Development

bFGF was commonly used in GRP cell culture, and it alone was sufficient to maintain the self-renewal of GRP cells *in vitro* (Gregori et al., 2002). Similarly, bFGF was also widely used in

OPC culture to suppress OPC differentiation and myelin gene expression (McKinnon et al., 1990; Bansal and Pfeiffer, 1997; Decker et al., 2000; Jiang et al., 2001). Thus, it seems that bFGF promotes progenitor maintenance by preventing the lineage progression of both GRP and OPC cells.

The present study has demonstrated that EGF is also effective in the maintenance of GRP and OPC culture. However, EGF appears to enhance the self-renewal of GRPs and OPCs by enhancing their responses to PDGFRα instead of directly inhibiting their differentiation. As a matter of fact, EGF treatment alone promotes the differentiation of both GRP cells and OPCs in culture. In the absence of PDGFaa and bFGF, EGF enhances GRP lineage progression to OPCs and OPC differentiation to mature OLs in synergy with T3 (**Figure 7**). Therefore, EGF and bFGF possess unique and distinct roles in oligodendroglial progenitor self-renewal and differentiation. Consistent with this notion, $bFGF^{-/-}$ knockout leads to a higher proportion of mature OLs (Murtie et al., 2005), whereas EGFR mutation reduces the number of mature OLs (Aguirre et al., 2007).

AUTHOR CONTRIBUTIONS

JY designed, performed experiments, collected, analyzed the data and wrote the manuscript; XC, BX, XZ and KZ performed experiments; ZZ analyzed data; MQ designed, supervised the experiments, collected, analyzed and discussed data and wrote the manuscript. All authors listed, have made substantial, direct and intellectual contribution to the work, and approved it for publication.

ACKNOWLEDGMENTS

This work is supported by the National Key Basic Research Program of China (Ministry of Science and Technology of the People's Republic of China; 2013CB531303, 2012CB910402), National Natural Science Foundation of China (81200961, 31372150, 31572224, 31471955), the Zhejiang Provincial Natural Science Foundation of China (Y2111122, LQ16C090004), and the Medical and Health Research Program of Hangzhou in China (20170533B10).

REFERENCES

Aguirre, A., Dupree, J. L., Mangin, J. M., and Gallo, V. (2007). A functional role for EGFR signaling in myelination and remyelination. *Nat. Neurosci.* 10, 990–1002. doi: 10.1038/nn1938

Alizadeh, A., Dyck, S. M., and Karimi-Abdolrezaee, S. (2015). Myelin damage and repair in pathologic CNS: challenges and prospects. *Front. Mol. Neurosci.* 8:35. doi: 10.3389/fnmol.2015.00035

Bansal, R., and Pfeiffer, S. E. (1997). Regulation of oligodendrocyte differentiation by fibroblast growth factors. *Adv. Exp. Med. Biol.* 429, 69–77. doi: 10.1007/978-1-4757-9551-6_5

Bansal, R., Stefansson, K., and Pfeiffer, S. E. (1992). Proligodendroblast antigen (POA), a developmental antigen expressed by A007/O4-positive oligodendrocyte progenitors prior to the appearance of sulfatide and galactocerebroside. *J. Neurochem.* 58, 2221–2229. doi: 10.1111/j.1471-4159.1992.tb10967.x

Barnett, S. C., Franklin, R. J., and Blakemore, W. F. (1993). *In vitro* and *in vivo* analysis of a rat bipotential O-2A progenitor cell line containing the temperature-sensitive mutant gene of the SV40 large T antigen. *Eur. J. Neurosci.* 5, 1247–1260. doi: 10.1111/j.1460-9568.1993.tb00910.x

Blank, T., and Prinz, M. (2014). NF-κB signaling regulates myelination in the CNS. *Front. Mol. Neurosci.* 7:47. doi: 10.3389/fnmol.2014.00047

Cai, J., Zhu, Q., Zheng, K., Li, H., Qi, Y., Cao, Q., et al. (2010). Co-localization of Nkx6.2 and Nkx2.2 homeodomain proteins in differentiated myelinating oligodendrocytes. *Glia* 58, 458–468. doi: 10.1002/glia.20937

Cao, Q., Xu, X. M., Devries, W. H., Enzmann, G. U., Ping, P., Tsoulfas, P., et al. (2005). Functional recovery in traumatic spinal cord injury after transplantation of multineurotrophin-expressing glial-restricted precursor cells. *J. Neurosci.* 25, 6947–6957. doi: 10.1523/JNEUROSCI.1065-05.2005

Chen, Y., Balasubramaniyan, V., Peng, J., Hurlock, E. C., Tallquist, M., Li, J., et al. (2007). Isolation and culture of rat and mouse oligodendrocyte precursor cells. *Nat. Protoc.* 2, 1044–1051. doi: 10.1038/nprot.2007.149

Chen, Y., Mei, R., Teng, P., Yang, A., Hu, X., Zhang, Z., et al. (2015). TAPP1 inhibits the differentiation of oligodendrocyte precursor cells via suppressing the Mek/Erk pathway. *Neurosci. Bull.* 31, 517–526. doi: 10.1007/s12264-015-1537-5

Cheng, X., Xie, B., Qi, J., Zhao, X., Zhang, Z., Qiu, M., et al. (2017). Rat astrocytes are more supportive for mouse OPC self-renewal than mouse astrocytes in culture. *Dev. Neurobiol.* doi: 10.1002/dneu.22476 [Epub ahead of print].

Chong, V. Z., Webster, M. J., Rothmond, D. A., and Weickert, C. S. (2008). Specific developmental reductions in subventricular zone ErbB1 and ErbB4 mRNA in the human brain. *Int. J. Dev. Neurosci.* 26, 791–803. doi: 10.1016/j.ijdevneu. 2008.06.004

Crang, A. J., Gilson, J. M., Li, W.-W., and Blakemore, W. F. (2004). The remyelinating potential and *in vitro* differentiation of MOG-expressing oligodendrocyte precursors isolated from the adult rat CNS. *Eur. J. Neurosci.* 20, 1445–1460. doi: 10.1111/j.1460-9568.2004.03606.x

Dadsetan, M., Knight, A. M., Lu, L., Windebank, A. J., and Yaszemski, M. J. (2009). Stimulation of neurite outgrowth using positively charged hydrogels. *Biomaterials* 30, 3874–3881. doi: 10.1016/j.biomaterials.2009.04.018

Decker, L., Avellana-Adalid, V., Nait-Oumesmar, B., Durbec, P., and Baron-Van, Evercooren, A. (2000). Oligodendrocyte precursor migration and differentiation: combined effects of PSA residues, growth factors, and substrates. *Mol. Cell. Neurosci.* 16, 422–439. doi: 10.1006/mcne. 2000.0885

Dincman, T. A., Beare, J. E., Ohri, S. S., and Whittemore, S. R. (2012). Isolation of cortical mouse oligodendrocyte precursor cells. *J. Neurosci. Methods.* 209, 219–226. doi: 10.1016/j.jneumeth.2012.06.017

Fernandez, M., Pirondi, S., Manservigi, M., Giardino, L., and Calzà, L. (2004). Thyroid hormone participates in the regulation of neural stem cells and oligodendrocyte precursor cells in the central nervous system of adult rat. *Eur. J. Neurosci.* 20, 2059–2070. doi: 10.1111/j.1460-9568.2004.03664.x

Galvez-Contreras, A. Y., Quiñones-Hinojosa, A., and Gonzalez-Perez, O. (2013). The role of EGFR and ErbB family related proteins in the oligodendrocyte specification in germinal niches of the adult mammalian brain. *Front. Cell. Neurosci.* 7:258. doi: 10.3389/fncel.2013.00258

Gard, A. L., and Pfeiffer, S. E. (1990). Two proliferative stages of the oligodendrocyte lineage (A2B5+O4- and O4+GalC-) under different mitogenic control. *Neuron* 5, 615–625. doi: 10.1016/0896-6273(90)90216-3

Gonzalez-Perez, O., Romero-Rodriguez, R., Soriano-Navarro, M., Garcia-Verdugo, J. M., and Alvarez-Buylla, A. (2009). Epidermal growth factor induces the progeny of subventricular zone type B cells to migrate and differentiate into oligodendrocytes. *Stem Cells* 27, 2032–2043. doi: 10.1002/stem.119

Gregori, N., Pröschel, C., Noble, M., and Mayer-Pröschel, M. (2002). The tripotential glial-restricted precursor (GRP) cell and glial development in the spinal cord: generation of bipotential oligodendrocyte-type-2 astrocyte progenitor cells and dorsal-ventral differences in GRP cell function. *J. Neurosci.* 22, 248–256.

Haas, C., Neuhuber, B., Yamagami, T., Rao, M., and Fischer, I. (2012). Phenotypic analysis of astrocytes derived from glial restricted precursors and their impact on axon regeneration. *Exp. Neurol.* 233, 717–732. doi: 10.1016/j.expneurol. 2011.11.002

Hart, I. K., Richardson, W. D., Heldin, C. H., Westermark, B., and Raff, M. C. (1989). PDGF receptors on cells of the oligodendrocyte-type-2 astrocyte (O-2A) cell lineage. *Development* 105, 595–603.

Herrera, J., Yang, H., Zhang, S. C., Proschel, C., Tresco, P., Duncan, I. D., et al. (2001). Embryonic-derived glial-restricted precursor cells (GRP cells) can differentiate into astrocytes and oligodendrocytes *in vivo*. *Exp. Neurol.* 171, 11–21. doi: 10.1006/exnr.2001.7729

Hill, C. E., Proschel, C., Noble, M., Mayer-Proschel, M., Gensel, J. C., Beattie, M. S., et al. (2004). Acute transplantation of glial-restricted precursor cells into spinal cord contusion injuries: survival, differentiation and effects on lesion environment and axonal regeneration. *Exp. Neurol.* 190, 289–310. doi: 10.1016/j.expneurol.2004.05.043

Hu, Q., Zhang, L., Wen, J., Wang, S., Li, M., Feng, R., et al. (2010). The EGF receptor-sox2-EGF receptor feedback loop positively regulates the self-renewal of neural precursor cells. *Stem Cells* 28, 279–286. doi: 10.1002/ stem.246

Ivkovic, S., Canoll, P., and Goldman, J. E. (2008). Constitutive EGFR signaling in oligodendrocyte progenitors leads to diffuse hyperplasia in postnatal white matter. *J. Neurosci.* 28, 914–922. doi: 10.1523/JNEUROSCI.4327-07.2008

Jiang, F., Frederick, T. J., and Wood, T. L. (2001). IGF-I synergizes with FGF-2 to stimulate oligodendrocyte progenitor entry into the cell cycle. *Dev. Biol.* 232, 414–423. doi: 10.1006/dbio.2001.0208

Liu, Y., Wu, Y., Lee, J. C., Xue, H., Pevny, L. H., Kaprielian, Z., et al. (2002). Oligodendrocyte and astrocyte development in rodents: an *in situ* and immunohistological analysis during embryonic development. *Glia* 40, 25–43. doi: 10.1002/glia.10111

McKinnon, R. D., Matsui, T., Dubois-Dalcq, M., and Aaronson, S. A. (1990). FGF modulates the PDGF-driven pathway of oligodendrocyte development. *Neuron* 5, 603–614. doi: 10.1016/0896-6273(90)90215-2

Morath, D. J., and Mayer-Pröschel, M. (2001). Iron modulates the differentiation of a distinct population of glial precursor cells into oligodendrocytes. *Dev. Biol.* 237, 232–243. doi: 10.1006/dbio.2001.0352

Murtie, J. C., Zhou, Y. X., Le, T. Q., and Armstrong, R. C. (2005). *In vivo* analysis of oligodendrocyte lineage development in postnatal FGF2 null mice. *Glia* 49, 542–554. doi: 10.1002/glia.20142

Neman, J., and de Vellis, J. (2012). A method for deriving homogenous population of oligodendrocytes from mouse embryonic stem cells. *Dev. Neurobiol.* 72, 777–788. doi: 10.1002/dneu.22008

Noble, M., Murray, K., Stroobant, P., Waterfield, M. D., and Riddle, P. (1988). Platelet-derived growth factor promotes division and motility and inhibits premature differentiation of the oligodendrocyte/type-2 astrocyte progenitor cell. *Nature* 333, 560–562. doi: 10.1038/333560a0

Phillips, A. W., Falahati, S., DeSilva, R., Shats, I., Marx, J., Arauz, E., et al. (2012). Derivation of glial restricted precursors from E13 mice. *J. Vis. Exp.* 64:3462. doi: 10.3791/3462

Qiu, M. (2013). Myelin in development and disease. *Neurosci. Bull.* 29, 127–128. doi: 10.1007/s12264-013-1325-z

Raff, M. C., Lillien, L. E., Richardson, W. D., Burne, J. F., and Noble, M. D. (1988). Platelet-derived growth factor from astrocytes drives the clock that times oligodendrocyte development in culture. *Nature* 333, 562–565. doi: 10.1038/333562a0

Rao, M. S., Noble, M., and Mayer-Pröschel, M. (1998). A tripotential glial precursor cell is present in the developing spinal cord. *Proc. Natl. Acad. Sci. U S A* 95, 3996–4001. doi: 10.1073/pnas.95.7.3996

Rao, S. N., and Pearse, D. D. (2016). Regulating axonal responses to injury: the intersection between signaling pathways involved in axon myelination and the inhibition of axon regeneration. *Front. Mol. Neurosci.* 9:33. doi: 10.3389/fnmol. 2016.00033

Rodríguez-Peña, A. (1999). Oligodendrocyte development and thyroid hormone. *J. Neurobiol.* 40, 497–512. doi: 10.1002/(SICI)1097-4695(19990915)40:4<497::AID-NEU7>3.0.CO;2-#

Sántha, P., Veszelka, S., Hoyk, Z., Mészáros, M., Walter, F. R., Tóth, A. E., et al. (2016). Restraint stress-induced morphological changes at the blood-brain barrier in adult rats. *Front. Mol. Neurosci.* 8:88. doi: 10.3389/fnmol.2015. 00088

Scafidi, J., Hammond, T. R., Scafidi, S., Ritter, J., Jablonska, B., Roncal, M., et al. (2014). Intranasal epidermal growth factor treatment rescues neonatal brain injury. *Nature* 506, 230–234. doi: 10.1038/nature12880

Shaw, G., Osborn, M., and Weber, K. (1981). An immunofluorescence microscopical study of the neurofilament triplet proteins, vimentin and glial fibrillary acidic protein within the adult rat brain. *Eur. J. Cell Biol.* 26, 68–82.

Sommer, I., and Schachner, M. (1981). Monoclonal antibodies (O1 to O4) to oligodendrocyte cell surfaces: an immunocytological study in the central nervous system. *Dev. Biol.* 83, 311–327. doi: 10.1016/0012-1606(81) 90477-2

Sun, J. J., Liu, Y., and Ye, Z. R. (2008). Effects of P2Y1 receptor on glial fibrillary acidic protein and glial cell line-derived neurotrophic factor production of astrocytes under ischemic condition and the related signaling pathways. *Neurosci. Bull.* 24, 231–243. doi: 10.1007/s12264-008-0430-x

Tang, J., Xu, H. W., Fan, X. T., Li, Z. F., Li, D. B., Yang, L., et al. (2007). Targeted migration and differentiation of engrafted neural precursor cells in amyloid β-treated hippocampus in rats. *Neurosci. Bull.* 23, 263–270. doi: 10.1007/s12264-007-0039-5

Tzahar, E., Waterman, H., Chen, X., Levkowitz, G., Karunagaran, D., Lavi, S., et al. (1996). A hierarchical network of interreceptor interactions determines signal transduction by Neu differentiation factor/neuregulin and epidermal growth factor. *Mol. Cell Biol.* 16, 5276–5287. doi: 10.1128/mcb.16.10.5276

Wu, Y. Y., Mujtaba, T., Han, S. S., Fischer, I., and Rao, M. S. (2002). Isolation of a glial-restricted tripotential cell line from embryonic spinal cord cultures. *Glia* 38, 65–79. doi: 10.1002/glia.10049

Yang, J., Cheng, X., Shen, J., Xie, B., Zhao, X., Zhang, Z., et al. (2016). A novel approach for amplification and purification of mouse oligodendrocyte progenitor cells. *Front. Cell. Neurosci.* 10:203. doi: 10.3389/fncel.2016.00203

Yang, J., Liu, X., Yu, J., Sheng, L., Shi, Y., Li, Z., et al. (2009). A non-viral vector for potential DMD gene therapy study by targeting a minidystrophin-GFP fusion gene into the hrDNA locus. *Acta Biochim. Biophys. Sin. (Shanghai)* 41, 1053–1060. doi: 10.1093/abbs/gmp080

Yoo, S., and Wrathall, JR. (2007). Mixed primary culture and clonal analysis provide evidence that NG2 proteoglycan-expressing cells after spinal cord injury are glial progenitors. *Dev. Neurobiol.* 67, 860–874. doi: 10.1002/dneu.20369

Zhang, Y., Guo, T. B., and Lu, H. (2013). Promoting remyelination for the treatment of multiple sclerosis: opportunities and challenges. *Neurosci. Bull.* 29, 144–154. doi: 10.1007/s12264-013-1317-z

Zhao, J. W., Raha-Chowdhury, R., Fawcett, J. W., and Watts, C. (2009). Astrocytes and oligodendrocytes can be generated from NG2$^+$ progenitors after acute brain injury: intracellular localization of oligodendrocyte transcription factor 2 is associated with their fate choice. *Eur. J. Neurosci.* 29, 1853–1869. doi: 10.1111/j.1460-9568.2009.06736.x

Permissions

List of Contributors

Yu-Feng Wang
Department of Physiology, School of Basic Medical Sciences, Harbin Medical University, Harbin, China
School of Basic Medical Sciences, Harbin Medical University, Harbin, China

Vladimir Parpura
Department of Neurobiology, The University of Alabama at Birmingham, Birmingham, AL, United States

Meetu Wadhwa, Garima Chauhan, Koustav Roy, Surajit Sahu, Satyanarayan Deep, Vishal Jain, Krishna Kishore, Koushik Ray, Lalan Thakur and Usha Panjwani
Defence Institute of Physiology & Allied Sciences (DIPAS), Defence Research and Development Organization (DRDO), New Delhi, India

Lisa Sevenich
Georg-Speyer-Haus, Institute for Tumor Biology and Experimental Therapy, Frankfurt am Main, Germany

Daxiao Cheng, Ting Zhang, Huifang Lou, Liya Zhu and Yijun Liu
Department of Neurobiology, Key Laboratory of Medical Neurobiology, Ministry of Health of China, Zhejiang Provincial Key Laboratory of Neurobiology, Zhejiang University School of Medicine, Hangzhou, China

Lunhao Chen
Department of Neurobiology, Key Laboratory of Medical Neurobiology, Ministry of Health of China, Zhejiang Provincial Key Laboratory of Neurobiology, Zhejiang University School of Medicine, Hangzhou, China
Department of Orthopedic Surgery, The First Affiliated Hospital, Zhejiang University School of Medicine, Hangzhou, China

Jiachen Chu
Department of Neurobiology, Key Laboratory of Medical Neurobiology, Ministry of Health of China, Zhejiang Provincial Key Laboratory of Neurobiology, Zhejiang University School of Medicine, Hangzhou, China
Department of Physiology, Johns Hopkins University School of Medicine, Baltimore, MD, United States

Zhuoer Dong
Middle School Attached to Northwestern Poly technical University, Xi'an, China

Jiao Wang, Jie Li, Qian Wang, Fangfang Zhou, Qian Li, Weihao Li, Yangyang Sun and Tieqiao Wen
Laboratory of Molecular Neural Biology, School of Life Sciences, Shanghai University, Shanghai, China

Yanyan Kong and Yihui Guan
Positron Emission Computed Tomography Center, Huashan Hospital, Fudan University, Shanghai, China

Yanli Wang
Institute of Nano chemistry and Nano biology, Shanghai University, Shanghai, China

Minghong Wu
Shanghai Applied Radiation Institute, School of Environmental and Chemical Engineering, Shanghai University, Shanghai, China

Xing-Shu Chen and Lan Xiao
Department of Histology and Embryology, Faculty of Basic Medicine, Collaborative Program for Brain Research, Third Military Medical University, Chongqing, China

Xu-Rui Jin
Department of Histology and Embryology, Faculty of Basic Medicine, Collaborative Program for Brain Research, Third Military Medical University, Chongqing, China
The Cadet Brigade of Clinic Medicine, Third Military Medical University, Chongqing, China

Ilias Kounatidis
Cell Biology, Development, and Genetics Laboratory, Department of Biochemistry, University of Oxford, Oxford, United Kingdom

Stanislava Chtarbanova
Department of Biological Sciences, University of Alabama, Tuscaloosa, AL, United States

Fangjie Shao and Chong Liu
Department of Pathology and Pathophysiology, Zhejiang University School of Medicine, Hangzhou, China

Justin Cohen, Luca D'Agostino, Joel Wilson, Ferit Tuzer and Claudio Torres
Department of Pathology and Laboratory Medicine, Drexel University College of Medicine, Philadelphia, PA, United States

Chih-Yen Wang, Kuan-Min Fang, Chia-Hsin Ho and Shun-Fen Tzeng
Institute of Life Sciences, College of Bioscience and Biotechnology, National Cheng Kung University, Tainan, Taiwan

Yuan-Ting Sun
Department of Neurology, National Cheng Kung University Hospital, College of Medicine, National Cheng Kung University, Tainan, Taiwan

Chung-Shi Yang
Institute of Biomedical Engineering and Nano medicine, National Health Research Institutes, Zhunan, Taiwan

Ulyana Lalo and Yuriy Pankratov
School of Life Sciences, University of Warwick, Coventry, United Kingdom

Alexander Bogdanov
Institute for Chemistry and Biology, Immanuel Kant Baltic Federal University, Kaliningrad, Russia

ylvian Bauer, Sandra Courtens, Manal Salmi, Nadine Bruneau and Pierre Szepetowski
INMED, French National Institute of Health and Medical Research INSERM U1249, Aix-Marseille University, Marseille, France

Robin Cloarec
INMED, French National Institute of Health and Medical Research INSERM U1249, Aix-Marseille University, Marseille, France
Neurochlore, Marseille, France

Natacha Teissier, Emilie Bois and Pierre Gressens
French National Institute of Health and Medical Research INSERM U1141, Paris Diderot University, Sorbonne Paris Cité, Paris, France
PremUP, Paris, France

Fabienne Schaller and Emmanuelle Buhler
INMED, French National Institute of Health and Medical Research INSERM U1249, Aix-Marseille University, Marseille, France
PPGI Platform, INMED, Marseille, France

Hervé Luche and Marie Malissen
Centre National de la Recherche Scientifique CNRS UMS3367, CIPHE (Centre D'Immunophénomique), French National Institute of Health and Medical Research INSERM US012, PHENOMIN, Aix-Marseille University, Marseille, France

Vanessa Pauly
Laboratoire de Santé Publique EA 3279, Faculté de Médecine Centre d'Evaluation de la Pharmacodépendance- Addictovigilance de Marseille (PACA-Corse) Associé, Aix-Marseille University, Marseille, France

Emilie Pallesi-Pocachard
INMED, French National Institute of Health and Medical Research INSERM U1249, Aix-Marseille University, Marseille, France
PBMC platform, INMED, Marseille, France

François J. Michel
INMED, French National Institute of Health and Medical Research INSERM U1249, Aix-Marseille University, Marseille, France
InMAGIC platform, INMED, Marseille, France

Thomas Stamminger
Institute for Clinical and Molecular Virology, University of Erlangen-Nuremberg, Erlangen, Germany

Daniel N. Streblow
Vaccine and Gene Therapy Institute, Oregon Health and Science University, Portland, OR, United States

Thomas Blank
Institute of Neuropathology, Faculty of Medicine, University of Freiburg, Freiburg, Germany

Michael Burnet
In Vivo Pharmacology, Synovo GmbH, Tübingen, Germany

Tobias Goldmann
Institute of Neuropathology, Faculty of Medicine, University of Freiburg, Freiburg, Germany
In Vivo Pharmacology, Synovo GmbH, Tübingen, Germany

Mirja Koch, Christian Schön, Johanna E. Wagner, Martin Biel and Stylianos Michalakis
Center for Integrated Protein Science Munich CiPSM and Department of Pharmacy, Center for Drug Research, Ludwig-Maximilians-Universität München, Munich, Germany

Lukas Amann and Shengru Pang
Institute of Neuropathology, Faculty of Medicine, University of Freiburg, Freiburg, Germany
Faculty of Biology, University of Freiburg, Freiburg, Germany

Michael Bonin
Institute for Medical Genetics and Applied Genomics Transcriptomics, University of Tübingen, Tübingen, Germany
IMGM Laboratories GmbH, Planegg, Germany

Marco Prinz
Institute of Neuropathology, Faculty of Medicine, University of Freiburg, Freiburg, Germany
BIOSS Centre for Biological Signalling Studies, University of Freiburg, Freiburg, Germany

Xuemei Xie, Li Peng, Jin Zhu, Yang Zhou, Lingyu Li, Yanlin Chen, Shanshan Yu and Yong Zhao
Department of Pathology, Chongqing Medical University, Chongqing, China
Key Laboratory of Neurobiology, Chongqing Medical University, Chongqing, China

Qi Chen
Institute of Neuroscience, State Key Laboratory of Neuroscience, Center for Excellence in Brain Science and Intelligence Technology, Chinese Academy of Sciences, Shanghai, China

Yu Zhang, Bing Xu and Xufei Du
Institute of Neuroscience, State Key Laboratory of Neuroscience, Center for Excellence in Brain Science and Intelligence Technology, Chinese Academy of Sciences, Shanghai, China
School of Future Technology, University of Chinese Academy of Sciences, Beijing, China

Yong Yan
Institute of Neuroscience, State Key Laboratory of Neuroscience, Center for Excellence in Brain Science and Intelligence Technology, Chinese Academy of Sciences, Shanghai, China
School of Life Science and Technology, Shanghai Tech University, Shanghai, China

Jiulin Du
Institute of Neuroscience, State Key Laboratory of Neuroscience, Center for Excellence in Brain Science and Intelligence Technology, Chinese Academy of Sciences, Shanghai, China
School of Future Technology, University of Chinese Academy of Sciences, Beijing, China
School of Life Science and Technology, Shanghai Tech University, Shanghai, China

Ping Wang
School of Basic Medical Sciences, Harbin Medical University, Harbin, China

Danian Qin
Department of Physiology, Shantou University, Shantou, China

Julie C. Savage, Katherine Picard, Fernando González-Ibáñez and Marie-Ève Tremblay
Axe neurosciences, Centre de Recherche du CHU de Québec – Université Laval, Québec City, QC, Canada
Département de médecine moléculaire, Université Laval, Québec City, QC, Canada

Nina Vardjan, Helena H. Chowdhury, Jelena Velebit, Špela G. Krivec, Saša T. Bobnar and Robert Zorec
Laboratory of Neuroendocrinology - Molecular Cell Physiology, Institute of Pathophysiology, Faculty of Medicine, University of Ljubljana, Ljubljana, Slovenia
Laboratory of Cell Engineering, Celica Biomedical, Ljubljana, Slovenia

Anemari Horvat and Marko Muhič
Laboratory of Neuroendocrinology - Molecular Cell Physiology, Institute of Pathophysiology, Faculty of Medicine, University of Ljubljana, Ljubljana, Slovenia

Maja Malnar
Laboratory of Cell Engineering, Celica Biomedical, Ljubljana, Slovenia

Marko Kreft
Laboratory of Neuroendocrinology - Molecular Cell Physiology, Institute of Pathophysiology, Faculty of Medicine, University of Ljubljana, Ljubljana, Slovenia
Laboratory of Cell Engineering, Celica Biomedical, Ljubljana, Slovenia
Department of Biology, Biotechnical Faculty, University of Ljubljana, Ljubljana, Slovenia

Katarina Miš and Sergej Pirkmajer
Laboratory for Molecular Neurobiology, Institute of Pathophysiology, Faculty of Medicine, University of Ljubljana, Ljubljana, Slovenia

Stefan Offermanns
Department of Pharmacology, Max Planck Institute for Heart and Lung Research, Bad Nauheim, Germany

Gjermund Henriksen
Nuclear and Energy Physics, Department of Physics, The Faculty of Mathematics and Natural Sciences, University of Oslo, Oslo, Norway
Norwegian Medical Cyclotron Centre Ltd., Oslo, Norway

Jon Storm-Mathisen
Division of Anatomy, Department of Molecular Medicine, CMBN/SERTA Healthy Brain Ageing Centre, Institute of Basic Medical Sciences, Faculty of Medicine, University of Oslo, Oslo, Norway

Linda H. Bergersen
Institute of Oral Biology, Faculty of Dentistry, University of Oslo, Oslo, Norway
Center for Healthy Aging, Faculty of Health and Medical Sciences, University of Copenhagen, Copenhagen, Denmark

Junlin Yang, Xuejun Cheng, Jiajun Qi, Binghua Xie, Xiaofeng Zhao, Kang Zheng and Zunyi Zhang
The Institute of Developmental and Regenerative Biology, Zhejiang Key Laboratory of Organ Development and Regeneration, College of Life and Environment Sciences, Hangzhou Normal University, Hangzhou, China

Mengsheng Qiu
The Institute of Developmental and Regenerative Biology, Zhejiang Key Laboratory of Organ Development and Regeneration, College of Life and Environment Sciences, Hangzhou Normal University, Hangzhou, China
Department of Anatomical Sciences and Neurobiology, University of Louisville, Louisville, KY, USA

Index

Printed in the USA
CPSIA information can be obtained
at www.ICGtesting.com
JSHW061054121023
49903JS00030B/321